HYPERSONIC INVISCID FLOW

Wallace D. Hayes
Princeton University

Ronald F. Probstein
Massachusetts Institute of Technology

DOVER PUBLICATIONS, INC.
Mineola, New York

Bibliographical Note

This Dover edition, first published in 2004, is an unabridged republication of *Hypersonic Flow Theory, Second Edition: Volume I: Inviscid Flows,* published by Academic Press, New York, in 1966. The first edition *Hypersonic Flow Theory,* was published by Academic Press in 1959. Since Volume II of the second edition never materialized, the present Dover edition is simply titled *Hypersonic Inviscid Flow.* A new Preface has been specially prepared for this volume.

Library of Congress Cataloging-in-Publication Data

Hayes, Wallace D. (Wallace Dean)
Hypersonic inviscid flow / Wallace D. Hayes, Ronald F. Probstein.
p. cm.
Originally published: 2nd ed. Hypersonic flow theory. Volume 1, Inviscid flows. New York : Academic Press, 1966.
Includes bibliographical references and indexes.
ISBN 0-486-43281-5 (pbk.)
1. Aerodynamics, Hypersonic. 2. Air flow—Mathematical models. I. Probstein, Ronald F. II. Title.

TL571.5.H34 2003
629.132'306—dc22

2003062540

Manufactured in the United States of America
Dover Publications, Inc., 31 East 2nd Street, Mineola, N.Y. 11501

This Dover edition is dedicated to the memory of
Wally Hayes (1918—2001)

PREFACE TO THE DOVER EDITION

It has been thirty-seven years since the Second Edition of this book was published. The book was intended to be the first of two volumes with the second volume treating hypersonic viscous and rarefied flows. Over the years I have been asked frequently, "Where is Volume II?" The answer is simply that the project was dropped because during the course of preparing the second volume our technical interests changed. Wally Hayes and I felt our detachment from the field and the concomitant lack of time to devote to the writing was inconsistent with the preparation of a quality text.

Numerous advances have taken place in the field since the initial publication of this volume. Despite these advances many colleagues who continued to work in hypersonics and many more who entered the field have informed me that the book remains relevant because of its fundamental approach with an emphasis on a unified and rational theory of hypersonic inviscid flow. Since its writing the most important advances in hypersonic flow theory have resulted from the application of high-speed computers to obtain solutions of the equations governing the flow fields and problems at hypersonic speeds, including chemical and physical nonequilibrium processes. But the conceptual insights to be obtained from the various hypersonic similitudes, constant-density solutions, small-disturbance theory, and thin-shock layer theory described in the book continue to contribute significantly to both understanding and design.

Many thanks are due to John Grafton, who felt that the book is a classic and should be made available once again to a wider audience through its publication in an affordable edition. The original edition is unaltered except for the correction of a few typographical errors and the inclusion of an erratum that points out an error made by the authors in the presentation of similar power-law solutions.

This Dover edition is dedicated to the memory of my close friend, colleague, and co-author, Wally Hayes. His contributions profoundly shaped the field and suffuse the book.

Ronald F. Probstein
Cambridge, Massachusetts
September, 2003

PREFACE TO VOLUME I

The field of hypersonic flow theory has expanded considerably since our first edition appeared. This expansion has forced the splitting of the second edition into two volumes. The material in the first seven chapters of the first edition has been expanded by a factor of more than two to form the present Volume I, devoted to inviscid hypersonic flows. The older material has been revised where appropriate, and extensive new material has been added. The additions appear in each of the seven chapters. This volume is thus less than a completely new text, though considerably more than the usual second edition.

In addition to reporting theories and results which have appeared in the literature since 1959, we have attempted to make the volume more comprehensive in three respects. First, we have distinguished nonequilibrium effects from viscous effects and have treated flows with nonequilibrium effects. Second, we have made the geometrical shapes of the bodies treated more general, considering asymmetric flows, conical flows, three-dimensional flows, and flows on blunted bodies to a far greater extent. Third, we have greatly extended the families of similitudes available in hypersonic flow theory and have outlined the relations between the various similitudes. As in the first edition, this volume serves as a vehicle for original work of the authors, otherwise unpublished.

The principles followed in writing were the same as those outlined in the original preface. This preface is here repeated essentially as given in the first edition.

Volume II, to be devoted to viscous and rarefied flows, is at present still in the planning stage. The subject matter of the last three chapters of the first edition has expanded even more than has the inviscid theory, and Volume II will have to be a new text. Its appearance is planned for 1969.

At the time of the first edition, no text on the companion subject of high-temperature gasdynamics had appeared. A few have now appeared, and are mentioned in Section 1.1. Of these the most comprehensive is that of Zel'dovich and Raizer. The authors have had the privilege of serving as the editors of the English translation of this book to be published by Academic Press at the end of 1966.

In the preparation of this volume, the authors are grateful for the extensive support of the Advanced Research Projects Agency through a contract technically administered by the Fluid Dynamics Branch of the Office of Naval Research with the Massachusetts Institute of Technology. The first author is also grateful for the support of the Office of Scientific Research, through the Gas Dynamics Laboratory of Princeton University. We wish to thank Miss Margaret M. Gazan for her skillful handling of most of the secretarial work. We thank also a number of others who have helped us in various ways, in technical discussions or in secretarial details.

Insofar as this volume is based upon our first edition, our acknowledgments for the first edition still apply. We shall not repeat these in this preface. However, we do wish to repeat one sentence from our earlier acknowledgments: "A particular debt is due to Roscoe H. Mills, Chief of the Fluid Dynamics Research Branch of the Aeronautical Research Laboratory, for his farsighted and vigorous support of research in American universities in the field of hypersonic flow." Without his activity during the period before hypersonic research became popular, its development would have been delayed by a number of years.

WALLACE D. HAYES

September, 1966 RONALD F. PROBSTEIN

PREFACE TO THE FIRST EDITION

Hypersonic flow theory is a branch of the science of fluid mechanics which is in active development at present. In this book we have endeavored to present the fundamentals of this subject as we understand them, together with a reasonably comprehensive report on the state of knowledge at this time. We feel that a book such as this one is needed now, even though numerous refinements and extensions of the theory will certainly be made later. In concentrating on the fundamental concepts of hypersonic flow, however, we hope to have produced a text which will be lasting as well as timely. The book is directed to students and research workers in the field of modern gasdynamics, and to hypersonic aerodynamicists. It should also be of interest to scientists and engineers desiring some insight into this relatively new field.

The scope of the book is indicated by the title. We have not included specific material on such aerodynamic subjects as the dynamics of hypersonic flight or hypersonic wing theory. Some of the material included is directly pertinent to these subjects, of course. We have not included any magnetohydrodynamic theory or any developments involving treatment of the Boltzmann equation. And we have generally taken the point of view of classical fluid mechanics and have not delved deeply in the field of high temperature gasdynamics.

This book serves as a vehicle for original work of the authors, otherwise unpublished. Most of this original work was done in the course of the preparation of the book to fill obvious gaps in the outlined subject matter. Some, of course, was done because specific questions suggested by our treatment of the subject invited further development.

In planning this book we set ourselves a number of guiding principles: The stress at all times is placed on the basic theory and on the related fundamental concepts. We have generally avoided empirical approaches and semi-empirical theories. Empirical results are mentioned only where they are so much in vogue as to demand attention or where they may contribute in some way to an understanding of hypersonic flow phenomena. Thus we present without apology theories which are correct but which cannot be applied accurately to hypersonic flows encountered in practice, provided they furnish fundamental concepts and lead to basic understanding. And theories which are incorrect or not rational we have ignored regardless of the excellence of their agreement with experiment. Experimental results have been included only for comparison with theoretical results, and not for their own interest. We feel that empirical approaches are certainly of value to the engineer, but would detract seriously from a book on theory.

We consider the material in this book to be essential to a hypersonic aerodynamicist. But we must emphasize that this book is not a handbook in any sense, and that we have made no attempt to present design information. The point here is that the understanding which comes from an appreciation of the theory is the soundest basis for engineering ability.

We hope the book will be useful as a text in graduate courses, in courses designed to introduce the student not only to hypersonic flow theory but also to modern approaches in theoretical aerodynamics in general. A course in gasdynamics or comprehensible fluid theory should be a prerequisite. Material from the book has been used in graduate classes of the authors.

Although the book is formally self-contained, the reader will find a background in the theory of compressible fluid flow most helpful. As to mathematical level, no effort to impose any artificial limit in this level was made. The requisite mathematical background is about what is needed for most compressible flow theory—primarily a knowledge of partial differential equations and vector analysis. Certain sections of Chapter III involve the concepts of dyadics or tensors. The reader will find an ability to appreciate approximations and their limitations most helpful.

Only directly cited references have been listed here. Although the list of references is thus governed by the plan of the book and is not intended as a general bibliography on hypersonic flow, it forms a reasonably comprehensive bibliography on hypersonic flow theory. An attempt has been made to include references for all results reported here except those which appear here for the first time.

We have endeavored to keep the notation as uniform as practicable throughout the text, while at the same time reasonably consistent with accepted usage. The principal symbols used have been listed in a symbol index, with which is included a brief discussion of our notation.

The book started as a projected 80-page contribution to *Advances in Applied Mechanics* undertaken by the senior author at the suggestion of Professor Theodore von Kármán. The second author joined the effort, and the concept of the contribution simply grew out of that of a short review paper into that of a reasonably comprehensive text. We are most grateful to the editors of *Advances in Applied Mechanics* for their release of our commitment for the review article and their encouragement of our expanding the work into a text. The writing of the book was mostly completed in 1957. Some of our original results have been duplicated independently by others, and these works have been cited herein.

This book is dedicated to Professor von Kármán, who was responsible for its inception. Both authors are pleased to acknowledge a personal debt to him, the senior author directly and the second author through Professor Lester Lees. Our debt is more than a personal one, however, and includes a more basic one. Our work rests heavily on the present state of development of the aeronautical sciences in many lands. Without the influence on these sciences of Professor von Kármán and his numerous able students of various generations our book could not have been written.

We hope we have caught in proof most of the miscellaneous inevitable errors which appear in the preparation of a technical book. We shall be grateful to readers who wish to inform us of errata or to comment on the content.

WALLACE D. HAYES
RONALD F. PROBSTEIN

February, 1959

CONTENTS

IV. CONSTANT-DENSITY SOLUTIONS

V. THE THEORY OF THIN SHOCK LAYERS

VI. NUMERICAL METHODS FOR BLUNT-BODY FLOWS

VII. OTHER METHODS FOR LOCALLY SUPERSONIC FLOWS

Erratum

Equation (2.6.21b) on page 62 is incorrect except in the case $j = 0$. The momentum balance was incorrectly applied, and the $p\ dA$ force on the streamtube neglected. This error carries through (2.6.22b), (2.6.26), and (2.6.47), which is not a solution of (2.6.7) except in the case $j = 0$. Impulse is, of course, a vector quantity and can be considered a scalar only if $j = 0$.

HYPERSONIC INVISCID FLOW

CHAPTER I

GENERAL CONSIDERATIONS

1. Introductory remarks

Within recent years the development of aircraft and guided missiles has brought a number of new aerodynamic problems into prominence. Most of these problems arise because of extremely high flight velocities, and are characteristically different in some way from the problems which arise in supersonic flight. The term "hypersonic" is used to distinguish flow fields, phenomena, and problems appearing at flight speeds far greater than the speed of sound from their counterparts appearing at flight speeds which are at most moderately supersonic. The appearance of new characteristic features in hypersonic flow fields justifies the use of a new term different from the well established term "supersonic".

These new characteristically hypersonic features may be roughly divided into those of a hydrodynamic nature which arise because the flight Mach number is large, and those of a physical or chemical nature which arise because the energy of the flow is large. If the gas involved is rarefied, so that the mean free path is not negligibly small compared with an appropriate characteristic macroscopic scale of the flow field, the same division applies to a certain extent if we include kinetic theory with hydrodynamics. Rarefied gas flows are encountered in flight at extreme altitudes.

The new features of a hydrodynamic nature are mostly of a kind allowing us to make certain simplifying assumptions in developing theories for hypersonic flow. However, certain important features which appear introduce additional complications over those met with in gasdynamics at more moderate speeds. In hypersonic flow the technique of linearization of the flow equations and the use of the mean-surface approximation for boundary conditions have a vanishing range of applicability. We find also that the entropy gradients produced by curved shock waves make the classical isentropic irrotational approach inapplicable. In many cases the boundary layer creates an important disturbance in the external inviscid flow field, and boundary layer interaction phenomena can be important in hypersonic flow. Generally, it is these hydrodynamic features of hypersonic flow which form the subject matter of the present volume and planned sequel.

The new features of a physical or chemical nature appearing in hypersonic flows are mostly connected with the high temperatures generally associated with the extremely strong shock waves present in such flows. At high temperatures in air or in other gases of interest vibrational degrees of freedom in the gas molecules may become excited, the molecules may dissociate into atoms, the molecules or free atoms may ionize, and molecular or ionic species unimportant at lower temperatures may be formed. In any of these processes there may be important time delays, so that nonequilibrium effects may appear. At sufficiently high temperatures the gas may radiate, giving a method for the transfer of energy which is negligible at lower temperatures. With the presence of different molecular or ionic species in large gradients of concentration, temperature, and pressure, the processes of diffusion become important. Finally, there are phenomena connected with the interaction of gas particles (or dust particles) with solid surfaces; here appear, for example, the accommodation coefficients of rarefied gas theory, catalytic recombination of dissociated atoms on the surface, and ionization of the surface material. These features of hypersonic flow belong to the field of high temperature gasdynamics, and, generally, are not treated in this book from a physical point of view. Material covering the subject matter of high temperature gasdynamics may be found in Zel'dovich and Raizer [1], Clarke and McChesney [1], and Vincenti and Kruger [1].

We must recognize, of course, that there is interplay between the hydrodynamics and the physics of hypersonic flow, that each affects the other. However, the influence of the physical phenomena on the flow is usually a local one, so that the principal features of the inviscid flow field may be obtained without a knowledge of the physical phenomena. This fact lends justification to our treatment of hypersonic flow from a hydrodynamic point of view. However, we must keep in mind that physical phenomena may not only strongly influence local details of hypersonic flow fields, but in extreme cases might control the nature of the entire flow. On the other hand, a knowledge of the hydrodynamic flow field is necessary for any estimation of physical effects. For the most part, though, only a rather rough picture of the flow field is needed, so that a treatment of high temperature gasdynamics independent of hypersonic flow theory is also justifiable.

In the present book we shall be concerned with the problem of determining the details of the flow field about a body placed in a high velocity gas stream. This gas stream is taken to be uniform with respect to all its basic properties, i.e., chemical composition, thermodynamic state, and velocity components. Insofar as possible, we shall treat the gas as a general fluid, and consider the perfect gas of constant ratio of specific heats as a special case. The Mach number of the free stream M_∞ is the ratio of the velocity of the free stream to the velocity of sound there, and is a basic parameter of the problem. In order that a flow may be termed hypersonic it is necessary that this parameter be large.

We naturally ask the question as to how large the free stream Mach number must be before we have a hypersonic flow. No direct answer may be given, as it depends upon the shape of the body, the particular gas involved, and upon the part of the flow field being considered. Some of the characteristic features of hypersonic flow appear on the forward parts of blunt bodies with M_∞ as low as three. Some features of hypersonic flow which some investigators consider essential do not appear unless M_∞ is about ten or larger. In short, we must recognize a certain arbitrariness in the term hypersonic which can be resolved only by reference to the particular flow and characteristic feature of immediate concern. The applicability of any part of hypersonic flow theory depends on the validity of the particular assumptions needed. Whether or not a flow is to be called hypersonic in the sense of a specific part of the theory must be assessed on the basis of this required validity.

2. General features of hypersonic flow fields

We shall begin our study of hypersonic flow theory by examining qualitatively the flow fields as they appear in observed hypersonic flows. Here we must make a distinction between the flow around a blunt body and that around a slender body (see Figs. 1–1 and 1–2). At the same time we must recognize that there exist bodies of intermediate shapes and that a slender body may be somewhat blunted at its nose. In all cases we observe that there is a strong fore-and-aft asymmetry in the flow pattern, and that the flow field is always completely undisturbed upstream of the body to within a very short distance of the nose of the body. The front of the body is enveloped by a shock wave, which extends downstream in the shape of a slightly flared skirt. The flow in front of this shock is undisturbed and the flow field of interest lies entirely behind the shock. Of principal interest to us is the flow field between the shock and the body. Here we notice that the inclination of surfaces in the flow field to the oncoming stream is very significant. The enveloping shock lies very close to body surfaces which have a sufficiently large positive inclination to the free stream direction. The region between the body and the shock here is termed the shock layer. No shock lies near body surfaces which have an appreciable negative inclination. The pressures on such surfaces are much less than those found in the thin shock layers, although usually greater than the pressure in the free stream. Far aft the shock wave becomes weak, and a wake is observed directly downstream of the body. The skirt-shaped relatively weak shock far downstream is termed the shock tail.

Within the shock layer the temperature and pressure are very much greater than in the free stream, with no limit on the ratios of these quantities across the shock. On the other hand, although the density is appreciably greater than in the free stream, the density ratio across the shock is limited to finite values.

If the temperature of the body is of the order of the temperature of the free stream, a large heat transfer takes place from the gas to the body. In this case the boundary layer may have densities higher than those found in the inviscid part of the shock layer. In general, the temperature of the body is an essential parameter in the determination of real-fluid effects, and even becomes an essential parameter for determining the forces exerted on the body if the gas is at low density, e.g., in a free molecule flow.

The shock waves enveloping the body are curved, and we observe large lateral entropy gradients in the flow. In accordance with the Crocco vorticity law, this flow is also highly rotational. The wake which extends behind the body is only partly attributable to viscous effects, and with no viscosity or heat conduction we should still observe an extensive wake behind the body as a result of the large entropy increase in the fluid which has passed near the body. Within the relatively wide entropy wake is observed the narrower viscous wake, often turbulent, and characterized by a decrease in total enthalpy if the body is cold.

As the shock grows weak in the shock tail far behind the body, the shock inclination approaches the free stream Mach angle $\sin^{-1}(1/M_\infty)$. In hypersonic flow this angle is very small. The entropy wake, or region of entropy increase, is formed behind the part of the shock wave which is relatively strong. This entropy wake has a lateral dimension which may be quite large but is limited.

Figure 1–1 gives a picture of a hypersonic flow on a blunt body, a right circular cylinder with its flat face traveling forward. Figure 1–1(a) is a free flight shadowgraph in air at a Mach number of 3. The dished appearance of the front face, the apparent thickness of the shock wave, and the bulged-out appearance of the sides are due to optical distortion. At a larger value of the Mach number the shock shape in front of the body would be but little altered, but the shock skirt and the other shock waves would have smaller inclination angles and would lie closer to the axis. This photograph was chosen because of the excellent picture it gives of the expansion about the rear corner of the body, the dead-water region behind the body, and the development of the highly turbulent wake. These features, of course, are all found in supersonic flows. The shock wave emanating from the side of the body probably results from recompression following overexpansion of the flow around the front corner of the body. The third shock is the rear shock from the recompression accompanying the necking-down of the dead-water region to form the wake. Figure 1–1(b) is a sketch showing qualitatively the characteristic features of this flow. Characteristic of all blunt-body flows is the subsonic region and the stagnation point behind the strong shock at the foremost point on the body. The flow in the shock layer on the front of this body is highly rotational and nonuniform.

Figure 1–2 gives a picture of a hypersonic flow on a slender pointed body

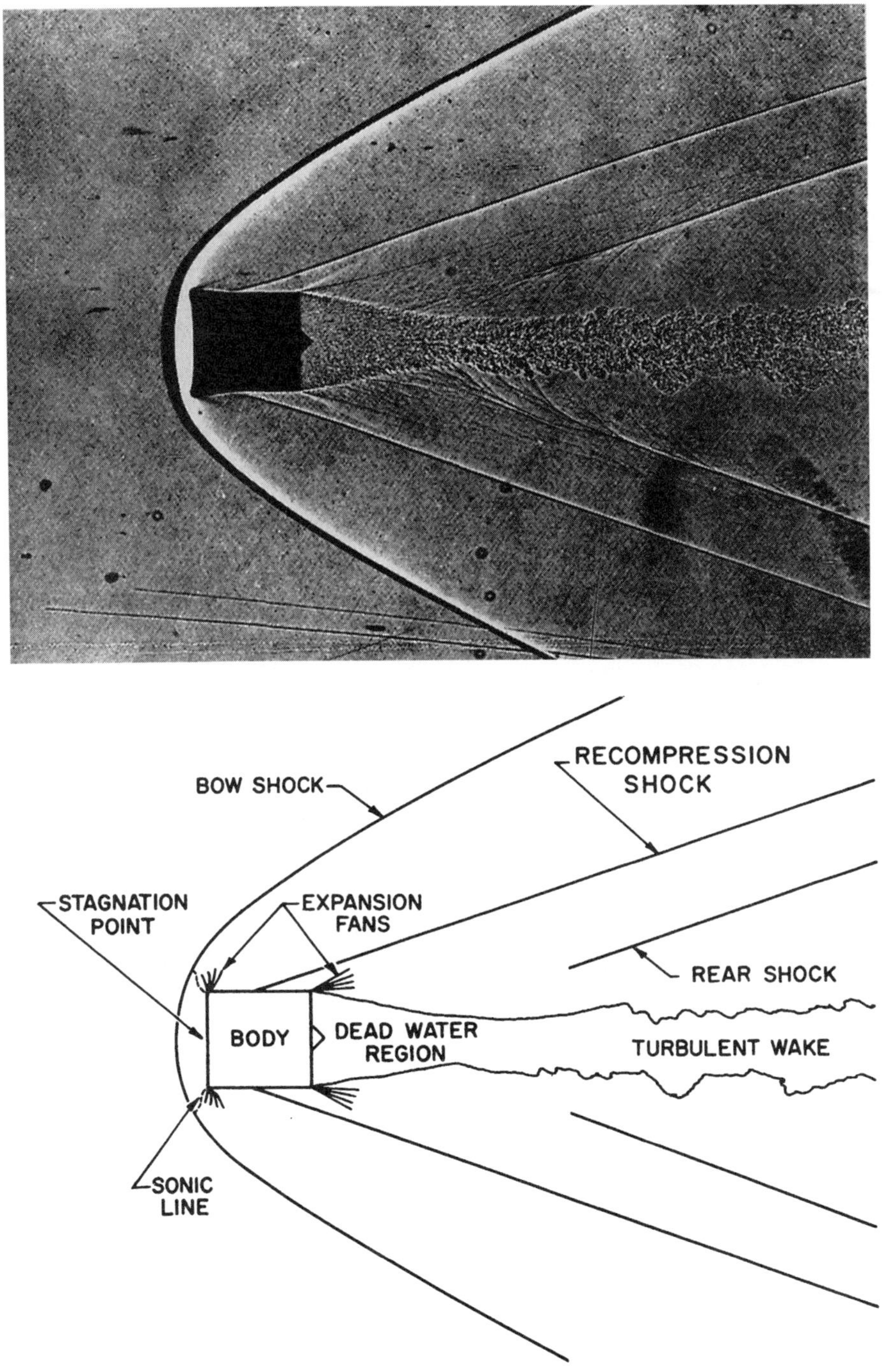

FIG. 1-1. Circular cylinder with flat face forward in air at $M_\infty = 3$. (a) Free flight shadowgraph. (Ames Research Center, courtesy of National Aeronautics and Space Administration.) (b) Sketch of flow field.

with a base flare. This is a free flight shadowgraph (with countercurrent air flow) at a Mach number of 9.6, Reynolds number of 10 million, and a free stream temperature of 290°R. The reader will observe the small inclinations of the shock waves, the boundary layer on the body, and the relatively weak rear shock. The body is at a slight angle of attack, with a resulting weak lateral

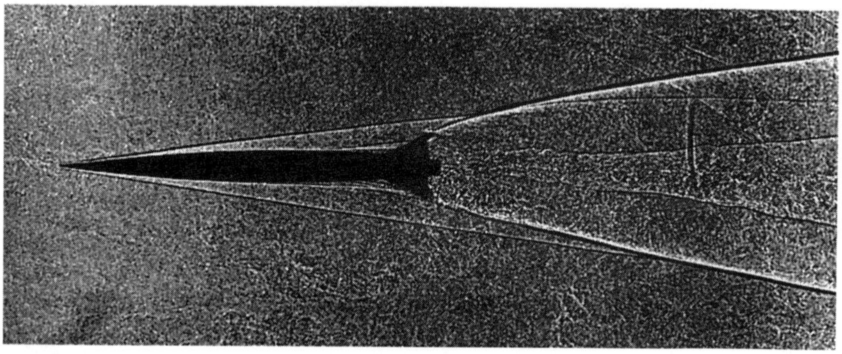

FIG. 1–2. Free flight shadowgraph in Free-Flight Wind Tunnel of a slender flared body in air at $M_\infty = 9.6$. (Ames Research Center, courtesy of National Aeronautics and Space Administration.)

asymmetry of the flow field. Laminar separation of the boundary layer occurs on the upper side, and a shock wave starts at the separation point and intersects the flare. Transition occurs in the separated boundary layer on the lower side. Characteristic of hypersonic slender-body flows is the fact that the velocity in the disturbed region is changed but very little from the velocity of the free stream, even though the other flow properties such as pressure, density, and speed of sound may be changed markedly. As long as the body is slender the speed of sound within the disturbed region remains low enough so that the entire flow field remains hypersonic. The concept of a shock layer may still be applied over the forward part of the body, but the concept is less appropriate for flows about slender bodies than for flows about blunt bodies. The shock waves observed with a slender body are much weaker than those with a blunt body, so that the entropy wake is less pronounced.

We must look a little more critically at conditions near the nose of a slender body in hypersonic flow. The remarks just made are based upon an idealized sharp tip on which a shock wave of small inclination may lie. In constructional practice it is next to impossible to provide a tip which is sharp enough to represent this idealization. In addition, the local heating in the vicinity of a very sharp tip may be so great that the sharp tip will rapidly melt away. Thus we must recognize the fact that slender bodies are really slightly blunted. At the tip of such a slender body a local blunt-body hypersonic flow is observed, with the attendant local strong shock wave and large entropy increase. The entropy wake from this local flow pattern lies in a layer next to the body which is generally initially thicker than the viscous boundary layer. Viscous effects play little or no part in determining the initial structure of the entropy wake, but this layer has an important effect on the development of the viscous boundary layer and vice versa. In fact, if the blunting of the slender body is slight it is generally difficult to identify an entropy wake distinct from the viscous boundary layer. If the blunting of the slender body is appreciable, the recognizable entropy wake is generally too extensive to be identifiable as a thin layer. The term entropy layer has been avoided here, as it is in common use for a quite different concept (Section 2.7).

Figure 1–3 is an interferogram of the flow over a flat plate at zero incidence at a Mach number of 12.7 in helium. The bluntness is a flat face normal to the flow direction of width 0.00093 inches, and the Reynolds number based

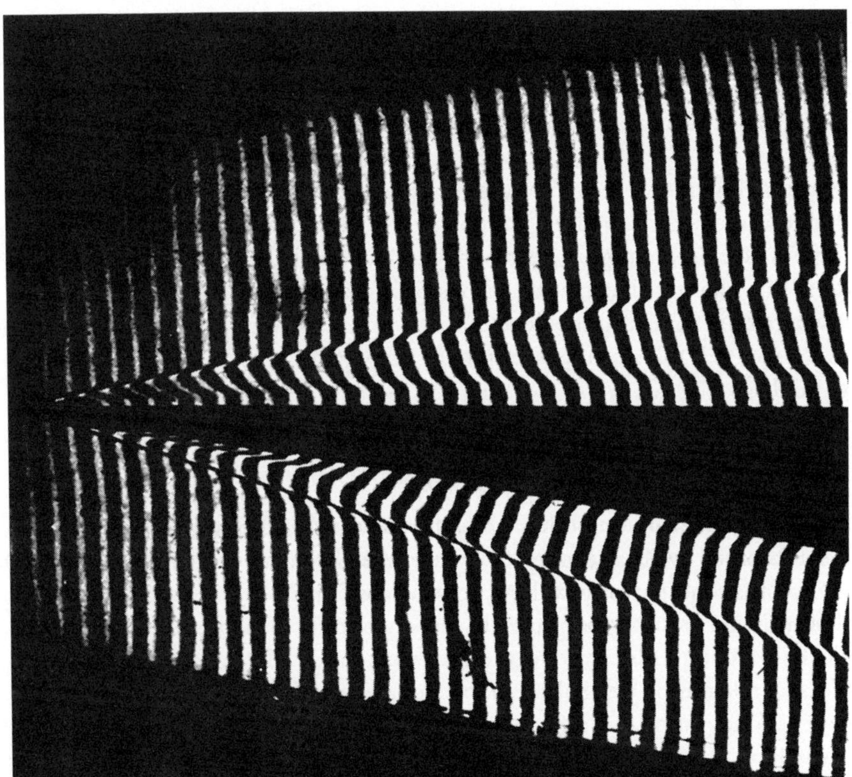

FIG. 1–3. Wind tunnel interferogram of a slightly blunted flat plate in helium at $M_\infty = 12.7$ and $Re_{d_\infty} = 808$. (Courtesy Gas Dynamics Laboratory, Princeton University.)

on this width and free stream conditions (Re_{d_∞}) is 808. The length of the plate shown in the photograph (about 35 fringes) is about 2 inches. The flow field below the plate is three-dimensional and should be disregarded. The shock wave above the plate is clearly visible, as is a layer of reduced density next to the plate. The decreased density in this layer is to a large extent attributable to the entropy jump across the strong shock very near the nose. However, it appears difficult to recognize a distinct viscous boundary layer within this layer of reduced density, and it is likely that viscosity plays a part in creating the identifiable outer boundary to this layer. Although the concept of a definable narrow entropy wake may not be a usable one in all cases, the recognition of the highly nonuniform entropy field behind a blunt nose is essential.

At sufficiently high free stream Mach numbers the value of the Mach number M_∞ is not a particularly important factor in determining the general shape of the shock wave near the body. The separation of the shock wave from the body depends primarily upon the density of the gas between them, which in turn depends upon the ratio of the density in the free stream to that just behind the shock. If this density ratio ϵ is small the shock lies close to the body; if it is not small the shock is farther from the body. Although this ratio is a variable dependent upon the inclination angle of the shock, it is useful to use it conceptually as a basic parameter. The density ratio across a shock will be treated in Section 1.4.

In these remarks we have not distinguished between, say, bodies of revolution and two-dimensional bodies. Important differences between bodies of these two types do exist, but appear primarily in fine details of hypersonic flow patterns and in the quantitative results. These differences will be pointed out in the text where they appear in the development. An understanding of these differences is not essential for an appreciation of the qualitative features of hypersonic flow fields.

3. Assumptions underlying inviscid hypersonic theory

In various parts of this book we shall make a number of assumptions which will be familiar to the reader with a general background in fluid mechanics. In this first volume we shall be developing inviscid hypersonic flow theory, with the basic assumption that all real-fluid effects such as viscosity may be neglected. Nonequilibrium effects (e.g. relaxation) will, however, be considered. In the second volume we shall treat viscous and rarefied gas flows. The distinction between nonequilibrium and viscous effects will be discussed in Section 1.8. We shall point out that nonequilibrium effects can be considered as inviscid even though they are dissipative.

The assumptions mentioned above are mostly quite standard, and not particularly characteristic of hypersonic flow theory. The reader as yet wholly unacquainted with hypersonic flow will encounter certain other assumptions here which are characteristically hypersonic. Some of these appear in the treatment of the interaction of shock waves and boundary layers, and thus lie outside the scope of this volume. Some of these underlie the inviscid theory, and deserve our attention at this point.

We shall pick out four basic assumptions which appear in inviscid hypersonic flow theory. Only one or two of these are needed for any particular development, and our main purpose in assembling and discussing them here together is to obtain a comprehensive picture of the assumptions used in inviscid hypersonic flow theory and their relation to each other. An assumption of the type being considered is always of the form that a particular quantity or parameter is small compared with one (or large compared with one). A particular theory based upon such an assumption is generally valid in an asymptotic sense as the chosen parameter is made to approach zero (or infinity) by a limiting process. Since products and ratios of parameters are themselves parameters the assumptions may appear in varying strength: for example, if a quantity $\epsilon^{1/2}$ is small, the quantity ϵ must be "very small"; or if $\sin \theta_b$ is small and $M_\infty \sin \theta_b$ is large, M_∞ must be "very large". In addition to the four hypersonic assumptions, the assumption needed for linearized supersonic or hypersonic flow is included for the sake of completeness.

The inviscid hypersonic assumptions are:

A.	$M_\infty \gg 1$	"Basic hypersonic"
B.	$\sin \theta_b \ll 1$	"Slender body"
C.	$M_\infty \sin \theta_b \gg 1$	"Strong shock"
D.	$\epsilon \ll 1$	"Small density ratio"
E.	$M_\infty \sin \theta_b \ll 1$	"Linearization"

Here θ_b is an appropriate maximum value of the inclination angle of the body or of a streamline with respect to the free stream direction. The limiting statement corresponding to an assumption may be designated by the letter followed by -lim, and a strong form of an assumption by the letter followed by -strong. Thus, for example, we designate the limiting process $M_\infty \to \infty$ by A-lim and the assumption $\epsilon^{1/2} \ll 1$ by D-strong. Note that assumption A involves only conditions in the free stream flow and that assumption B involves only the shape of the body. Assumptions C and E are mixed in nature, while assumption D primarily concerns the properties of the gas behind the shock.

By the definition of hypersonic flow, assumption A is required for all hypersonic flow theories. Briefly, the physical significance of this assumption is that the internal thermodynamic energy in the material in the free stream is small compared with the kinetic energy of this stream. Assumption A ensures that the Mach angle in the free stream is small. The physical significance of the other assumptions will appear naturally in the sections concerning the quantities involved in the assumptions (in particular ϵ in Section 1.4) or concerning the theories dependent upon the assumptions.

The concept which we shall term the Mach number independence principle (Section 1.6) depends on assumption C. Here we must notice that assumption C may not be applied to the entire shock tail if M_∞ is finite. In the vicinity of the shock tail the local flow inclination angle must be used in place of θ_b and this angle decreases toward zero as the shock grows weaker downstream. If we wish to apply the Mach number independence principle over the entire field we must use assumption A-lim.

The classical hypersonic similitude of Tsien and Hayes and the associated small-disturbance theory of Van Dyke require assumption B. The similar solutions of the small-disturbance theory require in addition assumption C and the assumption of a perfect gas of constant ratio of specific heats. The combination of assumptions B and C implies and requires assumption A-strong. Small-disturbance theory forms the subject of Chapter II.

Newtonian flow theory and various theories for thin shock layers related to Newtonian theory, treated in Chapters III to V, depend upon assumption D. The requirements for Newtonian theory are particularly stringent, as it is assumption D-strong which is needed in this case and there are restrictions on the body shape in order that the shock shape may be assumed known.

The application of supersonic linearized flow theory to hypersonic flow requires assumption E, which with A implies and requires assumption B-strong. This theory is not characteristic of hypersonic flow and, since assumption B-strong cannot be considered realistic, is not significant. Nonlinearity is an essential feature of hypersonic flow, and we shall not consider the linear theory further.

A word about the nature of the basic hypersonic limiting process A-lim is in order. The free stream Mach number is the ratio between the free stream velocity and the free stream sound speed. In the limiting process A-lim in which M_∞ approaches infinity we may consider as one possibility that the free stream velocity approaches infinity and that the free stream thermodynamic state remains fixed. However, such a process does not make physical sense, as then the energy of the gas and the temperatures in the shock layer increase without limit, and no true limiting state occurs. In the limiting process A-lim we may also consider that the free stream sound velocity is made to approach zero, while the free stream velocity and density are kept constant. Thus we consider the absolute temperature, pressure, and sound speed of the oncoming gas to approach zero. In such a limiting process a proper limiting state does appear. For a perfect gas with constant ratio of specific heats this distinction is unimportant, but for actual gases at elevated temperatures the distinction is usually an essential one.

4. The normal shock wave

Shock waves are an essential feature of any hypersonic flow, and we shall begin our analytical treatment of hypersonic flow with a study of them. The normal shock is treated first.

The subscripts ∞ and s will refer to conditions upstream and downstream, respectively, of the normal shock. The normal shock is governed by three basic conservation equations, corresponding to the three physical principles of conservation of mass, of momentum, and of energy. These equations are

$$\rho_\infty v_\infty = \rho_s v_s = m, \tag{1.4.1a}$$

$$p_\infty + \rho_\infty v_\infty^2 = p_s + \rho_s v_s^2 = P, \tag{1.4.1b}$$

$$h_\infty + \tfrac{1}{2}v_\infty^2 = h_s + \tfrac{1}{2}v_s^2 = H_n, \tag{1.4.1c}$$

where m, P, and H_n are constant. The quantity h is the specific enthalpy, defined with respect to the specific internal energy e by the relation

$$h = e + p/\rho, \tag{1.4.2}$$

and v is the flow velocity, directed normal to the shock. Both e and h are so defined as to be zero at zero absolute temperature. The quantity H is the total enthalpy, and the subscript n refers to the fact that the shock is considered normal. We shall generally know beforehand the properties of the gas in front of the shock, and shall want to know them behind. For this we must have an equation of state for the material behind the shock in order to relate p_s, ρ_s, and h_s. In this book, the term "equation of state" is used in a sense encompassing all the usual thermodynamic variables, and not in the limited sense specifying pressure as a function of volume and temperature. The equation of state required may be of the form

$$h = h(S, p), \tag{1.4.3}$$

where S is the specific entropy; T and ρ are immediately obtainable from (1.4.3) by differentiation, according to the well known thermodynamic formulas

$$T = \left(\frac{\partial h}{\partial S}\right)_p; \qquad \frac{1}{\rho} = \left(\frac{\partial h}{\partial p}\right)_S. \tag{1.4.4}$$

We should note that we are here assuming the existence of such an equation of state. This assumption is not always a valid one, and fails in particular if the gas is far from thermal equilibrium (but not frozen—see below).

A number of additional relations may be obtained from (1.4.1). Some of these are

$$v_\infty \pm v_s = m(1/\rho_\infty \pm 1/\rho_s) = v_\infty(1 \pm \epsilon), \tag{1.4.5a}$$

$$p_s - p_\infty = m(v_\infty - v_s) = \rho_\infty v_\infty^2(1 - \epsilon), \tag{1.4.5b}$$

$$h_s - h_\infty = \tfrac{1}{2}v_\infty^2(1 - \epsilon^2), \tag{1.4.5c}$$

$$\frac{p_s - p_\infty}{1/\rho_\infty - 1/\rho_s} = m^2 . \tag{1.4.5d}$$

Here ϵ is the density ratio across the shock, defined by

$$\epsilon = \frac{\rho_\infty}{\rho_s} . \tag{1.4.6}$$

We may eliminate v_∞^2 between (1.4.5b) and (1.4.5c) to obtain the Hugoniot relation

$$h_s - h_\infty = \frac{p_s - p_\infty}{2\rho_\infty}(1 + \epsilon) = \tfrac{1}{2}(p_s - p_\infty)(1/\rho_\infty + 1/\rho_s). \tag{1.4.7}$$

The importance of the Hugoniot relation lies in the fact that in it the velocities and the conservation constants of (1.4.1) have been eliminated. It provides a relation connecting the thermodynamic state quantities on the two sides of the shock. With the aid of the equation of state we may use the Hugoniot relation to plot a curve of the possible states of the gas behind the shock corresponding to a given state in front of the shock. In order to determine which of these states is actually obtained, some additional determining quantity or boundary condition must be given. For example, a specification of v_∞, of $v_\infty - v_s$, or of p_s will determine the shock. A more detailed investigation of the Hugoniot relation with sufficient conditions for uniqueness of a shock under various determining conditions may be found, for example, in Hayes [6, Arts. 1 and 2].

We now rewrite the Hugoniot relation in a form which expresses the density ratio ϵ explicitly

$$\epsilon = \frac{p_s/\rho_s}{(h_s + e_s) - (h_\infty + e_\infty) + p_\infty/\rho_s} . \tag{1.4.8}$$

We now consider the basic hypersonic limiting process (A-lim), in which the temperature and pressure before the shock approach zero and M_∞ approaches infinity. The terms $h_\infty + e_\infty$ and p_∞ in (1.4.8) are dropped, and we obtain

$$\epsilon_{\text{lim}} = \frac{p_s/\rho_s}{h_s + e_s} . \tag{1.4.9}$$

From this we see that the density ratio in the limiting case of a very strong shock depends only upon the thermodynamic state of the gas behind the shock, and that this limiting density ratio is finite.

In hypersonic flow theory we shall be interested primarily in a general fluid and shall consider the case of a perfect gas as a special case. The ratio of specific heats c_p/c_v, an important parameter for a perfect gas, is of essentially no significance in the gasdynamics of a general fluid such as a dissociating gas. Accordingly, we shall refer to the ratio of specific heats only with respect to a perfect gas, and shall not give this ratio a symbol *per se*. Instead, we shall use the symbol γ to refer to other dimensionless parameters which necessarily coincide with the ratio of specific heats only if the gas is perfect and this ratio is constant. Of these parameters the most important probably is the "isentropic exponent" or "effective ratio of specific heats" γ_e, defined as $\rho a^2/p$, where a is the speed of sound defined below. Except for flows such as those in shock tunnels or with particular fluids, the fluid in the free stream may be considered perfect, and we shall refer to γ_e in the free stream simply as γ.

In order to obtain an expression for ϵ in terms of the Mach number of the oncoming flow we introduce the notation for quantities before the shock

(1.4.10a) $$M_n = \frac{v_\infty}{a_\infty},$$

(1.4.10b) $$\gamma = \frac{\rho_\infty a_\infty^2}{p_\infty},$$

(1.4.10c) $$\epsilon_\infty = \frac{p_\infty/\rho_\infty}{h_\infty + e_\infty}.$$

The quantity a is the isentropic velocity of sound in the fluid medium, defined by the relation

(1.4.11) $$a^2 = \left(\frac{\partial p}{\partial \rho}\right)_S.$$

This quantity is necessarily identified with the actual speed of sound waves only for waves of sufficiently low frequency that real-fluid effects play no role. The quantity M_n is the Mach number of the oncoming flow normal to the shock, equal to M_∞ for a normal shock. In considering oblique shocks we shall define M_n as the normal component of the free stream Mach number M_∞, and thus make a distinction between the two quantities.

We now express the pressure ratio across the shock with the aid of (1.4.5b)

(1.4.12) $$\frac{p_s}{p_\infty} = 1 + \gamma M_n^2(1 - \epsilon).$$

If we treat ϵ_{lim} as defined in (1.4.9) as a constant and combine (1.4.12) with (1.4.8) we obtain a quadratic equation for ϵ. The solution of this may be expressed

$$\begin{aligned}(1.4.13)\quad \epsilon = {}& \tfrac{1}{2}(1+\epsilon_{\text{lim}})(1+\gamma^{-1}M_n^{-2}) \\ & - \tfrac{1}{2}\sqrt{[(1-\epsilon_{\text{lim}})-(1+\epsilon_{\text{lim}})\gamma^{-1}M_n^{-2}]^2+4(1-\epsilon_{\text{lim}}\epsilon_\infty^{-1})\gamma^{-1}M_n^{-2}}.\end{aligned}$$

The positive sign on the radical in (1.4.13) corresponds to the trivial solution $\epsilon = 1$. For M_n large (1.4.13) may be expanded as

$$(1.4.14)\qquad \epsilon = \epsilon_{\text{lim}}\left[1+\frac{\epsilon_\infty^{-1}-\epsilon_{\text{lim}}}{1-\epsilon_{\text{lim}}}\gamma^{-1}M_n^{-2}\right]+O(M_n^{-4}),$$

or with $|1-\epsilon_{\text{lim}}\epsilon_\infty^{-1}|$ small, as

$$\begin{aligned}(1.4.15)\quad \epsilon = {}& \epsilon_{\text{lim}}+(1+\epsilon_{\text{lim}})\gamma^{-1}M_n^{-2}-\frac{(1-\epsilon_{\text{lim}}\epsilon_\infty^{-1})\gamma^{-1}M_n^{-2}}{(1-\epsilon_{\text{lim}})-(1+\epsilon_{\text{lim}})\gamma^{-1}M_n^{-2}} \\ & +(1-\epsilon_{\text{lim}}\epsilon_\infty^{-1})^2O(M_n^{-4}).\end{aligned}$$

If the relation

$$(1.4.16)\qquad \epsilon_\infty = \epsilon_{\text{lim}}$$

holds, as is the case with a perfect gas of constant ratio of specific heats, we obtain simply

$$(1.4.17)\qquad \epsilon = \epsilon_{\text{lim}}+(1+\epsilon_{\text{lim}})\gamma^{-1}M_n^{-2}.$$

It is not necessary for the fluid to be a perfect gas for (1.4.16) to hold, but the relation should probably be considered accidental otherwise. However, it is possible to change ϵ_∞ by redefining e and h so that they have a value different from zero at absolute zero temperature. Hence it is possible to satisfy (1.4.16) for one particular shock, but it is not generally possible to satisfy this relation for all shocks with a given ρ_∞ and p_∞.

In the special case of a perfect gas with constant ratio of specific heats this ratio equals γ, and also $\epsilon_\infty = \epsilon_{\text{lim}}$. The limiting density ratio is

$$(1.4.18)\qquad \epsilon_{\text{lim}} = \frac{\gamma-1}{\gamma+1},$$

and (1.4.17) becomes

$$(1.4.19)\qquad \epsilon = \frac{\gamma-1}{\gamma+1}\left(1+\frac{2}{\gamma-1}M_n^{-2}\right).$$

The purpose of the foregoing calculations for the density ratio is to provide a basis for estimating the value of this quantity and for acquiring an understanding of its variations. We note first the consequences of different ways of changing M_n, corresponding to the different ways of applying the basic hypersonic limiting process $M_\infty \to \infty$ (A-lim) discussed in the previous section. If M_n is changed or made to approach infinity in such a way that the state properties behind the shock are unchanged, ϵ_{lim} is unchanged, and (1.4.13) to (1.4.15) above give explicit statements as to the effects of M_n on ϵ. If M_n is changed in such a way that ρ_∞ and v_∞ are unchanged, ϵ_{lim} changes but slightly; for large M_n the quantity ϵ_{lim} is constant within a relative error of $O(M_n^{-2})$, and (1.4.14) is of the correct form for an appropriate description of the effects of M_n. However, in this case the coefficient of M_n^{-2} in an expansion in powers of M_n^{-2} is different from that given in (1.4.14). But if the thermodynamic state in front of the shock is fixed, and M_n is changed by changing v_∞, the value of ϵ_{lim} may vary considerably; these equations for ϵ then give no explicit information on the variation of ϵ with M_n, except for the special case of a perfect gas. Unfortunately, the variations in M_n with the angle of an oblique shock in a given flow are of the latter type.

We next ask how close ϵ is to ϵ_{lim} if M_n is large. Equation (1.4.14) tells us that the relative error in using ϵ_{lim} for ϵ is of the order of $\epsilon_\infty^{-1} M_n^{-2}$. This means that in an analysis in which ϵ_{lim} is used for ϵ it is necessary not only that M_n^2 be large but that $\epsilon_\infty M_n^2$ be large. The quantity $\epsilon_\infty M_n^2$ or its equivalent appears in certain hypersonic analyses as a basic parameter.

What values can ϵ_{lim} have in actual gases? As long as the gas behind the shock is a perfect gas and is physically equivalent to the gas in front, (1.4.18) applies. In this case $\epsilon_{\text{lim}}^{-1}$ is equal to the number of classical degrees of freedom excited, plus one. All monatomic gases have $\epsilon_{\text{lim}} = \frac{1}{4}$, and diatomic gases at moderate temperatures have $\epsilon_{\text{lim}} = \frac{1}{6}$. At higher temperatures in polyatomic gases, vibrational degrees of freedom become excited, and ϵ_{lim} drops moderately.

A striking decrease in ϵ_{lim} occurs only if some physical mechanism appears which causes a large contribution to $h_s + e_s$ without a corresponding contribution to p_s/ρ_s. A mechanism which absorbs energy from the dynamic degr
of freedom of the gas is generally of this type; here the practically im the
examples are dissociation and ionization. The energy of dissociati the
as a potential energy contribution to $h + e$, which does not cor in p/ρ.
temperature or to p/ρ. With dissociation the number of m the effect
gas constant) in a diatomic gas doubles, and this results at elevated
However, the effect of the large energy of dissociati because of
of the increase in the number of molecules in prac ckel [2]). The
temperatures ϵ_{lim} may drop to a value of the ergy to or from a
the effect of dissociation (see, for example. sociation, nonequili-
effect of ionization is similar to that of

If there is a significant time dela
vibrational degree of freedom or

brium phenomena appear. If the gas is at a sufficiently high temperature, it may transfer a significant quantity of energy away from the region of the shock by radiation. Either with nonequilibrium or with radiation we may consider three different possibilities: First, all but a negligible portion of the energy transfer may be accomplished within a thin layer which may be considered as the shock wave itself. In this case only the structure of the shock is affected, except that if there is radiation present the energy lost must be accounted for by a corresponding decrease in the total enthalpy H_n, i.e., by a correction of the energy conservation equation (1.4.1c). Second, there may be only a negligible portion of the energy transferred within a clearly identifiable thickness of the shock wave. In this case the preceding analysis holds again, provided that the equation of state used takes into account the fact that any state variable involved in the nonequilibrium is frozen. The process of equilibration or radiation must then be taken into account in the flow field behind the shock. Third, the situation may be intermediate between the first two, and a significant portion of the energy transfer may occur both within the thickness of the shock and behind the shock. In this case it is difficult to define a thickness for the shock wave and the problem of determining the flow field becomes fundamentally more difficult.

If a nonequilibrium process takes place in the gas but with a characteristic delay time which is sufficiently short (as in the first possibility mentioned above), the departure at any instant of the state of the gas in the flow field from a thermodynamic equilibrium state will be small. In this case we may ignore the nonequilibrium phenomenon and use the "equilibrium" equation of state for the gas. If the characteristic time of the nonequilibrium is sufficiently large, the energy transfer may be negligible not only within the thickness of the shock (as in the second possibility mentioned above) but within the entire flow field of interest. In this case the gas is said to be in frozen equilibrium, and there exists a "frozen" equation of state which we may use to calculate the flow field of interest. In ideal fluid theory a single equation of state is assumed to ...d for the fluid, but it is immaterial whether this equation of state is based ...ermodynamic equilibrium or is a frozen equation of state. In the inter- ... case for which appreciable transfers of energy occur within the flow ...erest behind the shock, a simple equation of state with two independent va... ...s not apply.

... we ... librium ... zen equilibrium fits the requirements of inviscid flow theory present ... treat the fluid as though it were in thermodynamic equi- librium ... hich is of most practical interest to us is air, and Figs. 1–4 ... equilibri... form of a ... modynamic properties of argon-free air in thermodynamic ... n Feldman [1]. Each of these three figures is in the ... with specific enthalpy and entropy as the ordinate and absciss... ... ves of constant pressure, temperature, and the compressibility factor defined in (7.1.34) appear on Fig. 1–4a, and curves of constant density and speed of sound on Fig. 1–4b. On Fig. 1–4c are plotted curves of constant altitude and velocity in front of a normal shock for which the stagnation thermodynamic state behind the shock corresponds to the enthalpy and entropy given. For higher temperature data see Chance Vought [1]

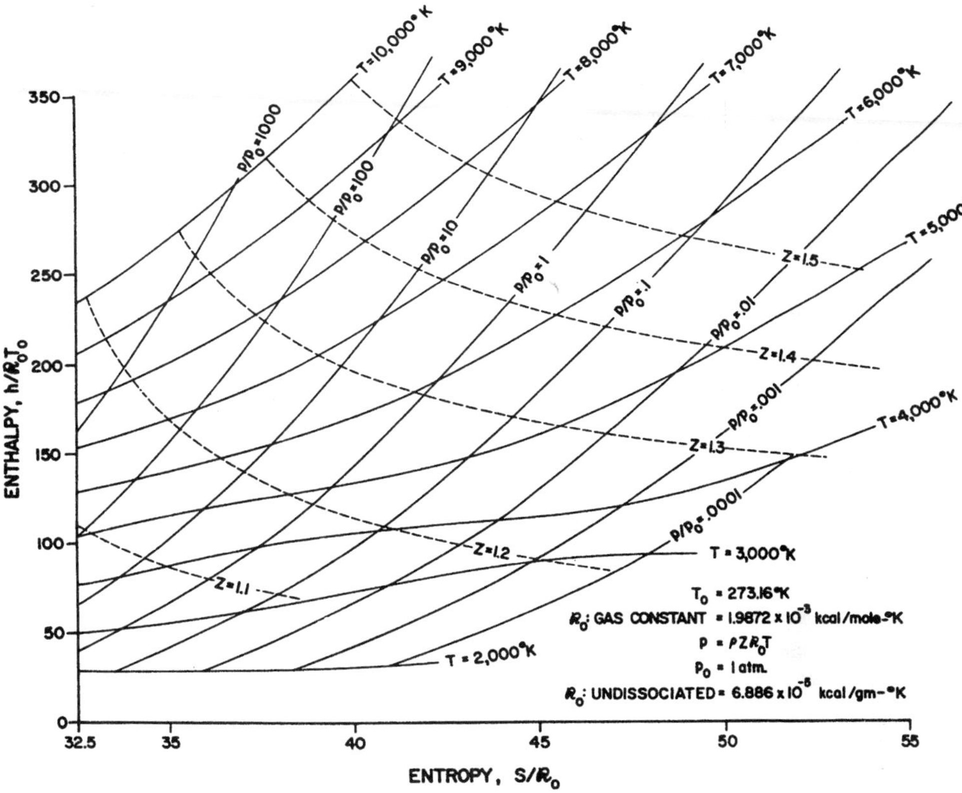

FIG. 1–4a. Mollier diagram for argon-free air. Pressure, temperature, and compressibility factor (Feldman [1]).

and Predvoditelev *et al.* [1; 2]. For a Mollier diagram of more extensive range see ARO [1].

One way of obtaining conditions behind a shock with the aid of such a set of diagrams or an equivalent set of tables is with a successive approximation procedure. Let us assume that the quantities v_∞, p_∞, h_∞, and ρ_∞ are known. An initial guess of ϵ is made, with which values for p_s and h_s are obtained from (1.4.5b, c). The corresponding value of ρ_s is taken from the plot of thermodynamic properties or interpolated from the equivalent set of tables. An improved value of ϵ is then obtained from (1.4.6) and the process is repeated. It may be shown that this procedure is convergent. For air, charts for ϵ and other thermodynamic properties behind normal and oblique shocks are available for various flight speeds and altitudes (see, for example, Feldman [1] and Moeckel [2]).

We turn finally to a discussion of the dimensionless parameters for which the symbol γ is used and which coincide with the ratio of specific heats for a

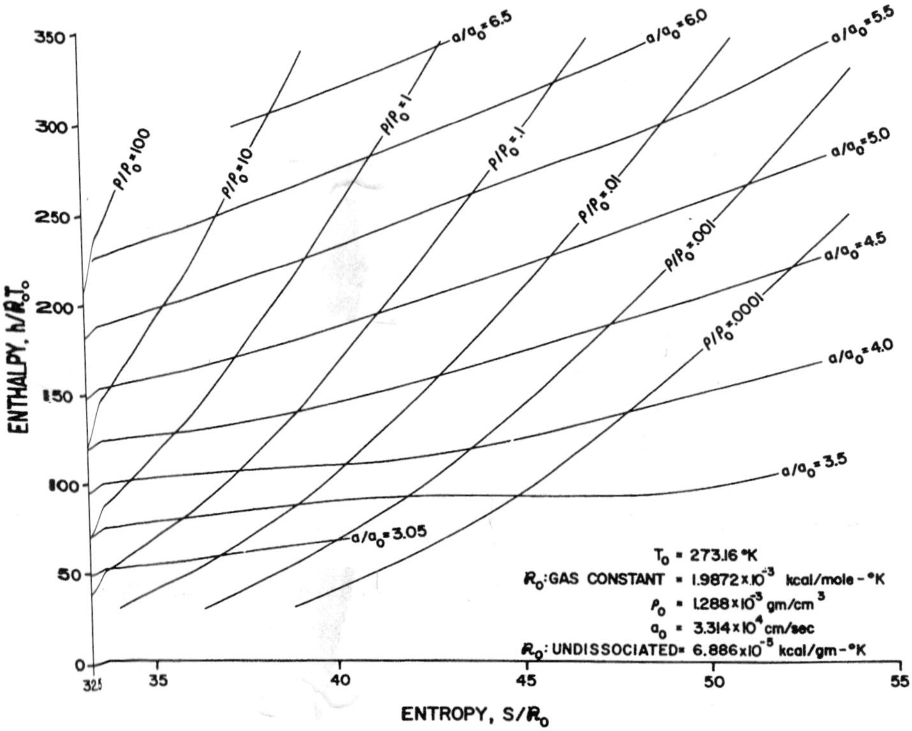

FIG. 1–4b. Mollier diagram for argon-free air. Density and speed of sound (Feldman [1]).

perfect gas when this ratio is constant. We shall define three of these, γ_ϵ, γ_e, and γ_*. The first of these is defined as the ratio of enthalpy to internal energy,

$$\gamma_\epsilon = \frac{h}{e}. \tag{1.4.20}$$

We may note that $\epsilon_{\lim}$ from (1.4.9) may be expressed

$$\epsilon_{\lim} = \frac{\gamma_\epsilon - 1}{\gamma_\epsilon + 1}, \tag{1.4.21}$$

γ_e or γ_* gives us no information with regard to the effects of changes in the entropy of the material.

5. Oblique and curved shocks

Over almost all of its extent, the enveloping shock on a body in hypersonic flow is oblique and curved. We shall extend our results on the normal shock to include the effects of obliquity, and shall examine briefly some of the effects of shock curvature.

Looking first at the case of the oblique shock pictured in Fig. 1–5, we imagine

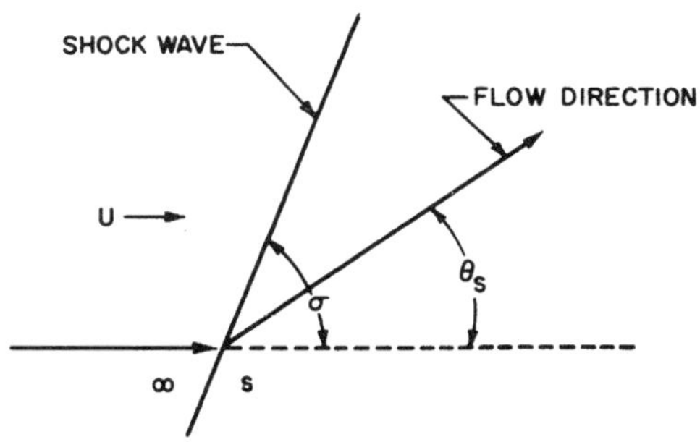

FIG. 1–5. Oblique shock.

an observer who travels along the shock with a velocity equal to

$$u_\infty = u_s = U \cos \sigma, \tag{1.5.1}$$

where U is the free stream velocity and σ is the inclination angle of the shock with respect to the free stream direction. With respect to such an observer the shock wave appears to be a normal one with an upstream velocity and Mach number equal to

$$v_\infty = \epsilon^{-1} v_s = U \sin \sigma, \tag{1.5.2a}$$

$$M_n = U \sin \sigma / a_\infty = M_\infty \sin \sigma. \tag{1.5.2b}$$

The total enthalpy H_n with respect to this observer is not the same as the total enthalpy H of the free stream, but is related to it by

$$H - H_n = \tfrac{1}{2} u_\infty^2 = \tfrac{1}{2} U^2 \cos^2 \sigma. \tag{1.5.3}$$

The thermodynamic properties of the gas on either side of the shock are the same for the moving observer as for a fixed observer. What is most important here is the basic idea that any oblique shock may be considered in terms of a normal one.

We may now calculate the angle between the shock and the flow direction behind the shock. The result is immediate:

$$\tan(\sigma - \theta_s) = v_s/u_s = \epsilon \tan \sigma, \tag{1.5.4}$$

where θ_s is the angle of deflection of streamlines passing through the shock. We may note that if ϵ is small and the shock is not nearly normal, the streamlines behind the shock must lie close to it. If we differentiate (1.5.4) we obtain

$$(1 + \epsilon^2 \tan^2 \sigma)\frac{d\theta_s}{d\sigma} = (1 - \epsilon)(1 - \epsilon \tan^2 \sigma) - \tan \sigma \frac{d\epsilon}{d\sigma}. \tag{1.5.5}$$

With σ equal to the Mach angle in the free stream, $\epsilon = 1$, $\theta_s = 0$, and $d\epsilon/d\sigma < 0$, so that $d\theta_s/d\sigma > 0$. For a normal shock with $\sigma = \frac{1}{2}\pi$, $\theta_s = 0$, and $d\theta_s/d\sigma = -\epsilon^{-1}(1 - \epsilon) < 0$. It is clear that the quantity $d\theta_s/d\sigma$ is zero for at least one intermediate value of σ, given by $\tan \sigma = \epsilon^{-1/2}$ if ϵ is constant. In general there is only one such angle, and this point on the curve of possible shock solutions is termed the detachment point. The flow deflection angle θ_s takes its maximum value at the detachment point. Shocks are termed weak or strong according to whether σ is less than or greater than the value of σ at detachment. However, except when we are considering this type of division, we shall be using the word "strong" to refer to a shock for which M_n is large and ϵ close to ϵ_{lim}.

The sonic point refers to a shock for which the velocity behind the shock is sonic, with the condition

$$u_s^2 + v_s^2 = a_s^2. \tag{1.5.6}$$

In general there is only one such point, and shocks are divided by this point into those termed supersonic and those termed subsonic, according to whether the velocity behind the shock is supersonic or subsonic. Normally the detachment point is subsonic and the sonic point is weak; this is the case in a perfect gas. However, Hayes in unpublished work has shown that an anomalous case is possible, with the detachment point supersonic and the sonic point strong. Such a case may only exist at high free stream Mach numbers. We shall implicitly assume that only the normal case occurs.

With the basic hypersonic limiting process $M_\infty \to \infty$ (A-lim) the sonic and detachment points generally remain distinct. In the special case of a perfect gas with constant ratio of specific heats these two points coalesce in the limit as M_∞ becomes very large.

In considering curved shock waves we note first that the entropy of the gas just behind the shock is a function of the shock inclination angle. The mass flow m across the shock is given by

$$m = \rho_\infty v_\infty = \rho_\infty U \sin \sigma. \tag{1.5.7}$$

If now (1.4.5d), (1.4.7), and (1.5.7) are differentiated with respect to the shock angle, and combined with the differential state equation

$$dh = T\,dS + \frac{1}{\rho}\,dp, \tag{1.5.8}$$

we arrive at a differential relation connecting the shock angle and the entropy behind the shock S_s

$$T_s\,\frac{dS_s}{d\sigma} = U^2 \sin\sigma \cos\sigma\,(1 - \epsilon)^2. \tag{1.5.9}$$

We shall consider here only the relatively simple case of a curved shock in two-dimensional flow, and shall be interested primarily in the vorticity. To consider a more general type of curved shock would only lead us into more complicated mathematics and would not give a more useful result for the vorticity. A left-handed cartesian coordinate system is set up as shown in Fig. 1–6 with origin at the point of interest on the shock. The entropy behind

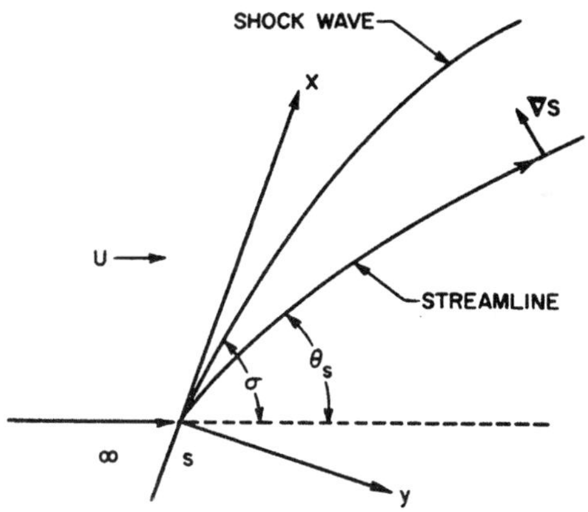

FIG. 1–6. Coordinate system (left-handed) for a curved shock.

the shock is constant along a streamline, and $|\nabla S|$ is used to designate the magnitude of the entropy gradient aligned as indicated in Fig. 1–6. Crocco's vorticity law as applied to a steady isocompositional two-dimensional isoenergetic (constant total enthalpy) flow states simply that the product of the vorticity and the speed is equal to T times the magnitude of the entropy gradient. If we now calculate the magnitude of the entropy gradient in terms of the derivative of entropy immediately behind the shock and express the vorticity law with correct sign we obtain

$$T_s\,|\nabla S_s| = \frac{T_s\,dS_s/dx}{\sin(\sigma - \theta_s)} = -\frac{U\cos\sigma}{\cos(\sigma - \theta_s)}\,\zeta_s\,, \tag{1.5.10}$$

where ζ is the vorticity defined here as

$$\zeta = \frac{\partial u}{\partial y} - \frac{\partial v}{\partial x} \tag{1.5.11}$$

in the left-handed coordinate system of Fig. 1–6. Combining (1.5.10) with (1.5.4) and (1.5.9) gives us

$$\zeta_s = - U\epsilon^{-1}(1 - \epsilon)^2 \cos \sigma \frac{d\sigma}{dx}. \tag{1.5.12}$$

The term $d\sigma/dx$ is simply the curvature of the shock, and (1.5.12) gives a direct relation between shock curvature and vorticity behind the shock.

The relation (1.5.12) was first found by Truesdell [1], and has been later rediscovered by others who were unaware of Truesdell's work. Lighthill [3, pp. 14 and 15] found an extension of the result to a general curved shock, obtaining the vorticity immediately behind the shock in terms of the principal curvatures of the shock surface. Hayes [5] obtained a generalized result with a purely dynamic derivation in which the constancy of total enthalpy and Crocco's vorticity law are not used. The derivation of (1.5.12) utilizing Crocco's law has been followed above in order to emphasize the connection between entropy gradient and vorticity.

The vorticity produced by a curved shock is clearly strongly dependent upon ϵ, and becomes very large as ϵ becomes small for a given shock inclination and curvature. However, behind a point on a shock for which the shock is normal, $\cos \sigma = 0$, and the vorticity is zero. Relation (1.5.12) may be readily shown to be valid in axisymmetric flow as well as in two-dimensional flow. If ζ_s is taken to be the component of vorticity perpendicular to both the streamline immediately behind the shock and the free stream direction, (1.5.12) is valid in general.

6. Mach number independence principle

The definition of hypersonic flow used by Oswatitsch [1; 2] involved the concept of the basic hypersonic limiting process $M_\infty \to \infty$ (A-lim). On the basis of his analysis of the flow of a perfect gas, he concluded that in the limit of very large values of M_∞, the flow pattern and pressure coefficients on a body were independent of the value of M_∞. The same result had been found by Sir Isaac Newton (see Section 3.1), for his model of a rarefied gas. We shall restate this important idea in a way which does not depend upon the gas being a perfect one. Chernyi [3] credits this principle to S. V. Vallander (1949), but gives no reference.

The word similitude usually is used to refer to an equivalence between two physical problems which are different from each other in some fundamental way. Oswatitsch stated his result as a similitude. An inviscid perfect gas with constant ratio of specific heats is a self-similar fluid which for any flow field permits the general Mach number similitude obtained from dimensional analysis. The general similitude expressed by Oswatitsch utilizes this property of a perfect gas. When the result is properly restated for a general ideal (inviscid) fluid, it really states that in the limit $M_\infty \to \infty$ two flows with different values of M_∞ are fundamentally identical, i.e., are not different from each other in any fundamental way. Although the result expressed for a general fluid may certainly be termed a similitude, it is preferable to consider it as an "independence principle". Such an independence principle is thus a special type of a similitude which is stronger than a general similitude.

In order that the flow field may be independent of the free stream Mach number M_∞ it is necessary that at any point on the shock the density ratio ϵ be independent of M_∞. From (1.4.14) it may be seen that M_∞ must be sufficiently high that M_n^{-2} or $(M_\infty \sin \sigma)^{-2}$ will be small compared with ϵ_∞ and that ϵ may be replaced by $\epsilon_{\lim}$. In order that $\epsilon_{\lim}$ be independent of M_∞ as M_∞ approaches infinity it is necessary that ρ_∞ and U be fixed in the basic hypersonic limiting process (A-lim) and that a_∞ approach zero. These considerations have already been briefly discussed in Section 1.4. The requirement that $(M_\infty \sin \sigma)^{-2}$ be small is equivalent to our strong shock assumption (assumption C), which is thus a basic assumption for the independence principle. If the quantity ϵ_∞ is itself small, we need a stronger form of the assumption (C-strong).

With the condition that M_∞ is sufficiently large that the strong shock approximation is valid, we may state the Mach number independence principle: The flow around a body and behind the bow shock depends only upon the density ρ_∞ and the uniform velocity U of the given gas in the free stream, and is independent of the free stream pressure p_∞, enthalpy h_∞, temperature T_∞, and speed of sound a_∞. A physical interpretation of the independence principle is that in a flow with ρ_∞ and U fixed the entire flow field becomes frozen as a_∞ is reduced to extremely small values, and a proper limiting flow field is approached.

The justification of this principle comes readily out of the results of the previous two sections. The flow of interest about the body is separated from the free stream by a shock wave which is everywhere strong and for which ϵ may be taken to be $\epsilon_{\lim}$. The pressure, enthalpy, and other quantities immediately behind the shock are dependent only upon ρ_∞, U, and σ, and may be considered independent of p_∞ and h_∞. Conditions immediately behind the shock may be considered to serve as boundary conditions for the flow field. Thus, a flow solution obtained for one sufficiently large value of M_∞ will serve for another large value of M_∞ if ρ_∞ and U are the same. With the assumption of uniqueness of the flow solution the independence principle follows immediately.

Since the flows behind the shock at two appropriate values of M_∞ are essentially identical under our conditions, it is clear that the independence principle applies also to real-fluid effects such as those due to viscosity, and to the physical gasdynamics effects mentioned in the first section. The only possible exception to this is the structure of the shock wave itself. However, the mean molecular collision time in the gas in the free stream is much greater than the transit time of the gas through the shock wave, and the vector velocity of individual molecules differs but negligibly from the free stream velocity. We may conclude that the independence principle applies also to the structure of the shock wave.

The Mach number independence principle as we have stated it thus holds for boundary layers in hypersonic flows provided that the exterior inviscid flow follows the independence principle. Where the Mach number is sufficiently large, as for what is termed the strong interaction region (see, for example, Dorrance [1], or our first edition), the independence principle holds even though the inviscid flow field is completely determined by the boundary layer. The independence principle does not hold on a flat plate for what is termed the weak interaction region. The independence principle does hold for rarefied gas flows and free molecule flows with $M_\infty \sin \theta_b$ large.

Mathematically phrased, the Mach number independence principle states that if ρ_∞ and U are fixed, the solution within a fixed finite domain approaches a limiting solution uniformly in the limit $M_\infty \to \infty$.

As we have already remarked, as long as M_∞ is finite, there is some point on the tail of the shock for which assumption C fails, and the independence cannot apply to the entire flow field including the shock tail. However, the body and the subsonic part of its wake are within the zone of action of only a relatively limited portion of the shock, and the flow on the body cannot be affected by the solution for the shock tail. Thus the failure of the independence principle on the shock tail is of no significance for the flow near the body.

7. General strip theory

There is another general principle which comes out of a well-known concept in aerodynamics; this concept goes under the name of "strip theory", and we use the same term for the principle. The principle is that if the shape of a body in a given flow field is altered in such a way that a suitably defined aspect ratio $\mathcal{R}$ approaches infinity, the flow about the body approaches a state which can be described as independent of the aspect ratio. This principle, although applying generally to all nontransonic flows, is particularly important in hypersonic flows. This is because with slender flat bodies the requirement for validity of strip theory is one which is not at all severe. We may note that in this principle the flow field is fixed and one parameter of the body geometry is made very large, while in the Mach number independence principle the body geometry is fixed and one parameter of the flow field is made very large.

To illustrate the principle we first consider a body of span b which lies along a z axis normal to the undisturbed flow direction. This axis serves as a basic axis, and the body occupies the interval $-\frac{1}{2}b \leqslant z \leqslant \frac{1}{2}b$ along it. The shape of the body is characterized by its cross sections normal to the z axis. These cross-sectional shapes are taken to depend smoothly only upon the variable $2z/b$, which varies between -1 and 1. The largest dimension of the cross sections is of the same order of magnitude as a reference chord length c, and the aspect ratio is defined by $AR = b/c$.

A limiting process is now considered, in which c is fixed and $b \to \infty$, so that $AR \to \infty$. In the equations of motion (with no restriction to inviscid flow) all partial derivative terms in $\partial/\partial z$ become negligibly small in this limiting process. Physically, this means that the flow locally approaches a two-dimensional flow. If the aspect ratio is large enough ($AR \gg 1$) that the flow closely approximates a two-dimensional flow locally, it may be considered to be independent of the actual value of AR. It can be shown that the total forces and total heat flow on the body divided by AR are then independent of AR.

This principle should be considered as an independence principle in the sense of the last section, rather than as a similitude. The local flows at corresponding points in two flows of sufficiently large but different values of AR are not different from each other in any fundamental way. Thus, the result applies with respect to all real-fluid effects.

An immediate generalization is obtained by allowing the basic axis to be yawed at a fixed angle relative to the free stream direction. The local flow then approaches that about an infinite yawed two-dimensional body. A requirement here is that the inclination angle α of the axis relative to the free stream direction (the complement of the yaw angle) is not too small, by which we mean $AR \sin \alpha \gg 1$. An additional generalization is obtained by letting the basic axis be a space curve which is required to expand affinely with b under the limiting process. The body is described in terms of cross sections locally normal to the axis curve.

Here the characteristic curvature K of the axis curve must not be too great, so that $cK \ll 1$, and the curve should be projectible on a (y, z) plane without multiple points. These statements, although susceptible to mathematical demonstration, are most easily understood physically.

With the curvature of an axis curve small enough we may simply ignore the curvature and apply the principle completely locally. We conclude that the local flow depends only on the local cross-sectional shape and the angle of yaw, within the arbitrariness permitted by the invariance of the free stream flow under translation or rotation about the flow axis. No result for total forces comes out of this localized strip theory principle, of course.

As is also true with the Mach number independence principle, the strip theory principle can be applied to unsteady flows. The formulation of the conditions needed for unsteady flows is omitted here.

In analogy with the case of the Mach number independence principle, as long as Æ is finite, there is some point downstream of the body where the principle fails, and the flow is not locally like that on a two-dimensional infinite body. In addition, failure of smoothness, such as discontinuities in cross section or corners in a general axis curve, results in a local failure of the principle which does not generally imply any overall failure of the principle. In addition we may note that in hypersonic flows the conditions on the magnitude of Æ needed for validity may be more stringent where the cross-sectional shape is blunted relative to the transverse flow than where the shape is sharp.

The independence principles of this and the last section can be applied simultaneously, as a combined strip theory-Mach number principle. Here both M_∞ and Æ need to be very large, with ρ_∞, U, c, and α (as appropriate) held fixed. The manner of passing to the limit in the conceptual double limiting process involved does not affect the result.

8. Dissipative effects

The flow of any fluid is influenced by a number of effects which are usually termed "real-fluid" effects, and which are neglected in ideal-fluid theory. These effects have their roots in the behavior of individual molecules, and the parameters that are important are the mean free path of kinetic theory and the mean free time between one collision and the next for a given molecule. A basic distinction must be made among the cases for which the mean collision time of the gas is small, is of the same order of magnitude, or is large compared with an appropriate characteristic macroscopic time of the flow field. In some cases it may be more convenient to make the corresponding distinction with respect to the mean free path. If chemical changes or transfers of energy from one mode to another are involved, a characteristic reaction or relaxation time may be involved. Such a characteristic reaction time may be far different from the mean collision time. An analogous distinction should be made among various cases with respect to the ratio of the characteristic reaction time and an appropriate characteristic macroscopic time of the flow field.

Real-fluid effects appear if the ratio of the characteristic molecular times (collision or reaction) to the characteristic macroscopic time is not negligible. A common property of all real-fluid effects is that they are dissipative. Dissipation is associated with irreversibility, and is usually identified with an increase in world entropy. We make no attempt here to define either local or world entropy in a general context, but note only that appropriate quantitative measures of dissipation are available with gas flows of practical interest. In any case, we shall consider "real-fluid" and "dissipative" as synonymous.

If the mean free path and the mean collision time are sufficiently small and the velocity and temperature gradients moderate we may consider the gas as a continuum in quasi-equilibrium for which a simple equation of state may be used. This equation of state may be somewhat more complicated than the usual equilibrium equation of state with two independent variables (with one phase and fixed composition), but should have a restricted number of independent variables. In this latter case additional determining equations beyond the basic hydrodynamic equations are required to determine the solution. The various characteristic molecular times are parameters in these additional equations.

If with small collision time a characteristic reaction time is extremely large, so that the ratio of macroscopic to reaction time is negligibly small, we can simply neglect the corresponding reaction. In this case the fluid is "frozen" with respect to that reaction. Locally, we can consider the flow as nondissipative with respect to that reaction. Eventually, the reaction must proceed toward equilibrium, and this process is then dissipative.

If, on the other hand, the reaction times as well as the collision times are all very small compared with the characteristic macroscopic time, the time delay in the process of equilibration may be neglected. In this case the fluid is considered to remain in thermodynamic equilibrium. Through most of this volume we assume this is the case.

If the collision times are not small the problem of defining the state of the fluid becomes more difficult. The state cannot be determined through a restricted number of independent variables but requires the specification of one or more functions, with functional equations to determine the solution. For example, in monatomic kinetic theory with neutral particles the distribution function is required, with the Boltzmann equation as the required determining functional equation. If the ratio of macroscopic time to collision time in this case is negligibly small we are in the free molecule regime. In analogy with the frozen case above, free molecule flow is locally nondissipative in the absence of wall interactions.

We next draw a distinction between nonequilibrium effects and viscous effects. This distinction is most easily formulated in the continuum case (with small collision times) but extends to the general case. The distinction is based upon the concept of spatial homogeneity. Spatial homogeneity is the property that spatial gradients of all state variables are all zero, that the ∇ operator applied to any dependent variable is zero. We define a nonequilibrium effect as a dissipative effect which is present with spatial homogeneity. We define a viscous effect as a dissipative effect which depends inherently upon spatial gradients.

A minor difficulty appears here, and is connected with the continuity equation. With the strict definition of spatial homogeneity the density of the fluid could not change with time. This limitation is far too restrictive, and we extend the concept of spatial homogeneity to permit isotropic changes of volume of a parcel of fluid with time.

With the continuum assumptions the only change appears with respect to the velocity gradient, by including cases for which

$$\nabla \mathbf{q} = \beta \nabla \mathbf{r} \tag{1.8.1}$$

within the scope of spatial homogeneity. The scalar β is connected with the change in the volume V by

$$3\beta = \nabla \cdot \mathbf{q} = \frac{d \ln V}{dt} \tag{1.8.2}$$

In the case of the Boltzmann equation, the distribution function $f(\mathbf{c}, \mathbf{r}, t)$ is permitted to have the property

$$\nabla_{\mathbf{r}} f = \beta \nabla_{\mathbf{c}} f. \tag{1.8.3}$$

A dissipative effect which is defined as a nonequilibrium effect because it arises out of a density change with time has a dual nature. Because of its connection with a spatial gradient it could also be considered viscous, even though it is connected only with the isotropic part of the gradient. Bulk or dilatational viscosity is one example of a dissipative effect of this type. Here we consider these effects nonviscous.

A nonequilibrium effect (nonviscous) in a continuum is generally connected with one of the additional determining equations, in which a characteristic reaction time t_r multiplied by a time derivative is equated to a term of differential order zero. Operationally we can write

$$-t_r \frac{D}{Dt} \approx 1, \tag{1.8.4}$$

where the operators act on one of the state variables. If t_r is small this yields a singular perturbation problem associated with an initial value problem. A singular perturbation problem is a problem in differential equations connected with the highest order derivative appearing with a vanishingly small parameter as a factor. Thin layers somewhat analogous to boundary layers may appear behind a wave surface such as a shock. An example of such a layer is a relaxation zone behind a shock. See, for example, Clarke and McChesney [1, Section 6.14], and Sections 5.10, 6.6, and 7.5.

A viscous effect in a continuum is generally connected with a determining equation, in which a characteristic kinematic "viscosity" ν multiplied by a Laplacian operator is equated to a total time derivative. Operationally we can write

$$\nu \nabla^2 \approx \frac{D}{Dt}. \tag{1.8.5}$$

The quantity ν is a parameter that has the same dimensions as the kinematic viscosity, i.e. L^2/T. It may be connected with shear viscosity, of course, or with other characteristically viscous effects such as heat conduction or diffusion. If ν is small we again have a singular perturbation problem, but this time associated with boundary conditions on a wall or boundary conditions arising from a contact discontinuity. The thin layers associated with such singular perturbation problems are the classical boundary layers, shear layers, and wakes.

The quantity ν is essentially a characteristic collision time multiplied by a characteristic velocity squared. This velocity is of the order of a velocity of sound. Thus we may write

$$\nu \approx t_c a^2. \tag{1.8.6}$$

Note that (1.8.4) and (1.8.5) are consistent in orders of magnitude if a wave equation of the operational form

$$a^2 \nabla^2 = \frac{D^2}{Dt^2} \tag{1.8.7}$$

holds, with t_r replaced by t_c.

In our discussion of viscous effects we have lumped together various classical continuum effects under the same name. Each of these is connected with the gradient of a particular dependent variable. Thus classical viscosity is connected with velocity gradient, heat conduction with temperature gradient, and diffusion with concentration gradient.

When can we consistently take one effect into account and not another? This evaluation must be based primarily upon the characteristic times or kinematic coefficients involved. We may neglect viscosity compared with a nonequilibrium effect only if the corresponding reaction time t_r is very much larger than the collision time t_c. In most cases this criterion does not hold for bulk viscosity considered as a nonequilibrium effect. We may neglect viscosity compared with heat conduction, for example, only if the kinematic viscosity is very much smaller than the corresponding ν for heat conduction.

In this volume viscous effects are not treated, and we shall consider only nonequilibrium effects. These effects, however, are generally considered separately, and the major portion of the volume is based upon the assumptions of an ideal fluid, with no dissipation except that engendered by shock waves. For the nonequilibrium effects considered, consistency with the neglect of viscosity requires that t_r not be too small.

CHAPTER II

SMALL-DISTURBANCE THEORY

1. Introduction and basic equations

The variety of hypersonic flow theory appropriate for slender bodies is termed the small-disturbance theory. Here the word small has been used because the velocity perturbations are small compared with the free stream velocity and the pressures are small compared with the free stream dynamic pressure. However, the velocity perturbations are not small compared with the free stream sound velocity, and the pressure perturbations are not small compared with the free stream static pressure. Hence the disturbances are not at all small in the sense usually associated with linearized theory, and the small-disturbance theory is an essentially nonlinear one.

The small-disturbance theory is inherently connected with classical hypersonic similitude; the similitude may be considered as a direct result of the theory, or the basic equations of the theory appear as a consequence of applying the similitude concept. The idea of hypersonic similitude is due to Tsien [1], who investigated the two-dimensional and axisymmetric irrotational equations of motion. By showing equivalence of a steady hypersonic flow on a slender body with an unsteady flow in one fewer space dimensions, Hayes [1] pointed out that the similitude should apply to three-dimensional slender bodies in rotational flow. Subsequently, hypersonic similitudes with viscosity, small blunting, and large angles of attack have been studied. In this volume we shall try to present a unified picture of hypersonic similitude in general.

The establishment of hypersonic similitude, however, does not provide hypersonic flow solutions. The appropriate equations of motion must be written down and solutions obtained. Although some solutions were obtained previously by Goldsworthy [1] and Linnell [1] for particular cases, the first comprehensive study made outside of Russia was that of Van Dyke [2]. The Russian work is discussed at the end of this section. Van Dyke also coined the term "small-disturbance theory". We shall here follow a development which is similar to his, but with two important differences. We shall use a development which is suitable for a general inviscid fluid instead of only for a perfect gas, and which is suitable for treating unsteady flows. We shall also be following more closely the original equivalence idea of Hayes [1], in which no restriction on the form of the equation of state was implied.

A parameter τ is introduced, whose purpose will be to serve as a measure of the maximum inclination angle of Mach waves in the flow field. In order to be able to define it in terms of known quantities, we shall require in the steady flow case that it be of the order of the maximum body inclination angle. Thus we define it as the thickness ratio or as the angle of attack of the body, whichever is larger. In unsteady flow we should use a maximum value of the lateral velocity of the surface divided by the free stream velocity if that quantity is of larger order than the thickness ratio. Thus τ is always defined in terms of the maximum inclination of individual particle paths near the body. If $M_\infty \tau$ is of the order of one or larger, the quantity τ as defined will serve as a measure of the maximum inclination angle of the Mach waves. If $M_\infty \tau$ is small, corresponding to linearized or almost linearized flow, we must replace τ by M_∞^{-1} where it is used as a measure of relative orders of magnitudes of various quantities. We shall assume in general that $M_\infty \tau$ is not small, and consider separately the modifications of the theory necessary when $M_\infty \tau$ is small. The slender-body assumption $\sin \theta_b \ll 1$ (B) which underlies the small-disturbance theory is equivalent to the requirement that τ is small. Although we shall be treating primarily the steady case, we shall keep the formulas in a form appropriate for discussing the unsteady case.

We assume that at the shock the inclination of individual particle paths is also of the order of magnitude τ. Considering the steady case, and assuming that the shock is not a nearly normal one, we may conclude from (1.5.4) or may cite the result (4.1.10) obtained later that

$$\tan \sigma = \frac{\tan \theta_s}{1 - \epsilon} + O\left(\frac{\epsilon \tan^3 \theta_s}{(1 - \epsilon)^3}\right). \tag{2.1.1}$$

Since $\tan \theta_s$ is of the order of magnitude τ so also is $\tan \sigma$ presuming only that $1 - \epsilon$ is not small.

With the shock inclination angle small, with $\tan \sigma$ of the order of τ, we conclude not only that the lateral velocity components behind the shock are of order $U\tau$, but that the axial perturbation velocity component is of order τ times this, or of order $U\tau^2$. This suggests the transformation

$$q_x = \tau q_x' , \tag{2.1.2}$$

through which a reduced axial perturbation velocity component is introduced which is of the same order of magnitude as the lateral velocity. The subscript on the velocity component q indicates its direction, and x is the axial cartesian coordinate.

The order of magnitude of the speed of sound behind the shock is $\epsilon^{1/2} U \sin \sigma$. With this result we note that the Mach angle behind the shock is of the order of magnitude $\epsilon^{1/2}\tau$, and that the flow there is still hypersonic. Since $\epsilon^{1/2}$ is less than one, the inclination of the characteristics (in steady flow) is of the order of magnitude τ. This suggests the transformation

$$x = \tau^{-1}x', \tag{2.1.3}$$

through which reduced derivatives of various quantities in the axial direction are of the same order of magnitude as lateral derivatives.

With the transformations (2.1.2) and (2.1.3) we are ready to write down the equations of motion for the flow. We shall use vector notation, with the understanding that the vectors are two-dimensional, defined with respect to the (y, z) space of the lateral coordinates. In accordance with the equivalence concept, the equations are expressed with respect to an observer who is fixed with respect to the fluid in the undisturbed free stream. In the coordinate system of such an observer, the axial velocity which appears is the perturbation, and the free stream velocity U appears only in the formulation of boundary conditions.

The equations of continuity, momentum, and entropy are now expressed for the flow field:

$$\frac{\partial \rho}{\partial t} + \nabla \cdot (\rho \mathbf{q}) = -\tau^2 \frac{\partial \rho q_x'}{\partial x'}, \tag{2.1.4a}$$

$$\frac{\partial \mathbf{q}}{\partial t} + \mathbf{q} \cdot \nabla \mathbf{q} + \frac{1}{\rho} \nabla p = -\tau^2 q_x' \frac{\partial \mathbf{q}}{\partial x'}, \tag{2.1.4b}$$

$$\frac{\partial q_x'}{\partial t} + \mathbf{q} \cdot \nabla q_x' + \frac{1}{\rho} \frac{\partial p}{\partial x'} = -\tau^2 q_x' \frac{\partial q_x'}{\partial x'}, \tag{2.1.4c}$$

$$\frac{\partial S}{\partial t} + \mathbf{q} \cdot \nabla S = -\tau^2 q_x' \frac{\partial S}{\partial x'}. \tag{2.1.4d}$$

The entropy equation here replaces the more usual energy equation. We are assuming the flow to be nondissipative, and the entropy equation is more convenient for our purposes.

The boundary conditions at infinity are simply that the velocity is everywhere zero, and that the thermodynamic state of the fluid is uniform and is given. Thus ρ_∞, p_∞, h_∞, and a_∞ are known. The location of the surface of the body is expressed by an equation of the form

$$f(x', y, z, t) = 0. \tag{2.1.5}$$

The boundary condition to be satisfied on the body is

$$\frac{\partial f}{\partial t} + \mathbf{q} \cdot \nabla f = -\tau^2 q_x' \frac{\partial f}{\partial x'}. \tag{2.1.6}$$

In linearized and transonic small-disturbance theories a useful and permissible procedure is to satisfy this boundary condition on a suitable cylindrical mean surface instead of on the body itself. In hypersonic flow this procedure is never permissible because of the large gradients of the flow variables normal to the surface, and the boundary condition (2.1.6) must be satisfied on the boundary surface proper. The original paper of Tsien [1] was in error on this small but essential point.

The shape of the shock wave is given by an equation similar to (2.1.5), by

$$g(x', y, z, t; \tau) = 0. \tag{2.1.7}$$

To write the shock conditions we need an expression for the normal component of the relative velocity of the fluid with respect to the moving shock surface. We choose a vector $\mathbf{q}_{\text{shock}}$ which is the velocity of a point moving so that it stays on the shock surface; $dg/dt = 0$ for such a point. We define the relative velocity $\mathbf{q}_r$ by

$$\mathbf{q}_r + \mathbf{q}_{\text{shock}} = \mathbf{q}. \tag{2.1.8a}$$

Expanding dg/dt and using (2.1.8a) we obtain

$$\mathbf{q}_r \cdot \nabla g = \frac{\partial g}{\partial t} + \mathbf{q} \cdot \nabla g + \tau^2 q'_x \frac{\partial g}{\partial x'} . \tag{2.1.8b}$$

The normal component of $\mathbf{q}_r$ is then

$$q_{rn} = \frac{1}{|\nabla g|}\left(\frac{\partial g}{\partial t} + \mathbf{q} \cdot \nabla g + \tau^2 q'_x \frac{\partial g}{\partial x'}\right). \tag{2.1.8c}$$

This quantity does not have the arbitrariness permitted in the choice of $\mathbf{q}_{\text{shock}}$. The component q_{rn} is to be substituted in place of v in the shock conditions (1.4.1). These conditions are all expressed in terms of the change in various quantities across the shock. Brackets are introduced here to mean the difference between the quantity inside the brackets after the shock and the same quantity before the shock. Thus, for example, the pressure jump is written as $[p] = p_s - p_\infty$. With this notation the shock conditions become

$$\left[\rho\left(\frac{\partial g}{\partial t} + \mathbf{q} \cdot \nabla g\right)\right] = -\tau^2\left[\rho q'_x \frac{\partial g}{\partial x'}\right], \tag{2.1.9a}$$

$$\begin{aligned}&\left[\rho\left(\frac{\partial g}{\partial t} + \mathbf{q} \cdot \nabla g\right)^2 + (\nabla g)^2 p\right] \\ &\qquad = -\tau^2\left[2\rho\left(\frac{\partial g}{\partial t} + \mathbf{q} \cdot \nabla g\right) q'_x \frac{\partial g}{\partial x'} + \left(\frac{\partial g}{\partial x'}\right)^2 p\right] + O(\tau^4),\end{aligned} \tag{2.1.9b}$$

(2.1.9c)
$$\left[\frac{1}{2}\left(\frac{\partial g}{\partial t} + \mathbf{q}\cdot\nabla g\right)^2 + (\nabla g)^2 h\right] = -\tau^2\left[\left(\frac{\partial g}{\partial t} + \mathbf{q}\cdot\nabla g\right)q_x'\frac{\partial g}{\partial x'} + \left(\frac{\partial g}{\partial x'}\right)^2 h\right] + O(\tau^4).$$

In addition we have a condition corresponding to the two components of momentum not included in (2.1.9b), that the direction of the velocity jump is normal to the shock wave

(2.1.10a)
$$[\mathbf{q}] = A\,\nabla g,$$

(2.1.10b)
$$[q_x'] = A\frac{\partial g}{\partial x'}.$$

In (2.1.10) the quantity A is a scalar proportionality factor, a function of position on the shock surface. Since A is initially undetermined, the two-dimensional vector equation (2.1.10a) expresses a single condition. With the values of A specified by (2.1.10a), equation (2.1.10b) expresses another single condition.

With τ sufficiently small we may neglect terms of order τ^2 in (2.1.4) to (2.1.9), so that the right-hand sides of all these equations may be taken to be zero. We are here making the slender-body assumption (B). Omitting (2.1.4c), equations (2.1.4) to (2.1.10a) are the correct equations governing two-dimensional unsteady flow, with the variable x' appearing only as a parameter. These equations give an analytic formulation of the principle that a three-dimensional hypersonic slender-body flow is equivalent to a two-dimensional unsteady flow. Once a solution to the two-dimensional unsteady problem has been obtained, (2.1.4c) and (2.1.10b) may be solved to obtain q_x'. The equations (2.1.4) to (2.1.10) with terms of order τ^2 dropped are referred to as the small-disturbance equations.

An observation of Van Dyke [2] on the order of error is of interest and is quoted here: "The error in the various first-order small-disturbance theories decreases progressively from $O(\tau^{2/3})$ in transonic flow to $O(\tau)$ in linearized supersonic flow to $O(\tau^2)$ in hypersonic flow. Therefore, under the plausible assumption (confirmed by later examples) that these mathematical order estimates give a reasonable indication of the actual physical magnitude of error, the practical need for a second-order solution is seen to be greatest for transonic flow and least at hypersonic speeds."

We turn now briefly to the case of linearized or almost linearized flow, with $M_\infty\tau$ small compared with one. The same analysis we have made above could then be made with M_∞^{-1} in place of τ, with the same arguments that q_x' is of the same order of magnitude as the other unreduced velocity components and that derivatives with respect to x' are of the same order of magnitude as derivatives with respect to the other unreduced coordinates. The terms to be dropped in order to obtain the small-disturbance equations would then be of order M_∞^{-2}. It is not necessary to change our analysis above, however, if we simply recognize that the right-hand terms bearing the factor τ^2 are really of order M_∞^{-2}, and that, for example, q'_x as defined by (2.1.2) is really of order $(M_\infty \tau)^{-1}$ times the other unreduced velocity components. Thus we are able to include the case of small $M_\infty \tau$ in our original analysis. But it is clear that we must assume not only that τ^2 is small but that M_∞^{-2} is small, and that the basic hypersonic assumption (A) is needed for the small-disturbance theory in addition to the slender-body assumption (B).

The concept of the equivalence of a three-dimensional physical problem to a two-dimensional unsteady problem with a parameter or a two-dimensional or axisymmetric physical problem to a one-dimensional unsteady problem with a parameter has been very fruitful. This is because this concept has permitted immediate adaptation of certain known unsteady gasdynamic solutions to hypersonic problems and because it has provided a reduction of one in the number of essential independent variables in many unsteady problems. In addition, it gives an immediate physical conceptual picture of the meaning of hypersonic similitude. We shall refer to this concept as the equivalence principle. This principle states, in essence, that the flow as viewed in any transverse plane is independent of the flow in any other transverse plane. In the two-dimensional physical case we may liken the flow in a given transverse plane to the flow in a cylinder driven by a piston.

In case the basic flow field is a steady one the dependence upon x' and t is always a dependence upon the single variable $x' + U\tau t$. We have then, for example, that

$$\frac{\partial f}{\partial t} = U\tau \frac{\partial f}{\partial x'} \tag{2.1.11}$$

and the basic equations may readily be changed to the steady-state form for an observer moving with the body. If the time-varying point of view is kept, the quantity $-(U\tau)^{-1}x'$ is a phase parameter indicating the delay or advance in time of the motion at the particular value of x' as compared with that at $x' = 0$. In this case x' is not a basic independent parameter, and the solutions with different values of x' are identical except for a shift in the time scale equal to the phase parameter as defined above. On the other hand, in the unsteady case x' is a basic independent parameter, and a different two-dimensional unsteady problem must be solved for each value of x'.

In the treatment above the quantity τ is considered as a constant which is small. To establish a similitude or to set up an expansion scheme to improve the small-disturbance solution we must consider τ as a parameter and the functional behavior with respect to τ of the various quantities occurring must be investigated. With the body shape given we may consider the function f

of (2.1.5) independent of τ, but the shock shape function g of (2.1.7) and the other variables of the problem are not. A direct procedure is to expand all quantities involved in power series in τ^2. Equations (2.1.4) to (2.1.10) yield successive sets of equations independent of τ by the process of equating like powers of τ^2. Investigations of Waldman [1] have shown that an impressive improvement may be obtained by this method for moderately thick bodies at high Mach number. An empirical way of taking finite thickness into account is to use the sine of the maximum inclination angle for τ in place of the tangent in the hypersonic similitude parameter (see Van Dyke [1]). Waldman found that the τ^2 expansion method did not succeed well in extending the range of validity of small-disturbance theory to lower supersonic Mach numbers. A better approach for this purpose is through the unified similitude described in Section 2.3.

The well known slender-body theory of subsonic, transonic, and supersonic flow (for bodies which have small aspect ratio as well as small thickness ratio) appears at first glance to be very similar to the hypersonic small-disturbance theory. In the slender-body theory the flow pattern is considered to be two-dimensional, with the axial variable acting as a parameter. This similarity is illusory, and there are significant differences between the two theories. Besides the requirement of small angles of incidence there are other very restrictive requirements for the validity of the slender-body theory, not only that the aspect ratio be small but also that the incidence angle vary smoothly along the body. The two-dimensionality of the slender-body theory is only a local property which holds sufficiently near the body, and does not exist over the entire flow field. And the pressure on the body is determined by the theory with an arbitrary additive function of the axial variable; this function can only be determined with the aid of the full three-dimensional equations, imposing conditions relatively far from the body. Thus the slender-body theory, useful in its place, lacks by far the generality of the hypersonic small-disturbance theory.

In Russia, some developments of the small-disturbance theory were made at an early date. Following the concept of the equivalence principle stated by Hayes [1], Bam-Zelikovich, Bunimovich, and Mikhailova [1] developed an appropriate theory and gave a more complete proof of the equivalence principle. Il'yushin [1] developed the theory independently, and proved the equivalence principle.

Il'yushin's term for the equivalence principle is the "law of plane sections". He presents a careful order-of-magnitude analysis not unlike that given above, but restricted to a perfect gas. The unification of linearized theory with that for which $M_\infty \tau$ is not small is accomplished by using $\tau + M_\infty^{-1}$ in place of τ as a measure of wave inclination, in his analogues of our (2.1.2) and (2.1.3). Il'yushin considers a large number of examples analytically, including the case of a cone at zero incidence for which the exact solution of Taylor and

Maccoll [1] was available but apparently not known to him. He also obtains the law of hypersonic similitude, referred to by him as the "method of affine models". In a footnote to his paper, which was published in 1956, it is stated that the paper was issued in a limited printing in 1948 and that it is reprinted without change. This unfortunate large delay in publication was the result of secrecy restrictions.

2. Hypersonic similitude

Before we express the classical hypersonic similitude, let us look briefly at the concept of similitude itself. A similitude or similarity rule expresses an equivalence between two physical situations between which there is some intrinsic dissimilarity and which are thus not identical in all essential respects. The description of a physical situation may be expressed in functional form, with some quantity of physical interest expressed as a function of all the parameters (and functions) which determine the physical situation. The result which a similitude accomplishes is a reduction in the number of independent parameters on which the function depends. In general, a regrouping of the parameters in the original functional formulation is necessary in accomplishing this reduction.

The classical dimensional similitude afforded by dimensional analysis fits into this general definition, with the necessary regrouping of parameters yielding new dimensionless parameters according to the Pi theorem. It must be emphasized that we do not generally have full dimensional similitude in fluid mechanics. Dimensional similarity between two different fluid flows depends either upon certain self-similar properties of the fluid or upon a particular correspondence between the equations of state of two different fluids. This correspondence must be regarded as accidental unless the two fluids are themselves self-similar, or unless the equations of state are otherwise idealized, as with the ideal dissociating gas of Lighthill [3, §2.4] or with a van der Waals gas. The self-similar fluid of most practical importance is the perfect gas of constant ratio of specific heats γ. Two perfect gas flows with the same value of γ can be dimensionally similar. However, our policy in this treatment is to consider such a perfect gas as a special case.

A calorically perfect gas (constant γ) may be considered to be fully self-similar. A perfect gas with γ not constant or a fluid whose equation of state can be expressed simply in terms of an arbitrary function of a single variable may be considered to be partially self-similar (cf. Wecker and Hayes [1]). Similarity results involving partially self-similar fluids and different fluids similar to each other have been given by Kochina and Mel'nikova [1; 2]. See also Korobeinikov, Mel'nikova, and Ryazanov [1].

The only general similitude for a general ideal fluid is the geometric one between flows of the same fluid which is afforded by the scale transformation, in which the basic distance coordinates and the time coordinates are all changed by the same constant factor. The similitude states that, with a given nondissipative flow, there can exist a similar one with a different scale but with the same values of the velocity and the thermodynamic state variables of the fluid at corresponding points. The functional dependence of any of these quantities may be expressed in a form independent of the absolute scale. The existence of steady conical flows depends upon the scale transformation, and thus such flows can exist in a general ideal fluid. This scale similitude is taken for granted in all the ideal fluid theory treated here. Thus, for example, the lateral coordinates y and z will be kept in an unreduced form in this section, with no loss of generality.

The establishment of a similitude must be based upon an analysis of the physical situation in which certain simplifying assumptions have been made. These assumptions which have been used in the analysis become the fundamental assumptions underlying the resulting similitude. In general, the greater are the number of assumptions made, the smaller are the number of independent parameters remaining in the functional formulation and the more powerful is the similitude in relating equivalent systems. Conversely, the fewer are the demands made of the number of different systems which must be related as equivalent, the fewer are the assumptions which need to be made and the more general is the similitude in the classes of systems to which it may be applied.

In inviscid fluid theory the assumption is made that the viscosity is so small it may be neglected. The generalized Prandtl-Glauert similitude of supersonic flow theory assumes that the equations of motion may be linearized and permits a functional specification of the results in which there is no dependence upon the Mach number *per se*. The more usual Prandtl-Glauert similitude requires also the mean-surface approximation and permits results expressed as independent of the parameter $\sqrt{M_\infty^2 - 1}\,\tau$ *per se*, where τ is the thickness ratio or angle of attack.

Hypersonic similitude is the similitude which arises from the small-disturbance analysis of the previous section. Thus the assumptions underlying hypersonic siuilitude are the same as those made for the small-disturbance theory, and the characteristic one of these is the slender-body assumption (B) that τ is small. With the geometric similitude given by the scale transformation not considered, hypersonic similitude considers two flows as equivalent if both flows are given by the same solutions of the small-disturbance equations. With a general fluid the fluid must be the same for both flows. We shall consider primarily the case of steady flow, but shall discuss the unsteady case briefly in Section 2.10. See Oswatitsch [3] for an alternative treatment of hypersonic similitude.

For classical hypersonic similitude we consider a family of bodies with related shapes, given by

$$f(\tau x, y, z) = 0 \tag{2.2.1}$$

in a system of coordinates fixed with respect to the body. The quantity τ remains defined as the thickness ratio (or angle of attack). Besides the angle of attack, various other geometric parameters must vary in proportion to τ, such as aspect ratio and gap ratio. The body is placed in a uniform steady hypersonic flow with U, ρ_∞, and p_∞ given in the free stream. Taking the pressure as typical of the dependent variables, we may express the functional dependence of p, in general, as

$$p = p(x, y, z; \tau, U, \rho_\infty, p_\infty). \tag{2.2.2}$$

Equation (2.2.2) is written for a particular fluid under consideration, and the functional dependence of p upon the equation of state is understood. If τ is small, so that the small-disturbance theory is applicable, we see from (2.1.11) that the functional dependence on x, τ, and U must be expressible as a dependence on $x' = \tau x$ and $U\tau$. Hence (2.2.2) may be reexpressed as

$$p = p(\tau x, y, z; U\tau, \rho_\infty, p_\infty). \tag{2.2.3}$$

The primary feature of hypersonic similitude is the reduction by one of the number of quantities upon which p depends, from (2.2.2) to (2.2.3). Thus, in (2.2.3), there is no dependence of p upon the parameter τ *per se*. Equation (2.2.3) may be re-expressed in terms of the pressure coefficient

$$C_p = \tau^2\, \Pi(\tau x, y, z; U\tau, \rho_\infty, p_\infty). \tag{2.2.4}$$

Similar expressions may be obtained for the other variables of the flow, with the lateral velocities proportional to $U\tau$ and the axial perturbation velocity to $U\tau^2$. The applicability of the concept of hypersonic similitude to a general fluid is evident from the statement of the concept in Hayes [1] and has been exploited independently by Cheng [1]. Cheng verified the similitude and presented correlations of wedge-flow calculations for equilibrium air.

For an example we shall consider a simple Prandtl-Meyer flow under hypersonic conditions, and shall use the result (7.1.15) of Chapter VII for the turning angle. We replace the term representing $\sqrt{M^2 - 1}$ by M, and replace $H - h$ except in the differential by $\frac{1}{2}U^2$. The result is a simplified form of (7.1.15) appropriate for hypersonic flow

$$\nu = -\frac{1}{U}\int \frac{dh}{a} = -\frac{1}{U}\int \frac{dp}{\rho a}, \tag{2.2.5}$$

in which the integral is taken along an isentropic path and the turning angle ν is defined to be zero in the free stream. Note that the integral is one taken only with respect to thermodynamic variables. With ρ_∞ and p_∞ fixed both the initial point and the path of the integral are determined, and the value of p will be given by the value of the integral. Thus we have

$$p = p(U\nu, \rho_\infty, p_\infty). \tag{2.2.6}$$

The ray angle is simply the Mach angle less the turning angle in a Prandtl-Meyer flow, or

$$\frac{y}{x} = \frac{a}{U} - \nu \tag{2.2.7}$$

in our hypersonic flow. The quantity a has a functional dependence of the same form as does p in (2.2.6). Hence we may write

$$\frac{Uy}{x} = a - U\nu = Y(U\nu, \rho_\infty, p_\infty). \tag{2.2.8}$$

Assuming the functional relation in (2.2.8) may be reversed, we may express $U\nu$ in (2.2.8) in terms of Uy/x and substitute in (2.2.6). The result is a new functional expression for p,

$$p = p\left(\frac{Uy}{x}, \rho_\infty, p_\infty\right), \tag{2.2.9}$$

which is a relatively very simple special case of (2.2.3). It may be readily shown that the quantities q_y and $U\tau q_x'$ have the same functional dependence as does p. Note that these results are for a general fluid.

If the fluid is a perfect gas with a constant ratio of specific heats γ, the single parameter γ characterizes the complete equation of state. The addition of γ as an independent parameter in (2.2.2) to (2.2.4) takes care of the functional dependence upon the equation of state. In addition, because of the self-similar properties of such a gas, we may drop the variables ρ_∞ and p_∞ provided that $U\tau$ is reduced to dimensionless form. To do this we need any quantity dependent upon the free stream state which has the dimensions of velocity. It is convenient to choose the free stream speed of sound for this purpose, already expressed for a general fluid (1.4.10b) by

$$a_\infty^2 = \frac{\gamma p_\infty}{\rho_\infty}. \tag{2.2.10}$$

The quantity $U\tau$ is divided by a_∞ and thus reduced to $M_\infty\tau$. This quantity $M_\infty\tau$ is the basic similarity parameter K of classical hypersonic similitude

$$K = M_\infty\tau. \tag{2.2.11}$$

Since a_∞ is a function of ρ_∞ and p_∞, K may be substituted for $U\tau$ in (2.2.3) or (2.2.4) if desired.

For a perfect gas, then, (2.2.4) may be reexpressed in the form

$$C_p = \tau^2\, \Pi(\tau x, y, z;\, K, \gamma). \tag{2.2.12}$$

The pressure coefficient of (2.2.4) or (2.2.12) may be expressed either relative to zero pressure or, more conventionally, relative to the free stream pressure. The free stream pressure coefficient relative to zero pressure must be of the form indicated by (2.2.12) but independent of the space coordinates, and is

$$\frac{p_\infty}{\frac{1}{2}\rho_\infty U^2} = \frac{2\tau^2}{\gamma K^2} \tag{2.2.13}$$

for any fluid. A lift coefficient based upon a lateral projected area has the same dependence as does C_p, except that the coordinate variables do not appear. The same is true for a drag coefficient which is based on a projected frontal area. Using a lateral projected area as reference for both coefficients, we have

$$C_L = \tau^2\, \Lambda(K, \gamma), \tag{2.2.14a}$$

$$C_D = \tau^3\, \Delta(K, \gamma). \tag{2.2.14b}$$

Analogous reductions of (2.2.4) with corresponding laws for force coefficients may be obtained for partially self-similar fluids.

The complete similarity of shape imposed by (2.1.5) requires that the angle of attack α and the aspect ratio $Æ$ of the body vary proportionally with τ, and (2.2.14a) is usually expressed with α^2 in place of τ^2. The similitude gives no information on a hypersonic flow in which α and τ are varied independently. The ratio α/τ is a basic parameter in hypersonic similitude.

We now turn to the Mach number independence principle or similitude of Oswatitsch discussed earlier, and ask what is the result of combining the two concepts. To use the independence principle, we must consider a limiting process in which p_∞ approaches zero with the other parameters of the problem held fixed. The combined result, expressed for the pressure, is

$$p = p(\tau x, y, z;\, U\tau, \rho_\infty) \tag{2.2.15}$$

to replace (2.2.3), or

$$C_p = \tau^2\, \Pi(\tau x, y, z;\, U\tau, \rho_\infty) \tag{2.2.16}$$

to replace (2.2.4). The functional dependence upon the equation of state is again understood.

For the perfect gas of constant ratio of specific heats, a_∞ goes to zero in the limiting process, and the basic similitude parameter K goes to infinity.

By the independence principle we may neglect dependence upon K, and obtain

$$C_p = \tau^2 \, \Pi(\tau x, y, z; \gamma) \tag{2.2.17}$$

in place of (2.2.12). The free stream pressure of (2.2.13) is zero in this limit. Oswatitsch [1, §5] has given suitable equations of motion for such a combined theory in two-dimensional and axisymmetric flow. We must note the severe limitations on such a combined theory. For its application not only must τ be small but K must be very large. In terms of our earlier discussion on assumptions, A-strong or A-lim is required.

Comparisons of exact theoretical results and experimental results with the predictions of hypersonic similitude indicate a wide range of validity for the similitude. An experimental check of the similitude for yawed cones is shown in Fig. 2–1, taken from Hamaker, Neice, and Wong [1]. Some comparisons

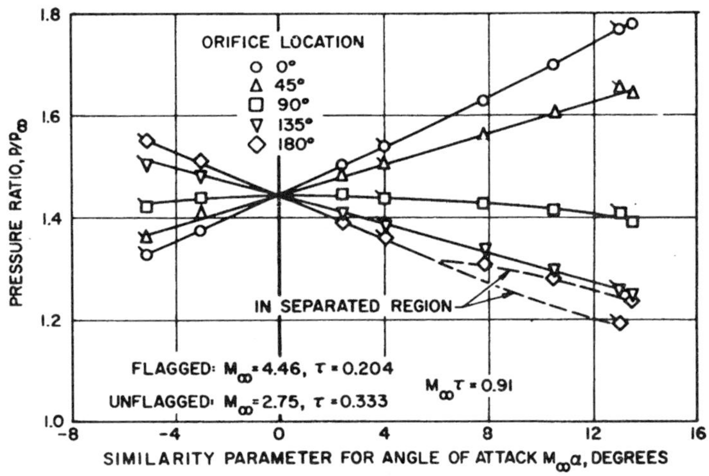

FIG. 2–1. Experimental check of hypersonic similitude for the pressure on yawed cones (Hamaker, Neice, and Wong [1]).

of exact theoretical results with results of the small-disturbance theory appear in Figs. 2–2 to 2–9 which accompany Sections 2.3 and 2.5; others appear in the references cited at the beginning of Section 2.1. This general agreement lends weight to the equivalence principle upon which hypersonic similitude is based. However, at moderately low Reynolds numbers on slender bodies the effects of the displacement thickness of the boundary layer may be large enough to invalidate the similitude. If the Reynolds numbers on similar bodies are controlled so that the displacement thickness varies in direct proportion to the body thickness, hypersonic similitude should still hold. But in applying the small-disturbance theory with the displacement thickness taken into account, this displacement thickness may not be known in advance because of interactions of the type discussed in Chapter IX of our first edition, or in Dorrance [1]. In comparing experimental results with the results of the small-disturbance theory, a correction for the boundary layer thickness may be advisable. Such corrections were made, for example, by Kubota [1] (see discussion under effect (f) in Section 2.7). Viscous hypersonic similitude, in which the displacement thickness is taken into account, is treated by Hayes and Probstein [1].

3. Unified supersonic-hypersonic similitude

The Prandtl-Glauert similitude for general bodies in linearized steady supersonic flow has the quantity $\beta\tau$ as a basic similitude parameter, where

$$\beta = \sqrt{M_\infty^2 - 1}\,. \tag{2.3.1}$$

In hypersonic flow β is very close to M_∞, and the similitude parameters merge. At moderate values of the Mach number the flow may be considered linearized if τ is small, and the classical hypersonic similitude as just presented is clearly incorrect. It has long been clear that the two similitudes for the two steady flow regimes may be joined by the simple process of replacing M_∞ by β in hypersonic similitude. An examination of the derivation of the small-disturbance theory shows that the error incurred in this replacement is of no greater order than that of the errors inherent in the original theory.

This idea was explored and justified heuristically by Van Dyke [1; 2] for the case of a perfect gas of constant ratio of specific heats. The correct form of the unified similitude is obtained by replacing the parameter $K = M_\infty\tau$ by a "unified" hypersonic similarity parameter

$$K' = \beta\tau \tag{2.3.2}$$

in the equations of functional dependence such as (2.2.12). In place of (2.2.13) we have

$$\frac{p_\infty}{\frac{1}{2}\rho_\infty U^2} = \frac{2\tau^2}{\gamma K'^2} - \frac{2}{\gamma\beta^2 M_\infty^2}\,. \tag{2.3.3}$$

It is convenient to define a density coefficient to measure the difference between the density in the flow field and in the free stream,

$$C_\rho = \frac{\rho - \rho_\infty}{\frac{1}{2}\rho_\infty U^2} \cdot \frac{p_\infty}{\rho_\infty}\,. \tag{2.3.4}$$

It may be noticed that the free stream density contribution to this density coefficient is the same as the free stream pressure contribution to the pressure coefficient and is given simply by the negative of (2.3.3).

We summarize here the unified similitude as given by Van Dyke for a perfect gas: The pressure coefficient relative to free stream pressure is given by (2.2.12) with K replaced by K', or by

$$C_p = \tau^2 \,\Pi(\tau x, y, z; K', \gamma). \tag{2.3.5}$$

The density coefficient C_ρ defined in (2.3.4) is given by an expression of exactly the same form. The velocity components q_y, q_z, and $q_x' = \tau^{-1} q_x$ are given by functional expressions of the form

$$q_y = U\tau\, Q(\tau x, y, z; K', \gamma). \tag{2.3.6}$$

Note that the free stream pressure coefficient and the free stream density coefficient do not fit (2.3.5) because of the term $-2/\gamma\beta^2 M_\infty^2$ in (2.3.3), and hence that the absolute pressure and density do not follow the unified similitude.

In case the fluid is a general one and not a perfect gas the unification replaces (2.2.4) by

$$C_p = \tau^2 \,\Pi(\tau x, y, z; K', \rho_\infty, p_\infty), \tag{2.3.7}$$

in which the understood hypersonic dependence on the particular equation of state is indicated by the presence of ρ_∞ and p_∞. If K' is small enough so that only second-order effects need to be taken into account, we may replace (2.3.7) by

$$C_p = \tau^2 \,\Pi(\tau x, y, z; K', \gamma_*), \tag{2.3.8}$$

where γ_* is given by

$$\gamma_* = \frac{1}{a^2}\left(\frac{\partial \rho a^2}{\partial \rho}\right)_S. \tag{2.3.9}$$

The quantity γ_* was defined previously in (1.4.24), and a discussion of its relation to second-order supersonic theory is included at the end of Section 1.4.

Finally, if K' is small enough so that only linear effects need be taken into account, the dependence on γ_* may be dropped from (2.3.8). The resultant relation is the one given by the general Prandtl-Glauert similitude.

This unified form of hypersonic similitude is much superior in practice to the original form, permitting the results of experiment to be correlated well over a much wider range of Mach numbers than if $K = M_\infty \tau$ is taken as the similitude parameter. A comparison of the two forms of hypersonic similitude is given for the case of the flow on a wedge in Fig. 2–2. Henceforth we shall consider hypersonic similitude in steady flow to have this unified form.

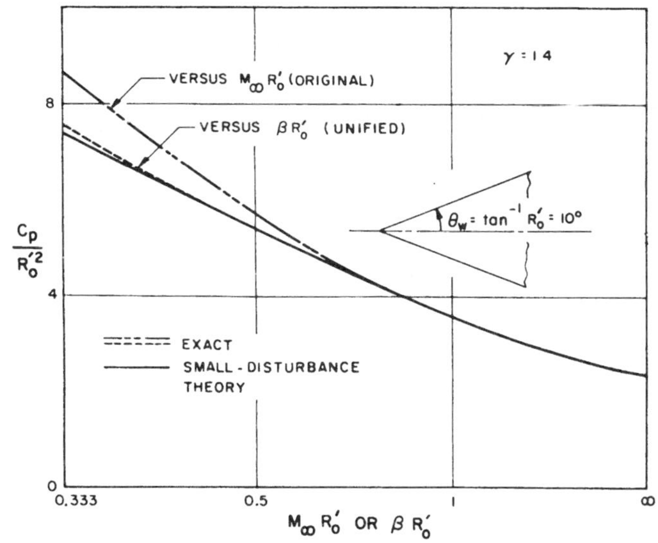

FIG. 2–2. Comparison of original and unified supersonic-hypersonic similarity law for a wedge of 10° half-angle (Van Dyke [2]).

4. Slender-body strip theory

The combination of the small-disturbance theory with the strip theory concept discussed in Section 1.7 leads to a strip theory for slender bodies which are flat. This slender-body strip theory has the essential features of a similitude, and corresponds to the more classical aerodynamic concept of strip theory. A significantly weaker condition than the condition $A\!\!R^{-1} \ll 1$ of Section 1.7 is required. The basic requirement is $\tau A\!\!R^{-1} \ll 1$.

Slender-body strip theory is not dependent upon the small-disturbance assumptions, and, for example, is applicable in supersonic wing theory with the basic requirement replaced by $\beta^{-1} A\!\!R^{-1} \ll 1$. Thus we can apply strip theory to a body without small-disturbance theory being invoked (or vice versa, of course). Here, for simplicity, we start with the small-disturbance equations already derived. We could just as well carry through a similar analysis without the small-disturbance simplification. The similitude we shall express is that labeled "piston theory" in Section 2.9 in Fig. 2–30, applied to the "flat normal" strip case of Fig. 2–31.

We follow a process of reduction similar to that used for the classical hypersonic similitude. The body is assumed to be a "flat" one, with a characteristic lateral scale determined by an aspect ratio $\mathcal{R}$. The meaning of flat will be that the body shape is close to a mean cylindrical surface, here considered to be the (x, z) plane. For this strip theory the restrictions on the shape of the body are very similar to those of the mean-surface or quasi-cylinder assumptions of linearized theory, though these assumptions remain invalid in hypersonic flow. The coordinate z is the lateral coordinate along the body, and y is the coordinate normal to the mean surface. The body shape is expressed by a relation of the form

$$f(\tau x, y, \tau \mathcal{R}^{-1} z, t) = 0 \tag{2.4.1}$$

instead of by (2.1.5), and the aspect ratio $\mathcal{R}$ is considered as a parameter of the problem. We shall be interested in the case for which $\tau \mathcal{R}^{-1}$ is small. By using the scale transformation discussed at the beginning of Section 2.2 we may put (2.4.1) into the more recognizable form

$$f\left(\frac{x}{c}, \frac{y}{\tau c}, \frac{z}{\mathcal{R} c}, \frac{t}{\tau c}\right) = 0, \tag{2.4.2}$$

where c is a reference axial dimension of the body.

The shape of the shock wave is assumed to be given by a relation of the same form as that for the body, and it is assumed that the shock forms a "flat" surface in the same sense as does the body. The value of q_z behind the shock is then of order $\tau^2 \mathcal{R}^{-1} U$ instead of τU as in the general case. This suggests the transformation analogous to (2.1.2) by letting

$$q_z = \tau \mathcal{R}^{-1} q_z', \tag{2.4.3}$$

with q_z' of the same order of magnitude as q_y. We complete the analogy with the transformation corresponding to (2.1.3) by letting

$$z = \tau^{-1} \mathcal{R} z'. \tag{2.4.4}$$

We may now complete the analysis and obtain equations analogous to (2.1.4) through (2.1.10). The terms on the right-hand side are now of order $\tau^2 \mathcal{R}^{-2}$ instead of τ^2.

The transformation (2.4.4) is clearly appropriate to produce derivatives with respect to z' which are of the same order of magnitude as derivatives with respect to y or x'. But we should justify that q_z', which is of order $U\tau$ behind the shock, will remain of that order throughout. We pick out the appropriate small-disturbance momentum equation

$$\frac{\partial q_z'}{\partial t} + q_y \frac{\partial q_z'}{\partial y} + \frac{1}{\rho} \frac{\partial p}{\partial z'} = -\tau^2 \mathcal{R}^{-2} q_z' \frac{\partial q_z'}{\partial z'} \tag{2.4.5}$$

and note that it is the pressure gradient term which controls changes in q'_z. That q'_z remains of order $U\tau$ is then justified by the fact that $\partial p/\partial z'$ must be of the same order of magnitude as $\partial p/\partial x'$ or $\partial p/\partial y$.

Finally, we must justify the assumption made that the shock shape is flat if the body is flat. The thickness of the shock layer and body together is of the order τc or $M_\infty^{-1}c$, whichever is greater, while the lateral dimensions of the shock are the same as those of the body, or Ac. For the shock shape to be flat both τA^{-1} and $M_\infty^{-1}A^{-1}$ must be small. The quantity τA^{-1} must be small anyway, and the requirement that $M_\infty A$ be large follows unless the similitude parameter $K = M_\infty \tau$ is small. As in the development of the small-disturbance theory, we should replace τ in our order-of-magnitude analysis by the Mach angle M_∞^{-1} if K is small, i.e., if we are near the linearized flow case.

For steady flow it is convenient to replace M_∞ by β, whereby the quantity $M_\infty A$ becomes the reduced aspect ratio βA of linearized supersonic theory. We note, incidentally, that the requirement for the applicability of strip theory in linearized supersonic flow is simply that βA be large.

If $\tau^2 A^{-2}$ (and also $M_\infty^{-2}A^{-2}$) is sufficiently small we may drop the right-hand sides in the reduced equations of which only (2.4.5) is given here. The resultant equations are the one-dimensional unsteady flow equations in y, with both x' and z' as parameters. To express the similitude we must first recognize that with A variable, the quantity τA^{-1} must appear in equations of the form (2.2.4). Thus to take the aspect ratio into account we should have expressed the functional dependence of the pressure coefficient in the form

$$C_p = \tau^2 \, \Pi(\tau x, y, z; \; U\tau, \tau A^{-1}, \rho_\infty, p_\infty) \tag{2.4.6}$$

for the general case of slender-body flow instead of in the form of (2.2.4). With the similitude afforded by slender-body strip theory, with τA^{-1} small, the dependence on z and τA^{-1} is one on $z' = \tau A^{-1} z$, and we may replace (2.4.6) by

$$C_p = \tau^2 \, \Pi(\tau x, y, \tau A^{-1} z; \; U\tau, \rho_\infty, p_\infty). \tag{2.4.7}$$

The strip theory similitude amounts to this elimination of dependence upon τA^{-1}. Further reduction for the case of a perfect gas or the development of a unified form to fit the supersonic case may be carried out exactly as was done above for classical hypersonic similitude. With strip theory valid, the analogy of a two-dimensional physical flow to flow in a cylinder driven by a piston applies also to three-dimensional flows.

As suggested in the generalizations presented in Section 1.7, slender-body strip theory is applicable to more general shapes, with the requirement that the body surface approximate a smooth reference cylinder or mean surface. The coordinate system must be constructed suitably, with one lateral coordinate normal to the mean surface and the other tangential to it. The reference lateral scale cannot be greater than a characteristic lateral dimension of the mean surface, such as the radius of an annular wing. See Section 2.9 for additional discussion.

5. Examples of small-disturbance solutions

In order to illustrate the small-disturbance theory we shall present the results of the application of the theory to a few simple problems. The cases we shall consider are those of the wedge, the plane ogive, the cone, and the ogive of revolution. Our results are all taken from the basic report of Van Dyke [2] (see also Van Dyke [3]), to which the reader is referred for details of the calculations. In accordance with our statement made at the end of Section 2.3 we have followed Van Dyke in using the unified form of the hypersonic similitude parameter in expressing the results. The assumption of a perfect gas of constant ratio of specific heats is made throughout, and the diatomic value of γ equal to 1.4 or 1.405 was used in computing the curves.

In either the two-dimensional case or the axisymmetric case the body shape is given by a function $R(x)$ in Van Dyke's notation, equal to y or r. The slope of the body at the nose is given by

$$R_0' = R'(0) = (\tan \theta_b)_0 \tag{2.5.1}$$

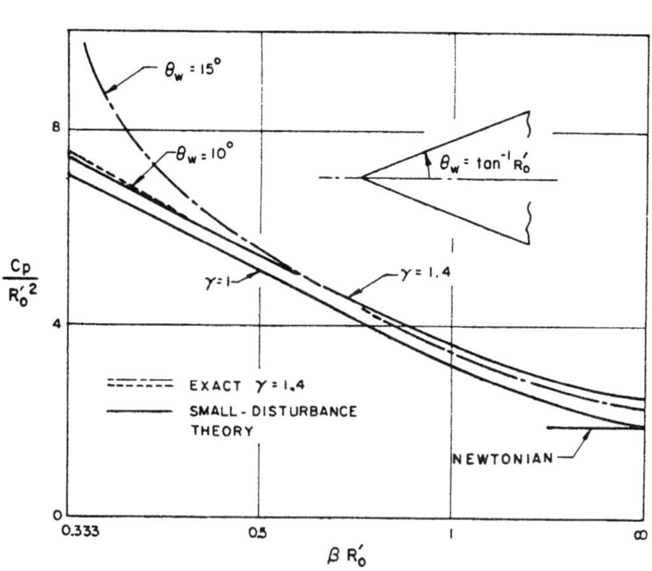

FIG. 2–3. Pressure coefficient on a wedge in small-disturbance theory (Van Dyke [2]).

with the prime denoting differentiation. The quantity R_0' is the quantity τ of our presentation of the small-disturbance theory, and the basic similitude parameter is thus $\beta R_0'$. In Figs. 2–3 through 2–9 taken from Van Dyke [2] are plotted various quantities against $\beta R_0'$. For convenience in including the limit $\beta R_0' = \infty$ the abscissas of these plots have been made linear in $(\beta R_0')^{-1}$.

For the wedge the small-disturbance result was found by Linnell [1] and by Bam-Zelikovich, Bunimovich, and Mikhailova [1] (see also Ivey and Cline [1]) and is

$$\frac{C_p}{R_0'^2} = \frac{2 \tan \sigma_0}{R_0'} = \frac{\gamma + 1}{2} + \sqrt{\left(\frac{\gamma + 1}{2}\right)^2 + \frac{4}{\beta^2 R_0'^2}} \tag{2.5.2}$$

with the unified parameter substituted for $M_\infty R_0'$. This result may be obtained from the results of Sections 1.4 and 1.5, and it may be noted that within the accuracy of the small-disturbance theory the quantity $\frac{1}{2}(\gamma + 1)$ expressed in (2.5.2) is simply $(1 - \epsilon_{\lim})^{-1}$. Note also that (2.5.2) expresses a result both for pressure and for shock position. This result with $\gamma = 1.4$ is plotted on Fig. 2–3, together with pressure curves for thick wedges for comparison. Also for comparison is the same curve for $\gamma = 1$; the ordinate of this curve at $\beta R_0' = \infty$ gives the Newtonian value, from the so-called Newtonian theory to be discussed later.

The body shape for the plane ogive is expressed in the form

$$y = R_0'x + \tfrac{1}{2}R_0''x^2 + O(x^3), \tag{2.5.3}$$

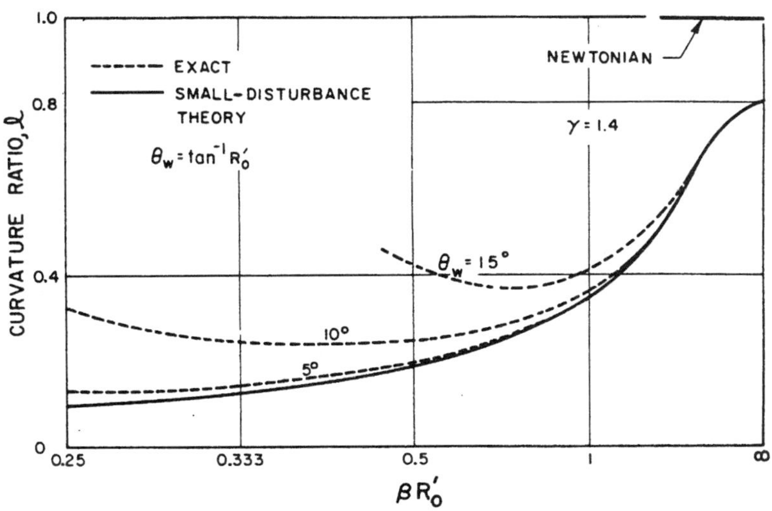

FIG. 2–4. Initial ratio of shock to body curvature for a plane ogive (Van Dyke [2]).

and R_0'' is the curvature of the body evaluated at the nose. The curvature of the shock $d \tan \sigma/dx$ evaluated at the nose divided by the body curvature R_0'' is denoted by l and is plotted for $\gamma = 1.4$ in Fig. 2–4. The pressure gradient along the body evaluated at the nose is given in Fig. 2–5. For comparison, some exact results of Kraus [1] for thick ogives have been shown, and also the

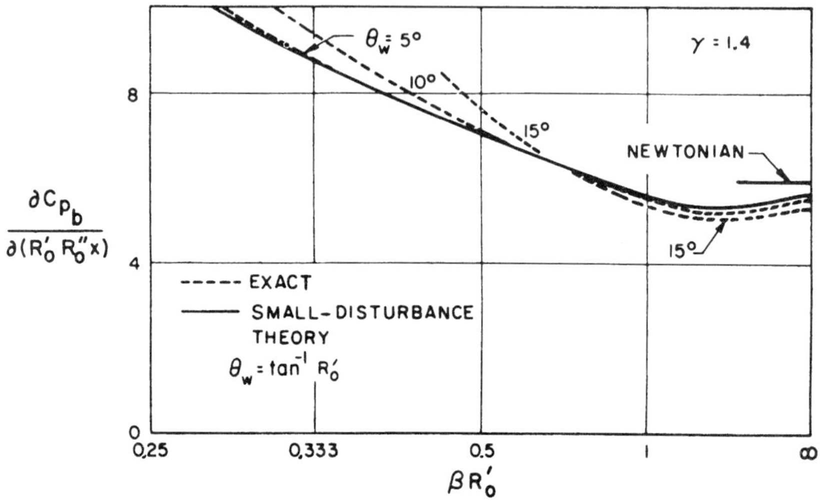

FIG. 2–5. Initial pressure gradient on a plane ogive (Van Dyke [2]).

values given by the Newtonian theory. It may be noted that for the "uncorrected Newtonian" pressure without the necessary centrifugal correction of Busemann (see Section 3.1), the ordinate in Fig. 2–5 would be 4.

For the cone and ogive of revolution the calculations were made for $\gamma = 1.405$ in order to permit a direct comparison with the solutions tabulated by Kopal [1]. For the cone, the ratio of cone angle to shock angle $R_0'/\tan \sigma_0$ is plotted in Fig. 2–6, and the pressure coefficient is plotted in Fig. 2–7. Also plotted in Fig. 2–7 are the corresponding linearized, second-order, and Newtonian results, and some exact results for thick cones taken from Kopal [1].

For the ogive of revolution, the ratio l of the shock curvature to body curvature at the nose is given in Fig. 2–8. Also presented here are some exact results of Shen and Lin [1] for thick ogives. The pressure gradient along the body evaluated at the nose is given in Fig. 2–9, with Newtonian, tangent-cone, and cone-expansion results also given for comparison. The tangent-cone approximation is discussed in Section 7.3. The cone-expansion method is the same as the shock-expansion method, applied to the three-dimensional flow field immediately behind the locally conical shock at the nose. The shock-expansion approximation is discussed in Section 7.2. As with the plane ogive, the ordinate in Fig. 2–9 for the incorrect "uncorrected Newtonian" pressure would be 4.

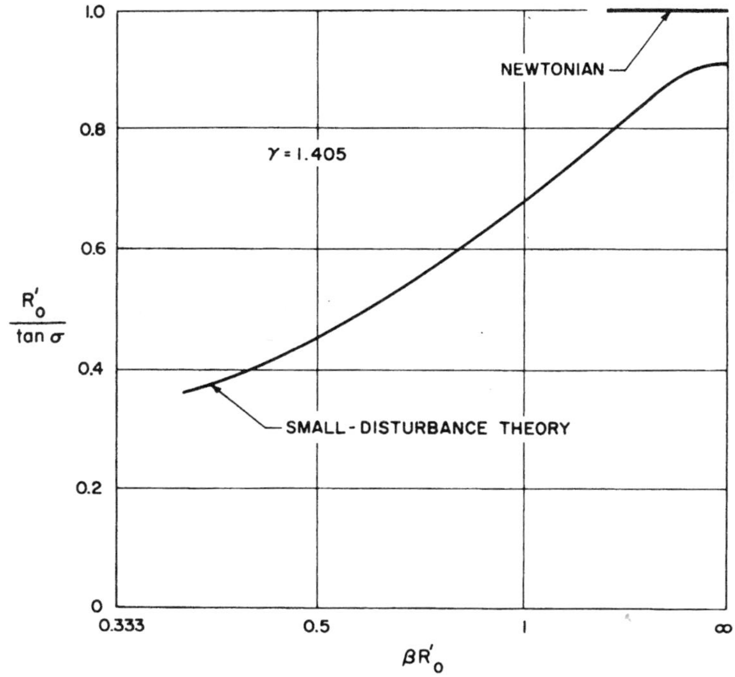

FIG. 2-6. Ratio of cone angle to shock angle (called δ/τ in Van Dyke [2]).

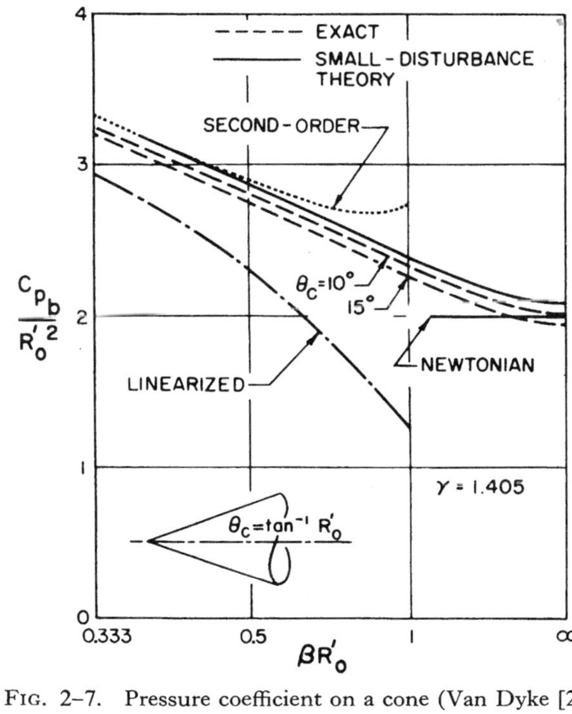

FIG. 2-7. Pressure coefficient on a cone (Van Dyke [2]).

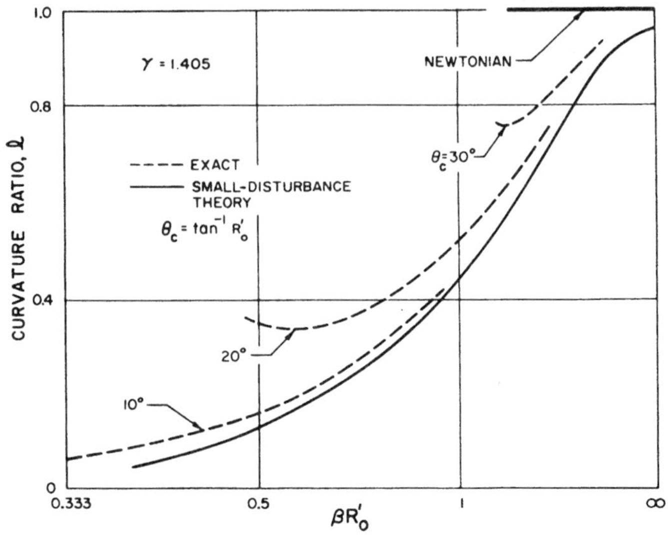

FIG. 2–8. Initial ratio of shock to body curvature for an ogive of revolution (Van Dyke [2]).

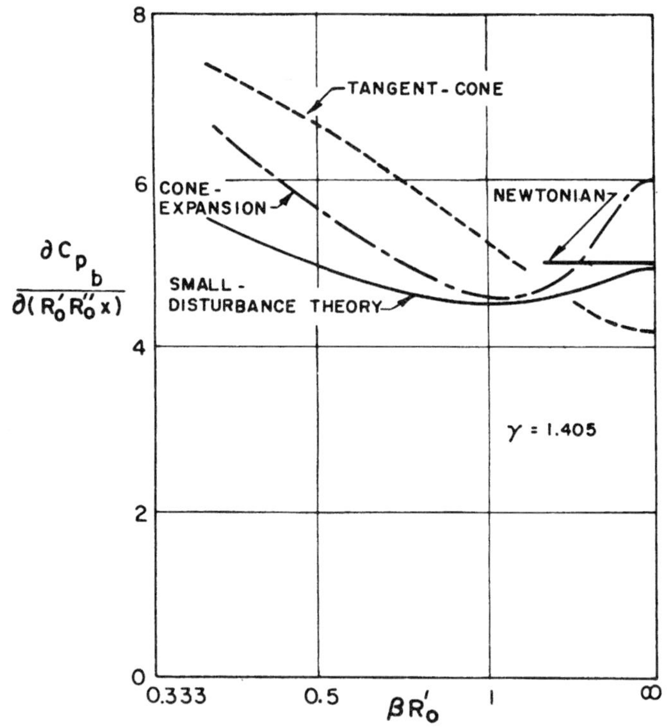

FIG. 2–9. Initial pressure gradient on an ogive of revolution (Van Dyke [2]).

6. Similar power-law solutions

We shall examine next a particular group of solutions to the small-disturbance equations, comprising those solutions for two-dimensional and axisymmetric steady flow which are self-similar. Such solutions are termed similar solutions, and have the property that the solution in the lateral variable y or r at one value of t (or of x for the corresponding hypersonic steady flow) is similar to the solution at any other value of t. This self-similar property permits a decrease in the number of essential independent variables from two to one, and yields a treatment using ordinary differential equations instead of partial differential equations.

Most of the investigations of similar solutions have been carried out with unsteady problems with certain symmetries in mind, although many of them are directly applicable to our related hypersonic steady flow problems. The unsteady problems of greatest practical interest have been those with spherical symmetry, for which no corresponding steady flow problems exist (barring, of course, hypersonic flows in a four-dimensional universe). Many well known solutions in gasdynamics are self-similar; examples include Prandtl-Meyer flows, supersonic attached flows on a circular cone, and the flows in idealized shock tubes. But the similar solutions of interest to us here are different from the more classical ones mentioned in that the gas must have self-similar thermodynamic properties (i.e. must be a perfect gas of constant γ), and a similar behavior must be required of shock waves of varying strength (ϵ must be constant).

The earliest work on similar solutions of the type we are considering was carried out by Bechert [1] and by Guderley [1]. Guderley derived the appropriate equations and boundary conditions for the problems we shall treat below, reduced the problem to a single first-order differential equation, and applied his theory to the problem of implosions. In his treatment Guderley gives a detailed study of the various singular points of the basic differential equation. Bechert's earlier work was mostly limited to a polytropic gas and is therefore of little practical interest. His main contribution lies in introducing the concept of such solutions.

During World War II, Taylor [1] solved the problem of a violent spherical explosion; publication of his results was delayed by security restrictions until 1950. Following Taylor's approach Sakurai [1; 2], S. C. Lin [1], Latter [1], Lees and Kubota [1], and Kubota [1] treated a number of related problems. Lin also noted that the result he had obtained for a violent cylindrical explosion would apply to an axisymmetric hypersonic steady flow. Cheng and Pallone [1] and Lees and Kubota developed the analogy further and presented comparisons of the theory with experiment. All these authors were apparently unaware of the extensive early work of various Russian authors.

In Russia the term used for these solutions is "automodel", etymologically equivalent to "self-similar". Two authors are predominant in the Russian development of similar solutions, Sedov and Stanyukovich. Sedov obtained the equations of Guderley in somewhat more general form [1], and obtained an analytic solution for the problem of a violent spherical explosion [2], the same problem for which Taylor obtained a numerical solution. Presentations of his theory appear in editions of his book on similitude [3], beginning with the second edition. Following Sedov's approach, Krasheninnikova [1] and others (see Chernyi [3]) treated a number of related problems. The analogy with steady hypersonic flows following the equivalence principle was developed by Grodzovskii [1] and Chernyi [2; 3]. Stanyukovich, e.g. in [1], has also investigated several types of similar solutions; a part of his book [2] is devoted to such solutions. Korobeinikov, Mel'nikova, and Ryazanov [1] present a good summary of recent Soviet work on the violent explosion and related problems. Soviet authors generally use the term point explosion for the violent explosion.

The approach we shall use is that of Sedov, with part of his notation unchanged and part altered to fit the remainder of this volume. The gas is assumed to be a perfect gas with constant ratio of specific heats γ, and the flow is assumed to have planar, cylindrical, or spherical symmetry. Uniformly in this book we shall use the integer j to denote one less than the number of lateral space dimensions involved. Thus j takes the value 0 in the planar case, the value 1 in the cylindrical case, and the value 2 in the spherical case.

Although other types of similar solutions may be considered (see, for example, Chernyi [3, p. 92]), we shall restrict ourselves to solutions of the power-law type. Similar solutions may be realized only with appropriate boundary conditions. Besides the basic text of Sedov [3], Chernyi [3] and Mirels [2] present expositions of the theory applicable to hypersonic flow theory.

The parameters s and m of Sedov are restricted to particular values by

$$s = 0, \tag{2.6.1a}$$

$$m = -1. \tag{2.6.1b}$$

No loss in generality is introduced through our imposition of conditions (2.6.1). Table 2–1 compares the notations of this book, Sedov, Chernyi, and Mirels for the basic parameters of the theory. It should be noted that some Soviet authors use ν for our j, and thus for Sedov's and Chernyi's $\nu - 1$.

TABLE 2–1

Parameter notation for similar solutions

Hayes-Probstein	Sedov	Chernyi	Mirels
$1 + j$	ν	ν	$1 + \sigma$
k	δ	$1 + n$	m
ω	$k + 3 = \omega$	0	0

The basic problem posed is one in one space dimension and time, for which the three equations of motion may be written

$$v_t + vv_r + \rho^{-1}p_r = 0, \tag{2.6.2a}$$

$$\rho_t + (\rho v)_r + jr^{-1}\rho v = 0, \tag{2.6.2b}$$

$$(p\rho^{-\gamma})_t + v(p\rho^{-\gamma})_r = 0, \tag{2.6.2c}$$

with subscripts denoting partial differentiation, with r used to denote the one spatial coordinate, and with v the velocity in the r direction. If $j = 0$, r is simply y. Dimensionless variables V, R, and P replacing the dependent variables above are defined

$$v = rt^{-1}V, \tag{2.6.3a}$$

$$\rho = Ar^{-\omega}R, \tag{2.6.3b}$$

$$p = Ar^{2-\omega}t^{-2}P, \tag{2.6.3c}$$

where A is a constant with suitable dimensions, and ω is a dimensionless parameter; ω will equal 0 for the solutions of most practical interest. These new variables are required to be functions only of a single dimensionless independent variable λ, which is defined by the relation

$$\lambda = Brt^{-k}, \tag{2.6.4}$$

where B is a constant with suitable dimensions and k is a dimensionless exponent. The dimensionless variable λ is constant on a path for which r is proportional to t^k, or in an equivalent steady hypersonic flow to x^k. Substituting (2.6.3) and (2.6.4) into (2.6.2) we obtain the set of ordinary differential equations

$$\lambda\left[(V - k)V' + \frac{P'}{R}\right] = -V(V - 1) - (2 - \omega)\frac{P}{R}, \tag{2.6.5a}$$

$$\lambda\left[V' + (V - k)\frac{R'}{R}\right] = -(1 + j - \omega)V, \tag{2.6.5b}$$

$$\lambda(V - k)\left[\frac{P'}{P} - \gamma\frac{R'}{R}\right] = -2(V - 1) - \omega(\gamma - 1)V, \tag{2.6.5c}$$

in which the primes indicate differentiation with respect to λ. Boundary conditions must be compatible with the assumed forms of the solution. Thus, a body or a shock wave in the flow field must lie on a line of constant λ.

In order to reduce the set of differential equations (2.6.5) to a single one, the new dependent variable z is introduced to replace the variable P,

$$z = \frac{\gamma P}{R}. \tag{2.6.6}$$

This variable is also dimensionless, and is equal to t^2/r^2 (or x^2/U^2r^2 for the equivalent steady problem) times the square of the speed of sound. From (2.6.5) may be obtained the single differential equation

$$\frac{dz}{dV} = \frac{z\left\{\begin{array}{l}[2(V-1)+j(\gamma-1)V](V-k)^2 \\ \quad +(\gamma-1)(1-k)V(V-k)-[2(V-1)+\kappa(\gamma-1)]z\end{array}\right\}}{(V-k)\{V(V-1)(V-k)-[(1+j)V-\kappa]z\}}, \tag{2.6.7}$$

in which P, R, and λ do not appear. Note that in the transformation to (2.6.7) the variable V has replaced λ as the basic independent variable. The solution of this equation under appropriate boundary conditions is the essential step in the problem. The parameter κ appearing in (2.6.7) is defined by

$$\kappa = \frac{2(1-k)+k\omega}{\gamma}. \tag{2.6.8}$$

Once a solution for $z(V)$ is obtained from (2.6.7), λ and R may be obtained by quadratures over V. The relations to be used are also obtainable from (2.6.5), and are

$$\frac{d\ln\lambda}{dV} = \frac{z-(V-k)^2}{V(V-1)(V-k)-[(1+j)V-\kappa]z}, \tag{2.6.9a}$$

$$\frac{d\ln(V-k)R}{d\ln\lambda} = -\frac{(1+j-\omega)V}{(V-k)}. \tag{2.6.9b}$$

Let us now examine the equations obtained. We shall call a line in the (r, t) space for which λ is constant, for which r is proportional to t^k, a "similarity line". The velocity of a point on this line is krt^{-1}. In view of (2.6.3a) the velocity of a fluid particle relative to a similarity line is $(V-k)rt^{-1}$. Thus $V = k$ is the boundary condition on a physical boundary which follows a similarity line, for which the similarity line is a particle path. The speed of sound equals $z^{1/2}rt^{-1}$, and hence the characteristics move with a velocity $(V-k\pm z^{1/2})rt^{-1}$ relative to a similarity line. A disturbance may be propagated only between the two characteristics. A disturbance may be propagated outward only if $V-k+z^{1/2}>0$, and inward only if $V-k-z^{1/2}<0$. At a point for which $z=(V-k)^2$ the similarity line is one of the two characteristics in the flow field. Except in certain singular cases the variable λ has an extremum at such a point, which therefore is a type of branch point. The similarity line in this case is a limiting line of the flow field, an envelope of the characteristics. In the flows which we consider $z-(V-k)^2$ is greater than zero within the region of interest. The fact that the velocity relative to a shock jumps from supersonic to subsonic across the shock may be translated into the fact that $z-(V-k)^2$ jumps from a negative to a positive value across a shock which lies on a similarity line. For further details see Sedov [3].

In order to obtain a similar solution we must provide boundary conditions with suitable similar properties. In this respect the shock conditions are most critical. In an undisturbed region of flow the pressure must be constant in both time and space; from (2.6.3c) either this requires $k = 2/(2 - \omega)$ with P proportional to $\lambda^{-2+\omega}$, or it requires $P = 0$. Except for this particular value of k, the undisturbed pressure in front of a shock must be negligibly small, so that we may take the ϵ_{lim} given by (1.4.18) in place of ϵ. In addition, the density in the undisturbed region must be independent of t; from (2.6.3b) this requires that R be a constant and that the density in front of the shock be proportional to a power $-\omega$ of the distance variable r. In the problems of interest to us the density in front of the shock will be constant, and we shall set $\omega = 0$ later. The shock trajectory must be a similarity line in order for the entire solution to be similar. With the gas in front of the shock at rest and the density there given by

(2.6.10) $$\rho_\infty = Ar^{-\omega}$$

the independent variables take on the values

(2.6.11a) $$V_\infty = 0,$$

(2.6.11b) $$R_\infty = 1,$$

(2.6.11c) $$P_\infty = 0,$$

(2.6.11d) $$z_\infty = 0,$$

in front of the shock. The shock conditions, with ϵ given by (1.4.18), give us the boundary values for the variables immediately behind the shock

(2.6.12a) $$V_s = \frac{2k}{\gamma + 1},$$

(2.6.12b) $$R_s = \frac{\gamma + 1}{\gamma - 1},$$

(2.1.12c) $$P_s = \frac{2k^2}{\gamma + 1},$$

(2.6.12d) $$z_s = \frac{2\gamma(\gamma - 1)k^2}{(\gamma + 1)^2}.$$

The subscripts ∞ and s here are equivalent to Sedov's 1 and 2. The boundary conditions (2.6.12) may be readily obtained from (2.6.11) and the principles that $R(V - k)$, $P + R(V - k)^2$, and $z/(\gamma - 1) + \frac{1}{2}(V - k)^2$ are conserved across the shock. These principles are the same as those underlying (1.4.1).

In the solutions of interest to us which contain shocks, the location of the shock is generally unknown in advance. The fact that this boundary is a floating one requires that we impose an additional condition to determine the solution. If this other condition is that the flow has a physical boundary which follows a similarity line, we impose

$$V_b = k \tag{2.6.13}$$

on the given boundary $\lambda = \lambda_b$. The solution of (2.6.7) near $V = k$ has the approximate form

$$z = (V - k)^\alpha, \tag{2.6.14}$$

where

$$\alpha = \frac{1 + \frac{1}{2}(\gamma - 1)\omega - k^{-1}}{1 + \frac{1}{2}(\gamma - 1)\omega + \frac{1}{2}\gamma(1 + j - \omega) - k^{-1}}. \tag{2.6.15}$$

Provided

$$\omega < 1 + j, \tag{2.6.16a}$$

$$k^{-1} < 1 + \tfrac{1}{2}(\gamma - 1)\,\omega + \tfrac{1}{2}\gamma(1 + j - \omega), \tag{2.6.16b}$$

then z approaches infinity, a constant, or zero as $V \to k$, as k^{-1} is greater than, equal to, or less than $1 + \frac{1}{2}(\gamma - 1)\,\omega$. The pressure and the quantity P remain finite, while the density and the quantity R behave as z^{-1}. In addition, it may be shown that in order that λ_b be positive, it is necessary that

$$k^{-1} < 1 + \tfrac{1}{2}(1 + j - \omega). \tag{2.6.17}$$

If $\omega \geqslant 0$, (2.6.17) implies (2.6.16b). These solutions, through the equivalence principle, represent bodies with power-law shapes in hypersonic flow. With $\omega = 0$ the case $k = 1$ corresponds to wedges and cones for $j = 0$ and 1, respectively; with $k < 1$ but subject to (2.6.17) the bodies are convex, while with $k > 1$ the bodies are pointed and concave. The case $k = 1$ here is the case for which p_∞ is not required to be zero.

Other types of solutions are of interest besides those corresponding to power-law shaped bodies. Of these the most important is that for a violent or point explosion treated numerically by Taylor [1] and analytically by Sedov [2; 3]. With $\omega = 0$ the value for k in this case is $2/(3 + j)$, while in general k equals the minimum value according to (2.6.17). In treating this case it is convenient to employ the general integrals of the conservation equations as given by Sedov [3].

Sedov deals with integrals over r or λ of various quantities which obey conservation laws. Thus we may examine integrals representing total mass, momentum, energy, and entropy. At first glance such integrals only serve to define new dependent variables. The new total mass variable is a useful one which can be interpreted in terms of a Lagrangian coordinate. The results for total momentum and energy are useful only in certain special cases in which the new total variables are eliminated. For the entropy, it is preferable not to introduce a total quantity but to use the principle that entropy is a function of Lagrangian coordinate alone. Through the equivalence law, total mass or Lagrangian variable can be interpreted in terms of the stream function.

The integrals are defined per unit area if $j = 0$, per unit angle times depth if $j = 1$, and per unit solid angle if $j = 2$. This normalization corresponds to setting Sedov's $\sigma_\nu = 1$. The integrals are

$$\mathscr{M} = \int \rho r^j \, dr = \frac{A}{1 + j - \omega} r_0^{1+j-\omega}, \tag{2.6.18a}$$

$$\mathscr{I} = \int v \rho r^j \, dr, \tag{2.6.18b}$$

$$\mathscr{E} = \int \left(\frac{p}{\gamma - 1} + \frac{\rho v^2}{2}\right) r^j \, dr, \tag{2.6.18c}$$

with integration limits which will be discussed later. The quantity r_0 is the Lagrangian radius, and corresponds to the lower limit in (2.6.18a) being zero and the upper limit variable. We now define the dimensionless quantities M, I, and E by

$$\mathscr{M} = AB^{-(1+j-\omega)} t^{k(1+j-\omega)} M, \tag{2.6.19a}$$

$$\mathscr{I} = AB^{-(2+j-\omega)} t^{k(2+j-\omega)-1} I, \tag{2.6.19b}$$

$$\mathscr{E} = AB^{-(3+j-\omega)} t^{k(3+j-\omega)-2} E. \tag{2.6.19c}$$

Our quantity M equals $\lambda^{1+j-\omega}$ times Sedov's M. On introducing the dimensionless variables defined by (2.6.3) and (2.6.4), we are led to the dimensionless relations

$$M = \int_{\lambda_1}^{\lambda_2} R \lambda^{j-\omega} \, d\lambda, \tag{2.6.20a}$$

$$I = \int_{\lambda_1}^{\lambda_2} RV \lambda^{1+j-\omega} \, d\lambda, \tag{2.6.20b}$$

$$E = \int_{\lambda_1}^{\lambda_2} \left(\frac{P}{\gamma - 1} + \frac{RV^2}{2}\right) \lambda^{2+j-\omega} \, d\lambda, \tag{2.6.20c}$$

with the integration carried out between $\lambda = \lambda_1$ and $\lambda = \lambda_2$.

We now keep λ_1 and λ_2 fixed, differentiate $\mathscr{M}$, $\mathscr{I}$, and $\mathscr{E}$ with respect to time, and write the standard conservation relations noting that the fluid velocity relative to a similarity line is $v - krt^{-1}$. The result is

$$\frac{d\mathscr{M}}{dt} + \rho r^j(v - krt^{-1})\Big| = 0, \tag{2.6.21a}$$

$$\frac{d\mathscr{I}}{dt} + [v\rho r^j(v - krt^{-1}) + pr^j]\Big| = 0, \tag{2.6.21b}$$

$$\frac{d\mathscr{E}}{dt} + \left[\left(\frac{p}{\gamma - 1} + \frac{\rho v^2}{2}\right) r^j(v - krt^{-1}) + pvr^j\right]\Big| = 0. \tag{2.6.21c}$$

In dimensionless form these become

$$k(1 + j - \omega)M + R(V - k)\lambda^{1+j-\omega}\Big|_{\lambda_1}^{\lambda_2} = 0, \tag{2.6.22a}$$

$$[k(2 + j - \omega) - 1]I + [RV(V - k) + P]\lambda^{2+j-\omega}\Big|_{\lambda_1}^{\lambda_2} = 0, \tag{2.6.22b}$$

$$[k(3 + j - \omega) - 2]E + \left[\left(\frac{P}{\gamma - 1} + \frac{RV^2}{2}\right)(V - k) + PV\right]\lambda^{3+j-\omega}\Big|_{\lambda_1}^{\lambda_2} = 0. \tag{2.6.22c}$$

These equations serve in general to give expressions for the quantities M, I, and E. In the particular case

$$\omega = 1 + j \tag{2.6.23}$$

we have

$$R(V - k) = \textit{constant}, \tag{2.6.24}$$

with no condition on the value of k. In the particular case

$$k = \frac{1}{2 + j - \omega} \tag{2.6.25}$$

we have

$$[RV(V - k) + P]\,\lambda^{2+j-\omega} = \textit{constant}, \tag{2.6.26}$$

while in the particular case

$$k = \frac{2}{3 + j - \omega} \tag{2.6.27}$$

we have

$$\left[\left(\frac{P}{\gamma - 1} + \frac{RV^2}{2}\right)(V - k) + PV\right]\lambda^{3+j-\omega} = \textit{constant}. \tag{2.6.28}$$

The three solutions (2.6.24), (2.6.26), and (2.6.28) correspond to constant rate of transfer across similarity lines of mass, momentum, and energy, respectively. When the constants on the right-hand sides of these equations are zero, the corresponding rates of transfer are zero and the corresponding total quantities defined between any two similarity lines are constant. Thus we shall refer to solutions satisfying (2.6.24), (2.6.26), and (2.6.28) with the right-hand constants zero as constant-mass, constant-momentum, and constant-energy solutions, respectively.

We turn now to the entropy relation. The entropy is constant along particle paths, and is thus a function of $\mathscr{M}$ alone if the lower limit is fixed at a physical boundary or at $\lambda_1 = 0$. In our calorically perfect gas p/ρ^γ is a function of entropy alone. Thus we may write

(2.6.29) $$\frac{p}{\rho^\gamma} = f(\mathscr{M}).$$

Substituting from (2.6.3) and (2.6.19a) we obtain

(2.6.30) $$A^{1-\gamma}PR^{-\gamma}\lambda^{2-\omega+\gamma\omega}t^{k(2-\omega+\gamma\omega)-2} = f(AB^{-(1+j-\omega)}Mt^{k(1+j-\omega)}).$$

Functional reasoning on the exponents of t immediately gives $f(\mathscr{M})$ proportional to $\mathscr{M}^{[k(2-\omega+\gamma\omega)-2]/k(1+j-\omega)}$, from which

(2.6.31) $$PR^{-\gamma}\lambda^{2-\omega+\gamma\omega}M^{-\frac{k(2-\omega+\gamma\omega)-2}{k(1+j-\omega)}} = \textit{constant},$$

provided $\omega \neq 1 + j$. With $(V - k)_1 = 0$ or $\lambda_1 = 0$, we may eliminate M between (2.6.31) and (2.6.22a) with λ_2 as dependent variable. We obtain

(2.6.32) $$P^{k(1+j-\omega)}\lambda^{2(1+j-\omega)}R^{2-k(2-\omega+\gamma+\gamma j)}(V - k)^{2-k(2-\omega+\gamma\omega)} = \textit{constant}.$$

The constant in (2.6.31) or (2.6.32) is discontinuous across a shock, of course. Isentropic solutions obey

(2.6.33) $$PR^{-\gamma}\lambda^{2-\omega+\gamma\omega} = \textit{constant}.$$

The case $\omega = 1 + j$ is rather special with respect to the entropy equation. Inspection of (2.6.30) shows that a similar power-law solution can exist only if

(2.6.34) $$k = \frac{2}{2 + (\gamma - 1)(1 + j)}.$$

The integral (2.6.24) applies in this case. In the constant-mass case we have

(2.6.35) $$V \equiv k,$$

and all the basic equations but (2.6.5a) are singular. The function $R(\lambda)$ is arbitrary, and the basic equation (2.6.5a) for P is linear if R is given.

There is one exception to the requirement (2.6.34) with $\omega = 1 + j$. This is the case

$$k = 1 \tag{2.6.36}$$

with p independent of λ and proportional to $t^{-\gamma(1+j)}$ instead of following (2.6.3c). The quantity R is arbitrary, and is constant in the isentropic case. These solutions correspond to source flows through the equivalence principle.

Returning to the constant-energy or point explosion solution, with k given by (2.6.27) and the constant in (2.6.28) zero, the solution may be written

$$z = \frac{(\gamma - 1)V^2(k - V)}{2(V - k/\gamma)}. \tag{2.6.37}$$

This solution satisfies the shock boundary conditions (2.6.12). The quantity z is infinite at $V = k/\gamma$, which may be shown to be at $(1 + j)V = \kappa$, with κ defined by (2.6.8). From (2.6.9) we obtain near $V = k/\gamma$ the approximate equations

$$\frac{d \ln \lambda}{d \ln (V - k/\gamma)} = \frac{\gamma - 1}{1 + j - \gamma\omega + 2(\gamma - 1)}, \tag{2.6.38a}$$

$$\frac{d \ln R\lambda^{-\omega}}{d \ln \lambda} = \frac{1 + j - \omega}{\gamma - 1} - \omega = \frac{1 + j - \gamma\omega}{\gamma - 1}. \tag{2.6.38b}$$

We require at this point that

$$\gamma\omega < 1 + j. \tag{2.6.39}$$

With condition (2.6.39) we conclude that we may set the inner boundary condition

$$\lambda_b = 0, \qquad V_b = k/\gamma \tag{2.6.40}$$

to replace (2.6.13). At this point the density, which is proportional to $R\lambda^{-\omega}$, goes to zero. The pressure, which is proportional to $P\lambda^{2-\omega}$, is finite at $\lambda = 0$.

In expressing the complete constant-energy solution it is convenient to introduce two parameters φ and Φ which serve to give us expressions of simpler form. We repeat (2.6.27) and write

$$k = \frac{2}{3 + j - \omega}, \tag{2.6.41a}$$

$$\varphi = k[\gamma + 1 + j(\gamma - 1)], \tag{2.6.41b}$$

$$\Phi = k\left(\frac{\gamma + 1 - \varphi}{\varphi} + \frac{\gamma - 1}{2\gamma - \varphi}\right); \tag{2.6.41c}$$

the quantity Φ is Sedov's α_1.

We now define the following functions of V, each of which may be shown from (2.6.12a) to take the value 1 immediately behind the shock:

$$F_1(V) = \frac{\gamma+1}{2k} V, \tag{2.6.42a}$$

$$F_2(V) = \frac{\gamma+1}{\gamma-1}\left(1 - \frac{1}{k} V\right), \tag{2.6.42b}$$

$$F_3(V) = \frac{\gamma+1}{\gamma-1}\left(\frac{\gamma}{k} V - 1\right), \tag{2.6.42c}$$

$$F_4(V) = \frac{\gamma+1}{\gamma+1-\varphi}\left(1 - \frac{\varphi}{2k} V\right). \tag{2.6.42d}$$

The function F_3 takes the value zero on the axis, as we may note from (2.6.40). We now define the constant B of (2.6.4) so that $\lambda_s = 1$. The quadratures of (2.6.9) together with the basic solution (2.6.37) then yield the results for a given value of the time

$$\frac{r}{r_s} = \lambda = F_1^{-k} F_3^{\frac{k(\gamma-1)}{2\gamma-\varphi}} F_4^{-\Phi}, \tag{2.6.43a}$$

$$\frac{v}{v_s} = \frac{\gamma+1}{2k}\lambda V = F_1^{1-k} F_3^{\frac{k(\gamma-1)}{2\gamma-\varphi}} F_4^{-\Phi}, \tag{2.6.43b}$$

$$\frac{\rho}{\rho_s} = \frac{\gamma-1}{\gamma+1}\lambda^{-\omega} R = F_1^{k\omega} F_2^{-\frac{k(2+2j-\gamma\omega-\omega)}{2-\varphi}} F_3^{\frac{k(1+j-\gamma\omega)}{2\gamma-\varphi}} F_4^{\frac{2(1+j)-\omega\varphi}{2-\varphi}\Phi}, \tag{2.6.43c}$$

$$\frac{T}{T_s} = \frac{(\gamma+1)^2}{2\gamma(\gamma-1)k^2}\lambda^2 z = F_1^{2(1-k)} F_2 F_3^{-\frac{k(1+j-\gamma\omega)}{2\gamma-\varphi}} F_4^{-2\Phi}, \tag{2.6.43d}$$

$$\frac{p}{p_s} = \frac{\gamma+1}{2k^2}\lambda^{2-\omega} P = F_1^{2(1-k)+k\omega} F_2^{-1-\frac{k(\gamma-1)(2+2j-\omega)}{2-\varphi}} F_4^{\frac{2(\varphi-1+j)-\omega\varphi}{2-\varphi}\Phi}, \tag{2.6.43e}$$

$$\frac{r_0}{r_s} = \left(\frac{F_1}{F_2}\right)^{-k} \left(\frac{F_3}{F_2}\right)^{\frac{k\gamma}{2\gamma-\varphi}} \left(\frac{F_4}{F_2}\right)^{\frac{\varphi}{2-\varphi}\Phi} \tag{2.6.43f}$$

These results were obtained by combining results taken from Sedov's book [3, Chapter IV, §§11 and 14]. The quantity T is the absolute temperature. The Lagrangian coordinate r_0 is defined by (2.6.18a) and is equal to the original value of r for a fluid particle before the shock wave reached it. Equations (2.6.43) are given in Sedov [3] only for the case with $\omega = 0$ and j general and for the case with $j = 2$ and ω general. Verification that (2.6.43) satisfy the appropriate basic relations is straightforward but tedious.

We have kept ω arbitrary above solely in order to preserve the generality of Sedov's constant-energy solution. We are actually interested only in the case for which ρ_∞ is constant, i.e. for which $\omega = 0$ and for which $k = 2/(3+j)$ for the constant-energy solution.

The limiting process $\gamma \to 1$, equivalent to the process $\epsilon \to 0$, is important in hypersonic flow theory. This limiting process for the similar solutions has been studied by Freeman [4], and in extensive detail by Brocher [1; 2]. In the constant-energy case we are interested primarily in the pressure near $\lambda = 0$. In this limit we have $F_1 = 1$, $F_2 = 2$, $F_4 = 1$, and we obtain from (2.6.43e)

$$\frac{p}{p_s} = \frac{1}{2}. \tag{2.6.44}$$

In the general case with a shock, the entire mass flow is concentrated in a thin layer behind the shock in this limit, with the physical boundary (body) close to the shock if (2.6.17) holds. If k is equal to or less than the value corresponding to the constant-energy solution, there is behind this layer near the shock a region of negligible density in which the pressure is a function of t alone. In more generality, we may show that the pressure on the body p_b or in the low-density region is given in the limit $\gamma \to 1$ by

$$\frac{p_b}{p_s} = \frac{2+j-\omega-k^{-1}}{1+j-\omega}, \tag{2.6.45}$$

of which (2.6.44) is a special case.

With $k^{-1} > \frac{1}{2}(3+j-\omega)$ singularities or anomalies of various types appear. As an example, the constant-momentum solution may be examined briefly, for which

$$k = \frac{1}{2+j-\omega}. \tag{2.6.46}$$

The solution for z comes immediately from (2.6.26), and is

$$z = \gamma V(k - V). \tag{2.6.47}$$

With this solution, which fits the shock conditions (2.6.12), a study of (2.6.9a) tells us that a limiting line appears at

$$V_l = \frac{k}{\gamma+1}, \tag{2.6.48}$$

at a finite value of λ_l less than λ_s. In general we may conclude that this solution is not physically realizable.

In the limit $\gamma \to 1$, however, we find that both density and pressure in the region behind the layer behind the shock approach zero. In this limit, then,

the limiting line of the flow finds itself in a vacuum and is of no physical importance. Thus in the limit the constant-momentum solution is of physical interest. The vanishing of the pressure is consistent with (2.6.45) and (2.6.46).

An interesting particular solution found by Sedov [3, p. 271] was pointed out by Nikol'skii [1] to correspond through the equivalence principle to hypersonic nozzle flow. This solution satisfies (2.6.23), (2.6.33), (2.6.34), and (2.6.35). The solution is

$$z = \lambda^{-2}[1 + K\lambda^2], \tag{2.6.49a}$$

$$P = A\lambda^{j-1}[1 + K\lambda^2]^{\frac{\gamma}{\gamma-1}}, \tag{2.6.49b}$$

$$R = \gamma A\lambda^{1+j}[1 + K\lambda^2]^{\frac{1}{\gamma-1}}, \tag{2.6.49c}$$

where

$$K = \tfrac{1}{4}(\gamma - 1)^2(1 + j)k^2. \tag{2.6.50}$$

In this flow, since each particle path is a similarity line, an outer physical boundary may be placed at any value of λ.

Finally, a word about implosion solutions. In the analysis above we have considered only explosion solutions, with the time t going from 0 to ∞; implosion solutions are solutions for which the time goes from $-\infty$ to 0. The planar implosion case $j = 0$ is trivial, and corresponds to the intersection of two equal uniform shocks. But the cylindrical case $j = 1$ is of interest to us, and corresponds to a focusing axisymmetric shock. Such a shock may be produced in a duct of circular cross section, for example; however, very near the convergence point of the implosion solution the equivalence principle must fail, and a lambda shock must appear in the steady hypersonic flow.

For similar implosion solutions we need to substitute $-t$ for t in the defining relations of this section. But if we also change the sign of v, so that a positive V corresponds to negative v, the entire analysis given above is unchanged, including the establishment of boundary conditions for flow boundaries and for shock waves with the gas in front undisturbed. The solution of Guderley [1] represents the asymptotic behavior of quite arbitrary implosions, and is particularly interesting. The solution is required to satisfy the boundary conditions

$$\lambda_b = \infty, \tag{2.6.51a}$$

$$z_b = 0, \tag{2.6.51b}$$

$$V_b = 0. \tag{2.6.51c}$$

Since the boundary point given above and the shock boundary point of (2.6.12) lie on different sides of the curve $z = (V - k)^2$ in the (z, V) space, the integral

curve must pass through a singular point in traversing this line. This singular point is a saddle point in this space, and corresponds to the numerators and denominators of (2.6.7) and (2.6.9a) vanishing simultaneously. An integral curve which possesses this property and which satisfies the boundary conditions may be found only with a particular value of k which depends upon the constant parameters of the problem.

With constant density in the undisturbed flow, i.e. with $\omega = 0$, Chisnell [1] and Culler and Fried [1] have obtained by independent and completely dissimilar analyses the same approximate formula

$$k = \frac{1}{1 + \dfrac{j\alpha_0}{(\alpha_0 + 1)(\alpha_0 + 2/\gamma)}} = \frac{1}{1 + \dfrac{j}{1 + 2/\gamma + \gamma\alpha_0}}, \tag{2.6.52}$$

where α_0 is the symbol used by Culler and Fried for

$$\alpha_0 = \sqrt{\frac{2}{\gamma(\gamma - 1)}}. \tag{2.6.53}$$

Chisnell notes that α_0 is the Mach number of the flow following a moving strong shock. This approximation is remarkably accurate. With $j = 1$, it gives a value of k which is too low by about 0.3 % at $\gamma = 1.1$, and which is in error by not more than about 0.05 % for γ between 1.4 and 2.0.

Actually, the boundary conditions (2.6.51) are not essential, and the essential boundary condition is that the integral curve pass through the saddle point. The solution for the inner regions with λ smaller than at the saddle point is completely independent of the physical solution for the outer region with λ larger than this critical value. The solution in the outside region need not be a similar one at all at finite values of $-t$, and the solution has an asymptotic validity for small values of $-t$ with quite general implosions with suitable symmetry. The reason for this independence is that no characteristic from the outer region can penetrate into the inner region. Guderley's solution continues for $t > 0$ with a solution of the explosion type, with different shock conditions and boundary conditions at $\lambda = 0$.

If the gas flow exists at finite pressure and $v = 0$ at $\lambda = 0$, a particular boundary condition for V applies. The pressure there is proportional to $t^{k(2-\omega)-2}$. The radius r of a gas parcel is then proportional to $t^{[2-k(2-\omega)]/\gamma(1+j)}$. The physical velocity v is dr/dt, and (2.6.3a) gives

$$V_0 = \frac{2 - k(2 - \omega)}{\gamma(1 + j)} \tag{2.6.54}$$

there. This is the desired condition. With k given by (2.6.27) it is consistent with (2.6.40). With (2.6.23) and (2.6.34) it is consistent with (2.6.35).

Through the equivalence principle, the similar time-dependent flows we have discussed may be interpreted as corresponding to steady hypersonic flows if $j = 0$ or $j = 1$. Besides the interpretations mentioned earlier, the constant-energy or point explosion solutions correspond to flows on a blunted flat plate ($j = 0$) or cylindrical rod ($j = 1$), and the constant-momentum solutions with $\gamma \to 1$ correspond to free layer solutions of Newtonian theory (Section 3.3). Implosions correspond to intersecting or focusing shocks. Table 2–2 presents a summary of the solution types we have treated or discussed.

TABLE 2–2

Similar power-law solutions

k	ω	Type ($j = 0$ or $j = 1$)
$\to\infty$	0	Exponential body (Chernyi [3, p. 92])
>1	0	Concave pointed power-law bodies
1	0	Wedge or cone
1	$1 + j$	Source
$\dfrac{2}{2 + (\gamma - 1)(1 + j)}$	$1 + j$	Nozzle
$\leqslant 1$	0	Implosions
$<1, > \dfrac{2}{3 + j}$	0	Convex power-law bodies
$\dfrac{2}{3 + j}$	0	Blunted flat plate or cylinder (point explosion)
$\dfrac{1}{2 + j}$	0	Free layer with $\gamma \to 1$

We present next a few graphs to illustrate those solutions of most importance in hypersonic flow theory. Where appropriate, we use here $x = Ut$ in place of t. For additional graphs and tables the reader is referred to Chernyi [3], Sedov [3], and Mirels [2].

Since we are concerned here with hypersonic small-disturbance theory rather than with explosion theory we shall confine our interest in the numerical solutions to the planar and cylindrical cases ($j = 0, 1$), corresponding to two-dimensional and axisymmetric steady flows. In Fig. 2-10 the functions v/v_s, ρ/ρ_s, and p/p_s are plotted versus y/y_s for the planar case with $\gamma = 1.4$. We have

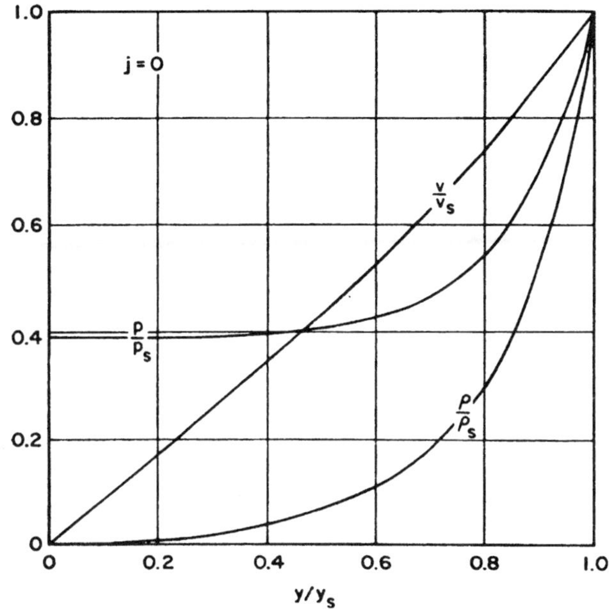

FIG. 2–10. Point explosion solution, two-dimensional case; $\gamma = 1.4$ (Sedov [3]).

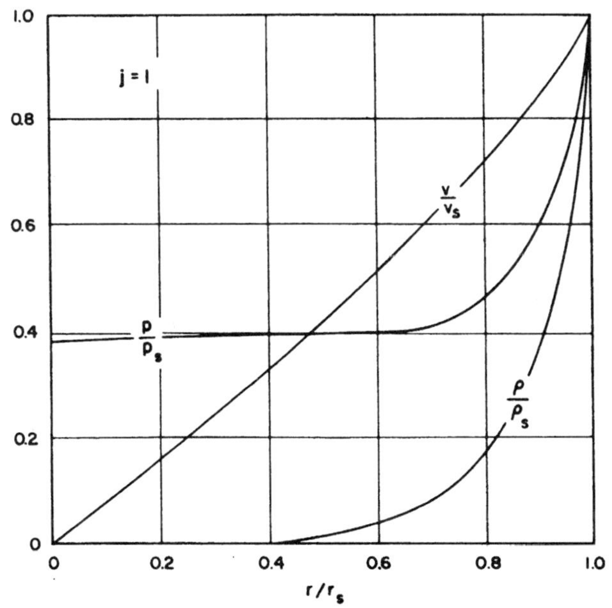

FIG. 2–11. Point explosion solution, axisymmetric case; $\gamma = 1.4$ (Sedov [3]).

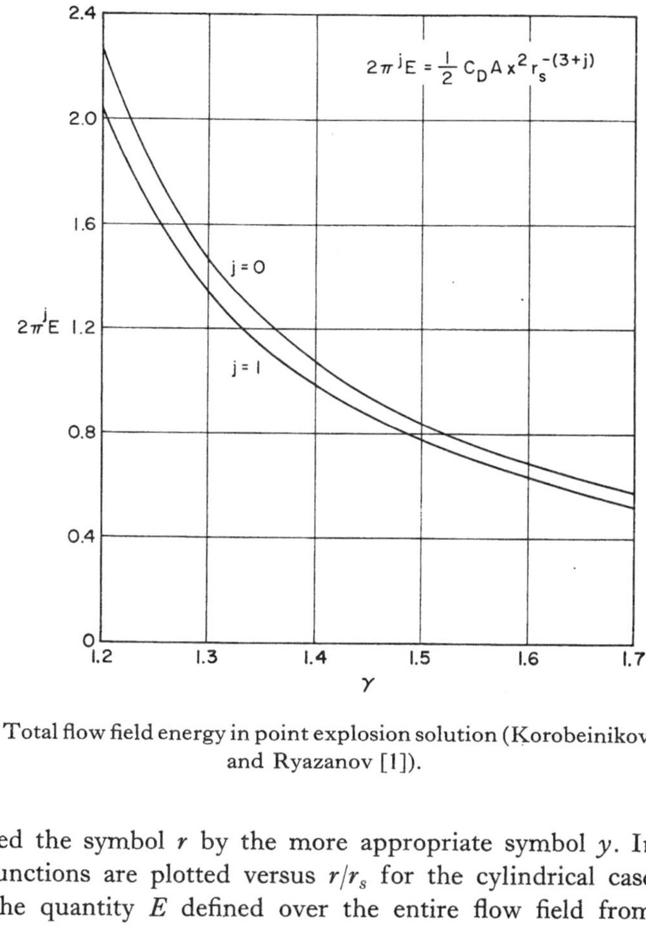

FIG. 2–12. Total flow field energy in point explosion solution (Korobeinikov, Mel'nikova, and Ryazanov [1]).

here replaced the symbol r by the more appropriate symbol y. In Fig. 2–11 the same functions are plotted versus r/r_s for the cylindrical case, also with $\gamma = 1.4$. The quantity E defined over the entire flow field from $\lambda_b = 0$ to $\lambda_s = 1$ is presented in Fig. 2–12. This quantity is multiplied by the normalization constant $2\pi^j$.

For the power-law bodies, plots of $\lambda_b = r_b/r_s$ and of dimensionless pressure are given in Figs. 2–13 for the two-dimensional case ($j = 0$) and in Figs. 2–14 for the axisymmetric case ($j = 1$). In these figures the pressure results for $\gamma = 1$ are obtainable from

$$C_{p_b} = 2\left(\frac{dr_b}{dx}\right)^2\left[1 - \frac{1-k}{k(1+j)}\right]. \tag{2.6.55}$$

This equation can be derived from the equations in this section but is most easily derived with Newtonian theory. The appropriate equations are (3.4.38) or (3.3.14) and (3.3.17) of Chapter III.

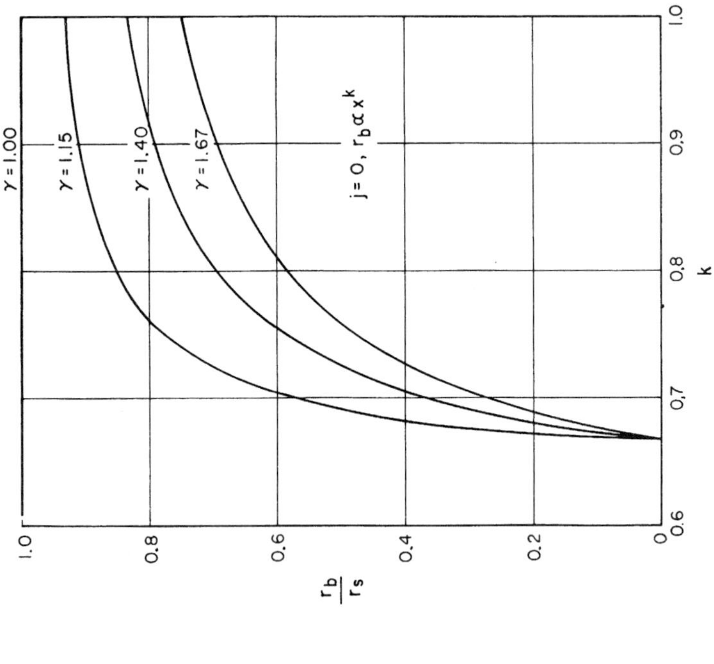

FIG. 2–13a. Ratio of body radius to shock radius in two-dimensional power-law solutions (Mirels [1]).

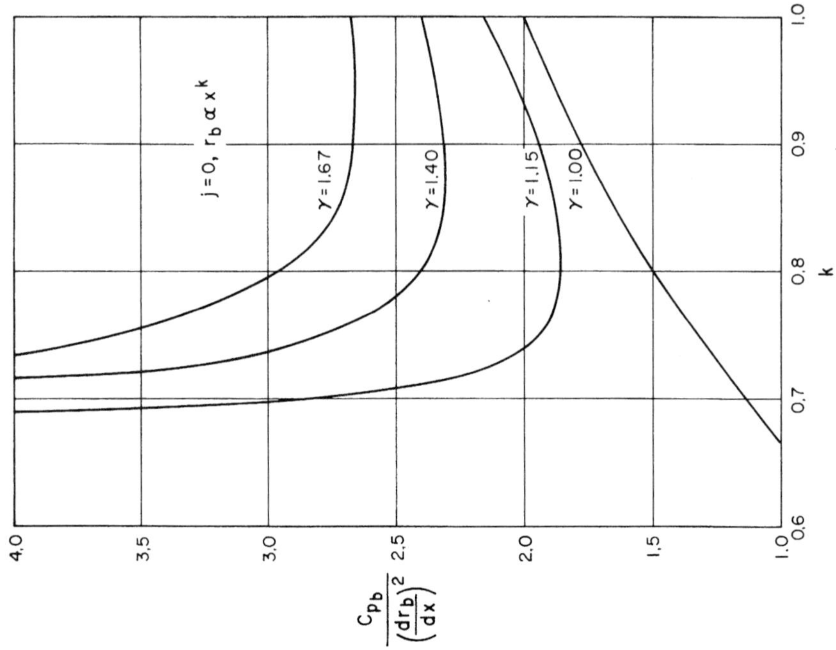

FIG. 2–13b. Pressure coefficient on body in two-dimensional power-law solutions (Mirels [1]).

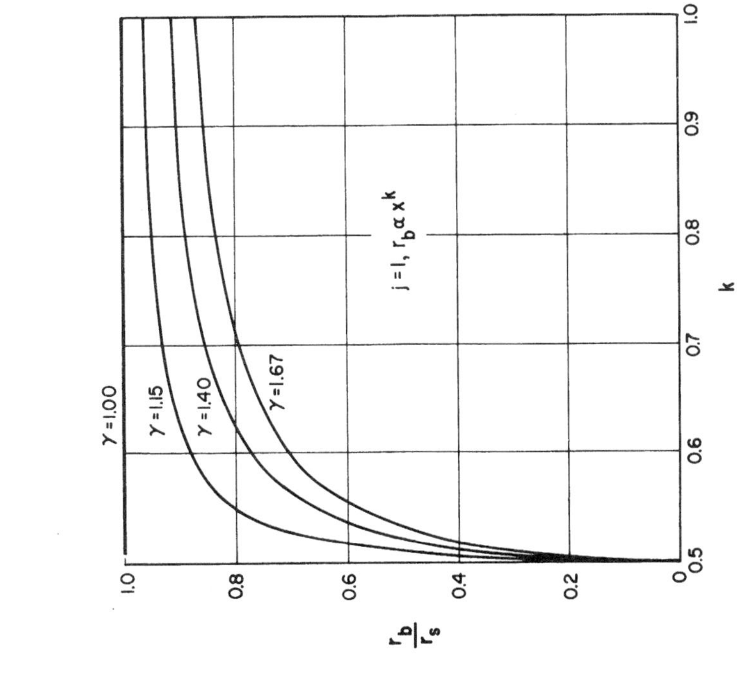

FIG. 2–14a. Ratio of body radius to shock radius in axisymmetric power-law solutions (Kubota [1]).

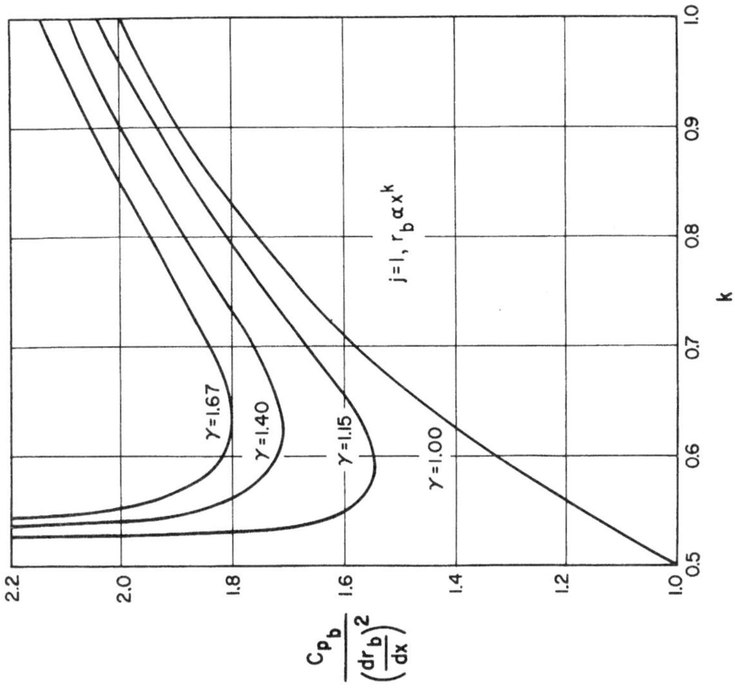

FIG. 2–14b. Pressure coefficient on body in axisymmetric power-law solutions (Kubota [1]).

The application of similar solutions to steady flows is valid generally within limitations which are discussed below. The validity has been checked experimentally and found to hold subject to the limitations. For power-law bodies, for example, the reader is referred to Kubota [1] and to Freeman, Cash, and Bedder [1]. Comparison with experiment and correlation procedures for the blast wave case may be found in Lees and Kubota [1], Chernyi [3], Lukasiewicz[1] and Kuehn [1] while comparison with numerical calculations may be found in Chushkin [1]. Additional references will be given in the next section where large deviations from the blast wave solution are considered.

Figure 2–15 is a Schlieren photograph of a flow over a blunted flat plate in a

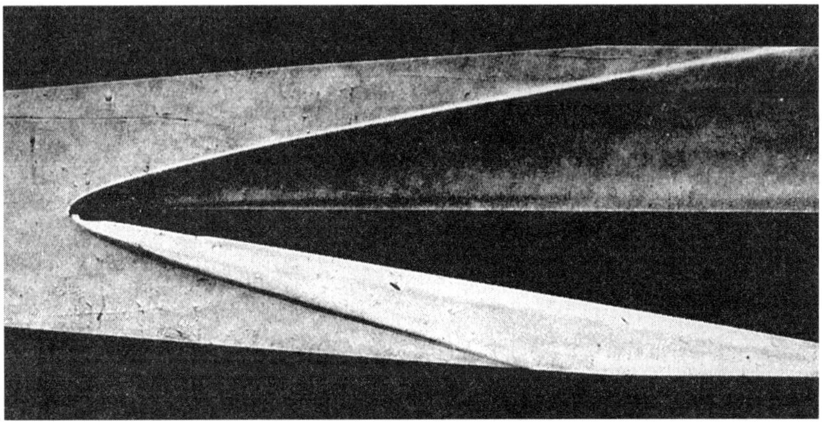

FIG. 2–15. Schlieren photograph of a blunted flat plate in helium at $M_\infty = 11.6$ and $Re_{d_\infty} = 16{,}500$. (Courtesy Gas Dynamics Laboratory, Princeton University.)

uniform flow of helium at $M_\infty = 11.6$. The leading edge has a flat face with a thickness of 0.017 inches, and the height of the part of the wind tunnel which is shown at the right of the photograph is $1\frac{1}{2}$ inches. As with Fig. 1–3, the flow on the bottom surface should be disregarded. The Reynolds number based on leading edge thickness Re_{d_∞} is about 16,500, and the effect of the boundary layer is negligible. The shock shape is close to a power-law shape, with a measured value of k from three tests of 0.66 ± 0.01. This may be compared with the theoretical value of 0.667. Although this agreement with the theory may be considered excellent for the shock shape, agreement for the pressure distribution on the plate is not too satisfactory in that the pressure does not follow a simple power law.

The limitations of the applicability of the theory are associated with various effects discussed below. In many cases we can consider one or more of these effects as being a perturbation on a given basic similar solution. For such an approach the effect must be weak, and a suitable parameter governing the

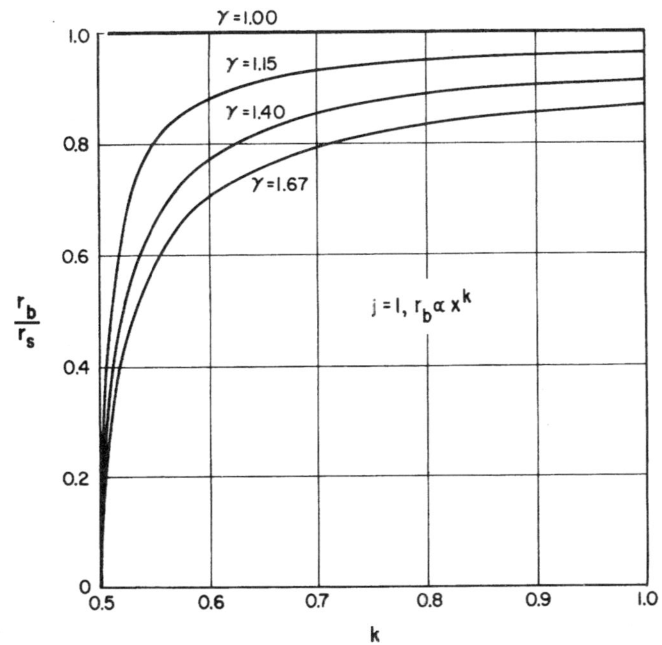

FIG. 2–14a. Ratio of body radius to shock radius in axisymmetric power-law solutions (Kubota [1]).

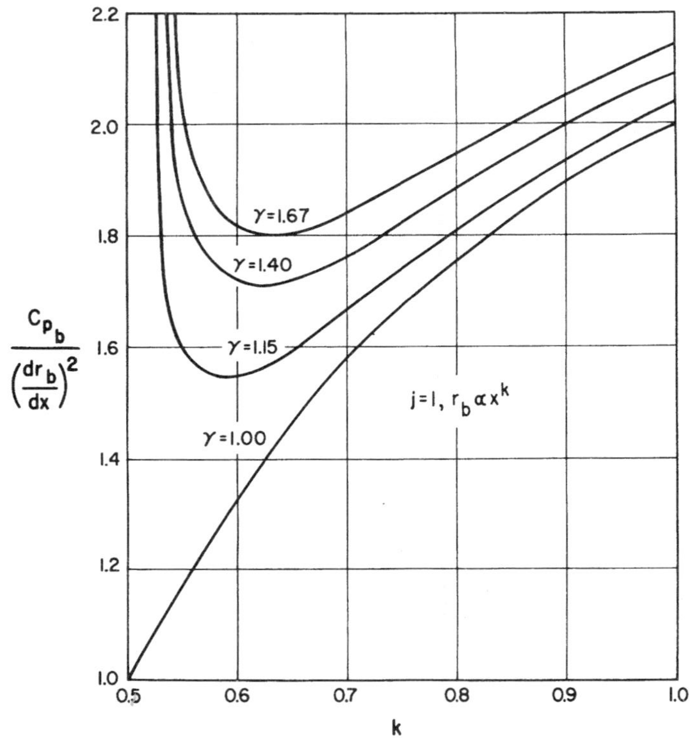

FIG. 2–14b. Pressure coefficient on body in axisymmetric power-law solutions (Kubota [1]).

7. Application of similar solutions to steady flows

In this section we discuss the application to steady flows of the similar power-law solutions treated in the last section. The solutions considered are limited to those with an outer shock wave moving into an undisturbed gas of constant density and negligible pressure ($\omega = 0$, $P_\infty = 0$). Through the equivalence principle, we replace t by x/U and write

$$A = \rho_\infty , \tag{2.7.1a}$$

$$B = x^k U^{-k} r_s^{-1} , \qquad \lambda_s = 1, \tag{2.7.1b}$$

$$\frac{dr_s}{dx} = \frac{kr_s}{x} . \tag{2.7.1c}$$

The pressure is given by

$$p = \rho_\infty U^2 r_s^2 x^{-2} (\lambda^2 P) \tag{2.7.2}$$

or by

$$C_p = \frac{2}{k^2}\left(\frac{dr_s}{dx}\right)^2 (\lambda^2 P) = \frac{4}{\gamma + 1}\left(\frac{p}{p_s}\right)\left(\frac{dr_s}{dx}\right)^2 . \tag{2.7.3}$$

For solutions corresponding to power-law bodies,

$$r_b = \lambda_b r_s , \qquad \frac{dr_b}{dx} = \lambda_b \frac{dr_s}{dx} . \tag{2.7.4}$$

The pressure coefficient on the body may be expressed

$$C_{p_b} = \frac{4}{(\gamma + 1)\lambda_b^2}\left(\frac{p_b}{p_s}\right)\left(\frac{dr_b}{dx}\right)^2 , \tag{2.7.5}$$

and the drag coefficient based upon frontal area may be shown to be

$$C_D = \frac{4k^2}{(\gamma + 1)\lambda_b^2}\left(\frac{1 + j}{3 + j - 2k^{-1}}\right)\left(\frac{p_b}{p_s}\right)\tau^2 , \tag{2.7.6}$$

where τ is defined by

$$r_b = \tau c \qquad at \qquad x = c, \tag{2.7.7}$$

and c is a reference axial length. The frontal area is $\pi\tau^2 c^2$ in the axisymmetric case ($j = 1$), and per unit depth is $2\tau c$ in the two-dimensional case ($j = 0$) with lateral symmetry. In the latter case ($j = 0$) the lateral projected area per unit depth is simply c, and the drag coefficient based on lateral area is 2τ times that given by (2.7.6).

The point explosion solutions are generally termed blast wave solutions when applied to steady flows. The energy released in the flow per unit distance is $2\pi^j\mathscr{E}$, and times the velocity U is the energy released in the flow per unit time. This quantity is identified with the rate DU at which work is done by a drag force D. Thus we have

$$2\pi^j\mathscr{E} = D. \tag{2.7.8}$$

From (2.7.1) and (2.6.19c) we obtain

$$2\pi^j E = \frac{C_{D_{\mathrm{fr}}} A_{\mathrm{fr}}\, x^2}{2r_s^{3+j}}, \tag{2.7.9}$$

where A_{fr} is the area on which this $C_{D_{\mathrm{fr}}}$ is based and has dimensions L^{1+j}. Equation (2.7.9) may be rewritten as

$$\frac{r_s}{x} = \left(\frac{C_{D_{\mathrm{fr}}} A_{\mathrm{fr}}}{4\pi^j E x^{1+j}}\right)^{\frac{1}{3+j}}. \tag{2.7.10}$$

The pressure coefficient may be expressed

$$C_p = \frac{4k^2}{\gamma+1}\left(\frac{p}{p_s}\right)\left(\frac{C_{D_{\mathrm{fr}}} A_{\mathrm{fr}}}{4\pi^j E x^{1+j}}\right)^{\frac{2}{3+j}} \tag{2.7.11}$$

with $k = 2/(3+j)$.

These blast wave solutions are identified as solutions for bodies with a blunted nose which causes a finite amount of drag. Behind the blunted nose the bodies have constant thickness and zero slope, and so bear no drag force except that caused by the nose. In (2.7.9) to (2.7.11) above, we generally choose A_{fr} as a measure of the frontal area of the blunted nose, with the nose drag coefficient $C_{D_{\mathrm{fr}}}$ then of the order of one. The prediction that the similar point explosion solution holds here applies only to the asymptotic behavior of the flow and not to the local solution near the nose.

The Lagrangian coordinate r_0 can be identified in terms of the stream function ψ by

$$\psi = \frac{\rho_\infty U}{1+j} r_0^{1+j}. \tag{2.7.12}$$

The stream function represents mass flow per unit depth in two-dimensional flows ($j = 0$), and mass flow per unit azimuthal angle in axisymmetric flows ($j = 1$). Note that frontal area is $2\pi^j r^{1+j}/(1+j)$, so that the initial frontal area of a portion of the flow field is identified with $2\pi^j\psi/\rho_\infty U$. In the two-dimensional case this is area per unit depth for a laterally symmetric solution.

The application of similar solutions to steady flows is valid generally within limitations which are discussed below. The validity has been checked experimentally and found to hold subject to the limitations. For power-law bodies, for example, the reader is referred to Kubota [1] and to Freeman, Cash, and Bedder [1]. Comparison with experiment and correlation procedures for the blast wave case may be found in Lees and Kubota [1], Chernyi [3], Lukasiewicz[1] and Kuehn [1] while comparison with numerical calculations may be found in Chushkin [1]. Additional references will be given in the next section where large deviations from the blast wave solution are considered.

Figure 2–15 is a Schlieren photograph of a flow over a blunted flat plate in a

FIG. 2–15. Schlieren photograph of a blunted flat plate in helium at $M_\infty = 11.6$ and $Re_{d_\infty} = 16{,}500$. (Courtesy Gas Dynamics Laboratory, Princeton University.)

uniform flow of helium at $M_\infty = 11.6$. The leading edge has a flat face with a thickness of 0.017 inches, and the height of the part of the wind tunnel which is shown at the right of the photograph is $1\frac{1}{2}$ inches. As with Fig. 1–3, the flow on the bottom surface should be disregarded. The Reynolds number based on leading edge thickness Re_{d_∞} is about 16,500, and the effect of the boundary layer is negligible. The shock shape is close to a power-law shape, with a measured value of k from three tests of 0.66 ± 0.01. This may be compared with the theoretical value of 0.667. Although this agreement with the theory may be considered excellent for the shock shape, agreement for the pressure distribution on the plate is not too satisfactory in that the pressure does not follow a simple power law.

The limitations of the applicability of the theory are associated with various effects discussed below. In many cases we can consider one or more of these effects as being a perturbation on a given basic similar solution. For such an approach the effect must be weak, and a suitable parameter governing the relative magnitude of the effect must be small compared with one. If such a parameter becomes large rather than small compared with one, it is generally appropriate to consider the flow in an entirely different regime, with a different interpretation of what solution may be basic and what effect may be a perturbation. If the parameter is neither large nor small we are in a transition zone in which no perturbation approach is applicable. Transition solutions linking well-defined regimes with known basic flow solutions must be sought by means other than perturbation methods. These other methods may be necessarily numerical, and include integral methods such as those considered in Section 5.8.

In our discussion of the pertinent effects we shall primarily consider each effect small, and so representable as a perturbation with a small parameter. We consider separately the various types of effects imposing limitations on similar solutions:

(a) Local equivalence failure. The small-disturbance theory and the consequent equivalence principle are based upon the smallness of a parameter τ^2, actually on the smallness of $\tau^2(x/c)^{-2(1-k)}$. Wherever the body slope or shock slope is not small compared with one the equivalence principle fails. This necessarily occurs sufficiently near the nose if $k < 1$ and sufficiently far downstream if $k > 1$. If $k = 1$ the body is a wedge or cone, and the body and shock slopes are independent of x. This type of limitation was discussed in Section 2.1.

(b) Counterpressure. If the undisturbed pressure p_∞ is not zero, the square of the Mach number $M_\infty^2 = \rho_\infty U^2/\gamma p_\infty$ is finite. If p_∞ is not small compared with $\rho_\infty U^2\tau^2(x/c)^{-2(1-k)}$, or $(M_\infty\tau)^{-2}$ is not small compared with one at $x = c$, the condition $P_\infty = 0$ fails. This failure occurs sufficiently far downstream if $k < 1$ and sufficiently near the nose if $k > 1$. If $k = 1$ the basic parameter is independent of x. It is in this $k = 1$ case that similar solutions are obtainable with P_∞ finite; for the purpose of our discussion we will consider finite P_∞ as giving failure of the $P_\infty = 0$ solution in this case too. First-order corrections for the effect of counterpressure have been given by Sakurai [1; 2], Sedov [3], and Kubota [1]. Good summaries of numerical results for the power-law solutions may be found in Mirels [1] and for the blast wave solutions in Korobeinikov, Mel'nikova, and Ryazanov [1, §3.1]. Counterpressure calculations for the blast wave solutions to second order have been presented by Swigart [1; 2], while results of complete numerical calculations are summarized in Belotserkovskii and Chushkin [2, Section 2.4].

The first two types of limitations involve basic assumptions without which (with $k \neq 1$) similar solutions are unobtainable. For the other limitations discussed, the equivalence principle and zero pressure in front of the shock are assumed. Thus similar solutions starting from an assumed similar shock shape do exist. These other limitations all involve inner boundary conditions and what we shall term displacement effects. Failure to satisfy inner boundary conditions can be interpreted as being caused by a displacement thickness of some kind, analogous to the displacement thickness of classical boundary layer theory. The displacement thickness or area resulting from a thin perturbed layer of flow in which mass is conserved is the integral of $1 - \rho u/\rho_\delta u_\delta$, over thickness or area, where $\rho_\delta u_\delta$ is the mass flow at the outer edge of the layer. An assumption underlying this concept is that in the absence of the perturbed layer the mass flow would take the constant value $\rho_\delta u_\delta$ across the range of stream function representing the layer.

(c) Entropy wake. If k is not too small, so that the corresponding body is quite sharp nosed, and the body is then slightly blunted, an entropy wake of the type discussed in Section 1.2 appears. This wake lies in a thin layer near the body, and is hotter and less dense than the flow near the body would be without the blunting. The result is an effective thickening of the body by a positive displacement thickness. This entity is often referred to in the literature as an entropy layer. Here we avoid this term to avoid confusion with the entity discussed below under (d). The entropy wake is a true wake of the blunt nose even though it lies as a layer on the body, and it passes with no essential change into the entropy wake behind the body.

We calculate for a calorically perfect gas the displacement area A_{ent} associated with an entropy wake on a power-law body. This area is defined by

$$A_{\text{ent}} = 2\pi^j \int \left(1 - \frac{\rho u}{\rho_\delta u_\delta}\right) r^j \, dr = 2\pi^j \int \left(\frac{1}{\rho u} - \frac{1}{\rho_\delta u_\delta}\right) d\psi , \tag{2.7.13}$$

with the integral taken over the entropy wake. We divide A_{ent} into two parts

$$A_{\text{ent}} = A_{\text{ent},\rho} + A_{\text{ent},u} , \tag{2.7.14}$$

where

$$A_{\text{ent},\rho} = \frac{2\pi^j}{u_\delta} \int \left(\frac{1}{\rho} - \frac{1}{\rho_\delta}\right) d\psi, \tag{2.7.15}$$

$$A_{\text{ent},u} = 2\pi^j \int \frac{1}{\rho} \left(\frac{1}{u} - \frac{1}{u_\delta}\right) d\psi . \tag{2.7.16}$$

The quantites ρ_δ and u_δ are assumed constant across the layer, and u_δ is immediately identified with U. The quantity $A_{\text{ent},u}$ is termed the velocity-defect displacement area, and is identically zero if the equivalence principle holds through the layer ($u = U$). The quantity $A_{\text{ent},\rho}$ is termed the constant-velocity displacement area.

With $p_\infty = 0$, we have immediately behind the shock $\rho_s = \rho_\infty/\epsilon$ and $p_s = \rho_\infty U^2(1 - \epsilon)\sin^2\sigma$, from (1.4.5b), (1.4.6), and (1.5.2a). Here ϵ is ϵ_{lim}, equal to $(\gamma - 1)/(\gamma + 1)$. Immediately behind the shock we have

$$\frac{p_s}{\rho_s^\gamma} = \rho_\infty^{-(\gamma-1)} U^2 \epsilon^\gamma (1 - \epsilon)(\sin\sigma)^2 . \tag{2.7.17}$$

From (2.7.5) and (2.7.7) we have for a power-law body

$$p_b = \rho_\infty U^2 (1 - \epsilon) \frac{1}{\lambda_b^2} \left(\frac{p_b}{p_s}\right) \tau^2 k^2 \left(\frac{x}{c}\right)^{-2(1-k)} . \tag{2.7.18}$$

The principle that p/ρ^γ is constant along a streamline yields, with the assumption that $p = p_b$,

$$\frac{1}{\rho} = \frac{\epsilon}{\rho_\infty} \left[k^2 \frac{p_b}{p_s}\right]^{-1/\gamma} \left(\frac{\tau}{\lambda_b}\right)^{-2/\gamma} \left(\frac{x}{c}\right)^{2(1-k)/\gamma} (\sin\sigma)^{2/\gamma} . \tag{2.7.19}$$

The fact that p is very close to p_b may be shown by a separate analysis (see Cheng [1] or Sychev [2]). The wake is thus in a region of essentially constant pressure. We now call $\sigma_{\text{bl}}(\psi)$ the shock angle which actually exists in the flow associated with a blunt nose, and ρ without subscript the local density in the layer in this case. The quantity $\sigma(\psi)$ without subscript is that associated with the reference shape without blunting, and ρ_δ is the local density in this case. The fact that $1/\rho$ or the area varies as $p^{-1/\gamma}$ was noted by Guiraud [2].

We define the reference area A by

$$A = \frac{2\pi^j}{\rho_\infty U} \left[k^2 \frac{p_b}{p_s}\right]^{-1/\gamma} \int_0^\infty [(\sin\sigma_{\text{bl}})^{2/\gamma} - (\sin\sigma)^{2/\gamma}]\, d\psi, \tag{2.7.20}$$

and obtain for the constant-velocity displacement area

$$A_{\text{ent},\rho} = A\epsilon \left(\frac{\tau}{\lambda_b}\right)^{-2/\gamma} \left(\frac{x}{c}\right)^{2(1-k)/\gamma} . \tag{2.7.21}$$

The integral in (2.7.20) should have negligible contribution outside a small range of ψ. We set the upper limit equal to infinity formally to indicate integration over the entire entropy wake. Note that $\tau/\lambda_b = \tau_s$ is the τ which would be defined in (2.7.7) using r_s instead of r_b. This area may be divided by the total cross-sectional area of the flow field

$$A_s = \frac{2\pi^j}{1 + j} c^{1+j} \left(\frac{\tau}{\lambda_b}\right)^{1+j} \left(\frac{x}{c}\right)^{k(1+j)} \tag{2.7.22}$$

to obtain a local parameter which measures the magnitude of the entropy wake effect. Entropy wakes on wedges and cones are included in the integral solutions of Chernyi [2; 3]; our results are consistent with the asymptotic values given by his solution. Entropy wakes have also been treated by Cheng [1].

To obtain the velocity-defect displacement area of (2.7.16) we first apply Bernoulli's equation to obtain

$$\frac{U}{u} = \left(1 - \frac{2\gamma}{\gamma - 1}\frac{p}{\rho U^2}\right)^{-1/2}. \tag{2.7.23}$$

Within the region in which the velocity defect is small, we may replace (2.7.23) by

$$\frac{U}{u} - 1 = \frac{\gamma}{\gamma - 1}\frac{p}{\rho U^2}. \tag{2.7.24}$$

Here we must be a little more careful on our choice of u_δ. We identify u with the shock angle σ_{b1}, and u_δ with the shock angle σ. The relations (2.7.18) and (2.7.19) give us the required approximations for p and ρ.

We may put the results of the calculation into a form similar to that of (2.7.21) and write

$$A_{\mathrm{ent},u} = A_{,u}\,\epsilon\left(\frac{\tau}{\lambda_b}\right)^{-2/\gamma}\left(\frac{x}{c}\right)^{2(1-k)/\gamma} \tag{2.7.25}$$

Now, $A_{,u}$ is not a constant, but is of the form

$$A_{,u} = \mathscr{A}\left[k^2\frac{p_b}{p_s}\right]^{\frac{\gamma-1}{\gamma}}\left(\frac{\tau}{\lambda_b}\right)^{\frac{2(\gamma-1)}{\gamma}}\left(\frac{x}{c}\right)^{-\frac{2(\gamma-1)(1-k)}{\gamma}} \tag{2.7.26}$$

where

$$\mathscr{A} = \frac{2\pi^j}{\rho_\infty U}\left[k^2\frac{p_b}{p_s}\right]^{-1/\gamma}\frac{2\gamma}{(1+\gamma)^2}\int_0^\infty (\sin\sigma_{\mathrm{b1}})^{2/\gamma}[(\sin\sigma_{\mathrm{b1}})^{2/\gamma} - (\sin\sigma)^{2/\gamma}]\,d\psi. \tag{2.7.27}$$

The quantity $\mathscr{A}$ has been defined in a form very similar to that of (2.7.20), so that (2.7.26) serves to give a comparison of the relative importance of the velocity-defect and constant-velocity contributions to the displacement area. If $\gamma - 1$ is not too small and $k < 1$, the effect of velocity defect should be small far enough downstream.

(d) Entropy layer. If $k < 1$ the shock shape is blunt at the nose without any extra blunting, and the equivalence principle fails there locally. Downstream of the nose the equivalence principle approaches correctness, but there is an effect of the failure near the nose which persists. The quantity p/ρ^γ behind the shock is given by (2.7.17). However, with the similar solution strictly interpreted through the equivalence principle, since $dr_s/dx = \tan \sigma$, the correct value for p/ρ^γ behind the shock should be

$$\frac{p_s}{\rho_s^\gamma} = \rho_\infty^{-(\gamma-1)} U^2 \epsilon^\gamma (1 - \epsilon)(\tan \sigma)^2 \tag{2.7.28}$$

in place of (2.7.17).

In the steady flow the streamlines near the body correspond to small values of $\cos \sigma$ and have an entropy jump far less than that indicated for the similar solution. This creates a layer near the body of fluid which, though representing the greatest values of entropy to be found in the flow field, has entropy far below that given by the similar solution. This layer then, though hot, is appreciably cooler than it ought to be. The result is an effective thinning of the body by a negative displacement thickness.

This layer has acquired the name entropy layer, and we shall adopt this terminology. The possibility of confusing this entropy layer with the layer we have termed an entropy wake is real and unfortunate. In the first edition of this book we used the designation "cool core" for the entropy layer to emphasize the contrast with the "hot core" of the similar solution. Many investigators refer to the entropy layer as a layer corresponding to high entropy and temperature, glossing over the fact that its essential nature lies in its relatively low entropy and temperature.

The choice of a name for an entity is not the significant point here. The important point is that the entropy wake and entropy layer concepts are fundamentally different, and the fact that the two entities are similar in many ways makes it even more important that we distinguish between them carefully.

In estimating the importance of the entropy layer, we may use the same approach as with the entropy wake. The negative displacement area $A_{\text{ent},\rho}$ is calculated by the same formula (2.7.21), where A is negative and is given by

$$-A = \frac{2\pi^j}{\rho_\infty U}\left[k^2 \frac{p_b}{p_s}\right]^{-1/\gamma} \int_0^\infty \left[(\tan \sigma)^{2/\gamma} - (\sin \sigma)^{2/\gamma}\right] d\psi. \tag{2.7.29}$$

In order to permit application of the theory to the blast wave or point explosion case, we replace τ by τ_s, the τ defined with respect to the shock shape instead of the body shape. They are related through

$$\tau_s = \tau/\lambda_b\,, \tag{2.7.30}$$

and we may simply replace τ/λ_b by τ_s in the equations above from (2.7.19) on. We write

$$\tan \sigma = \tau_s (x/c)^{-(1-k)} = \xi^{-\frac{1-k}{k}}, \tag{2.7.31}$$

where

$$\xi = r_s / \tau_s^{\frac{1}{1-k}} c. \tag{2.7.32}$$

The quantity $d\psi$ is expressed

$$d\psi = \rho_\infty U r_s^j \, dr_s = \rho_\infty U c^{1+j} \tau^{\frac{1+j}{1-k}} \xi^j \, d\xi. \tag{2.7.33}$$

We may now express $-A$ in the form

$$-A = 2\pi^j c^{1+j} \tau_s^{\frac{1+j}{1-k}} \left[k^2 \frac{p_b}{p_s}\right]^{-\frac{1}{\gamma}} \int_0^\infty \left[\xi^{\frac{-2(1-k)}{\gamma k}} - \left(1 + \xi^{\frac{2(1-k)}{k}}\right)^{-\frac{1}{\gamma}}\right] \xi^j \, d\xi. \tag{2.7.34}$$

For the velocity-defect effect we have from (2.7.27), with $\sin \sigma_{\text{bl}}$ replaced by $\sin \sigma$ and $\sin \sigma$ by 0, the expression

$$\mathscr{A} = 2\pi^j c^{1+j} \tau_s^{\frac{1+j}{1-k}} \frac{2\gamma}{(1+\gamma)^2} \left[k^2 \frac{p_b}{p_s}\right]^{-\frac{1}{\gamma}} \int_0^\infty \left(1 + \xi^{\frac{2(1-k)}{k}}\right)^{-\frac{2}{\gamma}} \xi^j \, d\xi. \tag{2.7.35}$$

The integrals appearing in (2.7.34) and (2.7.35) may be expressed in terms of Gamma functions. Figures 2–16 ($j = 0$) and 2–17 ($j = 1$) present results for the quantities $-A$ and $\mathscr{A}$ divided by the factor $c^{1+j} \tau_s^{(1+j)/(1-k)}$.

The quantity $A_{\text{ent},\rho}/A_s$ in the case of the entropy layer depends upon τ_s to the exponent $k(1+j)/(1-k) - 2/\gamma$, as was noted by Cheng [1] in the case $j = 0$ and by Sychev [2]. Small-disturbance theory itself involves errors of the order of τ_s to the exponent 2. Sychev set as a criterion for the entropy layer to be important that the first exponent be less than 2, or that

$$k < \frac{1}{1 + \dfrac{\gamma(1+j)}{2(\gamma+1)}}. \tag{2.7.36a}$$

For k large enough so that this condition does not hold, the effect of the entropy layer can be neglected. In our analysis this criterion applies for the constant-velocity displacement area. A similar argument applied to the velocity-defect displacement area leads to the more stringent condition

$$k < \frac{1}{1 + \dfrac{\gamma(1+j)}{4}} \tag{2.7.36b}$$

for the velocity-defect effect to have to be taken into account. This means that the velocity-defect effect is always negligible if the constant-velocity displacement effect is negligible.

at what problems can be solved without any perturbation of the similar solution.

The natural way to pose the problem of determining the flow on a body is to specify the body shape, with the unknown shock shape to be determined as part of the solution. Such a problem is termed a *direct* problem. An alternative way of posing the problem to a theoretician is to specify the shock shape, with the unknown body shape to be determined as part of the solution. Such a problem is termed an *inverse* problem. The distinction between these two types of problems is discussed further in Section 5.1, and is important in the treatments in Chapters V and VI.

In the present context, flows with the displacement effects (c) to (f) may be found with no perturbation simply by considering the problem to be solved as an inverse problem with a power-law shock wave. The altered body shape effect (e) is meaningless by itself here, but is used to cancel the displacement thickness effect of any of the other effects (c), (d), or (f). With a positive displacement thickness the physical body is made thinner, so that the combined real plus effective body shape follows the desired power law. With a negative displacement thickness, i.e. with an entropy layer, the physical body is made thicker.

A case that has received particular attention is the axisymmetric blast wave, with $j = 1$ and $k = \frac{1}{2}$. Sychev [2; 3] assumed a parabolic shock shape and calculated the shape of the required body numerically. In this case the entire body corresponds to minus the negative displacement area from the entropy layer. Yakura [1] used an analytic expansion scheme to investigate the same problem, including the two-dimensional ($k = \frac{2}{3}$) case, and found the asymptotic law for the shape of the Sychev body to be $r_b \propto x^{k/\gamma}$. Yakura's body is minus the negative constant-velocity displacement area. Figure 2–19 presents the body shapes calculated by Sychev and Yakura, and Fig. 2–20 presents the ratio of body to shock radius for the same shapes. The difference between the two shapes is primarily due to the velocity-defect effect, which is included in Sychev's calculations but is not included in the first-order theory of Yakura Subsequently, Mirels [2, Appendix] showed that Yakura's result follows simply from Mirels' general theory, and Sychev [4] has applied the PLK method to find a uniformly valid approximate analytic solution to his problem. Laval [4] studied the application of the PLK method to power-law shocks, and found it impossible to truncate the expansion except in the blast wave case. Figure 2–20 also gives results of Sychev [3] for a power-law body with $j = 1$, $k = 0.65$, showing that the entropy layer is of negligible importance in this case. Condition (2.7.36a) on k is just barely exceeded in this case. Studies similar to that of Yakura have been made by Freeman [7] and by Guiraud, Vallee, and Zolver [1]. The method of Yakura has been applied to power-law shocks by Lee [1].

The velocity-defect correction in the blast wave case gives a negative δr_b

to the Sychev body. From the previous relations in this section, the change in body radius can be shown to have a dependence upon x/c given by

$$-\delta r_b \propto (x/c)^{\frac{2(2-\gamma)(1-k)-jk}{\gamma}} \tag{2.7.43}$$

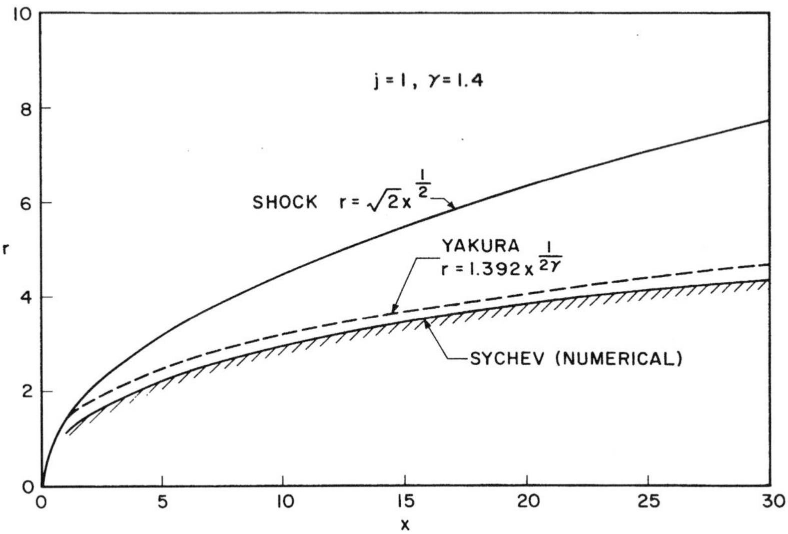

FIG. 2–19. Sychev body shape in axisymmetric blast wave case (Sychev [2; 3] and Yakura [1]).

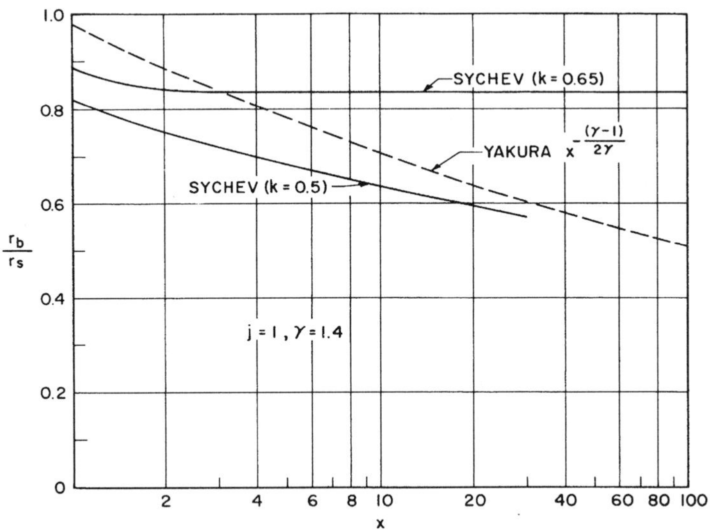

FIG. 2–20. Ratio of body to shock radius in axisymmetric flow, $k = 0.50$ and 0.65 (Sychev [2; 3] and Yakura [1]).

With $k = 2/(3 + j)$, this may be rewritten

$$-\delta r_b \propto (x/c)^{\frac{k(2+j-\gamma-\gamma j)}{\gamma}} \tag{2.7.44}$$

If $j = 1$ this exponent vanishes for $\gamma = 1.5$, and is only about 0.07 for $\gamma = 1.4$. Thus in the case studied by Sychev and Yakura δr_b should be approximately constant with x, and this conclusion is borne out in Fig. 2–19.

In correlating nose drag with shock shape for a Sychev body according to (2.7.10) it is important to recognize that the solution approaches the equivalent similar solution in an asymptotic manner downstream. The drag to be used is the drag of the entire Sychev body and not the drag of a restricted "nose" portion of it.

For the theory of perturbations of similar power-law solutions we depend upon the work of Mirels and Thornton [1] and of Mirels [1; 2]. The reader is referred to these references for details and calculations. Each dependent variable in a similar solution has the form

$$Q_i = x^{k_i} f_{i,0}(\lambda), \tag{2.7.45}$$

where k_i is a suitable exponent. Our λ is denoted η by Mirels. The perturbed solution is assumed to follow

$$Q_i = x^{k_i}[f_{i,0} + \alpha x^N f_{i,1}(\lambda)]. \tag{2.7.46}$$

where α is a small parameter and N is the exponent characterizing the perturbation. The differential equations governing the $f_{i,0}$ are the basic equations of the last section. With these quantities known, the basic hydrodynamic equations lead to a set of ordinary linear differential equations for the $f_{i,1}$. These equations may be solved by standard methods.

Appropriate perturbed boundary conditions must be satisfied at the shock and at the body (or axis in the blast wave case). These boundary conditions are translated into appropriate perturbation boundary conditions to be satisfied at the unperturbed shock ($\lambda = 1$) or at the unperturbed body ($\lambda = \lambda_b$ or $\lambda = 0$). The perturbation boundary conditions at the shock are sufficient to determine the solution of interest, at least within an undetermined multiplicative constant. At the body, the perturbation boundary conditions can be satisfied only with a perturbation in r_b following (2.7.46), and these conditions determine the multiplicative constant if appropriate. In the axisymmetric blast wave case (2.7.46) must be applied in terms of body area instead of body radius, with the corresponding $f_{i,0}(0) = 0$.

The perturbation effects discussed above all fit into this general form. Table 2–3 presents the appropriate small parameter and the appropriate exponent N for the various effects. Under entropy wake and entropy layer it is the constant-velocity displacement effect which is listed. The relatively weaker velocity-defect effect has been omitted. The perturbation solutions are asymptotically correct upstream if $N > 0$ and asymptotically correct downstream if $N < 0$. The perturbation is itself self-similar if $N = 0$.

TABLE 2-3

Perturbation parameters and exponents

Perturbation type	Small parameter	N	If $1 > k > \frac{2}{3+j}$:
(a) local equivalence failure	τ^2	$2(k-1)$	$N < 0$
(b) counterpressure	$(M_\infty \tau)^{-2}$	$2(1-k)$	$N > 0$
(c) entropy wake	$A\epsilon/c^{1+j}\tau^{1+j+2/\gamma}$	$\frac{2}{\gamma} - k\left(1 + j + \frac{2}{\gamma}\right)^*$	$N < 0$
(d) entropy layer	$-\epsilon\tau^{\frac{k(1+j)}{1-k} - \frac{2}{\gamma}}$	$\frac{2}{\gamma} - k\left(1 + j + \frac{2}{\gamma}\right)^*$	$N < 0$
(e) altered body shape	τ_0/τ	$k_0 - k$	$k_0 > k,\quad N > 0$
(f) boundary layer	$(\delta_0^*/\tau c)^2$	$\frac{3}{2} - 2k$	$k = \frac{3}{4},\quad N = 0$

* See discussion in text.

Although Mirels and Thornton are to be credited with putting all of these perturbations into one general scheme, a number of various authors have obtained solutions for various effects in terms of power-law perturbations. Waldman [1] and Mirels and Mullen [1] have thus studied local equivalence failure. Results for counterpressure have already been mentioned. The case $k_0 = 1$, with the alteration in body shape a constant change in slope so that the problem is one of a blunted flat plate at angle of attack, has been treated by Burke [1] and Kholyavko [1] as well as by Mirels and Thornton. Previous studies on the boundary layer case have been mentioned.

The case of the entropy layer, with the closely related case of the entropy wake, is rather special. Perturbation solutions satisfying the appropriate shock boundary conditions have a logarithmic behavior near $\lambda = \lambda_b$ or $\lambda = 0$, in addition to the expected power-law behavior. Physically, this difficulty comes from the fact that streamlines neighboring the body or axis deviate from the body or axis by an amount which corresponds to the desired body perturbation. It is for this reason that Mirels [2], who expresses boundary conditions at the body in terms of velocity rather than streamfunction, was not able to express a quantitative result and only expressed the exponent N of Table 2–3 and described the correct qualitative behavior. The particular difficulties of this problem have led to some controversy on the nature of the solution. There is no controversy on the existence of the logarithmic singularity and on the principle that a method using matched inner and outer expansions is required.

Physical reasoning of the type used with respect to the other perturbations leads to the value of N given in Table 2–3. A basic discrepancy involving terms in $\ln x$ appears to interfere with the matching of the inner and outer expansions. Guiraud [6] (see also Guiraud, Vallee, and Zolver [1]) has sought for alternative perturbation types which would permit the desired matching, but without success. He concludes that the constant-velocity displacement thickness must be zero, and finds that a matching without the logarithmic difficulty is then available with N replaced by $2N$. Vaglio-Laurin [2] presents an analysis using the same value of N found by Mirels [2] and obtains a matching. However, his analysis is not free of objection. The problem of matching is as yet unresolved.

In the authors' opinion, the value of N given must be essentially correct, and the displacement thickness is nonzero. It is likely that the form of the perturbation may have to be slightly altered. The perturbed solution would be of a form such as

$$Q_i = x^{k_i}[f_{i,0} + \alpha x^N f_{i,1}(\lambda, \ln \alpha x^N)] \tag{2.7.47}$$

instead of (2.7.46), with the condition that $f_{i,1}$ behave at worst algebraically in $\ln x$.

The general perturbation scheme may be carried to higher orders. To the bracket of (2.7.46) is added a term in $x^{2N}f_{i,2}(\lambda)$. Second-order perturbation equations may be found, and second-order boundary conditions formulated. However, asymptotic expansions to high order rarely provide a fruitful approach in applied mathematics, and we shall not pursue this subject. We may note that in the blast wave case, the perturbation corresponding to the velocity-defect effect appears proportional to x^{2N}, where N is the constant-velocity entropy layer exponent. Thus, velocity defect should naturally be taken into account in a second perturbation solution for the entropy layer.

If a perturbation solution is asymptotically correct upstream, with $N > 0$, then far downstream the flow will be in a different regime. For example, let us consider a finite Mach number flow with $k < 1$, in which counterpressure must be taken into account. Far downstream the flow will be quasi-linear and in a first-order wave structure regime (see, e.g., Hayes [2]). We note for a finite Mach number axisymmetric flow that the body shape generated by the original body plus the displacement thickness often results in an overexpanded flow locally, which then recompresses downstream with a resulting secondary

We next assume the equivalence principle, which in the present formulation amounts to taking $q_x = 0$. The result is

$$\frac{D(x)}{2\pi^j} = \int_{y_b}^{y_s} (e + \tfrac{1}{2}q^2)\rho y^j \, dy - \rho_\infty e_\infty \frac{y_s^{1+j}}{1+j} . \tag{2.8.2}$$

For this particular result, it is only necessary that equivalence hold at the station x. The nose drag D_N is identified as the limit of $D(x)$ as x approaches zero, so that

$$\frac{D_N}{2\pi^j} = \lim_{x\to 0} \int_{y_b}^{y_s} (e + \tfrac{1}{2}q^2)\rho y^j \, dy. \tag{2.8.3}$$

Thus the nose drag is identified with an instantaneous energy release; this result is consistent with (2.7.8).

We now consider the similitude connecting two flows, on different bodies at different velocities in the same fluid in the same undisturbed state. In consonance with the regular hypersonic similitude of Section 2.2 the bodies (except for the small bluntness) and the shock shapes are related affinely, with the same lateral body dimension τc. The body lengths and velocities are related so that the transit time c/U is the same; thus the two flows have the same value of $U\tau$. Under the equivalence principle the flows are similar, with the same flow variables p, ρ, q (and also q_x/τ) at the same values of x/c and of the lateral coordinates.

With the shock shape in one of the flows corresponding to an instantaneous energy release, with D_N in (2.8.3) not equal to zero, the shock shape in the second flow then also corresponds to an instantaneous energy release with the same value of D_N. Thus, with D_N an invariant, the two flows are similar with blunting. This observation gives the desired extension of hypersonic similitude in the form most closely associated with the equivalence principle.

Although Cheng and Chernyi considered only two-dimensional and axisymmetrical bodies, it can be shown that the above similitude may be still further extended to take into account three-dimensional shapes. The nose drag D_N may be considered as a concentrated single force ($j = 1$) or uniform concentrated force distribution ($j = 0$) on a leading tip or edge, respectively. In a three-dimensional extension we require that the concentrated force on the leading tips or the distribution on the edges be the same on two related bodies with the same lateral dimensions. In order to illustrate what is meant, we consider a flat three-dimensional wing with a blunted leading edge and sketch its planform in Fig. 2–23. The leading edge, which in this example does not extend across the span, bears a nose drag distribution which is not uniform.

In the example of Fig. 2–23 we have shown a leading edge which is not normal to the flow direction. Following the concept of cylindrical flow, we

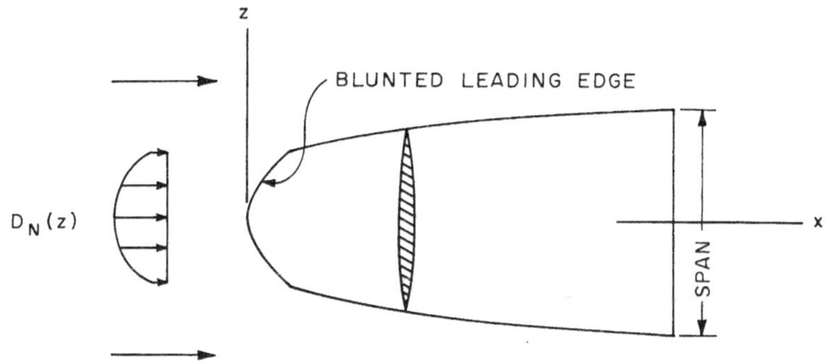

FIG. 2–23. Slightly blunted three-dimensional wing.

expect the actual force borne by the blunted leading edge to be normal to the leading edge. Thus a nose drag distribution on a swept-back leading edge has associated with it a nose side-force distribution as shown in Fig. 2–24. The nose drag force D_N per unit distance spanwise (in the z direction in Fig. 2–24)

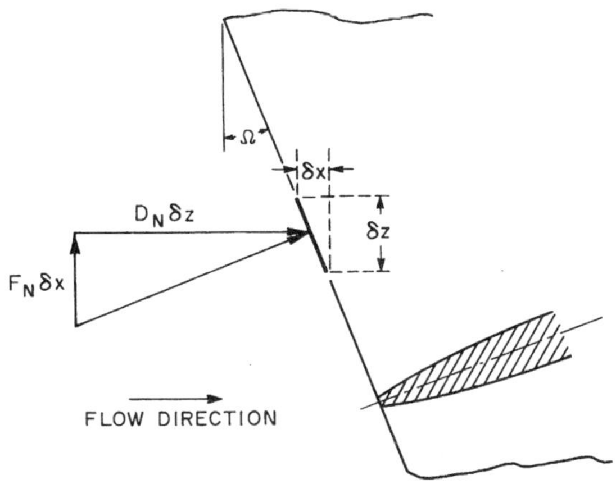

FIG. 2-24. Bluntness forces on a swept-back leading edge.

equals the normal force per unit of distance along the edge, and this equals the side force F_N per unit of distance in the free stream direction. Invariance of D_N entails invariance of F_N, and invariance of F_N is required by the similitude in order to yield a consistent rule for total side force.

Thus in the three-dimensional form of the similitude with slight blunting, leading edges need not be normal to the flow direction. In addition, a given body may have a combination of concentrated single nose drag forces on leading points and distributions of nose forces on leading edges.

The influence of small blunting appears in the extended similitude through an additional similarity parameter. A simple choice for this parameter is that of a quantity proportional to the ratio of D_N to the total drag. The nose drag in two-dimensional or axisymmetric flow is expressed

$$D_N = \tfrac{1}{2}\rho_\infty U^2 C_{D_j} d^{1+j}, \tag{2.8.4}$$

in which the A_{fr} of (2.7.9) has been set equal to d^{1+j}, with d the width or diameter of the blunted tip or leading edge. The drag coefficient $C_{D_{fr}}$ has been relabeled C_{D_j}. From the basic similitude results (2.2.3) or (2.2.4) the drag D is proportional to $\tfrac{1}{2}\rho_\infty U^2\tau^{3+j}c^{1+j}$. Since both D_N and D are invariants we choose their ratio

$$K_{N_j} = \frac{C_{D_j}(d/c)^{1+j}}{\tau^{3+j}} \tag{2.8.5}$$

as the basic similarity parameter for slight blunting in two-dimensional or axisymmetric flow.

To express the complete similitude in usual terms we must incorporate the scale transformation. The basic dependent variables are the thermodynamic variables p, ρ, h, S, etc. plus the lateral velocity $\mathbf{q}$ and the reduced axial perturbation velocity $q_x/\tau = q_x'$. The functional dependence of all these variables is on the same independent variables and parameters. We choose p again as typical of the dependent variables and write in place of (2.2.3) for the particular fluid under consideration

$$p = p(\tau x, y, z;\ U\tau, K_{N_j}, \rho_\infty, p_\infty). \tag{2.8.6}$$

To apply the scale transformation we simply leave the dependent variables unchanged and divide all the coordinate variables by τc. The result is

$$p = p\left(\frac{x}{c}, \frac{y}{\tau c}, \frac{z}{\tau c};\ U\tau, K_{N_j}, \rho_\infty, p_\infty\right). \tag{2.8.7}$$

This functional equation may be repeated for any of the other dependent variables ρ, h, S, $\mathbf{q}$, or q_x/τ.

As before, we may replace any of the dependent variables by a dimensionless equivalent obtained by dividing by a suitable combination of the independent parameters of the problem. Thus, we may replace p by $p - p_\infty$, and that in turn by $(p - p_\infty)/\rho_\infty U^2\tau^2 = \tfrac{1}{2}C_p/\tau^2$. We may replace ρ by ρ/ρ_∞ and $\mathbf{q}$ by $\mathbf{q}/U\tau$. Also, since a_∞ is a function of p_∞ and ρ_∞ alone, we may replace $U\tau$ as independent parameter by the more usual similarity parameter

$$K = U\tau/a_\infty = M_\infty\tau. \tag{2.8.8}$$

The shock shape, for example, may then be expressed in the form

$$\frac{y_s}{\tau c} = Y\left(\frac{x}{c}, \frac{z}{\tau c}; M_\infty\tau, K_{N_j}, \rho_\infty, p_\infty\right). \tag{2.8.9}$$

Note that in place of K_{N_j} we may use $K_{N_j}K^{3+j}$ as an alternative parameter in which M_∞ appears and τ does not.

Returning to the more general three-dimensional similitude, the nature of specification of nose blunting parameters depends on the particular case. For a blunted tip we must specify a parameter K_{N_j} as defined in (2.8.5) with $j = 1$ as in axisymmetric flow. For a blunted leading edge we must take the sweepback angle Ω into account. The free stream velocity component normal to the edge is $U \cos \Omega$, and the factor $\cos^2 \Omega$ appears in (2.8.4). We have $j = 0$ in this case, and defining K_{N_0} as before as a drag ratio leads to

$$K_{N_0} = \frac{\cos^2 \Omega \, C_{D_0}(d/c)}{\tau^3} = \frac{\sin^2 \Omega \, C_{D_0} d}{(\tau \tan \Omega)^2 \tau c} \tag{2.8.10}$$

in place of (2.8.5). Thus d is to be replaced by $d \cos^2 \Omega$ in (2.8.5). The quantity $\tau \tan \Omega$ is a geometric invariant in the similitude, so that

$$K'_{N_0} = \frac{\sin^2 \Omega \, C_{D_0} d}{\tau c} \tag{2.8.11}$$

may be used in place of K_{N_0}. If the sweepback angle is close to $\pi/2$, so that $\sin^2 \Omega$ may be replaced by one, the invariant K'_{N_0} is proportional to $d/\tau c$ and the blunting follows the affine law for the body shape.

We may check the consistency of (2.8.10) by examining the case of cylindrical flow. If a two-dimensional body is yawed with a yaw angle Ω, we can apply two-dimensional concepts to the normal flow. We replace U by $U \cos \Omega$, M_∞ by $M_\infty \cos \Omega$, τ by $\tau/\cos \Omega$, c by $c \cos \Omega$, and x by $x \cos \Omega$. The K_{N_0} calculated from (2.8.5) for the normal flow is

$$K_{N_0} = \frac{C_{D_0}(d/c \cos \Omega)}{(\tau/\cos \Omega)^3} = \frac{C_{D_0} \cos^2 \Omega (d/c)}{\tau^3}, \tag{2.8.12}$$

in accord with (2.8.10).

As an elementary check on the similitude with blunting we consider the τ_s characterizing the shock shape which would be obtained with a calorically perfect gas and no counterpressure from the blast wave theory. From (2.7.9) or (2.7.10) we can obtain

$$\tau_s = \left(\frac{C_{D_{fr}}(d/c)^{1+j}}{4\pi^j E}\right)^{\frac{1}{3+j}}. \tag{2.8.13}$$

The ratio of τ_s to the τ characterizing the body shape is then

$$\frac{\tau_s}{\tau} = \left(\frac{K_{N_j}}{4\pi^j E}\right)^{\frac{1}{3+j}}. \tag{2.8.14}$$

As expected, this quantity is invariant.

To examine the concept of the entropy wake displacement thickness let us again consider the two similar flows on bodies with the same lateral dimension, without blunting as well as with blunting. The displacement area with equivalence is given by (2.7.15), and is

$$A_{\text{ent},\rho} = 2\pi^j \int \left(\frac{1}{\rho_{\text{bl}}} - \frac{1}{\rho_0}\right) \frac{d\psi}{U}, \tag{2.8.15}$$

where the subscript 0 here indicates without blunting. For the entropy wake displacement thickness to fit the similitude it must have the same distribution with x/c for the two affinely related bodies. The density is given by an equation of state

$$\rho_{\text{bl}} = \rho(p_{\text{bl}}, S_{\text{bl}}), \tag{2.8.16a}$$

$$\rho_0 = \rho(p_0, S_0). \tag{2.8.16b}$$

The quantities p_{bl} and p_0 have distributions with x/c which are the same for the two bodies.

The entropy on a streamline is a function only of the normal velocity in front of the shock at the point the streamline crosses the shock. Thus we have

$$S = S(U \sin \sigma). \tag{2.8.17}$$

Under the similitude, however, it is not $U \sin \sigma$ which is invariant, but $U \tan \sigma$. Thus the requirement of assuming equivalence everywhere dictates that we consistently replace $\sin \sigma$ by $\tan \sigma$ in calculating velocity normal to a shock. In place of (2.8.17) we write

$$S_{\text{bl}} = S(U \tan \sigma_{\text{bl}}), \tag{2.8.18a}$$

$$S_0 = S(U \tan \sigma_0). \tag{2.8.18b}$$

With (2.8.18) the quantities S_{bl} and S_0 have distributions with ψ (independent of x/c) which are the same for the two bodies. Hence the displacement thickness of the entropy wake follows the similitude with equivalence assumed, and in fact may be considered the essential effect taken into account in the extension to include slight blunting.

The error from replacing $\sin \sigma_0$ by $\tan \sigma_0$ is of the order of τ^2 and properly neglected in hypersonic similitude. The error from replacing $\sin \sigma_{\mathrm{bl}}$ by $\tan \sigma_{\mathrm{bl}}$ is locally not small, and is the same as the error from neglecting the entropy layer effect (cf. 2.7.29)). This effect is to be compared with the entropy wake effect, which is the primary effect accounted for in the similitude. As we have shown in the previous section for a perfect gas, these two effects follow similar laws. The correctness of the similitude as we have stated it depends upon the ratio of entropy layer thickness to entropy wake thickness approaching zero in the limit $\tau \to 0$, $U \to \infty$.

To estimate the importance of the entropy layer in a perfect gas we may use (2.7.20) for A_{wake} and (2.7.29) for $-A_{\mathrm{layer}}$. We use σ_{bl} in place of σ in (2.7.29) and $\tan \sigma$ in place of $\sin \sigma$ in (2.7.20), and set $k = 2/(3 + j)$. The reference area which enters the assumed similitude is $A_{\mathrm{wake}} - A_{\mathrm{layer}}$, or

$$A = A_{\mathrm{wake}} - A_{\mathrm{layer}} \propto \int_0^\infty [(\tan \sigma_{\mathrm{bl}})^{2/\gamma} - (\tan \sigma)^{2/\gamma}]\, d\psi. \tag{2.8.19}$$

The error in the similitude from the entropy layer may be measured by the ratio

$$\frac{-A_{\mathrm{layer}}}{A} = \frac{\displaystyle\int_0^\infty [(\tan \sigma_{\mathrm{bl}})^{2/\gamma} - (\sin \sigma_{\mathrm{bl}})^{2/\gamma}]\, d\psi}{\displaystyle\int_0^\infty [(\tan \sigma_{\mathrm{bl}})^{2/\gamma} - (\tan \sigma)^{2/\gamma}]\, d\psi}. \tag{2.8.20}$$

In the similitude $U \tan \sigma_{\mathrm{bl}}$ with ψ (and τc) fixed is an invariant, and $\tan \sigma_{\mathrm{bl}} \to 0$ in the same manner as τ in the limit $\tau \to 0$. The denominator in (2.8.20) varies as $\tau^{2/\gamma}$, while the numerator approaches zero in the limit $\tau \to 0$ faster than $\tau^{2/\gamma}$, approximately as τ^{3+j}. Thus the effect of the entropy layer should be small if τ is small enough.

One way to get around the limitation on the extension imposed by the entropy layer is to consider that the physical body shape on a blunted slender body is described as a nominal body shape minus the negative displacement thickness from the entropy layer. In the similitude the nominal body shape follows the affine transformation law, and the actual body shape does not. As an example, if the Sychev body discussed in the previous section were an actual physical body, the corresponding nominal body would have zero thickness. Note that with this approach nominal bodies may well have negative thickness.

Another way to get around the limitation imposed by the entropy layer is available, with the velocity-defect displacement effect remaining as an error. With a perfect gas we simply require that A_{wake} vary as $\tau^{2/\gamma}(\tau c)^{1+j}$ in the similitude, where A_{wake} is defined as in (2.7.20), as a similarity condition to replace the invariance of K_{N_j}. In a general fluid this course must be replaced by the condition that $U \sin \sigma_{b1}$ or $\sin \sigma_{b1}/\tau$ is invariant for fixed ψ (and τc); this severe requirement dictates a specific blunting shape for each value of τ, obtainable only as a solution of an inverse blunt-body problem in each case. In each similitude with a general fluid there would be a minimum value of U for which the condition can be satisfied.

A simple example of the application of the similitude appears when we consider the case of a blunted two-dimensional or axisymmetric body of zero thickness at zero incidence. We consider a calorically perfect gas, so that p_∞ and ρ_∞ can be eliminated as independent parameters, and consider only pressure on the axis. In this case we rewrite (2.8.7) in a form in which τ has been eliminated,

$$\frac{p_b}{p_\infty} = f\left(\frac{x}{c} ; K_{N_j} K^{3+j}, \gamma\right). \tag{2.8.21}$$

In this problem the chord c is an inessential variable, and can also be eliminated. The similitude thus leads to the functional result

$$\frac{p_b}{p_\infty} = f\left(C_{D_j}\left(\frac{d}{x}\right)^{1+j} M_\infty^{3+j} ; \gamma\right) \tag{2.8.22a}$$

or, alternatively,

$$\frac{p_b - p_\infty}{p_\infty} = f\left(\frac{x}{d\, C_{D_j}^{\frac{1}{1+j}} M_\infty^{\frac{3+j}{1+j}}} ; \gamma\right). \tag{2.8.22b}$$

In the case $j = 0$ correlations of experimental data for pressure may be found in Chernyi [3, Fig. 5.4a] and Cheng [1], using the variables of (2.8.22b). Chernyi (Fig. 5.4b) also shows the corresponding correlation for shock shape, following (2.8.9) with the dependent variable multiplied by $K^{-2}K_{N_0}^{-1}$ in order to eliminate τ and c. In the case $j = 1$ we present in Fig. 2–25 the correlation given by Kuehn [1] from results of numerical calculations. Pressures close to the nose were excluded by Kuehn. For earlier calculations and correlations see also Van Hise [1]. In both the $j = 0$ and $j = 1$ cases the results are consistent with the similitude. Note that the main effect here beyond the blast wave effect is the counterpressure effect, which is not considered as a small perturbation.

As another example we consider the case of a blunted cone, with the free stream Mach number very high. Here $\tau = \theta_c$, and we take $M_\infty \theta_c \gg 1$. Eliminating M_∞ and c from (2.8.7) leads with $j = 1$ to

$$\frac{C_{p_b}}{\theta_c^2} = f\left(\frac{x\theta_c^2}{dC_{D_1}^{1/2}} ; \gamma\right). \tag{2.8.23}$$

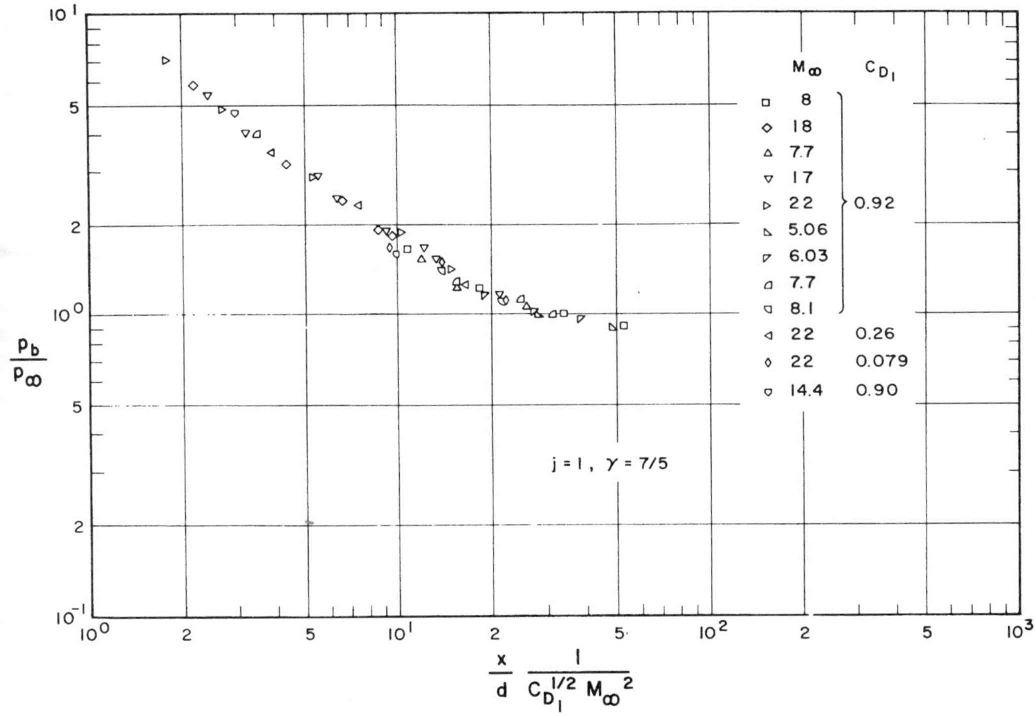

FIG. 2–25. Similitude correlation of calculated pressure distributions on blunted cylinders (Kuehn [1]).

A correlation of numerical characteristic solutions with $\gamma = \frac{5}{3}$ for cones with blunt conical noses is shown in Fig. 2–26. These results were taken from Wagner and Watson [1, Fig. 3]. For additional numerical calculations with $\gamma = 1.4$ see Chushkin [1] and Chushkin and Shulishnina [1]. In Fig. 2–26 the assumption $M_\infty\theta_c \gg 1$ fails for the slenderer cones. Nevertheless, the consistency with (2.8.23) is good over the entire range of θ_c. The departure from the correlation, which can be seen in Fig. 2–26, is associated with the breakdown of the similitude in the vicinity of the nose. Wagner and Watson also give correlated experimental results for pressure and shock shape.

As we have pointed out, the entropy wake is an essential part of the flow field on a blunted body. This layer has particular properties which permit certain conceptual simplifications and the formulation of an area rule. The entropy wake is a region of high entropy, and is also a region of high temperature and low density. It is thus a region of relatively high speed of sound; the local Mach number, though perhaps high enough to be considered hypersonic, is appreciably lower than the local Mach number in the flow outside the entropy

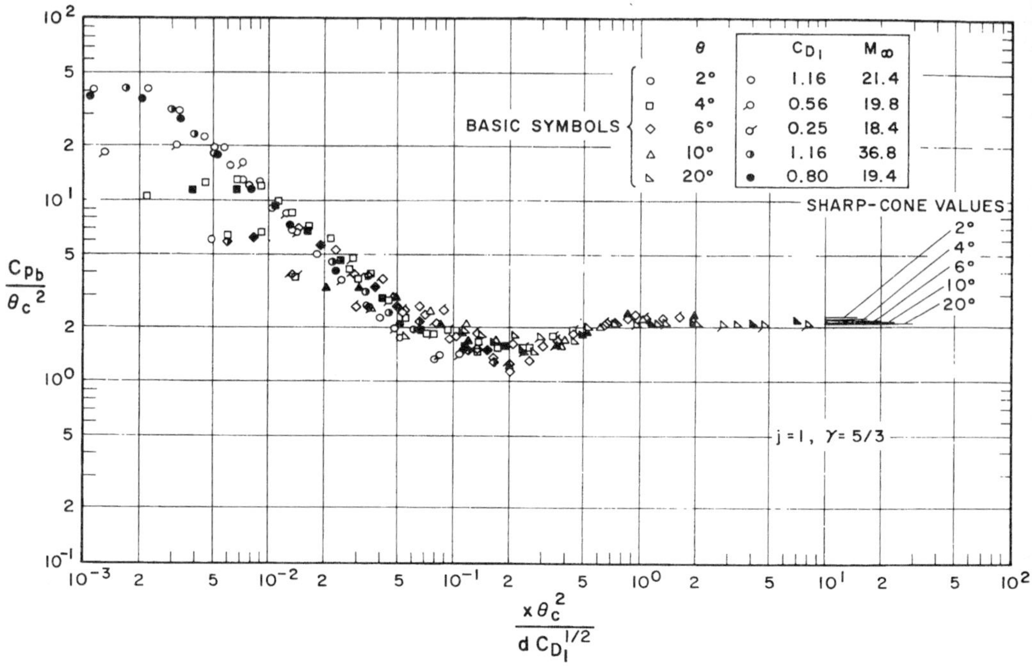

FIG. 2–26. Similitude correlation of calculated pressure distributions on slightly blunted slender cones (Wagner and Watson [1]).

wake. One possible result of this decreased local Mach number is to make the restrictions on slender-body strip theory more stringent with blunting.

The boundary between the entropy wake and the outer flow is not at all sharp. For conceptual purposes let us imagine that it is sharp, with a large discontinuity in entropy, density, and Mach number across the boundary. For such a sharply defined shear layer we may define a reflection coefficient which measures the strength of a disturbance reflected from such a layer to the strength of the disturbance incident on it. In Section 7.2 we derive an expression for such a reflection coefficient $\mathfrak{R}_v$, which is given by (7.2.19). For the case considered this coefficient is close to one, which means that a disturbance in the entropy wake is reflected from the boundary almost as from a solid wall. Therefore, the outer flow is influenced by the entropy wake only through the pressure distribution on the boundary since a single wave is only weakly propagated into the flow outside the wake.

If the changes of the shape of the body are sufficiently gradual and no part of the body extends laterally beyond the entropy wake, the lateral pressure gradients will remain small and we can consider the entropy wake as a constant pressure layer. Then the axial distribution of pressure is governed solely by the cross-sectional area distribution of the entropy wake. This distribution equals the area distribution determined by the boundary of the entropy wake less the area distribution of the body. Though formulated for a sharp boundary between entropy wake and outer flow, these concepts apply in essence to the actual case.

We are thus led to the area rule of Ladyzhenskii [2]. Two bodies which have blunted tips and are slender in both lateral directions will bear the same axial pressure distribution and will have the same drag in the same hypersonic free stream, provided:

(a) The nose drag on the blunted tips is the same for the two bodies.

(b) The two bodies have the same axial distribution of body cross-sectional area.

(c) No part of the body extends laterally beyond the confines of the entropy wake.

(d) The changes in shape of the body with distance downstream are sufficiently gradual.

The area rule implies that the entropy wake has approximately a circular cross section, that the outer flow is close to axisymmetric, and that the lateral forces on the body are negligible. Ladyzhenskii [1; 3] treats also a somewhat different type of area rule based upon $\gamma - 1$ being small. This will be taken up in Chapter V.

Ladyzhenskii's area rule is here discussed qualitatively rather than treated quantitatively simply because a comprehensive analytic investigation of its limitations is not available. To take one question as an example, we have not indicated how gradual a change in body shape needs to be to be sufficiently gradual. On this question we require an analysis similar in many respects to that in the classical slender-body theory of supersonic aerodynamics. Another question is that of the importance of velocity defect. This velocity defect can be significant if the entropy wake is itself not fully hypersonic. If the velocity defect is important, we may need to require in place of (a) above Ladyzhenskii's requirement that the two blunted tips coincide in shape.

9. Large incidence and correlation of similitudes

We next consider slender bodies at a large angle of incidence in a hypersonic flow, following the theory of Sychev [1]. In this case the perturbations produced by the body will not in general be small. If, however, the body is not only slender but also flat then the flow is a small-disturbance theory in the sense of this chapter and the equivalence principle may be applied. We consider this as a special case to be treated separately, and assume the body is not flat.

The body is taken to have a length c, and to have lateral dimensions of the order of δc in both lateral directions. The dimensionless parameter δ is a thickness ratio for the slender body, and we require $\delta \ll 1$. See Fig. 2–27.

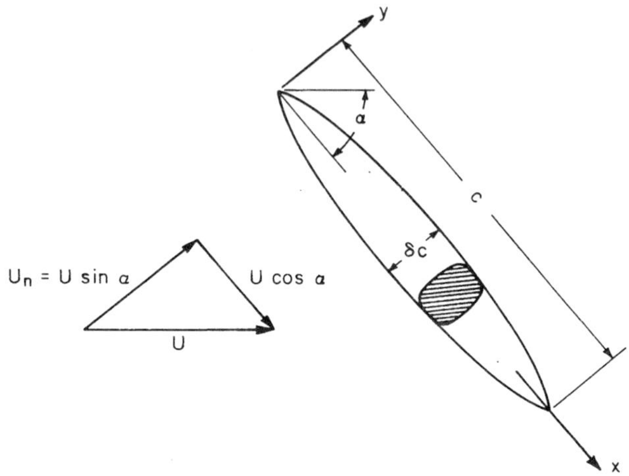

FIG. 2–27. Slender body at large incidence.

We set up a cartesian coordinate system aligned with the body rather than with the free stream direction, with x the axial coordinate along the body. The free stream velocity U lies in the (x, y) plane, at the incidence angle α with respect to the x axis. The lateral velocity components in the y and z directions are of the order of $U \sin \alpha$ unless α/δ is small, when they are then of the order $U\delta \cos \alpha$. Thus we may say that the lateral velocity is of the order of $U_n(1 + \delta \cot \alpha)$, where

$$U_n = U \sin \alpha \tag{2.9.1}$$

is the free stream velocity component normal to the axis of the body.

The analytic approach we follow is very close to that of Section 2.1, except that we shall keep to the steady equations in a reference frame fixed with respect to the body. Our approach is close to that of Sychev [1]. We again use the notation

$$\mathbf{q} = q_y \mathbf{j} + q_z \mathbf{k} \tag{2.9.2}$$

for the two-dimensional lateral velocity vector, and set

$$q_x = U \cos \alpha + \delta u. \tag{2.9.3}$$

The perturbation velocity in the x direction can be shown to be of the order of $\delta U_n(1 + \delta \cot \alpha)$, and the definition of u in (2.9.3) is in analogy with that of q'_x in (2.1.2). Directional derivatives in the x direction are of the order of δ times directional derivatives in the lateral directions, and the analogue of (2.1.3) is

$$x = \delta^{-1}x'. \tag{2.9.4}$$

The equations of continuity, momentum, and entropy may now be expressed

$$\delta \cot \alpha\, U_n \frac{\partial \rho}{\delta x'} + \nabla \cdot \rho \mathbf{q} = -\delta^2 \frac{\partial \rho u}{\partial x'}, \tag{2.9.5a}$$

$$\delta \cot \alpha\, U_n \frac{\partial \mathbf{q}}{\partial x'} + \mathbf{q} \cdot \nabla \mathbf{q} + \frac{1}{\rho} \nabla p = -\delta^2 u \frac{\partial \mathbf{q}}{\partial x'}, \tag{2.9.5b}$$

$$\delta \cot \alpha\, U_n \frac{\partial u}{\partial x'} + \mathbf{q} \cdot \nabla u + \frac{1}{\rho} \frac{\partial p}{\partial x'} = -\delta^2 u \frac{\partial u}{\partial x'}, \tag{2.9.5c}$$

$$\delta \cot \alpha\, U_n \frac{\partial S}{\partial x'} + \mathbf{q} \cdot \nabla S = -\delta^2 u \frac{\partial S}{\partial x'}. \tag{2.9.5d}$$

The analogy with (2.1.4) is completed when we note that the operator $\delta \cot \alpha\, U_n\, \partial/\partial x' = U \cos \alpha\, \partial/\partial x$ is equal to $\partial/\partial t$ for an observer who moves along the axis of the body with the constant velocity $U \cos \alpha$, that is, with $t = x/U \cos \alpha$.

Boundary conditions analogous to (2.1.6) are to be satisfied on the body, and shock conditions analogous to (2.1.9) and (2.1.10) are to be satisfied on the shock. We shall not express these here; for details see Sychev [1]. The one principal difference between the boundary conditions in this case and in the case of Section 2.1 lies in the outer undisturbed velocity condition. In the original small-disturbance theory the undisturbed gas in a transverse plane is at rest. In the analysis of this section the undisturbed gas has a lateral velocity given by

$$\mathbf{q}_\infty = U_n \mathbf{j}. \tag{2.9.6}$$

With $\delta \ll 1$ we drop the terms in δ^2 in the right-hand sides of (2.9.5). The resulting equations correspond to the extended equivalence law or law of plane sections of Sychev [1]. This law states that the three-dimensional steady flow on the inclined slender body may be calculated as a two-dimensional unsteady flow with free stream velocity U_n from the point of view of an observer who moves along the body with a velocity $U \cos \alpha$.

As with the regular equivalence law, this extended law leads immediately to a similitude. This similitude is exactly the same as that of Section 2.2, except that the independent parameter $U\tau$ is replaced by the pair of parameters

$U_n = U \sin \alpha$ and $\delta \cot \alpha$. In terms of the dimensionless similarity parameters, we replace $K = M_\infty \tau$ by any two of the three parameters

$$\delta \cot \alpha, \tag{2.9.7}$$

$$M_n = U_n/a_\infty , \tag{2.9.8}$$

and

$$K = M_n \delta \cot \alpha = M_\infty \delta \cos \alpha. \tag{2.9.9}$$

This Sychev similitude provides a firm bridge between classical hypersonic similitude for α small and strip theory (not of the slender-body type) for $\cos \alpha$ small. At small α, K is the usual hypersonic similarity parameter and $\delta \cot \alpha \approx \delta/\alpha$ is a geometrical invariant of the usual theory. At small $\cos \alpha$, $M_n \approx M_\infty$ is the free stream Mach number and $\delta \cot \alpha$ is negligibly small.

In examining equations (2.9.5) carefully, we note there are three regimes to be considered. Since $\delta \ll 1$, then we have $\delta \cot \alpha \ll 1$ unless α is small. The regime of hypersonic similitude is that in which $\delta \cot \alpha$ is of the order of one or larger. Here the first term in each of (2.9.5) is essential and not small. The regime of zero-yaw strip theory is that in which $\delta \cot \alpha/\delta^2 = 1/\delta \tan \alpha$ is of the order of one or smaller. Here the first term in each of (2.9.5) is of the order of the δ^2 terms already neglected, and must also be neglected in a consistent theory. In the intervening regime, or Sychev regime, $\delta \cot \alpha$ is small but need not be neglected in comparison to δ^2. Neglect of $\delta \cot \alpha$ in this regime gives a finite-yaw strip theory equivalent to combining strip theory with the cylindrical flow concept.

In the Sychev regime, then, there are two small parameters, $\delta \cot \alpha$ and δ^2, but with $\delta \cot \alpha \gg \delta^2$. Neglecting δ^2 while keeping $\delta \cot \alpha$ is permissible. But the resulting theory is a first-order theory rather than a zero-order theory in the sense that the influence of a small parameter is taken into account. It is not a first-order theory in the customary perturbation sense, inasmuch as the influence of the parameter $\delta \cot \alpha$ is taken into account to all orders. In other words, although $\delta \cot \alpha$ may actually be small, it is not treated as a small parameter in an expansion scheme in (2.9.5). But the effect considered, interpretable as the effect of nonzero sweepback angle $\Omega = \pi/2 - \alpha$ on strip theory, is a weak one unless α is small.

In the region downstream of the body, and particularly downstream of the leading tip, the assumptions underlying the Sychev theory may fail locally. This local failure has little or no effect on the flow near the body, and can be disregarded there. This may be seen from the fact that the pressure field in this downstream region is weak and its effect on the flow near the body will be small as a result of the hypersonic character of the normal flow.

If in (2.9.15) and the other reduced equations the terms in both τ^2 and $\delta^2\tau^2$ are dropped, the resulting equations for ρ and w are to be interpreted as the one-dimensional unsteady equations of the flow in a cylinder driven by a piston. Such an unsteady flow in a given transverse plane goes by the name piston theory and will be discussed further in the next section. The first two terms in (2.9.15) are the same as a time derivative term for an observer fixed in the undisturbed fluid. With the terms in τ^2 kept and those in $\delta^2\tau^2$ dropped, we obtain the flat-body version of the Sychev equations. The Sychev similitude applies, with τ a geometric invariant. Recall that M_n is invariant and $M_n\tau$ is the similitude parameter for cylindrical flow. In this case the error is far less, as $\delta^2\tau^2 \ll \tau^2$.

In treating these flat yawed bodies we have four approaches at our disposal. To compare these, we indicate these approaches and the corresponding parameter measuring the error in Table 2–4. In the Sychev regime, it is clear that

TABLE 2–4

Approaches for flat yawed bodies

Approach	Direction of neglected derivative	Error parameter
(a) Sychev	body axis	$\delta^2\tau^2$
(b) Piston theory	both in (x, y) plane	τ^2
(c) Slender-body strip	perpendicular to free stream	$\tau^2(\cos^2\alpha + \delta^2\sin^2\alpha)$
(d) Small-disturbance	free stream	$\tau^2\sin^2\alpha\,(1 + \delta^2\cot^2\alpha)$

neither slender-body strip theory or small-disturbance theory offers any improvement over the conceptually simpler piston theory. But the Sychev approach yields a significant improvement over all of these. A simple physical picture explains why the Sychev approach is better. Figure 2–29 shows the orientation of the fluid slabs for the three approaches in which slabs of fluid are assumed to

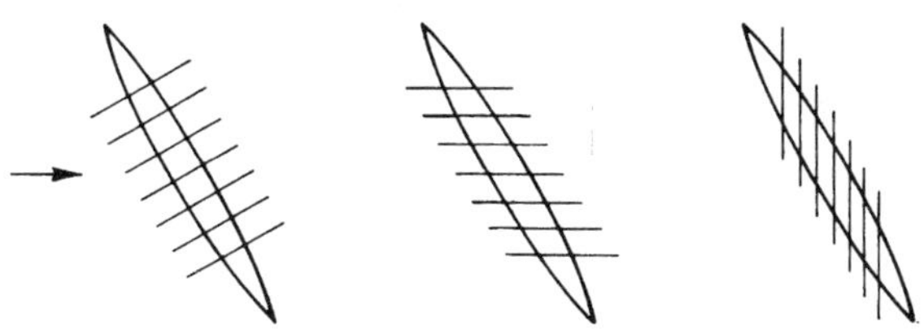

FIG. 2–29. Independent slabs according to various approaches.

behave independently. The errors are associated with gradients normal to the slabs. It is easy to see from Fig. 2–29 that these gradients may be expected to be least with the Sychev approach.

With a flat body, blunting along the leading edge may be taken into account following the cylindrical flow concept, as discussed in the last section. Thus the basic parameter is

$$K_{N_0} = \frac{C_{D_0}(d/\delta c)}{\tau^3}, \tag{2.9.16}$$

and is required to be an invariant function of x/c. However, τ is invariant, and this simply means $d/\delta c$ is invariant. In this case the entropy layer has no effect on the Sychev similitude.

At this point the reader will note that a large number of similitudes and independence principles have been introduced. We shall now consider the interrelationships between them, and shall attempt to present a comprehensive picture of the similitudes important in inviscid hypersonic flow theory.

The most basic similitude of all ideal-fluid (nondissipative) theory is that given by the scale transformation, discussed here in the first part of Section 2.2. This similitude we simply take for granted here, since it applies to all ideal-fluid flow. With the full freedom of the scale transformation invoked, the Mach number independence principle of Oswatitsch (Section 1.6) and the general strip theory (Section 1.7) become similitudes.

The similitudes we shall be considering all are expressed basically for finite Mach number, and all have a limiting form at infinite Mach number. Thus the entire pattern of similitude relationships without the Mach number independence principle is mirrored by a corresponding pattern with the principle in the limit $M_\infty \to \infty$. Thus no information is lost by omitting the Mach number independence principle from the detailed discussion.

Analogously, each similitude for a slender body with a sharp tip or leading edge has an extension to take slight blunting into account. This extension does not particularly shed light on the similitude relationships and can be included or not, as we please.

A similitude which we have not treated is the Prandtl-Glauert similitude of supersonic wing theory. We have rejected linearized flow as being inherently inconsistent with hypersonic flow theory, and accordingly do not consider the classical similitudes of nonhypersonic wing theory.

The relationships we are investigating are of the type in which one similitude may be considered to be a reduced form of another. This reduction is obtained through a specialization of the flow field, generally through a limiting process. We may thus look at the relationships in terms of various limiting processes. The omission of the Mach number independence principle means we are not considering the process $M_\infty \tau \to \infty$. The omission of blunting means we are not considering the process $K_{N_j} \to 0$. The rejection of the nonhypersonic similitudes means we have rejected the limit $M_\infty \tau \to 0$. When we consider Sychev similitude we must keep in mind its first-order nature in the sense discussed earlier. This means that we are not considering the limit $\delta \cot \alpha \to 0$ even though $\tan \alpha$ is of the order of one and $\delta \ll 1$. This special nature of Sychev similitude is indicated in the diagrams below by using dotted lines and boxes where it is concerned.

The limiting processes with which we are left are all geometrical ones, having to do with the shape of the body or its orientation in the undisturbed flow. There are four principal limiting processes, which we show in Table 2–5.

TABLE 2–5

Limiting processes for hypersonic similitudes

No.	Limit	Description	Type
①	$\delta \to 0$	body shape becomes slender	body shape
②	$\tau \to 0$	body becomes slender relative to flow	body shape
③	$AR \to \infty$ $(\sin \Omega = \cos \alpha \to 0)$	after ①, sweepback angle $\to 0$	orientation
④	$AR \to 0$ $(\cos \Omega = \sin \alpha \to 0)$	body becomes axially slender	body shape (orientation)

Processes ① $(\delta \to 0)$ and ② $(\tau \to 0)$ may appear at first glance to be similar, but the distinction between them is essential. Process ③ $(\Omega \to 0)$ may be considered as coming from a change in orientation, while the others involve changes in body shape. An exception is the case of process ④ $(AR \to 0)$ applied to a slender body at large incidence, which then comes from a change of orientation $(\sin \alpha \to 0)$. In the Sychev case, we must divide ① $(\delta \to 0)$ into two steps, a weak step ①$_w$ which goes to $\delta \ll 1$, and a limiting step ①$_l$ which completes the process. In addition, the limiting process $\delta\tau \to 0$ appears with respect to the Sychev similitude.

We start by diagramming the alternative approaches and corresponding similitudes for the flat slender body at large "incidence", which we term simply the yawed wing. These are shown in Fig. 2–30, with each arrow labeled according to the appropriate limiting process, and with each box indicating the appropriate similitude parameters. In this figure two of the arrows are unlabeled, indicating that no limiting process is involved in these cases in the Sychev regime with $\tan \alpha$ of the order of one (cf. Table 2–4).

Figure 2–31 presents the more comprehensive similitude picture, with the yawed wing case represented by a single box. Note that the corresponding theories for the flat normal (strip) case and the flat axial case have the same validity as the theory for the flat Sychev case. Application of the equivalence principle to the flat normal case or of the strip theory to the flat axial case leads to piston theory. The similitude presented as an example in Section 2.4 corresponds to piston theory applied in the flat normal case.

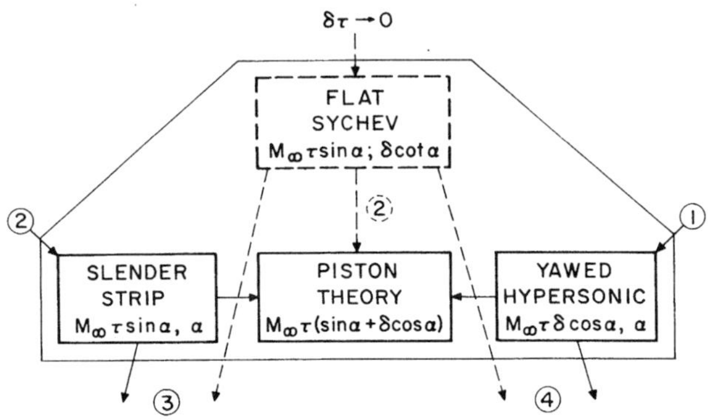

FIG. 2–30. Similitude correlation for yawed wing.

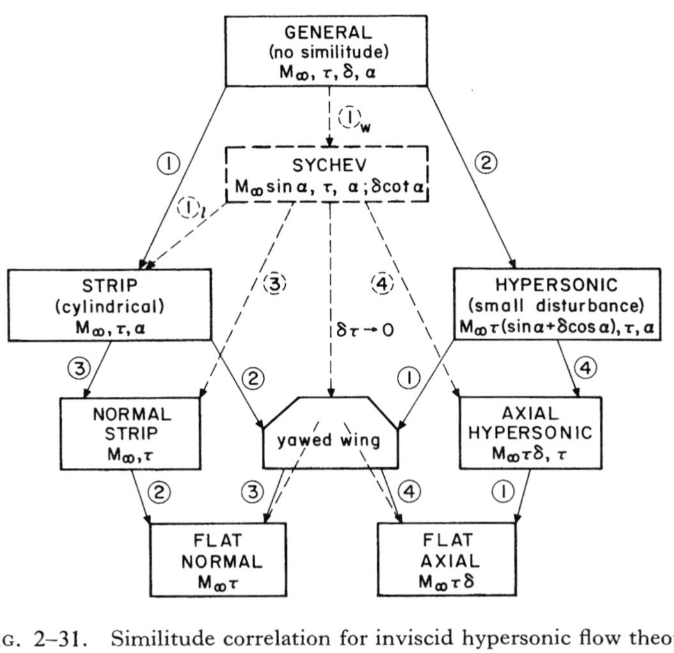

FIG. 2–31. Similitude correlation for inviscid hypersonic flow theory.

10. Unsteady flow theory

The question of the appropriate similitude for unsteady hypersonic flow was first examined by Lin, Reissner, and Tsien [1] in a paper which treated unsteady flow at all flight speeds. As with the earlier paper of Tsien [1] the investigation is limited to irrotational motions, and the error of Tsien of using the mean-surface approximation in the boundary conditions is repeated. The simplification obtainable by taking the point of view of an observer fixed in the undisturbed fluid was not pointed out by the authors, but is evident in their hypersonic equations. This paper does establish that Hayes' equivalence principle is applicable to unsteady flows, within the limitations inherent in the paper, although the authors did not specifically mention this fact.

The fact that the equivalence principle holds in unsteady flows was first noted by Hamaker and Wong [1], who demonstrated the equivalence analytically for potential flow and concluded that it should hold in general. Hamaker and Wong were primarily interested in the connection between unsteady hypersonic similitude and the dynamic similitude of bodies in free flight. They showed that complete dynamic similarity (with aerodynamic forces) can be obtained provided the body has certain symmetry and is not rolling.

The unsteady application of the equivalence principle was further exploited by Lighthill [2]. He likened the problem to that of the gas flow in a tube driven by a piston, and suggested the use of three terms in the isentropic expression for the pressure on a piston as a power series in its velocity. Within the accuracy of the first two terms in such a series we may neglect changes in the entropy, and Lighthill points out that the maximum error in the third term is small enough that this term is worth retaining for engineering purposes. Lighthill obtains analytic expressions for aerodynamic coefficients for a pitching symmetrical airfoil, with the assumption of a perfect gas. The accuracy of Lighthill's approximation in any particular case may be tested roughly by examining the magnitude of the next term in the series.

The relation which Lighthill used may be expressed

$$\frac{p}{p_\infty} = 1 + \gamma \frac{v}{a_\infty} + \frac{\gamma(\gamma+1)}{4}\left(\frac{v}{a_\infty}\right)^2 + \frac{\gamma(\gamma+1)}{12}\left(\frac{v}{a_\infty}\right)^3 \tag{2.10.1}$$

in our notation. This relation comes from the isentropic form of the Busemann expansion; compare (7.3.2) which includes the effect of a shock. Since v is of the order of $U\tau$, it is necessary that $M_\infty \tau$ be small for the application of Lighthill's formula. However, since the error is in the fourth term, acceptable accuracy may be obtainable with $M_\infty \tau$ of the order of about 1/3. Lighthill shows that his results are consistent with those of other investigators. Subsequently, Ashley and Zartarian [1] extended Lighthill's calculations and introduced the term "piston theory". This term is now generally applied to this branch of unsteady flow theory, and has been used previously in this chapter in connection with steady flows.

Studies of hypersonic unsteady flows have been made by Morgan, Runyan, and Huckel [1], Miles [1, Chapter 13; 2], and Zartarian, Hsu, and Ashley [1]. The approaches suggested by these investigators, besides the quasi-linear approach leading to (2.10.1), include unsteady analogues of the shock-expansion theory and of the tangent-wedge approximation. We shall discuss these approaches in steady flow in Chapter VII. Experimental results which can be used to compare with theoretical or empirical predictions appear to be quite limited. East [1] has obtained experimental results on oscillating airfoils with and without nose blunting, and has compared these with various predictions. He also gives a study of available theories.

Theoretical studies for an oscillating wedge have been made by Appleton [2] and by McIntosh [1]. The method used by both these investigators is to linearize the equations of motion with the assumption that the perturbations from a pure wedge flow are small. They both make the slender-body or piston-theory approximation without discussion of its validity. The primary aim of these investigations is to include the effect of reflection of acoustic waves from the shock. In both theories the neglect of these reflections provides a reduction to a "strong shock" piston theory of Miles [2] which is analogous to shock-expansion theory (cf. Section 7.2). McIntosh points out that an important effect of the reflections from the shock is a phase shift in the unsteady pressure distribution.

Landahl [1] has developed a theory for unsteady flow which takes into account the terms of order M_∞^{-2} which are neglected in piston theory. His theory is restricted to small values of $M_\infty \tau$, but its validity extends to smaller values of M_∞ than that of piston theory. Landahl's theory serves in particular to estimate the accuracy of piston theory when this is applied at moderate values of the Mach number. Good agreement is obtained with the results of Van Dyke [4] which include second-order effects.

In this book we are primarily interested in steady hypersonic flows, and here we shall only present heuristic arguments to delimit the validity of the small-disturbance theory in unsteady flow. In steady flow a parameter τ was defined, equal to the thickness ratio or the angle of attack of the body, whichever is greater. In unsteady flow τ must also be no less than the ratio of the maximum lateral velocity of any part of the body to the free stream velocity.

Provided the functional dependence of the body shape upon the distance and time is smooth and provided a suitably defined axial scale characterizing the time variations is of the order of the chord of the body or larger, the analysis we have given earlier or the somewhat more limited analysis of Lin, Reissner, and Tsien [1] is applicable, and the small-disturbance theory is correct to within an error of order τ^2. If the variations with time are slow, such as with a slowly accelerating body or with oscillating flow at low frequency, quasi-stationary theory applies and the equivalence principle remains valid, regardless of the distribution of the disturbances in the axial variable.

If the changes in body shape are not smooth in time and distance, failure of the equivalence principle may occur. To see this, let us imagine that at a point on the wing a local, violent disturbance to the flow occurs. According to the small-disturbance theory this disturbance should propagate normal to the body only, along a line fixed in the undisturbed fluid. But of course the disturbance propagates in all directions, and this diffraction effect is felt within a space of dimension of the order of τc, where c is the chord. For such an isolated disturbance the failure of the equivalence principle is felt only locally.

In general the unsteady flows of interest are not characterized by isolated violent disturbances; the flow patterns from smoother disturbances are themselves smoother, and clear-cut diffraction patterns do not appear. In order to see how failure of the equivalence principle may occur let us imagine that the unsteady disturbance may be broken up into superposable parts, each of which is sinusoidal both in time with frequency f and in distance along the body with wavelength λ. For a given combination of f and λ the disturbance may be divided into one representing a traveling wave going upstream with velocity $f\lambda$ and a wave going downstream with the same velocity. The latter appears to an observer moving downstream with velocity $f\lambda$ as a steady flow of velocity $|U - f\lambda|$. In order for the equivalence principle to apply it is necessary that the local Mach number corresponding to this velocity be high, or that

$$|U - f\lambda| \gg a. \tag{2.10.2}$$

Since a is of order $U\tau$ we may re-express this condition

$$\left|1 - \frac{f\lambda}{U}\right| \gg \tau. \tag{2.10.3}$$

Assuming we are considering only finite bodies the flow is primarily a transient one for the first wavelength or so, and this transient flow follows the equivalence principle. This transient flow will predominate over the entire body if the wavelength is large compared with the size of a diffraction pattern, or if

$$\frac{\lambda}{c} \gg \tau. \tag{2.10.4}$$

This condition is satisfied automatically if λ is of the same order as c.

In effect, then, the equivalence principle may be considered valid and the small-disturbance theory applicable for oscillatory unsteady flows provided either condition (2.10.3) or (2.10.4) holds, where λ is a suitably chosen axial scale of the disturbance and f is the frequency. Implicit in the investigations cited earlier is the assumption that the axial scale of the disturbance is of the same order of magnitude as the chord, hence that (2.10.4) is satisfied automatically.

With regard to the unsteady application of strip theory to slender bodies it is first necessary that the condition for slender-body strip theory in steady flow be satisfied, that $\tau Æ^{-1}$ be small. Then a condition equivalent to (2.10.4) must be satisfied, where λ is now a suitably chosen lateral scale for the unsteady motion. If the aspect ratio Æ is defined with respect to this lateral scale, the two conditions may be seen to be identical.

11. Nonequilibrium effects

The theories and similitudes presented so far in this chapter have all been given for an ideal fluid, one with no dissipative effects. To what extent are they valid with dissipative effects taken into account? With the exception of strip theory, formulated in Section 1.7 as an independence principle, the theories which yield similitudes are based upon the equivalence principle or on Sychev's extension of it. The equivalence principle requires that the velocity component along the axis of a slender body be nonzero and approximately constant.

Here viscosity interferes in the presence of a body. With viscosity, the boundary condition on the body in steady flow is that the velocity is zero. The assumption of approximately constant axial velocity necessarily fails. If the viscosity coefficient is small this failure is restricted to a thin layer near the body, the classical boundary layer for the velocity. If the boundary layer is thin enough the failure of equivalence may be neglected with respect to the flow outside the boundary layer. Viscosity does not interfere with equivalence in the asymptotic part of a wake or in the structure of a shock.

If the velocity boundary layer is sufficiently thin that it can be neglected, the equivalence principle may be applied if the ideal-fluid conditions for its validity hold. Nonequilibrium effects and viscous effects other than regular viscosity do not interfere. We shall show this in more detail later in this section after discussing briefly the physics and chemistry of nonequilibrium. The main difference which appears lies in the applicability of the scale transformation. In dissipative flows the determining equations and their solutions are not invariant under a scale transformation. Thus the scale transformation is not permitted. Hypersonic similitude applies then in the form of (2.2.3) or (2.8.6). The form of (2.8.7) was derived using the scale transformation and is not valid. In other words, in the dissipative hypersonic similitude the lateral scale of the slender body is an invariant.

Although other viscous effects besides viscosity do not interfere with equivalence, the effects are negligible outside the boundary layer unless the corresponding ν's of Section 1.8 are much larger than the kinematic viscosity. Thus, for example, heat conduction or diffusion can be included within hypersonic similitude only if the thermal or concentration boundary layers are much thicker than the velocity boundary layer. Although this case can occur in an ionized gas with respect to the heat conduction, the general case is that all viscous effects have ν's which are of the same order of magnitude.

If the thickness of the boundary layer is not negligible a hypersonic similitude for the outer flow is still available. For details see Hayes and Probstein [1]. This viscous hypersonic similitude requires boundary layer theory and thus lies outside the scope of this volume.

More than a brief introduction to the physics and chemistry of nonequilibrium effects would be out of place in a book devoted to flow theory. A most comprehensive treatment of the subject in the context of high temperature gasdynamics is to be found in Zel'dovich and Raizer [1], although the basic physics and chemistry may be found treated in a number of texts. Many of the articles in Rossini [1] are pertinent, and Hirschfelder, Curtiss, and Bird [1] is a standard reference on molecular gas theory and transport properties. Herzfeld and Litovitz [1] and Cottrell and McCoubrey [1] are good summaries on the transfer of energy to internal molecular modes. Penner [1] treats the chemistry of reacting flow systems. A valuable reference on the basic physics and chemistry of vibrationally relaxing and dissociating gases with application to gasdynamics is Clarke and McChesney [1], as is Vincenti and Kruger [1]. A basic treatise on ionization phenomena in gases is Massey and Burhop [1]. The compilation of articles edited by Bates [1] emphasizes the developments in the decade following the publication of Massey and Burhop.

Nonequilibrium effects may come from physical processes, such as vibrational excitation, or from chemical processes, such as formation or decomposition of NO in air. It is important to realize that there is no sharp boundary between physical and chemical processes. They can all be lumped together under chemical physics. With the gas a continuum in quasi-equilibrium the processes can be described by relatively simple reaction rates.

Nonequilibrium processes, particularly those involving energy transfer from one mode to another, are often describable as relaxation processes. A relaxation process is one governed by a differential equation of the form

$$t_r \frac{dx}{dt} + x = x_0(t), \tag{2.11.1}$$

for which the solution is

$$x = e^{-t/t_r}\left[x(0) + t_r^{-1} \int_0^t e^{t'/t_r} x_0(t')\, dt'\right]. \tag{2.11.2}$$

The quantity t_r (cf. (1.8.4)) is the characteristic relaxation time for the process. The quantity x is an appropriate dependent variable, which in the absence of the process ($t_r = 0$) would simply be equal to $x_0(t)$ at all times. The term relaxation thus refers to a mathematical behavior characterized by the factor e^{-t/t_r}.

If a gas is suddenly compressed or expanded, the new state is not immediately in thermal equilibrium. Equilibration of the translational degrees of freedom is very rapid, with the characteristic time the same as the collision time t_c. In a polyatomic gas equilibration of the rotational degrees of freedom is also rapid. The characteristic time is of the order of $300t_c$ for hydrogen at room temperature, and is of the order of $5t_c$ to $10t_c$ for nitrogen or oxygen at room temperature or above (Cottrell and McCoubrey [1, Chapter 5]). In molecular hydrogen at room temperature, the gas behaves immediately as though it were monatomic, with the transfer of energy to or from rotational degrees of freedom obeying approximately a relaxation law. In the case of nitrogen at room temperature this simplified picture is less accurate. However, the main point is that rotational equilibration is generally rapid enough so that rotational relaxation may be neglected. The main observable result of rotational relaxation lies in its contribution to classical bulk viscosity.

Vibrational equilibration is generally very slow. At low temperatures in a gas the question is moot, because the vibrational energy modes have negligible excitation. The characteristic time for vibrational relaxation in nitrogen falls from the order of $10^7\, t_c$ at 700°K to of the order of $10^4\, t_c$ at 4000°K (Cottrell and McCoubrey [1, Chapter 5]). At temperatures high enough so that dissociation begins to appear, the relaxation model loses its validity. A comprehensive treatment of vibrational and rotational energy exchange will be found in Herzfeld and Litovitz [1].

To illustrate vibrational relaxation we present the simple relaxation equations for a perfect diatomic gas. This gas has a specific internal energy

$$e = \tfrac{5}{2}\mathscr{R}T + e_{\text{vib}}\,, \tag{2.11.3}$$

with T the absolute temperature corresponding to the translational and rotational modes, both assumed always in equilibrium. If the gas as a whole were in thermal equilibrium, we would have

$$e_{\text{vib}} = e_{\text{vib},e}(T). \tag{2.11.4}$$

The relaxation equation is

$$t_{\text{vib}}\frac{de_{\text{vib}}}{dt} + e_{\text{vib}} = e_{\text{vib},e}\,. \tag{2.11.5}$$

The classical Landau-Teller theory for vibrational relaxation gives an approximate law for the dependence of t_{vib}/t_c on temperature. In its simplest form this law is

$$\frac{t_{\text{vib}}}{t_c} \propto T^{1/6} \exp\left(\frac{const.}{T^{1/3}}\right). \tag{2.11.6}$$

Equation (2.11.5) is the additional determining equation which must be added to the usual continuity, momentum, and energy equations. Equation (2.11.6) serves as an additional equation for the state variable t_{vib}.

So far we have implicitly understood that the gas of interest is a dilute gas of neutral particles. Before discussing chemical reactions we must be a little more explicit. In a gas of similar neutral particles the density is the mass per particle m times the number density n. Each particle has a collision cross section of area σ and a volume whose order of magnitude is $\sigma^{3/2}$. The parameter $n\sigma^{3/2}$, the ratio of the volume of a particle to the volume of space occupied per particle, is a basic diluteness parameter. We assume that it is negligibly small. The mean free path λ may be defined as the distance required for a particle to sweep out the volume n^{-1}. Thus we have

$$n\sigma\lambda = 1. \tag{2.11.7}$$

From the relation $\rho = mn$ we have

$$\rho\lambda = m/\sigma. \tag{2.11.8}$$

In such a dilute gas, the cross section does depend upon the temperature, but rather weakly as a rule. The characteristic velocity may be defined in a number of ways, but each definition gives a velocity nearly equal to $\sqrt{kT/m}$, where k is the Boltzmann constant. The important fact is that at a fixed temperature both the characteristic velocity and σ are fixed, and t_c is proportional to λ. Thus we may write, at fixed temperature,

$$\rho t_c = constant. \tag{2.11.9}$$

The diluteness parameter $n\sigma^{3/2}$ measures the time required by a collision divided by t_c, and thus represents the probability that any one particle is engaged in a collision. The probability that a particle is engaged in a three-body collision in a dilute gas is proportional to the square of this parameter, or to $n^2\sigma^3$.

The number of collisions per unit time per unit volume is simply n/t_c, and from (2.11.9) is proportional to ρ^2. The number of three-body collisions per unit time per unit volume is proportional to $n^2\sigma^{3/2}/t_c$, and is proportional to ρ^3. Processes in a dilute gas which depend upon two-body collisions will be termed second-order, or binary. Processes which depend upon three-body collisions will be termed third-order, or ternary. The classical continuum viscous effects in a dilute gas are all of second order.

In a dilute gas which is a mixture of different kinds of particles, the same general results hold. We must specify a different collision time for each category of collision and a different density for each species. The total pressure is the sum of partial pressures contributed by the several species present.

A chemical reaction, or process described chemically, is expressed through a chemical equation. The equation may or may not illustrate the details of the process in terms of what encounters between what chemical species are involved. In other words, the actual process may involve collisions with molecules not mentioned in the equation, or may involve several elementary processes in which species not appearing in the equation are created and destroyed. If when the density of each component is multiplied by the same factor K at fixed temperature, a reaction rate defined for change of an absolute concentration per unit volume is multiplied by K^N, the overall order of the reaction is defined as N. If a rate of reaction is proportional to a power N_i of the density of a particular species, the reaction is said to be of order N_i in that species. If all these orders exist, they are related by

$$N = \sum N_i , \tag{2.11.10}$$

where the sum ranges over the species whose densities affect the reaction rate. A chemical process need not have either orders in various species or an overall order. An overall reaction without an order may be composed of forward and reverse reactions with orders.

A chemical process may have an order within a range of temperature and density even though the detailed process is far more complicated than that described by the equation. It is possible to have overall processes which are of first order or of half-integral (e.g. $\frac{3}{2}$) order.

In hypersonic flow theory a process which is extremely important is that of dissociation of a diatomic gas. We consider nitrogen as an example, for which the chemical equation is

$$\mathrm{N_2} \rightleftarrows 2\,\mathrm{N}. \tag{2.11.11}$$

The actual process is described in terms of collisions by the equations

$$2\,\mathrm{N_2} \rightleftarrows 2\,\mathrm{N} + \mathrm{N_2}, \tag{2.11.12a}$$

$$\mathrm{N} + \mathrm{N_2} \rightleftarrows 3\,\mathrm{N}. \tag{2.11.12b}$$

The forward reaction, to the right in the equations, is of the second order. The reverse reaction, to the left in the equations, is of the third order. If both forward and reverse reactions are significant the net overall reaction has no order.

We let α be the fraction of N_2 molecules dissociated, and note that

$$\rho_{\text{N}} = \alpha\rho, \tag{2.11.13a}$$

$$\rho_{\text{N}_2} = (1 - \alpha)\rho. \tag{2.11.13b}$$

The rate of change of ρ_{N} is

$$\frac{d\rho_{\text{N}}}{dt} = \alpha\frac{d\rho}{dt} + \rho\frac{d\alpha}{dt}, \tag{2.11.14}$$

where $d\alpha/dt$ measures the rate of dissociation. We can calculate $d\alpha/dt$ by requiring the density ρ to remain constant in (2.11.14). The contribution to $d\rho_{\text{N}}/dt$ from the forward reaction from (2.11.12a) is proportional to $(1-\alpha)^2\rho^2$, and from (2.11.12b) is proportional to $\alpha(1 - \alpha)\rho^2$. The contribution from the reverse reaction from (2.11.12a) is proportional to $-\alpha^2(1 - \alpha)\rho^3$, while that from (2.11.12b) is proportional to $-\alpha^3\rho^3$. With $d\rho/dt = 0$, we may write

$$\rho\frac{d\alpha}{dt} = (1 - \alpha)\rho^2[A(1 - \alpha) + B\alpha] - \alpha^2\rho^3[C(1 - \alpha) + D\alpha], \tag{2.11.15}$$

where A, B, C, and D are functions of temperature alone. The classical equilibrium law

$$\frac{\alpha^2\rho}{1 - \alpha} = k(T) \tag{2.11.16}$$

must hold when $d\alpha/dt = 0$. We conclude that $AD = BC$ and that we may rewrite (2.11.15)

$$\frac{d\alpha}{dt} = (1 + a\alpha)f(T)[k(T)(1 - \alpha)\rho - \alpha^2\rho^2], \tag{2.11.17}$$

where $a(T) = (D - C)/C$. This expression is substituted into (2.11.14) to calculate $d\rho_{\text{N}}/dt$.

Lighthill [3] has suggested that a very realistic equilibrium model is obtained by setting

$$k(T) = \rho_D e^{-D/kT} \tag{2.11.18}$$

in (2.11.16), where ρ_D is a constant and D is the dissociation energy per molecule. At low temperatures this model corresponds to a perfect diatomic gas with vibrational modes half excited, so that $\gamma = \frac{4}{3}$. Freeman [2] has extended this simplified model to a consistent approximation for the dissociation and recombination rates. His model corresponds to choosing not only ρ_D to be a constant but $a = 0$ and

$$f(T) = CT^{-s}, \tag{2.11.19}$$

where s is a constant, chosen by him to be 0 or 2.5.

Another basic process of importance in hypersonic flow is that of ionization. We consider a hypothetical monatomic gas, which we term argon for convenience. The basic chemical equation is

$$\mathrm{Ar} \rightleftarrows \mathrm{Ar}^+ + \mathrm{e}^-. \tag{2.11.20}$$

With α the fraction of Ar ionized, (2.11.16) holds at equilibrium. If the principal forward reaction comes from atom-atom, electron-atom, and ion-atom collisions, the forward reaction follows (2.11.17) without the α^2 term. The rate in argon is actually controlled by the process

$$\mathrm{Ar} \rightarrow \mathrm{Ar}^*, \tag{2.11.21}$$

where Ar* indicates the lowest excited state of Ar.

An important alternative process is photoionization or its reverse, radiative recombination, expressed by

$$h\nu + \mathrm{Ar} \rightleftarrows \mathrm{Ar}^+ + \mathrm{e}^-. \tag{2.11.22}$$

The forward process is of first order in Ar and first order in the radiation field, the reverse process is of second order. In a physical situation in which emitted radiation escapes freely and the radiation field may be considered negligibly weak, thermal equilibrium cannot obtain, and the rate of ionization may be dominated by the forward collision process and the radiative recombination process. At sufficiently low densities the radiative recombination must predominate over any three-body recombination mechanism. In this case we would have

$$\rho \frac{d\alpha}{dt} = A(1 + a\alpha)(1 - \alpha)\rho^2 - B\alpha^2\rho^2, \tag{2.11.23}$$

with A and B functions of T. This may be rewritten

$$\frac{d\alpha}{dt} = f(T)[k'(T)(1 + a\alpha)(1 - \alpha) - \alpha^2]\rho. \tag{2.11.24}$$

This equation has a steady-state solution

$$\frac{\alpha^2}{(1 - \alpha)(1 + a\alpha)} = k'(T) \tag{2.11.25}$$

which does *not* correspond to thermal equilibrium. The above grossly simplified model serves to illustrate the effect of an ionization process in a hypersonic flow. The actual physical processes in argon are very much more complicated, and involve excited electronic states of neutral argon, radiation from de-excitation of Ar*, and at low enough temperature the ion Ar_2^+ (see Petschek and Byron [1], Harwell and Jahn [1], and Bray [2]).

Air at high temperature is a complicated mixture of species, including N_2, O_2, N, O, NO, electrons, and various positive and negative ions, besides traces of triatomic species. Many different processes must be considered simultaneously. We would note that for a density range 10 to 10^{-2} times standard atmospheric density the chemical composition of air remains essentially unchanged below about 2,500°K, although vibrational relaxation effects can be important there. Above this temperature O_2 begins to dissociate. The dissociation is essentially completed at about 4,000°K, when N_2 dissociation begins. At about 8,000°K the N_2 dissociation is essentially complete, the concentration of O atoms having remained practically unchanged. At a temperature of about 7,000 to 8,000°K, N and O atoms begin to ionize.

A summary of the various reactions and rate constants for air has been given by Wray [1]. The work of Hall, Eschenroeder, and Marrone [1] on the inviscid flow about a blunt nose is a good example of an analysis with rate-dependent dissociative reactions in air introduced into the flow calculations. Lin and Teare [1] present a rather complete series of calculations of the various effects on the reaction zone behind strong normal shock waves in air associated with spontaneous ionization rates.

We turn last to the most general case possible within our definition of nonequilibrium process. The density and the mechanical pressure (minus the isotropic part of the stress tensor) are always well defined, even if the fluid is far from being in thermal equilibrium. We assume the fluid is initially in a state with $p = p_\infty$ and $\rho = \rho_\infty$, in thermal equilibrium or in a well-defined quasi-equilibrium. Under the definition of nonequilibrium, the density as a function of time depends upon the time history of the pressure (or conversely). For a given gas, we may express this fact in terms of a functional in the form

$$\rho(t) = \mathop{\mathscr{F}}_{0}^{\infty}[\pi(\bar{t}); \rho_\infty], \tag{2.11.26}$$

using the notation of Volterra [1], with

$$\pi(\bar{t}) = p(t - \bar{t}). \tag{2.11.27}$$

Equation (2.11.26) merely says that if ρ_∞ and the time history of p from $t = -\infty$ to $t = t$ is given, $\rho(t)$ is given uniquely. The concept of a functional is a useful one; the basic reference on the subject is Volterra [1].

ones of boundary layer theory. The critical question is as to the magnitude of q'_x in the extended form of (2.1.4). With no body no difficulty appears. On a body the boundary condition on q'_x would be

$$q'_x = -U/\tau, \tag{2.11.37}$$

and the right-hand sides of (2.1.4) would then be of the same order as the terms on the left-hand sides. Because of the dissident boundary condition, (2.1.4c) no longer controls the magnitude of q'_x to the order of $U\tau$. Equivalence must fail, at least within the velocity boundary layer.

It is now apparent that there is still another similitude, which consists of an extension of the Sychev similitude of Section 2.9 to include nonequilibrium effects. The scale transformation is lost, and the lateral scale must be an invariant. No change need be made in any of the conditions. In strip theory no statement is necessary. Since strip theory is basically an independence principle it holds with all dissipative effects provided the lateral scale is fixed. Thus it is clear that the entire Fig. 2–31 applies with nonequilibrium effects and fixed lateral scale.

The loss of the scale transformation may be salvaged if the gas is a dilute one and the nonequilibrium processes have a single order N. First we note that a dilute gas which reacts with the same order in both directions has an equilibrium point which is a function of temperature alone and is a perfect gas. An example of such a reaction is $2\,HI \rightleftarrows H_2 + I_2$, which is second order in both directions. A perfect gas is self-similar in the sense of Section 2.2 and if ideal permits a similitude transformation in which T is fixed and ρ is changed. With nonequilibrium effects of a given order present, there is a transformation permitted in which both scale and density are changed.

To demonstrate this transformation, let us take a set of equations of the form

$$\frac{Dp_{A_i}/p}{Dt} = \rho^{N-1} f_i\left(\frac{p_{A_1}}{\rho}, \frac{p_{A_2}}{\rho}, \dots\right) \tag{2.11.38}$$

for the partial pressures p_{A_i}. Equations of this form would govern an N-th order reaction in which the species A_i participated. Along with the momentum, continuity, and energy equations, these equations are invariant under the transformation

$$p'_{A_i} = a p_{A_i}, \tag{2.11.39a}$$

$$\rho' = a\rho, \tag{2.11.39b}$$

$$t' = a^{1-N} t, \tag{2.11.39c}$$

$$\mathbf{r}' = a^{1-N}\mathbf{r}. \tag{2.11.39d}$$

Temperature and p_{A_i}/p are invariant. This transformation yields a similitude, in which the scaling similarity parameter is

$$K_{sc} = \rho_\infty^{N-1}\bar{c}, \tag{2.11.40}$$

where ρ_∞ is a characteristic density, and $\bar{c}$ is a characteristic scale.

Of the possibilities in this scale transformation for general order only one appears of practical interest. This is the case $N = 2$, with the scaling law termed second-order or binary. Viscous effects fall into this category, and the invariant $\rho_\infty c$ appears in Hayes and Probstein [1, Sect. 11] in treating rarefied gas flow. Many rates of reaction are controlled by two-body collisions, and flows with these reactions governing the nonequilibrium effects would obey binary scaling. Reviews of the applicability of binary scaling for reactions in air have been given by Hall, Eschenroeder, and Marrone [1] and by Gibson and Marrone [2]. Binary scaling can be applied where the reaction is of second order only in the forward direction, provided the reverse reaction may be neglected. Dissociation reactions sometimes fall into this category.

Scaling laws for molecular processes have been well known for a long time. A binary scaling law for gaseous discharges and general scaling laws in terms of reaction time for chemical kinetic processes have been in use since the 1880's. And the classical parameter *Re* of Reynolds is proportional to $\rho_\infty c$ at fixed temperature (and velocity), and thus corresponds to binary scaling.

With viscosity negligible and with the dissipative effects characterized by an order, the order scaling may be combined with hypersonic similitude. The characteristic scale is the lateral scale τc, and the scaling similarity parameter becomes

$$K_{sc} = \tau\rho_\infty^{N-1}c. \tag{2.11.41}$$

With K_{sc} invariant the scale transformation is permitted.

In establishing the equivalence principle under the functional formulation of (2.11.26), a different approach is needed. The physical idea underlying the equivalence is clear, but we have here consistently insisted on making the error terms explicit to order τ^2. Since the functional is written for a parcel of fluid, a Lagrangian formulation is suggested. If our original demonstration of the equivalence principle had been made using a Lagrangian approach the case using a functional would have fitted in naturally. Here we continue with the Eulerian formulation, and add Lagrangian *dependent* variables. We will not carry the analysis out in detail, but only outline the approach and obtain error orders.

In any inviscid hydrodynamic problem, the basic equations are the continuity and momentum equations. These equations involve the three dependent variables ρ, p, and $\mathbf{q}$, one more in number than the number of equations. The additional equation of state, equation of energy or entropy, and additional determining equations are needed specifically for the one purpose of establishing the relationship between p and ρ. This relationship is explicitly given by the functional, so only the continuity and momentum equations are needed.

The Lagrangian variable $\mathbf{r}_0$ is chosen as $\mathbf{r}$ in the undisturbed free stream. We can show that the relation between the velocity vector $\mathbf{q}$ and the vector $\mathbf{r}_0$ considered as a dependent variable in an Eulerian formulation is

$$\frac{D\mathbf{r}_0}{Dt} = \frac{\partial \mathbf{r}_0}{\partial t} + \mathbf{q} \cdot \nabla \mathbf{r}_0 = 0. \tag{2.11.42}$$

As with the momentum equation, this equation is divided into axial and transverse parts. We replace $\mathbf{r}_0$ by $\mathbf{r}_0 + \mathbf{i}(x + \tau\xi)$, $\mathbf{q}$ by $\mathbf{q} + \mathbf{i}\tau q_x'$, x by $\tau^{-1}x'$, and x_0 by $\tau^{-1}x_0'$. As with $\mathbf{q}$, the new $\mathbf{r}_0$ is two-dimensional. The equations for $\mathbf{r}_0$ and ξ take the form

$$\frac{\partial \mathbf{r}_0}{\partial t} + \mathbf{q} \cdot \nabla \mathbf{r}_0 = -\tau^2 q_x' \frac{\partial \mathbf{r}_0}{\partial x'}, \tag{2.11.43a}$$

$$\frac{\partial \xi}{\partial t} + \mathbf{q} \cdot \nabla \xi + q_x' = -\tau^2 q_x' \frac{\partial \xi}{\partial x'}. \tag{2.11.43b}$$

These equations replace (2.1.4d).

The functional is now expressed as in (2.11.26), but with

$$\pi(x_0') = p(t - \bar{t}, \mathbf{r}_0, x_0'). \tag{2.11.44}$$

We expand $\pi(x_0')$ in a Taylor series to order τ^2 and obtain

$$\pi(x_0') = \pi(x') + \tau^2 \xi\, \pi_{x_0'}(x') + O(\tau^4). \tag{2.11.45}$$

This expression substituted in the functional, with the functional expanded in a Taylor series (see Volterra [1]), gives an error term for ρ which is of the order of τ^2. This error treatment of the functional takes care of the error in the shock conditions, because the shock conditions are provided for in the functional. As before, the problem becomes an unsteady one in two dimensions when the terms in τ^2 are dropped. This completes the demonstration of the equivalence principle in this case.

CHAPTER III

NEWTONIAN THEORY

1. The gasdynamics of Sir Isaac Newton

The name of Sir Isaac Newton has been given to one branch of hypersonic flow theory, the branch based upon the zero density ratio assumption (D-lim), with the shock layer infinitesimally thin. Before we enter into an exposition of this theory let us inquire briefly as to what Newton's accomplishments in gasdynamics were, and as to why his name is attached to this body of modern hypersonic flow theory.

The investigations of Newton which are of interest to us were made in the last part of the seventeenth century, long before the concepts of thermodynamics, kinetic theory, viscous stresses, etc. were developed. Hence it is not surprising that some of Newton's analyses and results may appear to be in error in the light of today's knowledge. He had to invent a model for molecular interaction to serve in place of our present kinetic theory models, he made no distinction between the isentropic process in a sound wave and the isothermal process of a leisurely laboratory experiment, and his analysis of resistance in an incompressible fluid was based on an unrealistic model. In the present context, he did not have the notion of the shock wave or of the shock layer, and he apparently did not realize the magnitude of the flow velocities necessary for compressibility to play the dominating role in a fluid flow about a body. Newton's general approach was deductive, starting from a set of basic laws and a hypothetical model of a physical problem, leading to the logical solution of the problem based on the model. The results permitted some evaluation of the consistency of the model used, but not necessarily of its physical correctness.

A criticism of Newton's scientific results based upon our superior knowledge of two and a half centuries later is hardly of interest to us here. What is of interest is an appreciation for Newton's picture of the mechanics of matter, to serve as a framework in which to discuss his remarkable deductive accomplishments. The reference we shall use in our examination is Cajori's edition of the translation of the third edition of Newton's *Principia Mathematica* by Andrew Motte, Newton [1]. The portions of most direct interest are in Book II, Section VII, Propositions 32 through 35 (Theorems 27 and 28, Problem 7), and Note 35 in Cajori's appendix. According to Cajori, Proposition 34 of the second and third editions was Proposition 35 of the first edition.

Newton knew that gases such as air and steam are elastic, and that the relation between the density and the pressure ("compression") is what is now known as Boyle's law. He knew that liquids such as water and quicksilver are practically inelastic and have much higher densities than do gases. And he knew that water and steam were the same substance. He thus considered fluids to be either elastic, with variable density, or nonelastic ("compressed"), with the density an intrinsic property of the medium. In the absence of kinetic theory, Newton postulated repulsive ("centrifugal") forces between neighboring particles of an elastic fluid as a possible mechanism which would explain its elasticity. With this model the pressure of the elastic fluid arises primarily from these repellent forces. He finds that consistency with Boyle's law (the perfect gas law) is obtained with the assumption that the repellent force between two neighboring particles is inversely proportional to the distance between them (see Book II, Proposition 23). Newton also introduces the concept of a rare medium which consists of small particles with large distances between them, evidently with the additional assumption that the repellent forces between the particles may be neglected.

We shall examine Newton's conclusions on the basis of the model he postulated for a perfect gas, with an attempt to interpret his findings into present day terminology objectively. Thus we shall consistently interpret the interparticle forces inversely proportional to distance in terms of the pressure of a perfect gas. In assessing his results we should keep in mind that they depend deductively on a hypothetical model consistent with a realistic gas law, even though the details of the model would not be considered realistic today. We shall find that he clearly foreshadowed the concepts of Mach number similarity and of the Mach number independence principle, and that he obtained comprehensive and instructive results for hypersonic free molecule flows under hypothetical surface conditions.

In Proposition 32 Newton establishes the complete similarity of the motions of two similar systems of particles, provided, with the interpretation above, that the pressures are proportional to density times the squares of the velocities involved. His way of saying this is that the particles act on each other only "with accelerative forces that are inversely as the diameters of the correspondent particles, and directly as the squares of the velocities." The reader may note that acceleration times distance has the same dimensions as pressure divided by density. In Proposition 33 Newton extends this similarity to the resistance of a body in such a medium. His similarity condition may be re-expressed, that the velocity of the body must be proportional to the velocity of sound, with his definition of the velocity of sound. Thus his result may be expressed, that the drag coefficient of bodies of a given shape in similar elastic fluids is a function only of the Mach number. Newton did not specifically relate his similarity condition to the velocity of sound. We may note that Cranz [2, p. 45] derived Mach similarity from Newtonian similarity.

In Corollary II of Proposition 33 Newton first notes that "in a medium, whose parts when at a distance do not act with any force on one another, the resistance is as the square of the velocity, accurately." This case corresponds to our limiting case of infinite Mach number with the medium at absolute zero temperature. By the use of a similarity argument he then demonstrates that at very large velocity the repulsive forces (or pressure) may be neglected. He restates his conclusion in Corollary III: "Hence the resistance of a body moving very swiftly in an elastic fluid is almost the same as if the parts of the fluid were destitute of their centrifugal forces, and did not fly from each other; provided only that the elasticity of the fluid arise from the centrifugal forces of the particles, and the velocity be so great as not to allow the particles time enough to act." In these two corollaries Newton has demonstrated an equivalent of the Mach number independence principle or similitude of Oswatitsch for the perfect gas he is considering.

In Propositions 34 and 35 Newton considers a rare medium, in which it is evident he postulates no interaction between individual particles and considers only the force of impact between a body and the particles lying in the space swept out by the body. In Proposition 34 he assumes, but not explicitly, that the impact law for the transfer of normal momentum is independent of the angle of incidence of the body surface, and that there is no transfer of tangential momentum. He concludes, thus, that the impulse or the force of impact of a particle on the body is proportional to the sine of the angle of incidence and is directed normal to the body. The component of this impulse from a single particle contributing to the resistance is proportional to the square of the sine of the angle of incidence. In the Scholium to this proposition, Newton gives the result for the shape of the body of given length and base diameter which has minimum resistance according to his model. His calculation leading to the result has been found in his correspondence, and, according to Cajori, represents the earliest solution of a problem in the calculus of variations. A modern derivation of Newton's solution may be found in Eggers, Resnikoff, and Dennis [1], together with test results on such bodies.

In Proposition 35 Newton calculates the resistance of a sphere according to different impact laws. In his Case 1 he considers the impact to be completely elastic, while in Case 2 he considers the impact to be completely inelastic (for the normal component of momentum). The resistance for Case 1 is simply twice that for Case 2. In a free molecule flow, Newton's Case 1 corresponds to specular reflection, with all accommodation coefficients zero. Case 2 corresponds to an accommodation coefficient of zero for the tangential momentum or shear, and an accommodation coefficient of one for normal momentum or pressure if the body is very cold; if the body is not cold this normal accommodation coefficient must be greater than one. Except for normal momentum or pressure in Case 2, none of Newton's cases may be considered at all realistic for actual rarefied gas flows no matter what the body temperature is. The unrealistic model of specular reflection is sometimes treated as an interesting special case in modern theoretical investigations.

But for flows at high Mach number with a high density, essentially inviscid shock layer, Newton's analysis may be realistic. At very high Mach number the medium is essentially a rarefied one in front of the shock, and becomes a continuum only within the shock layer. With a very thin shock layer the shock has approximately the same angle of inclination as the body, and the normal momentum of an impinging molecule is lost inelastically and is transmitted to the body through the shock layer. With the shock layer inviscid the tangential component of the momentum of an impinging molecule is conserved. In order for the shock layer to be sufficiently thin, the strong form of the small density ratio assumption $\epsilon^{1/2} \ll 1$ (D-strong) discussed earlier is needed. This practical case corresponds to Newton's Case 2.

Epstein [1] was the first to obtain by a gasdynamic analysis the pressure result corresponding to Newton's Case 2

$$p_s - p_\infty = \rho_\infty U^2 \sin^2 \theta_b. \tag{3.1.1}$$

He obtained this result for the case of a two-dimensional wedge at velocities such that radiation energy losses would provide an infinitesimal density ratio across the shock. Epstein did not relate his result to the results of Newton. The connection between the shock wave result (3.1.1) and the analysis of Newton was noted independently by Busemann [2, pp. 276–277] and by Sänger [1, pp. 120–121] (the latter with an erroneous factor of $\frac{1}{2}$).

While the authors cited above were the first to obtain the sine-squared pressure law on the basis of the modern theory of oblique shocks, the sine-squared pressure law had been known previously for many years by ballisticians (cf. Cranz [1, § 12]). These earlier ballisticians used the law empirically, with an unknown multiplicative constant, and only considered the Newtonian law as one of several possible empirical laws. Newton was himself undoubtedly the most fundamental contributor to the science of ballistics, and it is probable that ballisticians have been well acquainted with Newton's results ever since the publication of the *Principia*. In any case the association of Newton's name with the sine-squared pressure law has existed for at least thirty years, and probably much longer.

The pressure result (3.1.1) as expressed is correct in the limiting case of zero density ratio provided the shock and body shapes may be assumed to be the same. Note that (3.1.1) has been given for the pressure immediately behind the shock. This pressure is equal to the pressure on the body only in case each particle follows an unaccelerated or essentially free path after impact with the shock layer, corresponding to the model for Newton's Case 2. This is the case for the flow on a wedge or on a cone at zero incidence. On a curved body, a particle is constrained within the continuum flow in the shock layer to follow a curved path, and the forces required to curve the particle paths must be taken into account. The result is a pressure difference across the shock layer equal to the momentum flow in the layer times the curvature of the layer. The necessity for this centrifugal (in present day sense, not in Newton's) force correction to obtain the pressure on the body was discovered by Busemann [2, pp. 276–277], who gave formulas for the correction. With a convex body the pressure on the body is less than that given by (3.1.1), and may drop to zero for a point on the body for which the surface still has positive incidence to the free stream.

In hypersonic flow theory, the inclusion of the centrifugal correction of Busemann is essential to a logical theory which is correct in a limiting sense. The use of a "modified Newtonian" sine-squared pressure formula without the centrifugal correction was proposed by Lees [1, Section (4)] on an empirical basis, and is common now for purposes of comparison with experiment. The modification of (3.1.1) consists of a multiplicative factor which makes the formula give the correct stagnation pressure, and the pressure is interpreted as the pressure on the body. This formula is valuable because it is easy to compute and gives a simple basis of comparison. It is not based on any rational theory, however, and its empirical basis should be kept in mind. This formula is discussed further at the end of Section 6.1.

With the recognition that the centrifugal correction to the pressure on the body must be included in a rational theory of hypersonic flow, we may turn our attention briefly to the question of terminology. The usual terms used are "Newtonian plus centrifugal" or "corrected Newtonian". A logical and appropriate possibility would be "Newton-Busemann". For simplicity we shall in this book use simply "Newtonian" to refer to the rational theory with the Busemann correction and the term "Newton-Busemann" for the correct pressure law. Since we are here primarily concerned with hypersonic flow *theory*, we must look on the uncorrected Newtonian pressure law applied to pressure on the body as being in error.

2. Two-dimensional and axisymmetric bodies

The basic assumption which underlies the Newtonian theory is the strong form of the small density ratio assumption, D-lim or D-strong in our terminology. Real gases, even those in which dissociation plays an essential role, are characterized by values of $\epsilon_{\lim}$ whose square roots cannot be considered, small, and Newtonian theory provides at best an extremely rough picture of actual hypersonic flows; it is extremely unlikely that many of the results we shall obtain in this chapter may be applied directly. The results of the succeeding four chapters will give us some understanding of the meaning of Newtonian theory in its relation to hypersonic flows. Before we get into our detailed analysis of Newtonian flow theory, it is appropriate to inquire as to whether this theory is of any more than academic interest in a study of hypersonic flow. There are two principal reasons for our studying Newtonian flow. The first is that one promising line of attack upon general blunt-body flows uses an expansion procedure in the density ratio ϵ, and the approximation of zeroth order in such an approach is a Newtonian one. We must, of course, understand thoroughly the zeroth approximation in order to exploit any expansion procedure. The second reason is that a pursuit of the Newtonian theory to its logical conclusions can shed light on phenomena in real gas flows. Many of the logical conclusions of Newtonian theory are anomalous, and involve infinite mass, zero pressure, concentration of force, and multiple condensation of the material. The presence of such anomalies, of course, depends on the limiting process $\epsilon \to 0$ (D-lim). These apparently completely unrealistic anomalous phenomena may have significant vestigial counterparts in hypersonic gas flows. An appreciation of the former is necessary for an understanding of the latter. Hence the anomalies of Newtonian theory are not to be avoided but rather to be sought out, in order to discover phenomena which may be important in hypersonic flow but which have no counterparts in flows at moderate speeds. In addition, the methods developed in Newtonian theory suggest analogous methods in more realistic theories. An example of this is the constant-density solution with cross flow given in Section 4.5, in which the method is suggested directly by the method of Section 3.6.

Newtonian flow theory is based upon the concept of an infinitesimally thin shock layer which coincides with the surface of the body, with the assumption that there is no friction between the layer and the body surface. On the basis of these assumptions and with no further specification as to the structure of the shock layer, we can obtain formulas for the steady-state pressures and forces on two-dimensional and axisymmetric bodies with suitable restrictions on the shape.

In general, however, some knowledge of the structure of the thin shock layer is desired or needed. For this purpose we need to specify in more detail the nature of the shock layer. We may in principle set up a number of models for the shock layer, or postulated rules of structure and behavior, all of which are consistent with the basic hypothesis that there is no friction between the layer and the body surface. We shall find it convenient here to consider two distinct conceptual models for the shock layer structure, which despite apparent dissimilarity are essentially identical. In our treatment of the shock layer structure we shall use the first model almost exclusively. The models considered are:

A. *The laminar layer model.* In this model, which is the one we shall consider as standard for Newtonian flow, the shock layer is considered to be composed of an infinite number of independent laminae. Each lamina is infinitesimally thin, and carries in it fluid which has entered the shock layer at some other point. This fluid has traveled between the point of entry and the point at which the shock layer is being examined along a path of zero lateral curvature on the body or shock layer surface. Such a path is known as a geodesic. With this model, the succession of laminae from the body side of the shock layer to the shock side furnishes us with a scale on which we must be able to identify the fluid with respect to its point of entry. In other words, we must be able to set up a one-to-one correspondence between the individual laminae within the layer and the points of entry into the shock layer of the corresponding streamlines. A given lamina can contain only material from one source. For any given point at which the shock layer structure is being investigated there will be a locus of the corresponding points of entry, called the "locus of entering streamlines". The one-to-one correspondence is a mapping of this locus onto the scale of laminae.

The above description of the model is for the steady flow case. In the unsteady case the additional time coordinate must be taken into account, and the specification of the analogue to the locus of entering streamlines is more involved. This case is treated in Section 3.9.

B. *The limiting perfect gas model.* According to this model we consider that the material is a perfect gas under the limiting processes $\gamma \to 1$ and $M_\infty \to \infty$, with zero viscosity. Such a model ought to be more realistic than other Newtonian models, as it does correspond to a gas flow and the only artificiality appears in the restrictions needed to make ϵ zero. On the whole this model will behave identically with the standard laminar layer model. The main purpose of considering such a model is to obtain suggestions as to whether the standard model should be modified when applied to unusual problems, and to obtain an idea as to when, if ever, the standard model necessarily diverges in behavior from the gas model.

The equivalence of this model with the laminar layer model may be obtained from the momentum equation, which states that the acceleration of a particle is minus the pressure gradient divided by the density. With the density approaching infinity in the limiting process considered, the particle acceleration approaches zero if the pressure gradient remains finite. In the direction normal to the shock surface the pressure gradient may approach infinity as the shock layer becomes infinitesimally thin, and the component of particle acceleration in this direction is not generally zero but is controlled by the geometry of the shock layer. In the directions parallel to the shock surface the pressure gradient is controlled by the shock pressure jump and must remain finite. Accordingly, the particle acceleration components tangential to the shock surface approach zero in the limiting process, and the particle path in the shock layer approaches a geodesic on the shock surface.

An assumed gas other than a perfect gas may equally be used in this conceptual limiting process as long as ϵ approaches zero and ϵ divided by the speed of sound in the shock layer also approaches zero. Such a gas would furnish us another model, also equivalent to the laminar layer model.

We now turn to the application to two-dimensional and axisymmetric bodies and introduce a stream function ψ which represents mass flow per unit depth for two-dimensional flows, and mass flow per unit azimuthal angle (in radians) for axisymmetric flows. In the undisturbed free stream this function is simply

$$\psi = \rho_\infty U y \tag{3.2.1a}$$

for two-dimensional flow, and

$$\psi = \tfrac{1}{2}\rho_\infty U r^2 \tag{3.2.1b}$$

for axisymmetric flow. It is convenient to combine these formulas, replacing r by y and introducing the integral parameter j, into the single formula

$$\psi = \frac{\rho_\infty U}{1+j} y^{1+j}. \tag{3.2.2}$$

Note that ψ is essentially frontal area times $\rho_\infty U$. The integer j is the integer of Table 2-1, with $j = 0$ for two-dimensional flow and $j = 1$ in axisymmetric flow. With $j = 1$ the coordinate y is to be read as r, the radial cylindrical coordinate.

We introduce also the quantity P which, times U, is the momentum in the shock layer per unit depth for two-dimensional flows, and per radian of azimuthal angle for axisymmetric flows. The change in the shock layer momentum between two points is simply the impinging mass flow times the tangential component of the free stream velocity, written as

$$dP = \cos\sigma \, d\psi. \tag{3.2.3}$$

In this equation σ is the shock inclination angle and a factor U has been cancelled. In integrated form this expression is

$$P = \int_0 \cos\sigma \, d\psi, \tag{3.2.4}$$

here expressed as an indefinite integral evaluated as zero at the point 0 at which the shock layer begins.

The pressure increase divided by $\rho_\infty U^2$ is one half the pressure coefficient C_p, defined by

$$\tfrac{1}{2}C_p = \frac{p - p_\infty}{\rho_\infty U^2}. \tag{3.2.5}$$

Immediately behind the shock, momentum considerations give us the formula

$$\tfrac{1}{2}C_{p_s} = \sin^2 \sigma. \tag{3.2.6}$$

On a convex body this is decreased by the Busemann centrifugal pressure correction equal to the momentum per unit depth $Uy^{-j}P$ times the curvature of the body $-\sin \sigma \, d\sigma/dy$ (see Fig. 3–1). The pressure on the body may thus be expressed

$$\tfrac{1}{2}C_{p_b} = \sin^2 \sigma - \frac{d \cos \sigma}{d\psi} P_1 , \tag{3.2.7}$$

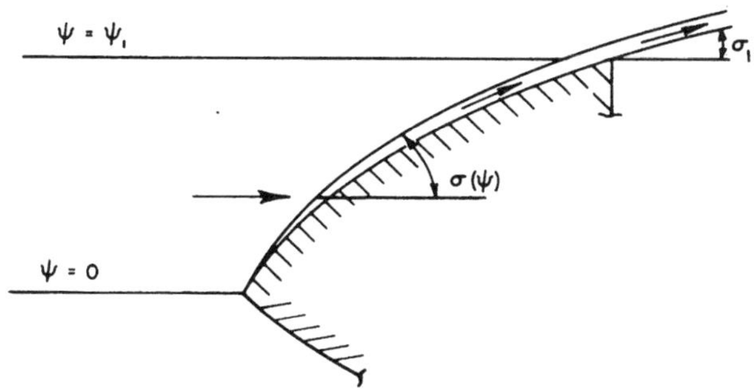

FIG. 3–1. Newtonian shock layer.

where the subscript 1 on P indicates that the integral (3.2.4) has been taken to the outer edge of the shock layer. The result (3.2.7) is termed the "Newton-Busemann" pressure law and is identical with the one originally given by Busemann [2] for the two-dimensional case.

The drag of the body can be calculated as an integral over projected frontal area. A simpler and more instructive approach is to use a momentum balance. The x-momentum entering is mass flow times U, while the momentum leaving is that in the shock layer times cos σ at the outer edge.

The drag of the body divided by U will be denoted $\tfrac{1}{2}\psi_1 C_D$, and is expressed

$$\tfrac{1}{2}\psi_1 C_D = \psi_1 - \cos \sigma_1 \, P_1 , \tag{3.2.8}$$

where the subscript 1 here denotes the outer edge of the body and ψ is defined to be zero where the shock layer begins. The drag coefficient here is based on frontal area. With a two-dimensional body the drag for the two sides should be calculated separately, and with an annular body of revolution the drag for the inner and outer portions should be calculated separately. The expression (3.2.8) was also given by Busemann. With two-dimensional bodies we may calculate the contribution to the lift of the flow on one side of the body. This is given, per unit depth of the body, by

$$\tfrac{1}{2}\psi_1 C_L = \pm \sin \sigma_1 \, P_1 \,, \tag{3.2.9}$$

and may also be obtained directly from momentum considerations or from appropriate integration of (3.2.7). This lift coefficient is also based on frontal area. A body of revolution (or a symmetric two-dimensional body) at zero incidence has zero lift, and (3.2.9) has no meaning applied to axisymmetric flow. We may note, though, that the axisymmetric theory can apply to a meridional section of a body of revolution. In this case a suitable integral will give a formula similar to (3.2.9) for a net lift on the body.

The formulas given above permit us to obtain by quadratures the pressures and forces on two-dimensional and axisymmetric shapes. Some examples of these calculations will be given later, and many may be found in the literature. See for example Ivey, Klunker, and Bowen [1] and Ivey and Morrissette [1], who developed the theory proposed by Busemann [2].

We shall now obtain a few results for the structure of the shock layer, with the laminar layer model. The mass per unit area in the shock layer is denoted by $\rho_\infty m$, and is given by

$$m = \frac{y^{-j}}{\rho_\infty U} \int_0 \frac{d\psi}{\cos \sigma} \,. \tag{3.2.10}$$

This quantity m has the dimensions of distance, and will appear as a fundamental independent variable for shock layer structure in Section 3.5. The velocity of the flow in the layer is simply $U \cos \sigma$, where σ is the angle of the body where the streamline of interest entered the shock layer. The quantity $\cos \sigma$ considered as a function of ψ thus describes the velocity distribution in the layer. If m in (3.2.10) is defined in terms of an integral over part of the layer, we may consider $\cos \sigma$ as a function of m. Distance along the layer is given by the integral of $dx/\cos \sigma$ or $dy/\sin \sigma$. The time that a particle has spent in the layer is thus given by

$$t = \frac{1}{U \cos \sigma_0} \int_0 \frac{dx}{\cos \sigma} \,, \tag{3.2.11}$$

where the subscript 0 on σ_0 here denotes conditions at the point of entry of the particle under consideration.

The Newtonian flow results for two-dimensional and axisymmetric shapes may be applied to bodies of considerably greater generality. Let us consider a body which is composed of a number of surfaces fitted together along sharp intersection lines or corners. Let us require that each individual surface be a piece of a two-dimensional surface (of a general cylinder with generatrix perpendicular to the flow direction) or a piece of a body of revolution whose axis is parallel to the flow direction. We require that all the corners of intersection be convex, and that each of these corners acts as a leading edge for both of its two defining surfaces. Then, on each surface we may set up strips (or meridional sections for pieces from bodies of revolution) which are of infinitesimal width. Within the assumptions of Newtonian flow theory the flow on each of these strips is independent from that on its neighbor, and the pressures and shock layer structure may be calculated for the strip on the basis of the two-dimensional or axisymmetric theory given in this section. It is possible to construct a more general convex body as a limiting case of such a composite surface, to which our simple Newtonian theory could be applied. Because of the strong restriction that none of the corners may be a trailing edge for a surface, we are limited in this concept to very special bodies. As a class, these bodies may be termed "Newtonian bodies without cross flow".

A further generalization is made possible by introducing the classical concept of cylindrical flow, with pieces from general cylinders whose generatrices are not perpendicular to the flow direction permitted in the composite body surface. With such surfaces the calculation of pressures and structure becomes more complicated and the restrictions on orientation of the corners become more involved. We shall investigate these cylindrical flows in Section 3.6, but shall not go into their application in composite bodies.

3. Simple shapes and free layers

We now continue our treatment of two-dimensional and axisymmetric flows, considering the impulse P_1 as the basic dependent variable, but with the subscript 1 dropped. We first consider P to be a function of σ, with $\psi(\sigma)$ given by

$$\psi = \int \frac{dP}{d\sigma} \frac{d\sigma}{\cos \sigma}. \tag{3.3.1}$$

The pressure may be expressed

$$\tfrac{1}{2}C_{p_b} = \sin^2 \sigma + \sin \sigma \cos \sigma \, P \frac{d\sigma}{dP}, \tag{3.3.2}$$

and the drag as before in (3.2.8). With this formulation we shall now define two families of shapes which we term "simple"; the simplicity of these bodies lies in their pressure expressions rather than in their geometries.

A "simple cosine shape" is defined by the relation

$$P(\sigma) = \rho_\infty U A (\cos \sigma)^n, \tag{3.3.3}$$

where A and n are constants. The pressure coefficient for this shape is

$$\tfrac{1}{2}C_{p_b} = 1 - \left(1 + \frac{1}{n}\right)\cos^2\sigma. \tag{3.3.4}$$

The integral (3.3.1) may be evaluated explicitly to give

$$\psi = \frac{nP}{(n-1)\cos\sigma} = \frac{n\rho_\infty UA}{n-1}(\cos\sigma)^{n-1}. \tag{3.3.5}$$

The shape with ψ chosen to be zero at $y = 0$ is determined by relation (3.2.2) between ψ and y, and by the integral for x

$$x = \int \cot\sigma\, dy = -\frac{n-1}{1+j}\left[\frac{(1+j)nA}{n-1}\right]^{\frac{1}{1+j}} \int (\cos\sigma)^{\frac{n-1}{1+j}}\, d\sigma. \tag{3.3.6}$$

For a general exponent this integral may be expressed in terms of an incomplete Beta function. Note that we may express

$$-\int_{\pi/2} (\cos\sigma)^m\, d\sigma = \tfrac{1}{2}B\left(\frac{m+1}{2}, \tfrac{1}{2}; \cos^2\sigma\right), \tag{3.3.7a}$$

$$\int_0 (\sin\sigma)^m\, d\sigma = \tfrac{1}{2}B\left(\frac{m+1}{2}, \tfrac{1}{2}; \sin^2\sigma\right), \tag{3.3.7b}$$

where

$$B(\alpha, \beta; t) = \int_0 t^{\alpha-1}(1-t)^{\beta-1}\, dt \tag{3.3.8}$$

is the incomplete Beta function. This function may itself be expressed as a hypergeometric function. The integrals in (3.3.7) and (3.3.8) are indefinite integrals which are specified to be zero at the lower limit indicated. A special case of interest is that for which

$$n = 2 + j, \tag{3.3.9}$$

which gives

$$x = [(2+j)A]^{\frac{1}{1+j}}(1 - \sin\sigma), \tag{3.3.10a}$$

$$y = [(2+j)A]^{\frac{1}{1+j}}\cos\sigma. \tag{3.3.10b}$$

This shape is a circle, corresponding to a circular cylinder in the two-dimensional case and to a sphere in the axisymmetric case. Note that the pressure difference from stagnation is 3/2 times the uncorrected Newtonian value (the correct value immediately behind the shock) for the cylinder, and 4/3 times for the sphere.

With $n > 1$ all these simple cosine shapes are finite and convex. At a point on the body the pressure is zero, and if the solution were continued the pressure would become negative. What must happen is that the shock layer must separate at this point and fly free. Such "free layers" are discussed below. Where $\cos \sigma$ is small, we have $y \propto (\cos \sigma)^{(n-1)/(1+j)}$ and $x \propto (\cos \sigma)^{(n+j)/(1+j)}$. With $n > 1$ this occurs near the nose. Near the nose of one of these bodies the shape has the form

$$y \propto x^k, \tag{3.3.11}$$

where

$$k = \frac{n-1}{n+j}. \tag{3.3.12}$$

If $n < -j$ we obtain bodies which are infinite in extent and concave. A body of this type has $P = \rho_\infty UA$ at $\sigma = 0$, and cannot start with $P = 0$. Some sort of forebody is needed on which an initial impulse in the shock layer of the correct magnitude is developed. Equations (3.3.11) and (3.3.12) still apply, but describe the asymptotic shape of the body at large distances (where $\cos \sigma$ is small).

The other family of simple shapes, denoted "simple sine shapes", is defined by the relation

$$P = \rho_\infty UA(\sin \sigma)^n, \tag{3.3.13}$$

where A and n are again constants. For these shapes the pressure is given by

$$\tfrac{1}{2}C_{p_b} = \left(1 + \frac{1}{n}\right) \sin^2 \sigma. \tag{3.3.14}$$

and the stream function by

$$\psi = n\rho_\infty UA \int_0 (\sin \sigma)^{n-1}\, d\sigma, \tag{3.3.15}$$

in terms of another integral reducible to an incomplete Beta function. Only in the two-dimensional case is the quadrature for x explicitly obtainable. We obtain with $j = 0$

$$x = \frac{nA}{n-1} (\sin \sigma)^{n-1}. \tag{3.3.16}$$

For $n = 2$ the two-dimensional body is again a circular cylinder, but in this case the concave side of a quadrant of a circle faces the stream.

With $n > 1 + j$ the simple sine shapes are all finite and concave. The local behavior near the nose again follows (3.3.11), but in this case the exponent k is given by

$$k = \frac{n}{n - 1 - j}. \tag{3.3.17}$$

With n negative we obtain bodies which are infinite and convex and, if the pressure is to remain nonnegative, we must have $n \leqslant -1$. As with the simple cosine shapes of negative n, the body cannot start with $P = 0$, and again we must have a forebody to furnish an initial impulse in the shock layer of correct magnitude. The case $n = -1$ corresponds to a free layer with $k = 1/(2 + j)$ describing the asymptotic shape at large distances, and with the pressure behind the layer equal to zero. The case $n = -2$ gives a pressure on the body half that immediately behind the shock, with $k = 2/(3 + j)$ for the asymptotic shape. This case corresponds to the constant-energy similar solution of the small-disturbance theory in the limit $\gamma \to 1$. It may be noted that the values of k in these two cases correspond with those given by (2.6.25) and (2.6.27), respectively, with $\omega = 0$.

For slender bodies derived from simple sine shapes we may calculate the drag coefficient in terms of the thickness ratio and the parameter n or k. With σ small (3.3.15) and (3.3.6) may be integrated to give the approximate integral

$$x = \frac{ky}{\sin \sigma}, \tag{3.3.18}$$

where k is to be taken from (3.3.17). This integral is valid only at small distances with $n > 1 + j$, and is valid only at large distances with $n < 0$. We define the thickness ratio τ as the ratio of y to x at the base of the body, and obtain

$$\tau = \frac{y_1}{x_1} = \frac{\sin \sigma_1}{k}. \tag{3.3.19}$$

The drag coefficient based on frontal area may be expressed

$$\begin{aligned} \tfrac{1}{2}C_D &= \frac{1}{\psi_1} \int_0^{\psi_1} (\tfrac{1}{2}C_{p_b})\, d\psi \\ &= \frac{n+1}{n+2} \sin^2 \sigma_1 . \end{aligned} \tag{3.3.20}$$

This result may also be obtained from (3.2.8) by an appropriate approximate analysis. Expressing n in terms of k from (3.3.17), the drag coefficient takes the form

$$\tfrac{1}{2}C_D = \frac{(2 + j)k - 1}{(3 + j)k - 2}\, k^2 \tau^2. \tag{3.3.21}$$

This result was obtained by Cole [1]. He notes that for the two-dimensional case $C_D/2\tau^2$ takes on the minimum value of 0.918 at $k = 0.864$, and that for the axisymmetric case it takes on the minimum value of 2/3 at $k = 2/3$. He notes also the condition for finite drag that k must be greater than $2/(3 + j)$, equivalent to the condition $n < -2$.

We now turn to a different formulation of the problem with ψ considered as the independent variable. With primes used to denote differentiation with respect to ψ, the relation determining σ is

$$\cos \sigma = P'. \tag{3.3.22}$$

The pressure is expressed by

$$\tfrac{1}{2}C_{p_b} = 1 - P'^2 - PP'', \tag{3.3.23}$$

and the drag of the part of the body out to ψ by

$$\tfrac{1}{2}\psi C_D = \psi - PP'. \tag{3.3.24}$$

The subscript 1 is dropped here. For determining the body shape with this formulation we note first that

$$dx = \cot \sigma \, dy = \frac{P' \, dy}{\sqrt{1 - P'^2}}. \tag{3.3.25}$$

Differentiation of (3.2.2) gives

$$dy = (\rho_\infty U)^{\frac{-1}{1+j}}[(1 + j)\psi]^{\frac{-j}{1+j}} \, d\psi, \tag{3.3.26}$$

whence we obtain

$$dx = (\rho_\infty U)^{\frac{-1}{1+j}}[(1 + j)\psi]^{\frac{-j}{1+j}} \frac{P' \, d\psi}{\sqrt{1 - P'^2}}. \tag{3.3.27}$$

With $P(\psi)$ given the quadrature of (3.3.27) gives us the shape. We may recognize the simple cosine shapes in this formulation as being those for which

$$P = C\psi^{\frac{n}{n-1}}, \tag{3.3.28}$$

with C a constant, and the simple sine shape with $n = 2$ as one for which

$$P = \psi \left(1 - \frac{\psi}{4\rho_\infty UA}\right). \tag{3.3.29}$$

A Newtonian free layer is a Newtonian shock layer whose position is determined not by the geometry of a body but by the condition that the pressure behind it is zero. As in the special case of the simple sine shape with $n = -1$, there must be a forebody to develop an initial impulse. If the curvature of the body is sufficiently large or if there is a discontinuity in slope, the shock layer separates from the body and becomes a free layer. The concept of a free layer is due to Busemann [2]. He did not present calculations for free layer shapes, but did point out the existence of a separated layer with zero pressure between the layer and the body. Lighthill [3] obtained the free layer shape in the axisymmetric case. The derivation of the first edition is that given below and was obtained independently.

The general shape of a free layer in two-dimensional or axisymmetric flow may be obtained following various approaches. Probably the most direct is the one using ψ as an independent variable, and we shall follow this approach. The pressure coefficient in (3.3.23) must be set equal to zero. This is equivalent to setting the drag in (3.3.24) equal to a constant, or

$$PP' = \psi - \psi_0 , \tag{3.3.30}$$

with ψ_0 a constant equal to $\frac{1}{2}\psi C_D$. The integral of (3.3.30) gives

$$P = \sqrt{(\psi - \psi_0)^2 + P_0^2} = P_0/\sin \sigma, \tag{3.3.31}$$

where P_0 is another constant, and shows that P takes the minimum value P_0 at $\psi = \psi_0$. We evaluate also the quantity

$$\frac{P'}{\sqrt{1 - P'^2}} = \frac{\psi - \psi_0}{P_0} = \cot \sigma, \tag{3.3.32}$$

and note that $\sigma = \pi/2$ at $P = P_0$. We may now evaluate x from (3.3.27), keeping in mind the relation (3.2.2) between ψ and y, as

$$x = \frac{\rho_\infty U}{(1 + j)P_0}\left[\frac{y^{2+j}}{2 + j} - y_0^{1+j} y\right] + x_0 . \tag{3.3.33}$$

Here y_0 is the arbitrary value of y at $\psi = \psi_0$ and x_0 is an arbitrary constant. In the axisymmetric case ($j = 1$) y_0^{1+j} cannot be negative, as this would require a value of P larger than can be developed; the free layer shape is a cubic curve. In two-dimensional flow ($j = 0$) y_0 cannot be negative if the body which develops P is restricted to $y \geqslant 0$; the free layer shape is a parabola.

The equation (3.3.33) may also be derived from (3.3.13) and (3.3.15) for the simple sine shape with $n = -1$. The asymptotic shape of the free layer obeys (3.3.11) with $k = 1/(2 + j)$. The Newtonian free layer is depicted in Fig. 3–2a.

We must now inquire as to whether this free layer may be considered in any way realistic for actual gas flows. An obvious step is to compare the asymptotic shock shape of a free layer with the asymptotic similar shapes in gas flows considered in Section 2.6. The limiting case of these similar solutions gives us the shape associated with a constant-energy similar flow, for which the parameter k is distinctly greater than that for a free layer. Other similar gas flows with finite bodies have still larger values of k. From this inconsistency we might

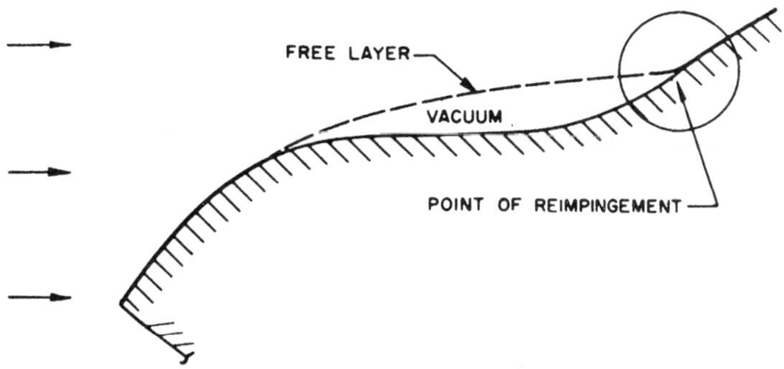

FIG. 3–2a. Free layer and reimpingement.

cursorily conclude that the free layer is unrealistic; we have noted in Section 2.6, however, that the constant-energy similar solution is itself unrealistic for steady hypersonic flows with γ close to 1. And it is only for a gas with γ very close to 1 that any Newtonian flow can be expected to be realistic. We may expect that the effect of an entropy layer would be to give a lower effective value of k. Thus we should not conclude that the free layer is unrealistic on the basis of a comparison with the constant-energy similar flow.

The correctness of the concept of a Newtonian free layer depends on the maintenance of the integrity of the layer, without appreciable thickening of the layer or loss of material to the vacuum region behind. We outline here the justification of the fact that the layer remains intact in a limiting sense, following the limiting perfect gas model for Newtonian flow. We consider that the body shape is given, that $M_\infty = \infty$ in the flow, and that the gas is a perfect one with γ very close to 1. And we consider first that the body is a pointed one, so that the velocity in each lamina is of the order of U. Conservation of total enthalpy yields that the velocity of sound is of the order $(\gamma - 1)^{1/2}U$. With finite pressure through the layer, its thickness is roughly proportional to $\gamma - 1$. We imagine that the layer shoots over a sharp edge and suddenly becomes free. A sound signal is propagated through the layer in a time proportional to the thickness divided by the sound speed; this time is proportional to $(\gamma - 1)^{1/2}$ with varying γ. In the limit as $\gamma \to 1$ this time approaches zero, and in the limit the entire shock layer is affected immediately by the fact that the layer has become free.

The material on the back side of the shock layer immediately begins to expand into the space between the shock layer and the body behind the corner. The Mach number within the layer is of the order $(\gamma - 1)^{-1/2}$, and we may apply the equivalence principle locally to the shock layer itself. Thus we may liken this expansion to a sudden one-dimensional expansion following a piston which is being withdrawn. If $\gamma - 1$ is very small this unsteady flow is one at constant velocity of sound. The unsteady characteristic equation at constant $a^2 = p/\rho$ may be written in the form $dp/p + dM = 0$. The pressure on the piston is expressible in terms of the initial pressure p_b by

$$p = p_b \exp(-M), \tag{3.3.34}$$

where M is the velocity of the piston divided by the constant velocity of sound. This Mach number is of the order of the Mach number of the primary flow in the layer times the angle of divergence, and if this angle is finite, M is of order $(\gamma - 1)^{-1/2}$. From (3.3.34) we see that the pressure on the body in the space approaches zero as $\gamma \to 1$. Even though the rarefaction wave sent from the corner to the shock reflects as a pressure wave (see Table 7–1), it may be shown that this reflected wave does not affect our limiting result. If the slope of the body is continuous at the Newtonian separation point, i.e. if there is no corner, the process described is delayed so that the pressure drop does not take place suddenly. However, this delay is of the order $(\gamma - 1)^{1/2}$ in time or distance, and an effective separation point defined by a chosen low body pressure will approach the Newtonian separation point in the limit $\gamma \to 1$.

The considerations above are for a pointed body, for which the shock layer may be considered itself hypersonic. With a blunt body the fluid nearest the body is at a relatively low velocity, at zero velocity at the body within the Newtonian approximation. With γ very close to one, we can express the Mach number M_b of the flow nearest the wall by a direct application of Bernoulli's law. At constant velocity of sound Bernoulli's law may be written $dp/p + M\,dM = 0$, and we obtain

$$M_b^2 = 2 \ln (p_0/p_b), \tag{3.3.35}$$

where p_0 is the stagnation pressure. If the Newtonian separation point occurs in a region of continuous body slope or at a corner with p_b/p_0 appreciably less than $\exp(-\frac{1}{2})$, then M_b^2 is appreciably greater than one although still of order one. The equivalence principle does not hold for the expansion, and the pressure on the body immediately downstream of the corner or Newtonian separation point is not vanishingly small. However, the rarefaction wave sent from the back of the layer toward the shock is rapidly reflected as a rarefaction almost fully from the strong vorticity in the lower part of the shock layer. This conclusion is based on (7.2.19) and the discussion following it. The end effect of these reflections is to make the presence of the low velocity flow near the wall unimportant, and to cause the pressure to be determined primarily by the high velocity flow. The case of Newtonian separation at a corner which is far enough forward on a blunt body so that p_b/p_0 is greater than exp $(-\frac{1}{2})$ $(M_b^2 < 1)$ is considerably more complicated. In any case, we may expect the pressure behind the corner to be primarily determined by the high velocity flow. Thus we may expect the pressure behind the Newtonian separation point to approach zero as $\gamma \rightarrow 1$ for a blunt body as well as for a pointed body, although this approach may be slower than with a pointed body.

Although the foregoing arguments justify that the free layer may be considered to be realistic, they also indicate the limitations inherent in the concept. We first note that the quantity $\gamma - 1$ we have been considering is essentially the

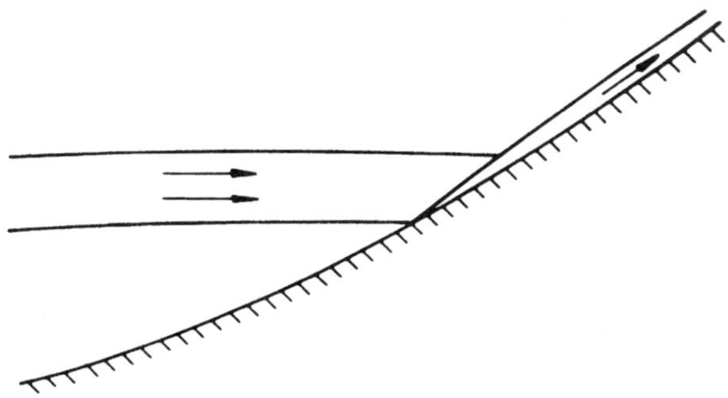

FIG. 3–2b. Reimpingement detail.

same as 2ϵ. The requirement that the layer remain thin is then that $\epsilon^{1/2}$ be small. This requires the strong form of the small density ratio assumption (D-strong); however, as we shall see later in Chapter IV, this assumption is required for the validity of most results in Newtonian flow. We have also assumed that Newtonian separation occurs at a corner or at least with the body curvature initially appreciably larger than the free layer curvature, so that there is an appreciable space between the free layer and the body. Even with $\epsilon^{1/2}$ small, this space should be amply large for the free layer to be realistic. If $\epsilon^{1/2}$ is finite and the space behind the free layer is constrained, we should expect an important effect of finite back pressure in this space. This effect would involve a thickening of the layer, and would be more pronounced with a blunt body. Freeman [1], in investigating flows on blunt bodies with ϵ small, has noted a singular behavior of his theory near the Newtonian separation point. This behavior is connected with the limitations on the concept of Newtonian separation, and also indicates a non-analytic dependence upon ϵ of the solution in the separation region. In Section 5.9 is reported the investigation of Freeman [5] on the nature of this singular point on a sphere. This investigation supports the reality of Newtonian separation as a limiting process, but also indicates clearly the severe limitations involved. Also reported in Section 5.9 is the investigation of Freeman [6] on transition far downstream of a body from a free layer solution to a blast wave or constant-energy solution.

The shape of the Newtonian free layer after it leaves the body is independent of the shape of the body downstream of the separation point, provided that a space is left between them. If the body shape is such that it intersects the course of the free layer, the layer will reimpinge on the body (see Figs. 3–2). The free layer is itself hypersonic, and we should not expect phenomena in any way similar to those observed when a free jet of incompressible fluid strikes a wall. Instead, we should expect that the Newtonian impact process be essentially repeated, and that the free layer lose its momentum component normal to the wall and preserve the component tangential to the wall. This transfer of momentum results in a concentration of forces along the line of impingement and a further condensation of the material. This force distribution is a force per unit distance and not a pressure; a similar force would be exerted by an attached Newtonian layer at a concave corner of the body. In the idealization of Newtonian flow such singular or anomalous force distributions are not uncommon. Although they may appear unreasonable they are logical consequences of the Newtonian assumptions and models. If such singular behavior appears in a Newtonian flow we may expect some corresponding local behavior in a real-gas flow with ϵ small. In the case of the phenomenon of impingement, however, the shock layer in a real gas may be closer to one of the constant-pressure type discussed next.

The Newtonian free layer is a special case of a more general family of layers characterized by constant pressure on the body side of the layer. Except for the free layer (zero pressure case), such constant-pressure layers may be either convex or concave. A constant-pressure shape is asymptotic to a straight line of slope $\tan \sigma$ equal to $\sqrt{\frac{1}{2}C_p/(1 - \frac{1}{2}C_p)}$, and may be obtained with the methods used for investigating the free layer. Note that the asymptotic value of $\frac{1}{2}C_p$ must be $\frac{1}{2}C_{p_s} = \sin^2 \sigma$. A constant-pressure layer may reimpinge on a body just as may a free layer. The space between a constant-pressure layer and the body must be filled with "dead" fluid at the specified pressure, at a high density which in the Newtonian limit approaches infinity. Since this space normally would have finite volume, we have the anomalous result that an infinite mass of fluid is stored therein.

With constant-pressure layers permitted the Newtonian flow on a general body becomes nonunique, as the possible constant-pressure layers on a given two-dimensional or axisymmetric body form a two-parameter family. For this very general family we permit the possibility that the pressure jumps discontinuously behind the layer at a separation point. If the flow separates at a corner, or if we require the pressure on the body to be continuous where the flow separates at a point which is not a corner, there is a reduction in the freedom of choice of solution and we are left with a one-parameter family. With the slope of the body continuous and its curvature finite, a constant-pressure layer may only separate under a pressure which is greater than that which would occur without separation. If the body curvature is continuous and the pressure is required to be continuous, a constant-pressure layer can separate only in a region of decreasing pressure or at a leading edge.

A constant-pressure layer other than a free layer which extends to infinity in the free stream must be rejected because of unacceptable downstream boundary conditions. If now we also reject solutions in which the layer reimpinges on the body the possible solutions form a zero-parameter family. In this case the solution is not necessarily unique, but we would have at most a finite number of solutions to choose from. In real gas flows the thin but finite viscous boundary layer does provide a mechanism for upstream influence in the shock layer, and a separation point far upstream from a pressure-increasing obstacle is quite generally observed in experiments. Hence we suggest here that a configuration with a Newtonian constant-pressure layer without discontinuities in the pressure (except at a corner) and without reimpingement of the layer may be an appropriate limiting configuration to represent real gas flows, in case the attached solution reaches zero pressure and the solution with a free layer involves reimpingement. The condition of contact with the body without reimpingement is somewhat analogous to the well known Kutta condition of airfoil theory, in that it is an ideal-fluid condition with an origin in real-fluid effects, and which provides uniqueness within inviscid theory. The constant-pressure layer without reimpingement is depicted in Fig. 3–3.

Another interesting generalization of the Newtonian free layer appears when the Newtonian theory is applied to the problem of a two-dimensional hypersonic sail. We define a "Newtonian sail" as a membrane which is a surface in a Newtonian flow. A membrane has the property that it may support no

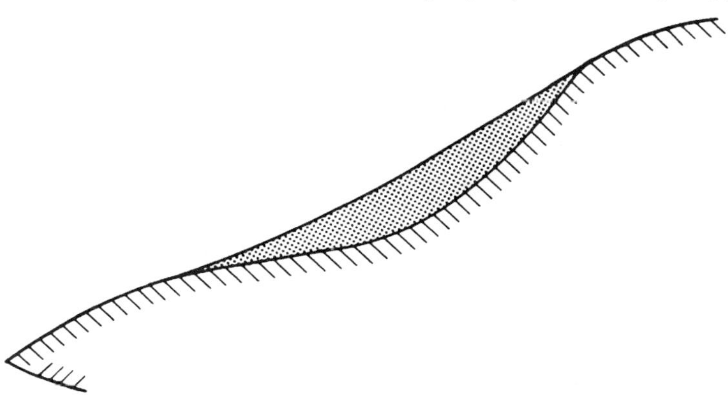

FIG. 3–3. Constant-pressure layer without reimpingement.

bending loads but will, in general, experience a two-dimensional stress in its surface. On the reverse side of the sail we specify a constant pressure. We shall restrict ourselves to a generalization of the free layer and not consider the analogous generalization of the constant-pressure layer. Accordingly, we specify this constant back pressure to be zero. And finally, we require that the pressure on the front of the sail be nonnegative. Each differential element of the sail must be in equilibrium under the forces exerted on it.

The stress in the membrane is a symmetric two-dimensional tensor, whose principal axes will be aligned with the natural coordinate system for either a two-dimensional or an axisymmetric sail. For a two-dimensional sail the principal stress component in the direction of the cylindrical axis of the sail exerts no forces on any element of the sail and may be completely disregarded. From the requirement of equilibrium in the tangential direction we may conclude that the other principal stress component is constant. We denote this stress by UT and note that the pressure difference across the sail is UT times the curvature. We may then derive an expression for the pressure on the back of the sail which is the same as (3.3.23) but with P replaced by $P - T$. Setting this back pressure equal to zero we obtain

$$0 = 1 - P'^2 - P''(P - T). \tag{3.3.36}$$

This relation we may integrate immediately to obtain an equation analogous to (3.3.30),

$$(P - T)P' = \psi - \psi_0. \tag{3.3.37}$$

When we integrate this expression with the requirement of nonnegative pressure on the front of the sail we obtain in place of (3.3.31)

$$T - P = \sqrt{(\psi - \psi_0)^2 + P_0^2}\,. \tag{3.3.38}$$

The succeeding development proceeds exactly as for the free layer, but with a change in sign in (3.3.32) and (3.3.33). The resulting shape ($j = 0$) is again a parabolic cylinder, but with the vertex of the parabola pointed downstream instead of upstream. Which shape of the family of possible shapes the sail assumes may be determined by the ratio of the length of the sail to its chord, by the tension in the sail, or by an elastic relation between the length and the tension.

An analysis of a two-dimensional hypersonic sail using the uncorrected Newtonian pressure law instead of the Newton-Busemann pressure law has been made by Daskin and Feldman [1]. With the uncorrected pressure law the shape of a two-dimensional sail is a catenary with its vertex pointed downstream. Daskin and Feldman did not consider the case of a sail with the complete pressure law.

Another two-dimensional sail configuration is possible, involving the constant-pressure layer discussed above. The shock layer is a constant-pressure layer which just grazes the rear of the sail; with no forebody this shape is simply the straight line chord from the leading edge of the sail to the rear edge. Behind the dead-water region the sail takes a circular arc shape. Which of the family of possible circular arc shapes the sail assumes is again determined by the ratio of length to chord, by the tension, or by an elastic relation between length and tension.

For an axisymmetric sail we must consider the azimuthal or meridional principal stress component in the membrane. We let UT be the longitudinal force per unit radian in the membrane; if the azimuthal stress component is nonzero, T is not a constant but a function $T(\psi)$. When we include the effect of the azimuthal stress in the condition for equilibrium of an element of the sail, we again arrive at (3.3.36) but with an additional term $P'T'$ on the right-hand side. We conclude that the first integral (3.3.37) is valid also for an axisymmetric sail, whether T is constant or not.

For an axisymmetric sail with zero azimuthal stress and constant T, the analysis above may be followed again and yields (3.3.33) with $j = 1$ and with reversed sign. Again the sail shape is the mirror image in a transverse plane of a free layer shape. For a sail with nonzero azimuthal stress we must inquire as to how the stress components and sail shape are specified or related. If the sail shape itself is specified, i.e. if the unstressed membrane shape is given and the membrane is inextensible, the sail behaves as though it were rigid and we may solve the problem completely with the methods of Section 3.2.

The problem of an elastic axisymmetric sail is nontrivial in the sense of this section, however. We assume that the unstressed shape of the sail is known, and that its complete elastic behavior under an arbitrary axisymmetric stress system is specified. A Lagrangian coordinate identifying position on the unstressed sail is defined and is used as the independent variable. Two elastic stress relations and (3.3.37) yield three ordinary differential equations for the three unknown dependent variables ψ, P, and T. With three boundary conditions from the conditions of the mounting of the sail we may obtain the complete solution. We shall not investigate the elastic sail further.

Finally, we introduce the concept of an "isotropic sail". An isotropic sail is a Newtonian sail for which the stress tensor in the membrane is isotropic, or for which the membrane has zero rigidity in shear. The condition of tangential equilibrium of a sail element requires that the stress, which now may be considered a scalar, be a constant over the entire sail. Thus we could refer jocularly to such an isotropic sail as a "Newtonian soap bubble". In the axisymmetric case T is r/U times this constant stress, and we may rewrite (3.3.37) as

$$(P - A\psi^{1/2})P' = \psi - \psi_0 , \tag{3.3.39}$$

where A is a constant. This differential equation may be solved numerically and the profile shape obtained from (3.2.1b) and (3.3.27).

4. Optimum shapes

The problem of determining the shape of a body so that its drag is a minimum is an old problem, one which occupied Newton in the seventeenth century. Newton's results and those of Eggers, Resnikoff, and Dennis [1] are based on analyses which omit the centrifugal correction to the pressure on the body. We set as our present task the determination of optimum minimum-drag shapes using the complete pressure relation (3.2.7). We have available an explicit formula for the drag (3.2.8), but note that this formula is somewhat different in form from that obtained using the uncorrected pressure expression. We shall find it convenient to distinguish between two different types of optimum shapes which we shall develop, designated by the terms "absolute optimum" and "proper optimum". Eggers, Resnikoff, and Dennis did estimate the effect of the centrifugal correction and obtained approximate optimum shapes with the effect included.

We shall concern ourselves primarily with the classical problem in which the quantities specified are the length and thickness of the body. The approach used here will be directly applicable or adaptable to problems with different isoperimetric conditions. In considering bodies of revolution we shall not include annular bodies; if desired, such bodies may be treated by the same methods. We shall work within the formulation of the problem in which $y(x)$ is the basic function treated. The formulations using the function $x(y)$ or the function $P(x)$ could also have been used, but the use of $y(x)$ appears direct and the most conventional.

Equation (3.2.8) for the drag is fundamental to any optimization considerations and is repeated here,

$$\tfrac{1}{2}\psi_1 C_D = \psi_1 - \cos\sigma_1\, P_1. \tag{3.4.1}$$

The subscript 1 refers to conditions at the rear of the body. With the thickness of the body given the quantity ψ_1 is fixed, and a minimum to the drag is obtained by requiring the product $\cos\sigma_1\, P_1$ to be a maximum. The two factors in this product are of different types, as $\cos\sigma_1$ depends only upon the shape of the body at its base, while P_1 is an integral taken over the entire body. We now make the apparently naive assumption that the two factors are independent and that we may maximize the product by maximizing the factors separately. The maximum of $\cos\sigma_1$ is simply 1, and is obtained by setting $\sigma_1 = 0$. The maximum of P_1 will be obtained by using the calculus of variations. With $\cos\sigma_1 = 1$ the requirement that the thickness of the body be given is inessential, and our analytical treatment may be directly applied to problems in which the body thickness is not specified. However, then variations in the quantity ψ_1 appearing on the right side of (3.4.1) must be taken into account.

The step of setting $\cos \sigma_1 = 1$ is a critical one, which will have to be examined carefully later. With $\cos \sigma_1 = 1$ we may express the drag with the aid of (3.2.4) as

$$\tfrac{1}{2}\psi_1 C_{D_{\sigma_1=0}} = \int_0^{\psi_1} (1 - \cos \sigma)\, d\psi = 2 \int_0^{\psi_1} \sin^2 \frac{\sigma}{2}\, d\psi. \tag{3.4.2}$$

For comparison, the corresponding expression for the drag using the uncorrected expression for the pressure is

$$\tfrac{1}{2}\psi_1 C_{D_{\text{uncorr}}} = \int_0^{\psi_1} \sin^2 \sigma \, d\psi. \tag{3.4.3}$$

For a slender body the approximate result is immediate,

$$C_{D_{\sigma_1=0}} = \tfrac{1}{2} C_{D_{\text{uncorr}}}\,. \tag{3.4.4}$$

This result was obtained by von Kármán [1], who did not, however, discuss the limitations involved.

We now express P_1 in terms of the function $y(x)$ with $\cos \sigma = 1/\sqrt{1 + y'^2}$ as

$$P_1 = \rho_\infty U \int_0^{x_1} \frac{y' y^j \, dx}{\sqrt{1 + y'^2}}, \tag{3.4.5}$$

where x_1 is the length of the body. Two auxiliary integrals of interest in various isoperimetric conditions are the integral for the volume per unit depth or radian

$$(1 + j)V = \int_0^{x_1} y^{1+j}\, dx, \tag{3.4.6}$$

and the corresponding integral for the surface area

$$S = \int_0^{x_1} \sqrt{1 + y'^2}\, y^j \, dx. \tag{3.4.7}$$

With the Lagrangian multipliers λ_V and λ_S the problem posed for the calculus of variations (compare Eggers, Resnikoff, and Dennis [1]) is the maximization of the integral of F over x, where

$$F = \left[\frac{y'}{\sqrt{1 + y'^2}} + \lambda_V y + \lambda_S \sqrt{1 + y'^2}\right] y^j. \tag{3.4.8}$$

Since F is a function of y and y' alone and does not depend explicitly upon x, we may apply immediately the first integral of the Euler equation of the calculus of variations. This first integral is

$$F - y' \frac{\partial F}{\partial y'} = A, \tag{3.4.9}$$

where A is a constant. With the F in (3.4.8) this takes the form

$$\frac{y'^3}{(1 + y'^2)^{3/2}} + \lambda_V y + \frac{\lambda_S}{\sqrt{1 + y'^2}} = Ay^{-j}. \tag{3.4.10}$$

Equation (3.4.10) is a first-order differential equation in x and y whose solution gives us the desired shapes. Note that three undetermined constants appear in (3.4.10) and that a fourth must appear in the general solution of the differential equation. This freedom is necessary to satisfy the various boundary and isoperimetric conditions. Of course, in special cases (3.4.10) is simplified: If there is no specification on the volume, λ_V is zero; if there is no specification on the surface area, λ_S is zero; if there is no specification on the length of the body A is zero (compare Eggers, Resnikoff, and Dennis [1]). If there is no specification on the thickness of the body, the quantity ψ_1 of (3.4.1) is no longer constant. This can be taken into account by subtracting $y'y^j$ from F. The condition $F_{y'}(x_1) = 0$ then applies, and the boundary condition $y_1' = 0$ must be satisfied at the base of the body, except with $\lambda_S > 0$. Exceptions to normal behavior and other difficulties will appear in the most general case, and a complete study will require use of the parts of the theory of the calculus of variations dealing with wall conditions, discontinuities in slope, and second variations. See, for example, Courant and Hilbert [1].

The simpler solutions of (3.4.10) are given in Table 3–1. Note that with $A = 0$ and the body length unspecified the solutions are the same for two-dimensional as for axisymmetric flow. With λ_V and λ_S both zero in axisymmetric flow, the solution may be expressed parametrically as

$$x = \tfrac{3}{4}y_0[\xi(\xi^2 - \tfrac{1}{2})\sqrt{\xi^2 - 1} - \tfrac{1}{2}\cosh^{-1}\xi], \tag{3.4.11a}$$

$$y = y_0\xi^3, \tag{3.4.11b}$$

where the parameter ξ goes from 1 to ∞. In axisymmetric flow with body length specified the body generally has a flat portion on the nose, as do the corresponding solutions of Newton and of Eggers, Resnikoff, and Dennis. The radius of this flat portion in solution (3.4.11) is y_0. For ξ large (3.4.11) approaches the form

$$y = y_0\left(\frac{4x}{3y_0}\right)^{3/4}. \tag{3.4.12}$$

Eggers, Resnikoff, and Dennis obtain a solution analogous to (3.4.11) with an asymptotic solution of a form similar to (3.4.12). They note that their asymptotic solution is a very accurate approximation to their complete solution in the case of reasonably slender bodies. It can be shown that this is even more true with our solution (3.4.11). The fact that the two theories give the same asymptotic

TABLE 3–1

Solutions to the variational equation

A	j	λ_V	λ_S	Solution
0	0 or 1	$\frac{1}{y_m}$	0	$x = y_m - (y_m^{2/3} - y^{2/3})^{3/2}$
0	0 or 1	0	$-\frac{a^3}{1+a^2}$	$y = ax$
$\frac{b}{\sqrt{1+a^2}}$	0	0	$b - \frac{a^3}{1+a^2}$	$y = ax$
y_0	1	0	0	Equations (3.4.11)

shapes is a consequence of the relation (3.4.4). The solution (3.4.11) was obtained independently by Gonor and Chernyi [1]. This paper did not correctly interpret and apply the result. This deficiency was repaired in Chernyi [3], where additional calculations may be found.

The solutions we have obtained correspond to optimum bodies only if we may set $\cos\sigma_1 = 1$. In general, however, the shapes obtained have finite slope at the base, and the question naturally arises as to whether our solutions have meaning. Mathematically, there is no objection to a discontinuity in slope, and we may require that the slope be zero at the base proper although the slope remains finite to within an infinitesimal distance of the base. Physically, such a solution requires that the direction of the shock layer be changed quite suddenly, and this would require a negatively infinite pressure acting over an infinitesimal part of the body at the base. Of course, we must reject such a mechanism as being unrealistic even by the rather liberal standards of Newtonian flow theory.

Another solution is available, however, and consists of a mechanism by which the direction of the shock layer is changed by means of a positively infinite pressure acting over an infinitesimal part of the outside of the layer at the base. This leads to the concept of a "thrust cowl" which will turn the flow in the shock layer smoothly into the proper direction. In order that this be done without significant losses, it is necessary that the chord of the cowl be large compared with the thickness of the Newtonian layer. But since this thickness is infinitesimal, the chord of the thrust cowl is itself infinitesimal in the Newtonian limit. This Newtonian thrust cowl is depicted in Fig. 3–4.

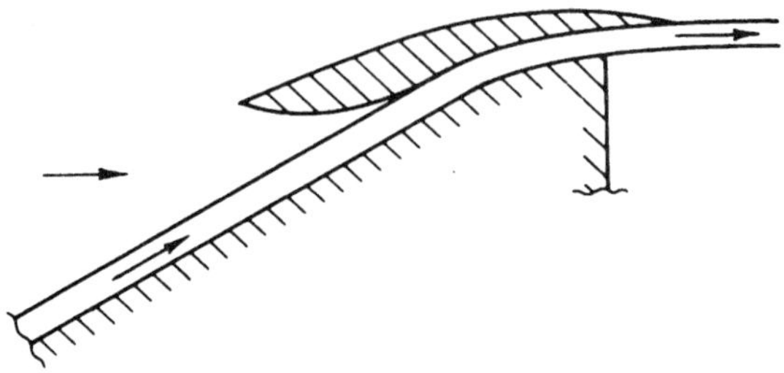

FIG. 3–4. Newtonian thrust cowl.

If the body in question has positive pressure everywhere the one thrust cowl at the base is sufficient. If the body has a region of negative pressure we must replace the body with a number of positive-pressure segments separated by corners fitted with thrust cowls. If in the Newtonian limit the number of these segments and cowls is increased without limit the body may be made to approach equivalence to a body with negative pressure. The realism of this picture is admittedly more dubious than that of a body with a single thrust cowl, but the picture is useful in permitting us to complete conceptually the class of bodies without detachment of the layer and with the condition $\sigma_1 = 0$ satisfied.

With the condition $\sigma_1 = 0$ satisfied and with thrust cowls provided as needed, then, we can find optimum shapes under a variety of isoperimetric conditions. We shall refer to these shapes as being "absolute optimum" shapes. In a real gas flow with a thin shock layer a thrust cowl may be used to decrease the net drag of a body. The advantage of such a cowl will be greater the thinner is the shock layer, and the closer we are to being able to utilize the Newtonian absolute optimum shapes.

We now turn to a different kind of optimum shape for minimum drag, for which we shall use the expression "proper optimum". For a proper optimum shape we set up different rules from those applicable to absolute optimum shapes. In particular, the possibility of using a thrust cowl is ruled out, and the possibility of having a discontinuity in the slope of the body is eliminated by the requirement that the pressure on the body be everywhere positive or zero. The pressure is itself expressed as an indefinite integral of the shape and, to the writers' knowledge, the classical calculus of variations does not provide a method for this type of isoperimetric condition. We shall use an attack based upon intuition, with the help of general results from the calculus of variations. We shall concentrate on the simplest problem, that of a two-dimensional body with given chord length and thickness; we suggest as an interesting problem the extension of the method given here to other isoperimetric conditions and to general axisymmetric shapes. We shall, however, include the axisymmetric case for a slender body of given chord and thickness.

We first note that our nonnegative pressure condition is violated for the absolute optimum shape only by the negatively infinite pressure at the base. Personifying the optimum body shape, we note that it is doing its best to have very low pressure far aft. Within the rules of the proper optimum, the best that the optimum shape can do in this regard is to have zero pressure over a finite section adjoining the base of the body. This suggests that a proper optimum body should consist of a forebody bearing positive pressures and an afterbody which is quite arbitrary (see, for example, Fig. 3–6). The afterbody lies within a free layer which separates from the forebody and which returns to the body at the base. The free layer must pass by the base unobstructed into the region behind the body.

If the shape of the free layer were predetermined we could draw its trace on a plot of tan σ versus x. Minimum drag would still be obtained by maximizing P at the end of the forebody and we should solve a problem in the calculus of variations with the end point not fixed but required to lie on this trace. A basic result of the theory of the calculus of variations is that such a boundary condition does not affect the basic differential equations obtained; in our case the same Euler equations and solutions would therefore apply for the forebody as the equations and solutions appropriate to the absolute optimum problem. The shape of the free layer is not predeterminable, but we guess that the forebody should have an absolute optimum shape anyway. We shall not attempt a complete proof that this guess is justified but shall outline a partial proof for the two-dimensional body of given chord and thickness. This is the simplest problem, as the absolute optimum shape is simply a straight line in this case. After presenting this justification we shall present examples.

Gonor [4] showed that a proper optimum body must have a finite free layer section adjoining the base. For the narrow class of variations ($\delta r_c' = 0$ in his notation) corresponding to fixed forebody slope at the free layer separation point, he proves that the forebody shape is an absolute optimum shape. Kraiko [1] extended the results of Gonor to include isoperimetric conditions. For the mathematically simpler case of a slender body, Miele [1] gave a complete proof.

Our method will consist of assuming the contrary and showing that a contradiction results. Our comparison shapes will, however, be limited to those consisting of a forebody followed by a free layer; hence our proof is only a partial one. We need certain properties of the free layer solution. The derivation of these properties is straightforward and will not be given here. We are interested in the change in the lateral coordinate y of a two-dimensional free layer in traversing a given distance in the axial direction. The properties we shall need are that the change in the lateral coordinate varies monotonically with the initial impulse with given initial angle, and that if the rear part of the forebody is shortened so that the free layer starts earlier, the value of y at the base must decrease monotonically with the amount of the body cut off.

We now suppose that the proper optimum shape has a forebody which is not the absolute optimum shape (a straight line in this case). The free layer shape is continued upstream from the separation point, and a straight line is drawn from the vertex tangent to the free layer (see Figs. 3–5). A shape for which this point of tangency lies aft of the base clearly has greater drag than does a simple wedge and is not considered. Also, it may be shown that the vertex must lie on the convex or upstream side of the free layer and that the point of tangency must exist. Two cases now exist, that for which the point of tangency lies in the part of the free layer downstream of the separation point (Fig. 3–5a), and that for which the point of tangency lies on the extension of the free layer upstream of the separation point (Fig. 3–5b). We consider the two cases separately.

In the first case we consider comparison body A to consist of the straight line to the point of tangency (Fig. 3–5a). Since this body has maximum impulse between the vertex and the tangency point and the initial angles are the same, the free layer from A will overshoot the base. Now we consider the body to be shortened to give comparison body B, with the property that the free layer just reaches the base. The drag of the original body is clearly greater than that of body A, and the drag of body A is greater than that of body B. Since body B satisfies the conditions required of a proper optimum body, the original body could not have been the proper optimum body as was assumed.

In the second case we consider comparison body A to consist of the straight line extended beyond the tangency point 3 to its intersection with the original body 4, followed by a thrust cowl at this intersection point to direct the layer along the original body, followed by the remaining part of the original body to the separation point 2. The numbers here refer to points labelled on Fig. 3–5b. Comparison body B consists simply of the straight line to the point of tangency, while B' is the same plus the free layer section between the point of tangency 3 and the separation point 2. The impulse at the intersection point 4 for body A is greater than that for the original body, because between the vertex and point 4 body A has an absolute optimum shape. The contribution of the segment 4–2 to the impulse is the same for both bodies, and the impulse at point 2 is greater for body A. We may write this result as

$$P_{A2} > P_{O2} \,. \tag{3.4.13}$$

We must now compare the increments in impulse developed along the extended free layer segment 3–2 and the corresponding part of body A, 3–4–2. We first must note that minus the Euler differential expression is the kernel of the first variation integral in the theory of the calculus of variations (see Courant and Hilbert [1]), equal to $3y'y''/(1+y'^2)^{5/2}$ in this case. This kernel has the same sign as does the curvature $y''/(1+y'^2)^{3/2}$ because y' is positive, and is negative for a concave shape (corresponding to a convex body) such as that for a free

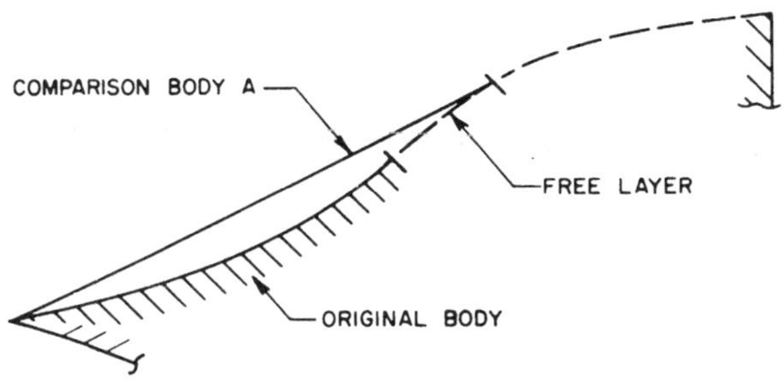

FIG. 3–5a. Proper optimization—first case.

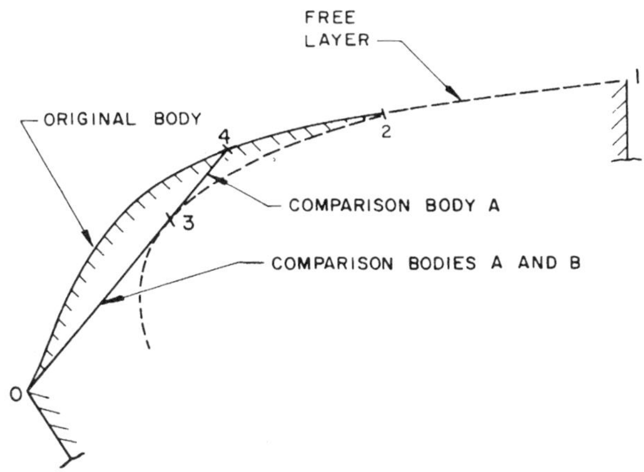

FIG. 3–5b. Proper optimization—second case.

layer. An infinitesimal increase of y over a concave section of a shape decreases the impulse P developed over that section. It is possible to construct a sequence of nested shapes each of which is concave where it does not coincide with the segment 3–4–2, and which provides a continuous transition between the free layer segment 3–2 and the segment 3–4–2. The impulse decreases monotonically on this sequence of shapes, and we may conclude that the impulse at point 2 on body B' is greater than for body A, or

$$P_{B3} + P_{3-2} > P_{A2}\,. \tag{3.4.14}$$

From (3.4.13) and (3.4.14) we obtain immediately

$$P_{B3} > P_{O2} - P_{3-2}\,. \tag{3.4.15}$$

But the quantity $P_{O2} - P_{3-2}$ is the impulse which the original free layer extended upstream would have at point 3, and the drag of the original body may be calculated from (3.4.1) at point 3, using this impulse. Hence the drag of body B is less than that of the original body. Also, from (3.4.15) the free layer which will come from body B separating at point 3 will overshoot the base. We consider body B shortened to give comparison body C with the property that the free layer from C just reaches the base. Since body C has less drag than does body B and satisfies the conditions required of a proper optimum body, our assumption that the original body was the proper optimum body must have been false.

Recognizing that our proof is a partial one and that a neater proof of the result (3.4.14) would be desirable, let us accept the result that the forebody must have an absolute optimum shape. We are left with a one-parameter family of shapes satisfying the specifications and need only to minimize the drag with respect to the parameter. Let a be the slope of the forebody and sc its chord, where c is the chord of the entire body. Let τ be the thickness ratio of the entire half-body on one side of the vertex (see Fig. 3–6). The change in y in the free

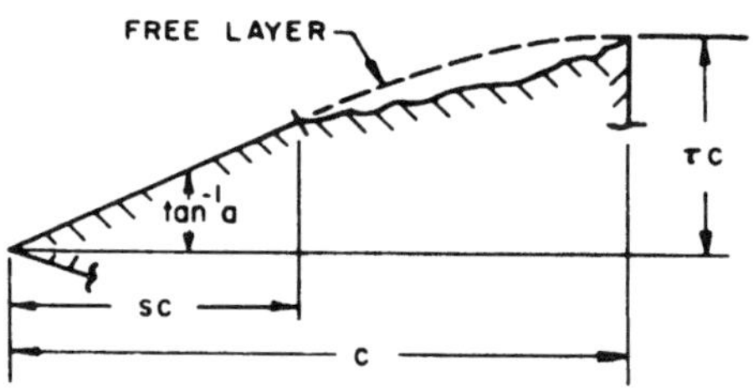

FIG. 3–6. Two-dimensional proper optimum shape.

layer is thus $(\tau - as)c$, and the change in x is $(1 - s)c$. Putting in the condition which comes from a determination of the initial impulse in the free layer, we may write the condition that the free layer just reach the base in the two-dimensional case as

$$\tfrac{1}{2}(\tau - as)^2 + \frac{as(\tau - as)}{1 + a^2} = \frac{a^2 s(1 - s)}{1 + a^2}\,. \tag{3.4.16}$$

This equation can be obtained from (3.3.33), with $j = 0$ and with a factor c^2 divided out.

The drag of the body equals the drag of the forebody, and can be shown to be

$$\tfrac{1}{2}\tau C_D = as\left(1 - \frac{1}{1 + a^2}\right) = \frac{a^3 s}{1 + a^2}\,. \tag{3.4.17}$$

The quantity s may be eliminated from (3.4.16) by using (3.4.17), and we obtain

$$\tau^2\left[1 - C_D + \frac{(1 + a^2)^2}{4a^4} C_D^2\right] - \frac{\tau C_D}{a} = 0. \tag{3.4.18}$$

The quantity τ is kept constant, and an extremum of the drag is obtained by setting the derivative of (3.4.18) with respect to a equal to zero. This procedure yields

$$\frac{\tau C_D}{a^2}\left[1 - \frac{1 + a^2}{a^3} \tau C_D\right] = 0. \tag{3.4.19}$$

That a minimum to the drag is obtained by setting the bracket in (3.4.19) equal to zero is readily verified, and a comparison with (3.4.17) tells us that

$$s = \tfrac{1}{2} \tag{3.4.20}$$

for minimum drag. A peculiar simplicity of this result lies in the fact that it is independent of thickness ratio, and no simple reason for this is apparent to the authors.

Results for the drag in terms of thickness ratio are best expressed parametrically, for example, in terms of a using (3.4.17) and (3.4.18). We may avoid the solution of a quadratic in a parametric representation by introducing a new parameter κ, which varies from $\frac{1}{2}$ for very thick bodies to $\frac{1}{2}\sqrt{3}$ for very thin bodies. In terms of κ the results may be expressed

$$\tau = a\kappa, \tag{3.4.21a}$$

$$a = \frac{\sqrt{3 - 4\kappa^2}}{2\kappa - 1}, \tag{3.4.21b}$$

$$\tfrac{1}{2}C_D = \frac{3 - 4\kappa^2}{8\kappa(1 - \kappa)}, \tag{3.4.21c}$$

$$\frac{1}{2\tau^2} C_D = \frac{(2\kappa - 1)^2}{8\kappa^3(1 - \kappa)}. \tag{3.4.21d}$$

For slender bodies the results are simpler. Besides the basic result of (3.4.20) we have, setting $\kappa = \frac{1}{2}\sqrt{3}$ in (3.4.21a and d),

$$a = \frac{2}{\sqrt{3}}\tau = 1.155\tau, \tag{3.4.22a}$$

$$\frac{1}{2\tau^2} C_D = \frac{4\sqrt{3}}{9} = 0.770. \tag{3.4.22b}$$

We shall compare this drag result with others after looking at the axisymmetric case.

The methods and basic approach presented are clearly applicable to two-dimensional problems with different isoperimetric conditions and to corresponding axisymmetric problems. It is clear, however, that both the justification arguments and the algebra of computing the optima will be even more involved and detailed. Moreover, we would have to go into such questions as whether the free layer surface area should be included in an isoperimetric condition on surface area. We shall limit ourselves here with the introduction of the concept of the proper optimum, and with the relatively simple illustrative examples presented.

However, we shall give the axisymmetric solution corresponding to the two-dimensional one above, but only for the slender-body case. The quantity s is defined as before, and a is defined by the equation giving the shape of the forebody

$$y = \tfrac{4}{3}asc\left(\frac{x}{sc}\right)^{3/4}, \tag{3.4.23}$$

which is equivalent to (3.4.12). The values of y and y' at the separation point 2 with $x = sc$ are given by

$$y_2 = \tfrac{4}{3}asc, \tag{3.4.24}$$

$$y_2' = a. \tag{3.4.25}$$

A new parameter β is introduced as the ratio between the thickness at the base and y_2, or

$$\tau c = \beta y_2, \tag{3.4.26}$$

from which we obtain

$$\tau = \tfrac{4}{3}a\beta s. \tag{3.4.27}$$

The drag of the forebody is given by the approximate expression

$$\tfrac{1}{2}(\tfrac{1}{2}\tau^2c^2)C_D = \int_0^{y_2} (1 - \cos\sigma)y\,dy + \tfrac{1}{2}(1 - \cos\sigma_2)y_2^2, \tag{3.4.28}$$

from which we may calculate

$$\tfrac{1}{2}\tau^2C_D = \tfrac{20}{9}a^4s^2. \tag{3.4.29}$$

Within the body of approximations which we are using, the term in y_0 in (3.3.33) does not appear, and the free layer shape may be expressed as

$$\tfrac{1}{3}y^3 = \tfrac{16}{9}a^3s^2c^2(x - x_0), \tag{3.4.30}$$

where x_0 is a constant. The coefficient on the right-hand side has been evaluated by matching the impulse from the forebody with the impulse in the free layer. The condition that the free layer just reach the base is thus

$$\frac{1}{3}\left(\frac{4}{3}\right)^3 a^3 s^3 c^3(\beta^3 - 1) = \frac{16}{9} a^3 s^2 (1 - s) c^3. \tag{3.4.31}$$

We then solve (3.4.31) for s and obtain

$$s = \frac{9}{4\beta^3 + 5}. \tag{3.4.32}$$

Next we divide (3.4.29) by τ^4 from (3.4.27), eliminate s using (3.4.32), and obtain another expression for the drag

$$\frac{1}{2\tau^2} C_D = \frac{5}{4^3 3^2}\left(\frac{4\beta^3 + 5}{\beta^2}\right)^2. \tag{3.4.33}$$

The drag expressed in (3.4.33) has a minimum at

$$\beta = \left(\tfrac{5}{2}\right)^{1/3} = 1.3572, \tag{3.4.34}$$

with a minimum value of

$$\frac{1}{2\tau^2} C_D = \frac{5}{16}\beta^2 = 0.5756. \tag{3.4.35}$$

The value of s at this minimum is

$$s = \tfrac{3}{5}. \tag{3.4.36}$$

Gonor [4] carried out the analogous calculation without the restriction to slender bodies, and gave numerical results for s and C_D. For the range $0 \leqslant \tau \leqslant 0.65$ he found that $0.6 \leqslant s < 0.7$.

The numerical results for two-dimensional and axisymmetric bodies are presented in Table 3–2. The values of the absolute optimum drag which are used as reference in the last column of Table 3–2 are

$$\frac{1}{2\tau^2} C_{D_{\text{abs. opt.}}} = \frac{1}{2} \tag{3.4.37a}$$

in the two-dimensional case, and

$$\frac{1}{2\tau^2} C_{D_{\text{abs. opt.}}} = \frac{27}{64} \tag{3.4.37b}$$

in the axisymmetric case. These values are readily obtained from (3.4.2) or (3.4.5) with (3.4.1).

TABLE 3–2

Comparative drags of optimum slender bodies

Shape	$\frac{1}{2\tau^2} C_D$	$\frac{C_D}{C_{D_{\text{abs. opt.}}}}$
Two-dimensional:		
Wedge	1.000	2.000
0.864 Power (Cole [1])	0.918	1.836
Proper Optimum	0.770	1.540
Absolute Optimum (wedge)	0.500	1.000
Axisymmetric:		
Cone	1.000	2.370
3/4 Power, uncorrected	0.844	2.000
3/4 Power	0.703	1.667
2/3 Power (Cole [1])	0.667	1.580
Proper Optimum	0.576	1.364
Absolute Optimum (3/4 power)	0.422	1.000

For the slender-body case, an alternative approach is available, which we shall briefly outline without considering isoperimetric conditions. For slender bodies the basic Newton-Busemann pressure law (3.2.7) may be rewritten (cf. Cole [1])

$$\tfrac{1}{2}C_{p_b} = y'^2 + \frac{1}{1+j}\, yy''. \tag{3.4.38}$$

The drag may be represented by the integral

$$\tfrac{1}{2}\psi_1 C_D = \rho_\infty U \int_0^1 F(y, y', y'')\, dx, \tag{3.4.39}$$

where

$$F = y^j y'^3 + \frac{1}{1+j}\, y^{1+j} y' y''. \tag{3.4.40}$$

We may minimize (3.4.39) using the usual technique, with the fact that the first integral to the Euler equation for (3.4.39) is

$$y'' F_{y''} + y' F_{y'} - F - y' \frac{d}{dx}(F_{y''}) = A, \tag{3.4.41}$$

where A is a constant. Putting F from (3.4.40) into (3.4.41) we obtain

$$y'^3 = A y^{-j}, \tag{3.4.42}$$

which we recognize as equivalent to (3.4.10) with the Lagrange multipliers zero. But while discontinuities in slope were permitted in our absolute optimum analysis, they are not permitted here, as the pressure given by (3.4.38) must remain finite.

Miele [1] has followed this scheme with success, formulating the problem as a Mayer problem and including isoperimetric conditions. The inequality $p \geqslant 0$ is included by making p proportional to the square of an additional real independent variable. The Lagrange multiplier corresponding to the defining equation for this variable is a switching function for the problem, and is zero on the body at nonzero pressure. He shows that, with the switching function zero at the free layer separation point, it cannot be zero at any point downstream. The conclusion is that the shape can have at most two subarcs, one for the forebody and one for the free layer. Thus for slender bodies Miele substantiates the results for the proper optimum given above.

We now ask what these results can teach us as to what optimum bodies should be like in real gas flows. The process of optimization is one by which momentum losses are minimized. These losses are of two kinds, the loss in momentum of the flow in entering the shock layer, and the loss in the axial component of the momentum from inclination of the momentum leaving the body. The thrust cowl of an absolute optimum body has as its aim the elimination of the second loss.

In a real gas flow with a shock layer of finite thickness a penalty in pressure and frictional drag must be paid for a thrust cowl. In addition, there will be practical difficulties in structural support and aerodynamic heating which may eliminate the thrust cowl for the designer. The practical realization of a design based upon the proper optimum concept, however, appears feasible especially when we realize that frictional resistance should be very low on the after section of the body corresponding to the free layer section.

We first note that our absolute optimum shape which appears on the proper optimum forebody (without isoperimetric conditions) is essentially identical to the shape which appears as optimum in the theory of linearized supersonic flow (for the axisymmetric case, see von Kármán [1]). We may guess, then, that an uncowled optimum shape should have a forebody with our absolute optimum shape, a wedge in the two-dimensional case and a $\frac{3}{4}$ power shape body in the axisymmetric case. This forebody should be followed by an afterbody designed to have low pressures on it. The chord ratio s should be intermediate between our Newtonian value ($\frac{1}{2}$ or $\frac{3}{5}$) and one, and should be closer to the former if the shock layer is very thin and closer to the latter if the shock layer is thicker. In any case, with the shock layer very thin, the optimum shape will surely not be realized with a simple power-law shape.

These conclusions cannot be expected to apply unless ϵ is small. Shmyglevskii [1; 2] has made numerical optimum shape calculations for a two-dimensional body with a perfect gas of $\gamma = 1.4$ and low hypersonic Mach numbers. His optimum shapes are close to straight lines, and show no discernible vestige of the proper optimum shapes.

We have concerned ourselves so far in this section exclusively with the problem of minimizing the drag of bodies for which no specification is made as to lift; the axisymmetric bodies are at zero incidence and provide no lift anyway. Let us now look at a simple optimization problem involving lift in two-dimensional Newtonian flow. The problem of maximizing only the ratio of lift to drag is not a properly posed problem in inviscid flow, as this ratio is simply the inverse of the angle of attack for a flat plate and may be made as large as we please. We shall consider a body of given chord c with the lift specified but with no other isoperimetric conditions, and shall minimize the drag. Following our absolute optimum concept we shall not require that the shock angle σ be continuous at the trailing edge.

We use (3.2.8) or (3.4.1) for the drag and (3.2.9) for the lift, and first consider that the shock angle at the trailing edge σ_1 and the quantity ψ_1 be fixed. We note that an increase in P_1 would increase the lift and decrease the drag. An accompanying decrease in σ_1 to return the lift to its prespecified value would further decrease the drag. Hence any optimum shape must be one for which P_1 is maximum, or one of the absolute optimum shapes we have considered. In the absence of other isoperimetric conditions this shape is a straight line, and we conclude that our body must be a flat plate with an angle of attack which we denote α. Of course, the quantity α is not necessarily the same as σ_1.

Defining c as the actual width of the flat plate we have

$$\psi_1 = \rho_\infty U c \sin\alpha \tag{3.4.43}$$

and

$$P_1 = \psi_1 \cos\alpha = \tfrac{1}{2}\rho_\infty U c \sin 2\alpha. \tag{3.4.44}$$

The lift and drag coefficients for the body based on c are simply $\sin\alpha$ times the corresponding coefficients based on frontal area. The lift coefficient based on c is thus

$$C_L = \sin 2\alpha \sin\sigma_1\,, \tag{3.4.45}$$

while the corresponding drag coefficient is

$$C_D = 2\sin\alpha - \sin 2\alpha\cos\sigma_1\,. \tag{3.4.46}$$

We next require that both C_L and C_D be stationary under a variation in α and σ_1, or that the Jacobian determinant vanish:

$$\begin{vmatrix} 2\cos 2\alpha\sin\sigma_1 & \sin 2\alpha\cos\sigma_1 \\ 2\cos\alpha - 2\cos 2\alpha\cos\sigma_1 & \sin 2\alpha\sin\sigma_1 \end{vmatrix} = 0. \tag{3.4.47}$$

We obtain immediately the condition

$$\cos \sigma_1 = \frac{\cos 2\alpha}{\cos \alpha}. \tag{3.4.48}$$

Putting condition (3.4.48) back into (3.4.45) and (3.4.46) yields for our optimum conditions

$$C_L = 2 \sin^2 \alpha \sqrt{3 - 4 \sin^2 \alpha} \tag{3.4.49}$$

and

$$C_D = 4 \sin^3 \alpha. \tag{3.4.50}$$

Combining (3.4.49) and (3.4.50) gives the ratio of lift to drag

$$\frac{C_L}{C_D} = \frac{\sqrt{3 - 4 \sin^2 \alpha}}{2 \sin \alpha}. \tag{3.4.51}$$

These results may also be obtained by eliminating σ_1 between (3.4.45) and (3.4.46) and minimizing C_D with C_L fixed.

The analysis above is only valid for α between 0 and $\pi/4$. Within this range σ_1 from (3.4.48) is always greater than α. Since the Newtonian shock layer must be deflected from its inclination angle α on the flat plate to the inclination angle σ_1, the flat plate must be fitted with a narrow strip which we term a "Newtonian chine strip". Such a strip is analogous to a Newtonian thrust cowl. The chine strip has a chord of larger order of magnitude than the thickness of the shock layer but which is still infinitesimal in the Newtonian limit (see Fig. 3–7).

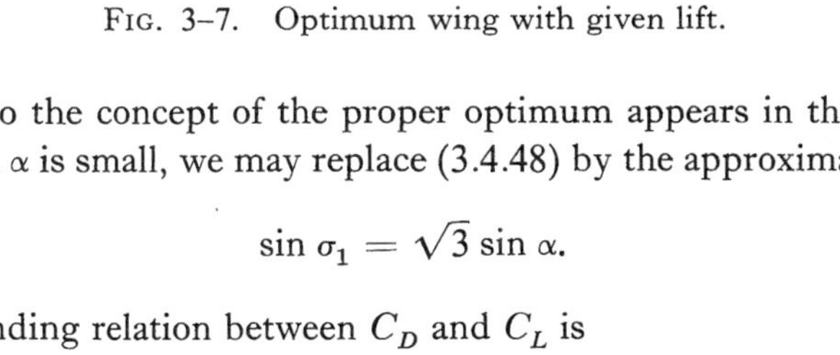

FIG. 3–7. Optimum wing with given lift.

No analogue to the concept of the proper optimum appears in this case. If the angle of attack α is small, we may replace (3.4.48) by the approximate expression

$$\sin \sigma_1 = \sqrt{3} \sin \alpha. \tag{3.4.52}$$

The corresponding relation between C_D and C_L is

$$C_D = \sqrt{2}(3)^{-3/4} C_L^{3/2} = 0.620 C_L^{3/2}. \tag{3.4.53}$$

The coefficient in this expression was minimized in our optimum procedure; for a flat plate the coefficient has the value $\frac{1}{2}\sqrt{2}$ or 0.707. We may note, incidentally, that a wing with this geometry flying in trim (i.e. with balanced moments) under these optimum loading conditions would be statically unstable.

A simpler problem related to the above is the following: Given the average angle of attack $\bar{\alpha}$ of a profile, what is σ_1 and the distribution of α for maximum L/D? As before, $\alpha = \bar{\alpha}$ and the profile is a flat plate. An easy calculation yields $\sigma_1 = \bar{\alpha}$ also, so that no chine strip is present.

The Newtonian chine strip appears in another problem, that of a Newtonian body of maximum drag. For a body of maximum drag P must again be a maximum, and the chine strip must turn the shock layer so that it shoots almost directly forward, with $\cos \sigma_1$ in (3.4.1) equal to minus one. The effect is similar to that which is the basis for the Pelton wheel. For a slender body the maximum drag is given approximately by

$$\tfrac{1}{2}C_{D_{\max}} = 2, \tag{3.4.54}$$

with the drag coefficient based on frontal area.

The process discussed in Section 3.3 in which a constant-pressure layer is formed which just grazes the rear edge of the body would destroy the results we have obtained requiring Newtonian chine strips. It appears plausible, though, that with ϵ low enough and the Reynolds number large enough it might be possible to obtain a practical realization of the deflection process envisioned in the concept of the Newtonian chine strip.

In looking back over the results of this section, we should note particularly that they all depend upon the maximization of the impulse in the shock layer. In fact we can look on this maximization of the impulse as a central optimization principle. The primary aerodynamic forces, lift and drag, are expressed directly in terms of the impulse in the shock layer and its direction as the layer leaves the body. Accepting the principle that we have full control over this direction, we have the greatest control over lift and drag if the impulse is at a maximum.

The problem of determining optimum body shapes is an extremely important one in practical aerodynamics. It is appropriate, then, to comment on the applicability of the results and methods of this section to practical hypersonic aerodynamics in a fluid in which we may expect reasonably thin shock layers. We may expect that hypersonic flows as met with in actual flight will occur primarily at extremely high altitudes and at low densities, and that we shall always be faced with strong viscous effects. This means that the frictional drag must be taken into account in the optimization procedure, and should have an importance roughly equal to that of the inviscid hypersonic drag. Thus the detailed results of this section are probably not applicable. However, much of the philosophy of this section and the methods developed should be applicable.

And with certain simple assumptions as to the dependence of the frictional drag on the parameters of the problem, certain parts of our results should be applicable. In particular, this should be the case if the frictional drag is a function only of those quantities which are held fixed in an optimization.

5. Shock layer structure and cross flow phenomena

In the analysis we have made of Newtonian flow on two-dimensional and axisymmetric bodies we have not had to concern ourselves with the structure of the shock layer. However, in analyzing Newtonian flow on general bodies in which the flow in the shock layer is not unidirectional we must analyze this structure. Before turning to the general problem let us first look briefly at the shock layer structure for certain two-dimensional and axisymmetric problems. The basic formula (3.2.10) needed has already been given.

In our investigation of shock layer structure we shall consistently use a notational convention to distinguish between the point at which a particle entered the shock layer and the point at which it is being observed. The subscript 1 will appear on quantities to designate their value at the point at which the structure is being investigated. Also, we shall use the subscript 1 to refer to a quantity at the top of the shock layer. These two usages are consistent because it is for a particle at the top of the shock layer that the point of entry and the point of investigation coincide. The lack of subscript 1 will generally refer to a particle within the shock layer and will generally designate the value of a quantity at the point of entry of that particle.

We look first at the case of the sphere for which the solution is given by (3.3.10). Letting R_s be the radius of the sphere, we express first the relation between ψ and σ

$$\psi = \tfrac{1}{2}\rho_\infty U R_s^2 \cos^2 \sigma. \tag{3.5.1}$$

The quantity m of (3.2.10) may be calculated as an indefinite integral for a point designated by the subscript 1

$$m = \frac{1}{\rho_\infty U y_1} \int_0 \frac{d\psi}{\cos \sigma} = R_s \frac{\cos \sigma}{\cos \sigma_1}. \tag{3.5.2}$$

The quantity m at $\psi = \psi_1$ is given for this case,

$$m_1 = R_s \tag{3.5.3}$$

to be independent of position. The quantity m_1, which always has the dimensions of distance, equals the mass per unit area in the shock layer divided by the free stream density ρ_∞. The variable m is analogous to the Howarth-Dorodnitsyn variable $\int_0 \rho \, dy$ which is used in boundary layer theory, and will be used in the remainder of this chapter as a basic independent variable.

Our reasons for using m as a basic variable are simple. We need a clearly defined scale normal to the shock layer with which to identify individual laminae of our laminar layer model. The intuitive distance scale is unavailable because the Newtonian shock layer is infinitesimally thin. But the variable m is clearly defined within the laminar layer model and is available to serve as the desired scale. If the Howarth-Dorodnitsyn variable is used as a scale in a non-Newtonian hypersonic flow, it will approach $\rho_\infty m$ in the Newtonian limit. The quantity m is the only variable which is suitable in the general Newtonian problem with cross flow present.

We denote the velocity in the shock layer by Uq; the quantity q is thus a dimensionless velocity. We may express the distribution of q in the layer for the case of the sphere either in terms of ψ or of m as

$$q = \cos \sigma = \frac{\sqrt{2\psi/\rho_\infty U}}{R_s} = \frac{m \cos \sigma_1}{R_s} \tag{3.5.4}$$

or

$$\frac{q}{q_s} = \frac{\cos \sigma}{\cos \sigma_1} = \sqrt{\frac{\psi}{\psi_1}} = \frac{m}{m_1}. \tag{3.5.5}$$

We notice that the distribution of q in terms of m is triangular. The velocity at the bottom of the layer, with $m = 0$ there, is zero for this case according to Newtonian theory.

We turn now to the case of the circular cylinder, with

$$\psi = \rho_\infty U R_s \cos \sigma, \tag{3.5.6}$$

and

$$q = \frac{\psi}{\rho_\infty U R_s}. \tag{3.5.7}$$

In this case the integral (3.2.10) for m does not converge and we cannot express q in terms of m. The mass per unit area is infinite everywhere on the body in this case. This anomaly will be resolved later when we obtain improved solutions for blunt two-dimensional bodies. We shall use m as the basic independent variable despite the anomaly which appears in this case; no solution exists for the structure of the Newtonian shock layer on a blunt two-dimensional body.

We shall look at one other simple shape, the two-dimensional concave quadrant of a circle, corresponding to (3.3.16) with $n = 2$. The quantity m is given by

$$m = -R_s \ln \cos \sigma, \tag{3.5.8}$$

and the velocity profile may be expressed

$$q = 1 - \frac{\psi}{\rho_\infty U R_s} = \exp\left(-\frac{m}{R_s}\right). \tag{3.5.9}$$

In this case the velocity is greatest at the bottom of the layer.

As we have indicated, the variable m serves as a scale which may be used to identify the laminae which make up the shock layer. The one-to-one correspondence which we shall require of this layer is between points on this scale of m and points on the "locus of entering streamlines" (see Section 3.2). We shall require that this mapping be continuous along any connected linear segment of the locus of entering streamlines. We also require that this mapping be continuous under a continuous change in the point at which the structure is being investigated. This requirement is illustrated in two examples given below. The uppermost value of m must correspond to the streamline which enters the layer at the point investigated. If these requirements are not met we must either abandon the laminar layer model or must accept new and usually anomalous phenomena in the flow field. The latter choice appears physically more sound.

To begin our discussion of these concepts we shall consider a two-dimensional problem, termed the "roller coaster problem" (see Figs. 3–8). In this problem a developed Newtonian shock layer lies on a body for which the angle θ_b increases to a value greater than $\pi/2$ and then decreases. The value of $\cos\theta_b$ thus decreases and becomes negative, then becomes positive again.

With the laminar layer model a basic difficulty appears. Considering the shock layer structure at a point at the bottom of the trough for which $\cos\theta_b = 0$, the locus of entering streamlines extends from the point in both directions. A continuous one-to-one correspondence between points in the layer and entering streamlines is impossible at this point as well as for other points (see Fig. 3–8a). Here in a simple example, then, we have a failure of the laminar layer model and may expect to find a new phenomenon which we must accept in order to resolve the problem. This phenomenon is a "pool" of infinitely high-density fluid at zero velocity and at the pressure $\rho_\infty U^2$. The resolution of the problem with the laminar layer model is shown in Fig. 3–8b. Newtonian anomalies in this solution are an infinite mass of fluid in the pool and zero velocity of the flow coming out of the pool.

Referring to Fig. 3–8a, we note that a general point has a locus of entering streamlines which has two branches, one connected to the point being investigated and the other disjoined from it. The point at the bottom of the trough is a singular point, for which the two branches of the locus of entering streamlines are connected, with the point itself the branch point. The configuration or topology of the two branches is different for a point on one side of the singular point from that for a point on the other side. And in the final solution with the pool, a point on the body above the pool is "shielded" by the pool from its second branch of entering streamlines on the other side of the trough. This shielding process permits us to keep the laminar layer model for points above the pool. These features all have their analogues in more general flows.

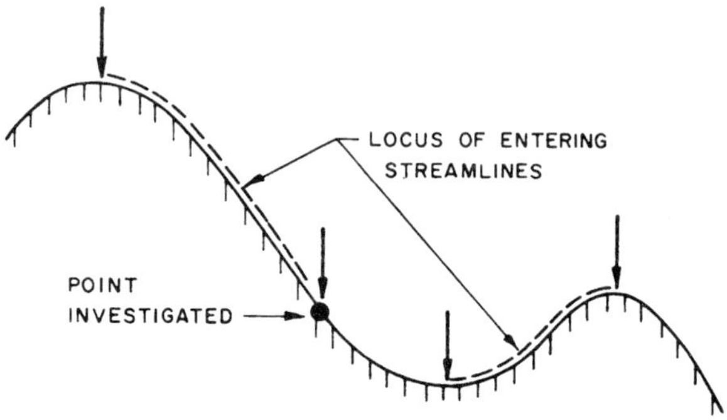

FIG. 3–8a. Roller coaster problem—locus of entering streamlines.

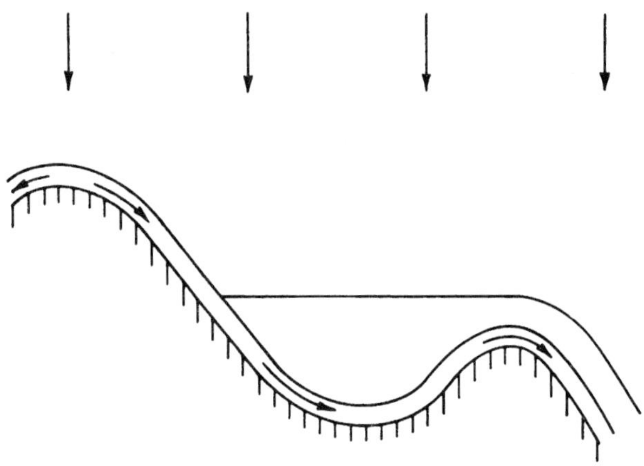

FIG. 3–8b. Roller coaster problem—laminar layer model.

On a general body in steady flow we can draw lines for which the axial coordinate is constant. With the flow axis aligned vertically these lines are contour lines. The orthogonal trajectories of these contour lines on the body surface are the paths of steepest descent, and we shall refer to the direction of such a path at any point as the "fall line" direction. This fall line direction is determined at any point by the intersection of the body with a plane through the external streamline going into the point and the normal to the body surface at the point.

A particle striking a surface which loses all of the normal component of its momentum without change in the tangential components assumes the fall line direction. Within our laminar layer model this is what happens to a particle in Newtonian flow. The particle subsequently experiences no forces except for those forces normal to the surface required to make the particle follow the surface. Under these conditions the particle follows a path on the surface which has zero lateral curvature. Such a path is termed a geodesic, and the reader should consult a text on differential geometry if he wishes a detailed treatment of geodesics and their properties. Here we only note, for example, that a geodesic on a sphere is a great circle, and that a geodesic on a developable surface such as a general cone is a straight line on the developed surface. The observation that the trajectories are geodesics was made by Grimminger, Williams, and Young [1].

The entering streamlines form a two-parameter family, as do the subsequent geodesic trajectories on the surface. The trajectories which pass over a given point on the surface form a one-parameter family, and the locus of points at which the trajectories start is our locus of entering streamlines (see Fig. 3–9).

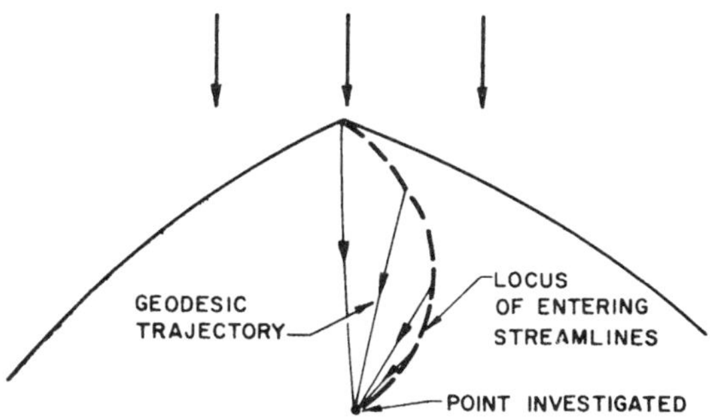

FIG. 3–9. Locus of entering streamlines.

The locus begins at a stagnation point or at a leading point or edge of the body and ends at the point in question.

We now turn to another problem in which the assumption of a continuous one-to-one correspondence is violated, which we term the "drain trough problem". The problem considered is that of a surface which is inclined to the free stream and whose principal characteristic is that it is laterally concave (see Figs. 3–10). At a sufficient distance from the leading edge there will be singular points whose loci of entering streamlines are branched, such as point A in Fig. 3–10b. These singular points themselves form a locus on the body. The locus of entering streamlines for a neighboring point, such as point B in Fig. 3–10b, consists of two distinct branches. A one-to-one correspondence established between m and entering streamlines for an ordinary point cannot be continuous as the point is made to cross the locus of singular points, and our basic requirement is violated.

Physically, we can say that the difficulty arises because material arrives at a point from various points on the surface in such a way that a continuous solution would be possible only if two streams were allowed to occupy the same space, only if a lamina is allowed to contain material from two sources. The alternative is that a collision or shock process ensues. Within the limiting perfect

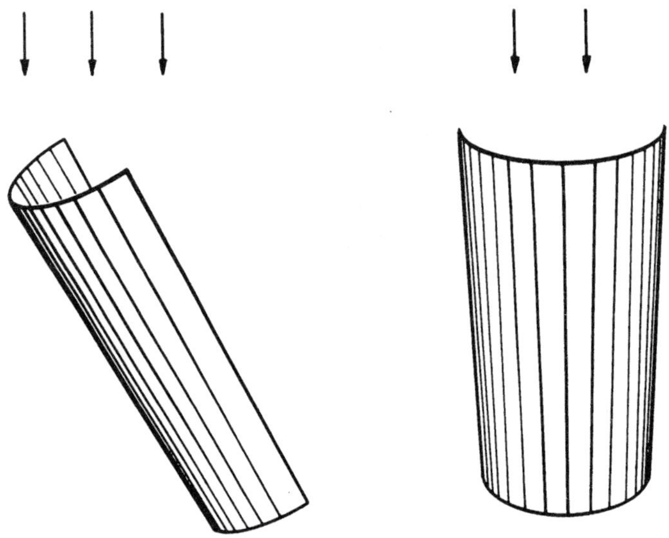

FIG. 3–10a. Drain trough problem—sketch.

gas model the flow in the shock layer is still hypersonic and at infinite Mach number, and with a second shock the Newtonian process is repeated and a second condensation takes place. The original Newtonian process reduced a spatial distribution of mass, mass flow, and momentum to a surface distribution. The second Newtonian process we are here concerned with reduces a surface distribution of these quantities to a linear distribution. The resultant flow is termed a "Newtonian shock line", and is illustrated in Fig. 3–10c. The Newtonian shock line begins at the first singular point on the body and follows a path on the body for which the lateral curvature of the path times the momentum flow in the shock line is balanced by lateral momentum fed into the line from the sides. In general, this path is not the locus of singular points. This shock line is the new and anomalous phenomenon which we must accept in order to resolve the problem.

If the path of the shock line along the body is concave with respect to the body, a force per unit distance is exerted on the body by the shock line. This force is simply equal to the momentum flow in the shock line times the component of the curvature of the shock line normal to the surface. If the path were convex with respect to the body a negative force per unit distance would be required. This possibility must be rejected and the shock line must fly free, in a manner analogous to that of a free layer. But in the case of a free shock line, the free stream cannot exert the singular forces needed to make the shock line curved, and its path is a straight line. If at the first singular point the surface is convex

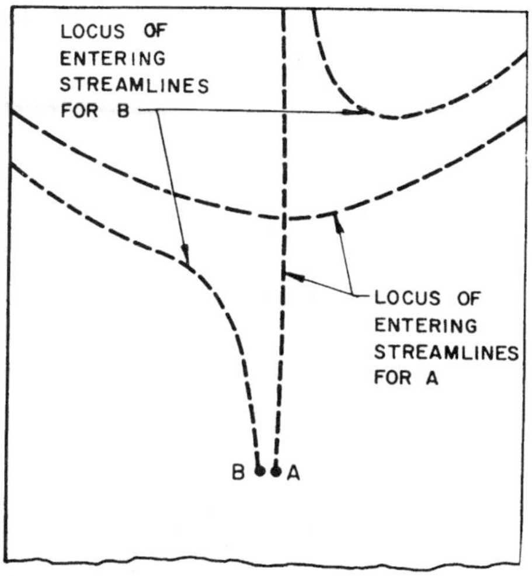

FIG. 3-10b. Drain trough problem—loci of entering streamlines.

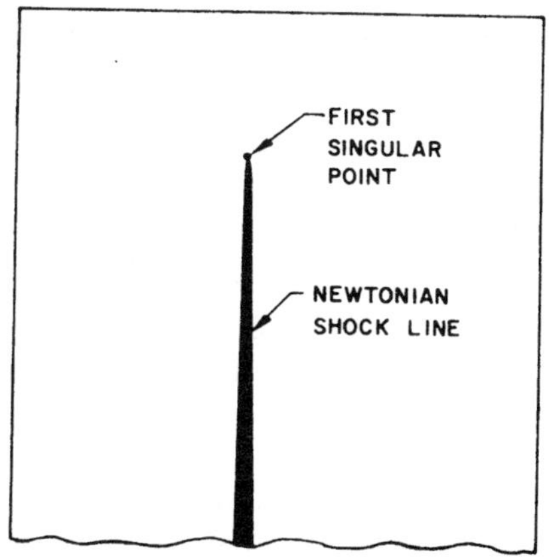

FIG. 3-10c. Drain trough problem—Newtonian shock line.

in the direction a shock line would otherwise take, a shock line is not formed. Instead, a Newtonian free layer is formed which erupts into the free stream. And if immediately after a shock line separates and becomes free the conditions on the surface are such that a regular Newtonian layer still cannot exist, a free shock layer is formed which erupts from the body and is connected to the free shock line like a web.

In either the solution with a Newtonian shock line or that with an erupting shock layer a point on the surface on one side of the line or layer is shielded by it from certain parts or branches of its locus of entering streamlines. The structure of the shock layer at this point is determined by the flow entering at those parts of the locus which are not shielded, which correspond to entering particles which actually reach the point in question. Thus the original difficulty of multiple correspondence between the structure of the layer and the locus of entering streamlines is removed through this shielding effect, and we are permitted to keep the laminar layer model.

Other even more anomalous behavior may be deduced for body shapes related to that for our drain trough problem. For example, we may imagine a body with a geometry such that a free shock line will reimpinge on the body.

Assuming a Newtonian process in this reimpingement, there will result an isolated force on the body at the reimpingement point. Such an isolated force represents a two-fold condensation with respect to the force distribution represented by conventional aerodynamic pressure.

We return now to our study of the regular Newtonian layer, assuming that singular behavior of the type just discussed does not appear. In our development, we shall always express pressures in terms of the dimensionless pressure coefficient. In defining this coefficient we have reduced the pressure using the known quantities ρ_∞ and U describing the free stream. With respect to the new dynamic quantities we shall introduce, we shall use ρ_∞ and U in their definitions in order to eliminate the dimensions of mass and of time. We wish to make no specification as to body dimensions, and accordingly do not choose to complete the reduction to dimensionless form by using a reference length in our definitions. Thus our basic variables will all have dimensions involving length. The reader will recognize that we have already carried out this plan in the definition of m, as given in (3.2.10), or of q, as defined earlier in this section. We have used dynamic forms for ψ and P in the first four sections of Chapter III primarily in order to have a consistent definition of ψ throughout the book.

All our results will be invariant with respect to the scale transformation discussed in the third paragraph of Section 2.2, and this property will be evidenced by dimensional consistency of the results. Reduction of our quantities to dimensionless form by use of a specified reference distance dimension would be a trivial process.

At each point in a regular Newtonian layer a scale in terms of m exists which we may use in describing the structure of the shock layer. The structure of the shock layer is described in terms of the velocity $U\mathbf{q}$ of the material in the layer, expressed functionally

$$\mathbf{q} = \mathbf{q}(m). \tag{3.5.10}$$

This vector is a two-dimensional vector in the curved two-dimensional space of the surface of the shock layer or body, and is dimensionless. The total mass per unit area is represented by the quantity

$$m_1 = \int_0^{m_1} dm, \tag{3.5.11}$$

which is a scalar in this two-dimensional space with the dimensions of length. We define the total mass flow vector

$$\mathbf{M}_1 = \int_0^{m_1} \mathbf{q}\, dm; \tag{3.5.12}$$

this vector has two components. And we define the total momentum

$$\mathfrak{P}_1 = \int_0^{m_1} \mathbf{qq}\, dm; \tag{3.5.13}$$

this quantity is a symmetric dyadic or tensor in this space, and has three independent components. Both $\mathbf{M}_1$ and $\mathfrak{P}_1$ have the dimensions of length. The subscript 1 indicates that the integral of (3.5.12) or (3.5.13) is taken through the entire layer. We shall use the same quantities without the subscript to indicate the corresponding indefinite integrals evaluated as zero at $m = 0$, representing quantities that are functions of m. Thus, for example, we have $\mathbf{M}_1 = \mathbf{M}(m_1)$. In two-dimensional or axisymmetric flow $\mathbf{M}$ and $\mathfrak{P}$ each have only one nonzero component. The nonzero components of $\mathbf{M}$ and $\mathfrak{P}$ are related to the ψ and P we have used in the first four sections by the relations

$$|\mathbf{M}| = \frac{\psi}{\rho_\infty U y^j}, \tag{3.5.14}$$

and

$$|\mathfrak{P}| = \frac{P}{\rho_\infty U y^j}. \tag{3.5.15}$$

The curvature of the surface is also a symmetric dyadic or tensor in the two-dimensional space of the shock layer, and has three independent components. We denote this quantity as $\mathfrak{K}$. As with any symmetric dyadic or tensor there exist principal axes with respect to which the cross component vanishes. The inverses of the two principal components of the curvature tensor are termed the principal radii of curvature. We define $\mathfrak{K}$ to have positive components for a convex body; with this definition the customary convention for curvature requires us to define the unit vector normal to the body $\mathbf{n}$ as directed into the body.

The inclination of the shock layer surface σ is the angle of the surface with the free stream direction, measured in the plane defined by the free stream direction and the normal to the surface. The quantity $\sin \sigma$ is the direction cosine between the surface normal and the free stream direction. The pressure immediately behind the shock is given as before by (3.2.6). The centrifugal correction needed to obtain the pressure on the body is obtained by integrating the normal pressure gradient from the shock.

The normal pressure gradient in a finite density flow with velocity Uq is

$$\frac{\partial p}{\partial y} = \rho U^2 q^2 K, \tag{3.5.16}$$

where $-K$ is the streamline curvature in the normal direction. In our case we may write

$$K = \frac{\mathbf{q}}{q} \cdot \mathfrak{K} \cdot \frac{\mathbf{q}}{q}. \tag{3.5.17}$$

We replace $\rho\, dy$ by $\rho_\infty\, dm$, and obtain

$$\frac{d\frac{1}{2}C_p}{dm} = \mathbf{q} \cdot \mathfrak{K} \cdot \mathbf{q} = \mathfrak{K} : \mathbf{qq}, \tag{3.5.18}$$

where the double dot product of two dyadics is the same as the double contracted product of the corresponding tensors. We integrate (3.5.18) and obtain

$$\tfrac{1}{2}C_p = \sin^2 \sigma - \mathfrak{K} : (\mathfrak{P}_1 - \mathfrak{P}) \tag{3.5.19}$$

for the pressure at any value of m and

$$\tfrac{1}{2}C_{p_b} = \sin^2 \sigma - \mathfrak{K} : \mathfrak{P}_1 \tag{3.5.20}$$

for the pressure on the body.

In a three-dimensional Newtonian sail, the stress is a general two-dimensional symmetric tensor in the surface of the sail. We denote this stress tensor or dyadic by $U\mathfrak{T}$. The condition of equilibrium of an infinitesimal element of the sail may be given

$$\nabla \cdot \mathfrak{T} = 0. \tag{3.5.21}$$

This is a vector equation, and expresses two conditions on the quantity $\mathfrak{T}$. With zero pressure on the back side of the sail, the equation analogous to (3.3.36) for equilibrium of an element in the normal direction is

$$0 = \sin^2 \sigma - \mathfrak{K} : (\mathfrak{P}_1 - \mathfrak{T}). \tag{3.5.22}$$

If the sail is isotropic the stress dyadic may be expressed

$$\mathfrak{T} = \mathcal{T}\mathfrak{I}, \tag{3.5.23}$$

where $\mathfrak{I}$ is the idemfactor. Condition (3.5.21) requires that $\nabla\mathcal{T}$ be zero, and hence that $\mathcal{T}$ is constant.

It is clear that in a general problem with cross flow any of the various phenomena discussed for two-dimensional and axisymmetric bodies may arise, including free layers, constant-pressure layers, and reimpingement. Many new phenomena must exist in a general flow, however, besides those connected with the Newtonian shock line which we have already discussed. For example, we can make a free shock layer reimpinge on a surface on which there is already a regular shock layer; we can make two free layers collide and coalesce in midstream; or we can control the pressure behind a constant-pressure layer by ventilation from the sides. An extended deductive exploration of such anomalous phenomena appears of dubious value at the present time, and we shall turn to other problems. We shall next investigate how the structure of a regular layer with cross flow is to be obtained.

6. Shock layer structure with cross flow

The analysis given below is essentially that of the first edition. Independently of the authors and of each other Guiraud [3] and Maikapar [1] presented methods for obtaining the structure and pressure distribution of three-dimensional Newtonian flows. Their methods, although of course equivalent to the authors', are different in detail.

In calculating the structure of a Newtonian flow, the first step is to calculate the particle trajectories as indicated in the last section. If we designate points on the surface in terms of some system of curvilinear coordinates (ξ, η), then for each point at which a streamline enters (ξ, η) there will be a trajectory on the surface satisfying an equation of the form $g(\xi_1, \eta_1) = 0$, where (ξ_1, η_1) designates a point on the trajectory. The quantities (ξ, η) are parameters in this equation, so we can rewrite it in general in the form

$$g(\xi, \eta; \xi_1, \eta_1) = 0. \tag{3.6.1}$$

This equation is termed the trajectory equation.

If we now hold (ξ_1, η_1) fixed, we observe that (3.6.1) gives all the entering points (ξ, η) whose trajectories pass over (ξ_1, η_1). Therefore the trajectory equation (3.6.1) with ξ, η as independent variables and ξ_1, η_1 as parameters gives the locus of entering streamlines. Thus the trajectory equation is the basic one in the three-dimensional case. Note that the locus of entering streamlines is defined implicitly by this equation rather than explicitly.

For the structure calculation we need not only the locus of entering streamlines for points on the surface, but also for each point an ordering of the points along the locus consistent with the ordering of points on the m scale. In other words, we need to know the order in which the fluid laminae lie on the surface. In addition, we need to know $\mathbf{q}$ at the point of interest for each entering streamline.

The magnitude of $\mathbf{q}$ is $\cos \sigma$, of course, but its direction depends upon the geometry of the surface. The direction may be obtained from the trajectory function through the relation $\mathbf{q} \cdot \nabla_1 g = 0$. The method is based on conservation of mass.

In addition to the point of interest 1 we need a neighboring point 1′. Figure 3–11 shows the loci of entering streamlines for these two points. The total

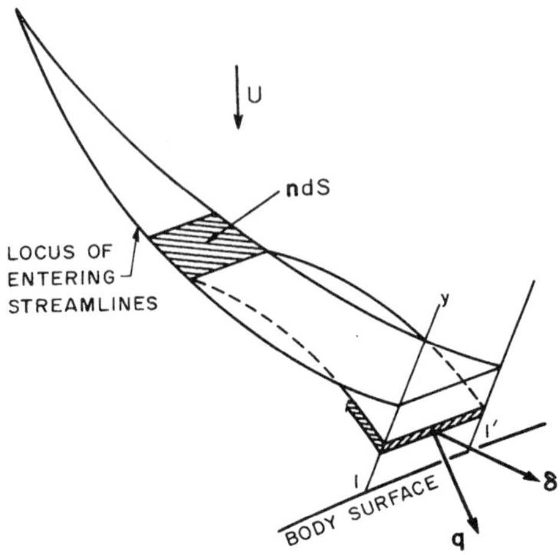

FIG. 3–11. Calculation of m with cross flow.

mass flow entering the shock layer between the two loci of entering streamlines must pass between the normals to the points 1 and 1′. The mass flow entering through the area element $\mathbf{n}\, dS$ is $\rho_\infty U \sin\sigma\, dS$, and is equal to the mass flow $\rho U \mathbf{q} \cdot \boldsymbol{\delta}_1\, dy$ through the area element $\boldsymbol{\delta}_1\, dy$. We replace $\rho\, dy$ by $\rho_\infty\, dm$ and obtain

$$dm = \frac{\sin\sigma\, dS}{\mathbf{q} \cdot \boldsymbol{\delta}_1}. \tag{3.6.2}$$

Here $\boldsymbol{\delta}_1$ is a vector in the surface normal to the segment between 1 and 1′ and equal in magnitude to its length.

There is a certain arbitrariness in this procedure in that we may choose any orientation for $\boldsymbol{\delta}_1$ that we please, and our final result must be independent of this choice. We may take advantage of this freedom in order to choose an orientation which will make the calculation simpler. As another possibility, we may check our calculation by repeating it with a different orientation for the two points.

Our first example is that of cylindrical flow, which consists of a constant velocity in the direction of the axis of a general cylinder superposed on a two-dimensional flow about the cylinder. Cylindrical flows are generally obtained by placing a cylinder in a uniform flow at an angle of yaw or sweepback, and we shall take this point of view. The coordinate x is taken along the surface normal to the generatrix of the cylinder, and y is taken along the axis, with $\mathbf{j}$ a unit vector in the y direction. The two-dimensional solution for the unyawed cylinder is presumed known, and the two-dimensional quantities m_1, $\psi_1/\rho_\infty U$, and $P_1/\rho_\infty U$ are denoted in vector form as m_{2-d}, $\mathbf{M}_{2-d}$, and $\mathfrak{P}_{2-d}$. In this notation we recognize that the latter two quantities are a two-dimensional vector and tensor, respectively, each with only one nonzero component.

In the yawed flow the velocity U normal to the cylinder is reduced by the factor $\cos \Omega$, where Ω is the angle of yaw, and the constant superposed axial velocity is $U \sin \Omega$. However, in our analysis we shall use the complement of Ω, termed σ_0; in the case of a blunt body σ_0 is the inclination angle of the surface at the two-dimensional stagnation point. Thus we shall have

$$\sigma_0 = \tfrac{1}{2}\pi - \Omega, \tag{3.6.3a}$$

$$\sin \sigma_0 = \cos \Omega, \tag{3.6.3b}$$

$$\cos \sigma_0 = \sin \Omega. \tag{3.6.3c}$$

Our reason for using σ_0 in place of Ω is to permit a more direct comparison of the example of cylindrical flow with the example of the flow near an axis of symmetry considered later.

We may now directly calculate the quantities m_1, $\mathbf{M}_1$, and $\mathfrak{P}_1$ for the yawed body, using the fact that $\mathbf{q} = \mathbf{q}_{2-d} \sin \sigma_0 + \mathbf{j} \cos \sigma_0$. The results are

$$m_1 = m_{2-d}, \tag{3.6.4a}$$

$$\mathbf{M}_1 = \mathbf{M}_{2-d} \sin \sigma_0 + \mathbf{j} m_{2-d} \cos \sigma_0, \tag{3.6.4b}$$

$$\mathfrak{P}_1 = \mathfrak{P}_{2-d} \sin^2 \sigma_0 + (\mathbf{M}_{2-d}\mathbf{j} + \mathbf{j}\mathbf{M}_{2-d}) \sin \sigma_0 \cos \sigma_0 + \mathbf{jj} m_{2-d} \cos^2\sigma_0. \tag{3.6.4c}$$

Since the curvature tensor has but one nonzero component, only the $\mathfrak{P}_{2-d}$ component of (3.6.4c) comes into the expression for the pressure. The inclination of the body is governed by the relation

$$\sin \sigma = \sin \sigma_0 \sin \sigma_{2-d} \tag{3.6.5}$$

obtainable from the law of cosines of spherical trigonometry, and the pressure on the body by the relation

$$\tfrac{1}{2} C_{p_b} = \sin^2 \sigma_0 (\sin^2 \sigma_{2-d} - \mathfrak{K} : \mathfrak{P}_{2-d}). \tag{3.6.6}$$

In the natural coordinates of the cylinder $\mathfrak{K}$ has only the one component not involving $\mathbf{j}$.

Thus the pressure coefficient is simply that for the two-dimensional flow times $\cos^2 \Omega$. It is evident that this principal result for the flow on a yawed infinite cylindrical body could have been readily obtained without the use of our cross flow theory.

We now look at the locus of entering streamlines, shown in the plane of the developed cylinder in Fig. 3–12.

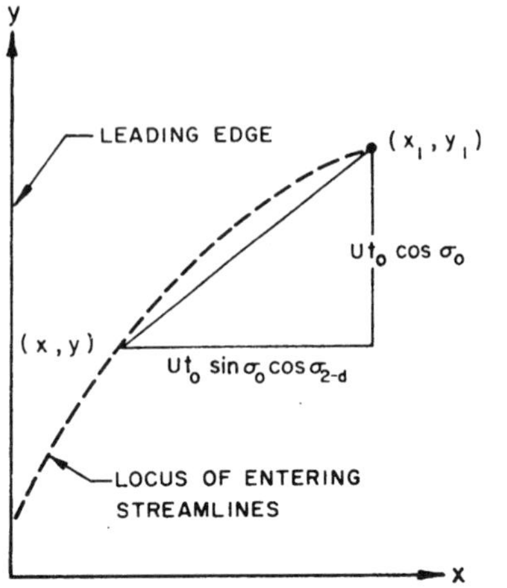

FIG. 3–12. Developed cylinder with cylindrical flow.

The geometry of this locus is determined by the trajectory equation

$$\frac{y_1 - y}{\cos \sigma_0} = \frac{x_1 - x}{\sin \sigma_0 \cos \sigma_{2-d}}, \tag{3.6.7}$$

where $\cos \sigma_{2-d}$ is a function of x. Note that the quantity expressed in (3.6.7) is essentially the time the particle has spent in the layer before reaching point 1. With allowance for the change in the definition of x and of σ_0, this expression is in agreement with (3.2.11). If the neighboring point 1′ is chosen to have the same value of x as does point 1, the area element between the two loci of entering streamlines is simply $\delta_1\, dx$ and the cylindrical structure of the shock layer is easily obtained by our general method.

So far, our cylindrical analysis has given us very little new. A new aspect appears when we permit the cylindrical body to be finite rather than infinite, with part of the leading edge a curve in the (x, y) plane different from the $x = constant$ curve representing the leading edge or stagnation line of the basic cylinder. If the locus of entering streamlines for a point starts at this leading edge rather than at the leading edge of the cylinder, the shock layer structure for the point will lack part of the complete cylindrical solution. This part is on the bottom of the layer and corresponds to the part of the locus of entering streamlines cut off by the leading edge. Such a solution is not cylindrical in the proper sense, but is easily obtainable from the two-dimensional results and equation (3.6.7) for the locus of entering streamlines.

If the two-dimensional body is sharp-nosed, with $\cos \sigma_{2-d}$ nonzero at the leading edge, the loci of entering streamlines are finite in length and the trajectory from a point on the leading edge has a nonzero angle on the (x, y) plane.

If the body is sufficiently long, then, a portion of the body will have a shock layer with the complete cylindrical structure. If the two-dimensional body is blunt, on the other hand, the fact that the body is finite is felt over the entire body. With the body blunt, and $\cos \sigma_{2-d} = 0$ at the leading edge, each locus of entering streamlines is infinite in extent and must be cut off by the leading edge if the body is finite. One result of this effect is that on a blunt two-dimensional body such as the circular cylinder the anomaly of infinite m_1 disappears if the body is finite and is yawed. The structure on a finite blunt cylindrical leading edge will be investigated as a special case of our next problem.

We consider now the problem of the flow near a straight axis of symmetry. A surface is considered which is tangent to the (x, y) plane along the y axis (see Fig. 3–13). The (x, y, z) coordinate system is aligned so that the (x, y)

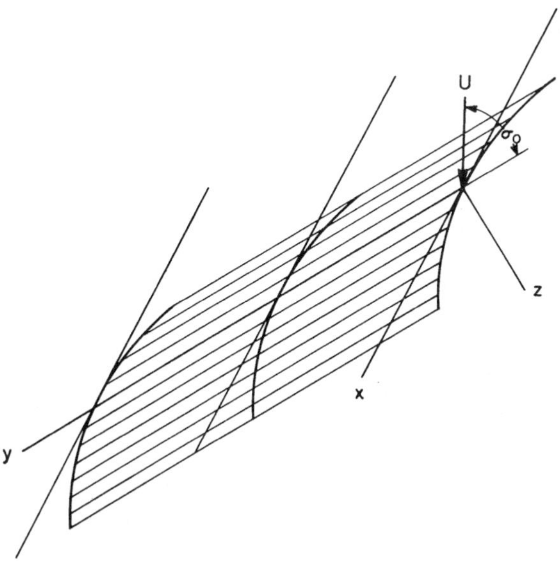

FIG. 3–13. Straight axis of symmetry.

plane is at an angle of incidence σ_0 to the flow direction, with direction cosines $(0, \cos \sigma_0, \sin \sigma_0)$. The surface is assumed to be given by the equation

$$z(x, y) = \tfrac{1}{2}K(y)x^2 + O(x^4), \tag{3.6.8}$$

and is locally symmetric with respect to x. The unit normal vector to the surface is approximately

$$\mathbf{n} = -\mathbf{i}Kx - \mathbf{j}\tfrac{1}{2}K'x^2 + \mathbf{k}(1 - \tfrac{1}{2}K^2x^2) \tag{3.6.9a}$$

and the curvature tensor is given by

$$\mathfrak{K} = -\nabla \mathbf{n} = \mathbf{ii}K + (\mathbf{ij} + \mathbf{ji})K'x + \tfrac{1}{2}\mathbf{jj}K''x^2. \tag{3.6.9b}$$

The orientation of the surface is such that with positive curvature the shape is convex, and a streamline striking the body at $x = 0$ has the line $x = 0$ as its subsequent trajectory. The body is assumed to have a definite leading edge at $y = 0$. If K is a constant the body is a section of a yawed blunt cylindrical body. If K is inversely proportional to y the body represents the neighborhood of a symmetry axis on a yawed cone. Our investigation will be a purely local one, with terms of higher order in x neglected, and with geodesics approximated by straight lines on an (x, y) plot.

The locus of entering streamlines is given by a trajectory equation similar to (3.6.7), with Kx in place of $\cos \sigma_{2-d}$. The trajectory equation is

$$x = \frac{x_1}{1 + K(y)(y_1 - y)\tan \sigma_0}. \tag{3.6.10}$$

The neighboring point to point 1 is placed at the same value of y, and the component of $\mathbf{q}$ parallel to $\boldsymbol{\delta}_1$ is simply $\cos \sigma_0$. The ratio δ/δ_1 is equal to x/x_1, and the area element is $\delta\, dy$. From (3.6.2) we obtain

$$dm = \frac{\tan \sigma_0\, dy}{1 + K(y)(y_1 - y)\tan \sigma_0}, \tag{3.6.11}$$

and the quadrature of this equation gives us m_1 on the body, which is locally independent of x.

The flow direction in this coordinate system has direction cosines $(0, \cos \sigma_0, \sin \sigma_0)$, and the dot product with $\mathbf{n}$ of (3.6.9a) gives $\sin \sigma$, the local body slope. The slope of the body is then

$$\sin \sigma = \sin \sigma_0 (1 - \tfrac{1}{2}K^2x^2) - \cos \sigma_0 (\tfrac{1}{2}K'x^2), \tag{3.6.12}$$

and this quantity in (3.2.6) gives the pressure immediately behind the shock. The velocity $\mathbf{q}$ has components $(Kx \sin \sigma_0, \cos \sigma_0)$, and the quantities $\mathbf{M}_1$ and $\mathfrak{P}_1$ may be calculated from (3.5.12) and (3.5.13). The result is

$$\mathbf{M}_1 = \mathbf{i}M_1x_1 \sin \sigma_0 + \mathbf{j}m_1 \cos \sigma_0, \tag{3.6.13a}$$

$$\mathfrak{P}_1 = \mathbf{ii}\mathfrak{P}_1x_1^2 \sin^2 \sigma_0 + (\mathbf{ij} + \mathbf{ji})M_1x_1 \sin \sigma_0 \cos \sigma_0 + \mathbf{jj}m_1 \cos^2 \sigma_0, \tag{3.6.13b}$$

where

$$M_1 = \frac{1}{x_1}\int_0^{m_1} Kx\, dm = \tan \sigma_0 \int_0^{y_1} \frac{K\, dy}{[1 + K(y_1 - y)\tan \sigma_0]^2}, \tag{3.6.14a}$$

$$\mathfrak{P}_1 = \frac{1}{x_1^2}\int_0^{m_1} K^2x^2\, dm = \tan \sigma_0 \int_0^{y_1} \frac{K^2\, dy}{[1 + K(y_1 - y)\tan \sigma_0]^3}. \tag{3.6.14b}$$

The relation between (3.6.13) and (3.6.4) is evident. The pressure on the body is given by (3.5.20), using (3.6.12) and the relation

$$\mathfrak{K}:\mathfrak{P}_1 = x_1^2[K\mathfrak{P}_1 \sin^2\sigma_0 + 2K'M_1 \sin\sigma_0 \cos\sigma_0 + \tfrac{1}{2}K''m_1 \cos^2\sigma_0], \tag{3.6.15}$$

in which K and its derivatives are evaluated at $y = y_1$. The bracket in (3.6.15) is a function of y_1 alone.

For a simple example we consider the finite yawed cylindrical body mentioned earlier. The quantity K is a constant, equal to R_s^{-1}. Quadrature of (3.6.11) gives

$$m_1 = R_s \ln\left(1 + \frac{y_1 \tan\sigma_0}{R_s}\right), \tag{3.6.16}$$

and of (3 6.14b) gives

$$\mathfrak{P}_1 = \frac{1}{2R_s}\left[1 - \left(\frac{R_s}{R_s + y_1 \tan\sigma_0}\right)^2\right]. \tag{3.6.17}$$

The pressure on the body, expressed as a function of x_1 and y_1, is

$$\tfrac{1}{2}C_{p_b} = \sin^2\sigma_0\left[1 - \frac{3x_1^2}{2R_s^2} + \frac{x_1^2}{2(R_s + y_1 \tan\sigma_0)^2}\right]. \tag{3.6.18}$$

As y_1 increases without limit, the third term in the bracket of (3.6.18) goes to zero. The remaining part of (3.6.18) is equivalent to (3.3.4) with $n = 2$, with account taken of the difference in geometry.

We turn next to the problem of a conical body. As we have defined it, a conical body will have a symmetry axis along a ray where the body is tangent to a right circular cone with axis in the free stream direction. The semiangle of this right circular cone is σ_0. Along the symmetry axis the conical body will osculate a different right circular cone of semiangle β. The basic dimensionless parameter of the problem, denoted κ, is the ratio of the tangents of these two semiangles. In this case we express the quantity K of (3.6.8) as

$$K(y) = \frac{1}{y \tan\beta} = \frac{\kappa}{y \tan\sigma_0} \tag{3.6.19}$$

in an equation which also defines κ. We have, therefore,

$$K' = \frac{-\kappa}{y^2 \tan\sigma_0}, \tag{3.6.20}$$

and

$$\tfrac{1}{2}K'' = \frac{\kappa}{y^3 \tan\sigma_0}. \tag{3.6.21}$$

The pressure immediately behind the shock, calculated with the help of (3.6.12), is

$$\tfrac{1}{2}C_{p_s} = \sin^2 \sigma_0 - \frac{x_1^2 \cos^2 \sigma_0}{y_1^2} \kappa(\kappa - 1). \tag{3.6.22}$$

The quantity x_1/y_1 in (3.6.22) is simply the angle from the symmetry axis measured on the cone surface.

With K given from (3.6.19) we may calculate m_1, M_1, and $\mathfrak{P}_1$ from (3.6.11) and (3.6.14). The results of these quadratures are

$$m_1 = \frac{y_1 \tan \sigma_0}{(\kappa - 1)^2} [\kappa \ln \kappa - (\kappa - 1)], \tag{3.6.23a}$$

$$M_1 = \frac{\kappa}{(\kappa - 1)^2} [(\kappa - 1) - \ln \kappa], \tag{3.6.23b}$$

$$\mathfrak{P}_1 = \frac{\kappa}{2y_1 \tan \sigma_0}. \tag{3.6.23c}$$

Applying these quantities to (3.6.15), we may obtain the pressure on the body

$$\tfrac{1}{2}C_{p_b} = \sin^2 \sigma_0 - \frac{3x_1^2 \cos^2 \sigma_0}{2y_1^2} \frac{\kappa^2}{(\kappa - 1)^2} [(\kappa - 1)(\kappa - 3) + 2 \ln \kappa]. \tag{3.6.24}$$

In the limit of κ very large (3.6.24) approaches the same form that (3.6.18) does with y_1 very large, with allowance for the different functions $K(y)$ in the two cases.

The problem of Newtonian flow on a general cone will be treated in a separate section. We have treated the conical case of the axis of symmetry to illustrate the general approach. A conical approach for the conical symmetry axis problem leads to the same results. In fact, the result (3.6.24) was first obtained following the conical approach. The approach we have chosen illustrates the centrifugal pressure correction in its general form.

Our final example is that of a general stagnation point, intermediate between the two-dimensional stagnation point on a body such as a circular cylinder and the axisymmetric stagnation point on a body such as a sphere. Again, we shall seek only a local solution, within a region small enough that from a front view a geodesic appears as a straight line. No solution for an entire body without two-dimensional or axial symmetry has been found which has the simplicity found for the circular cylinder and sphere. The system of coordinates is one in a projection of the surface in a plane normal to the flow direction, with the origin at the stagnation point. The axes are aligned in the directions of principal curvature. The basic parameter in this problem, again denoted κ, is the ratio of principal curvatures

$$\kappa = \frac{K_1}{K_2} = \frac{R_2}{R_1}, \tag{3.6.25}$$

where K_1 is the curvature in the x direction and K_2 the curvature in the y direction. There is a possibility of confusion here with respect to our use of the subscript 1. We use the subscripts 1 and 2 to indicate the x and y directions only upon the quantities K, $R = K^{-1}$, and a dimensionless pressure correction term π to be defined later. On all other quantities the subscript 1 will indicate, as before, the point under investigation or the top of the shock layer. Our results will all have to remain unchanged if we interchange x and y, κ and κ^{-1}, and the subscripts 1 and 2 on K, R, and π. This invariance gives us a ready method of checking our results. No solution is obtainable with κ negative, and we consider only positive values of κ.

The velocity of a particle in the shock layer is given by

$$\mathbf{q} = \mathbf{i}K_1 x + \mathbf{j}K_2 y \tag{3.6.26}$$

in terms of the coordinates of the point of entry. The equation for the trajectory of the particle is

$$\frac{x_1 - x}{K_1 x} = \frac{y_1 - y}{K_2 y}, \tag{3.6.27}$$

where x_1 and y_1 are the coordinates of a point on the trajectory. With x_1 and y_1 fixed and with x and y variable, (3.6.27) is the equation for the locus of entering streamlines. We rewrite (3.6.27)

$$\left(x + \frac{x_1}{\kappa - 1}\right)\left(y - \frac{\kappa y_1}{\kappa - 1}\right) + \frac{\kappa x_1 y_1}{(\kappa - 1)^2} = 0, \tag{3.6.28}$$

in which form the locus may be recognized as an equilateral hyperbola (see Fig. 3–14). The paths of steepest descent here are of the form $x = Ay^{\kappa}$.

We now choose the neighboring point 1′ to have the same value of x_1 as does point 1. The area element between the two loci is $\delta\, dx$ with δ their separation in the y direction. The two loci are related through an affine transformation, and the ratio δ/δ_1 is equal to y/y_1. Applying (3.6.2) we obtain

$$dm = \frac{y\, dx}{K_1 y_1 x} = \frac{\kappa R_1\, dx}{(\kappa - 1)x + x_1}, \tag{3.6.29}$$

for which the integral giving the correspondence between m and the coordinates of the point of entry is

$$m = \frac{\kappa R_1}{\kappa - 1} \ln \left[1 + \frac{(\kappa - 1)x}{x_1}\right] = -\frac{R_2}{\kappa - 1} \ln \left[1 - \frac{(\kappa - 1)y}{\kappa y_1}\right]. \tag{3.6.30}$$

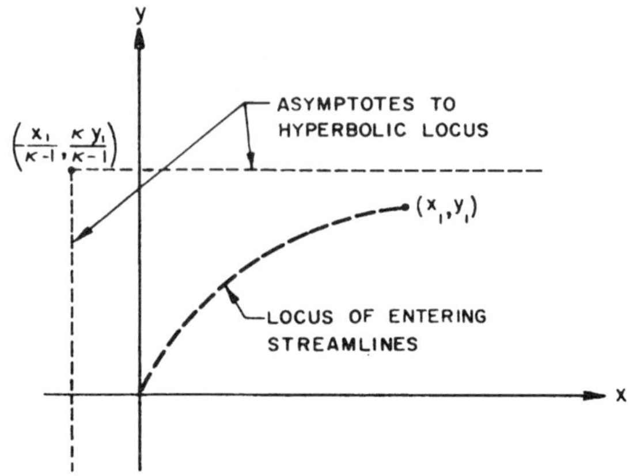

FIG. 3–14. General stagnation point.

The total quantity m_1 is simply

(3.6.31) $$m_1 = \frac{\kappa R_1}{\kappa - 1} \ln \kappa = \frac{R_1 R_2}{R_2 - R_1} \ln \frac{R_2}{R_1},$$

and it may readily be verified that this reduces to (3.5.3) for $\kappa = 1$ and that m_1 approaches infinity as κ is made indefinitely large.

The velocity profile in the shock layer may be described in terms of the velocity components

(3.6.32a) $$q_x = \frac{K_1 x_1}{\kappa - 1}\left[\exp\left(\frac{(\kappa - 1)m}{\kappa R_1}\right) - 1\right],$$

(3.6.32b) $$q_y = \frac{\kappa K_2 y_1}{\kappa - 1}\left[1 - \exp\left(-\frac{(\kappa - 1)m}{\kappa R_1}\right)\right].$$

The velocity profile for one component is concave and for the other is convex, except for $\kappa = 1$. With $\kappa = 1$, both components of the velocity profile become triangular, in agreement with (3.5.4).

The impulse tensor may be expressed in the form

(3.6.33) $$\mathfrak{P}_1 = \mathbf{ii}K_1 x_1^2 \pi_1 + \mathbf{jj}K_2 y_1^2 \pi_2 + (\mathbf{ij} + \mathbf{ji})\mathfrak{P}_{xy},$$

in terms of which the pressure on the body is expressed

(3.6.34) $$\tfrac{1}{2}C_{p_b} = 1 - (1 + \pi_1)K_1^2 x_1^2 - (1 + \pi_2)K_2^2 y_1^2,$$

with π_1 and π_2 constants to be determined, and $\mathfrak{P}_{xy}$ a cross component of $\mathfrak{P}_1$ which does not influence the pressure. Using (3.5.15) and the relations given for this problem, we may calculate the coefficients appearing in (3.6.34). The results are

(3.6.35a) $$1 + \pi_1 = (\kappa - 1)^{-3}[(\kappa - 1)(3\kappa - 1)(\tfrac{1}{2}\kappa - 1) + \kappa \ln \kappa],$$

(3.6.35b) $$1 + \pi_2 = (\kappa - 1)^{-3}[(\kappa - 1)(\kappa - 3)(\kappa - \tfrac{1}{2}) + \kappa^2 \ln \kappa].$$

It may be verified that $1 + \pi_1$ approaches 3/2 as κ approaches infinity, and approaches the value 4/3 for $\kappa = 1$, in agreement with our two-dimensional and axisymmetric results.

Additional information on this case will be found in Section 4.5, in which the density is assumed constant. Curves will be given for $\sqrt{1 + \pi_1}$ and $\sqrt{1 + \pi_2}$, and the velocity gradient at the body will be obtained.

We close with a formula due to Guiraud [5] for the total force on a surface. We consider a finite surface element, and note that the total force contributed by momentum entering the shock layer is $\rho_\infty U^2\mathbf{U}$ times the projected area of the surface, where $\mathbf{U}$ is a unit vector in the free stream direction. The momentum leaving the surface is an integral around the boundary of the surface of the scalar product of $\mathfrak{P}_1$ and $\boldsymbol{\nu}\, dl$, where $\boldsymbol{\nu}$ is a unit vector in the surface normal to the boundary and dl is distance along the boundary. Applying the principle of the conservation of momentum we can express the force

(3.6.36) $$\mathbf{F} = \rho_\infty U^2 \left[\mathbf{U} \iint \sin \sigma \, dS - \int \mathfrak{P}_1 \cdot \boldsymbol{\nu}\, dl\right].$$

This expression must be equivalent to that obtained by expressing the force as an integral of pressure over the surface. To show this we must apply the divergence theorem in the surface to the second integral, to obtain a surface integral of $\nabla \cdot \mathfrak{P}_1$. The unit vector $\mathbf{U}$ is taken into the integral and decomposed at each point according to

(3.6.37) $$\mathbf{U} = \mathbf{U}_t + \mathbf{n} \sin \sigma.$$

We may now rewrite (3.6.36) as

(3.6.38) $$\mathbf{F} = \rho_\infty U^2 \iint (\mathbf{n} \sin^2 \sigma + \mathbf{U}_t \sin \sigma - \nabla \cdot \mathfrak{P}_1)\, dS.$$

The quantity $\nabla \cdot \mathfrak{P}_1$, because of the curvature of the surface, has both normal and tangential components. We can identify the normal component as

(3.6.39) $$\mathbf{n} \cdot (\nabla \cdot \mathfrak{P}_1) = \mathfrak{K} : \mathfrak{P}_1.$$

The tangential component is simply given by the tangential component of momentum added per unit area per unit time, or

(3.6.40) $$(\nabla \cdot \mathfrak{P}_1)_t = \mathbf{U}_t \sin \sigma.$$

Thus (3.6.38), using (3.5.20), may be rewritten

$$\mathbf{F} = \tfrac{1}{2}\rho_\infty U^2 \iint \mathbf{n} C_{p_b}\, dS. \tag{3.6.41}$$

7. Conical flow

Conical flow is an inviscid steady flow which is invariant with respect to a scale transformation. The origin of coordinates is termed the vertex, and straight lines through the vertex are called rays. The principal property of conical flow is that the vector velocity and all thermodynamic quantities are constant along rays, and thus are functions only of two independent coordinate variables describing the rays. These coordinates are here chosen as coordinates in the surface of a unit sphere. The concept of conical flow was found and developed by Busemann [1; 3], who also characterized conical flows as flows having degenerate hodograph plots (plots with velocity components as independent variables).

In a conical flow, all natural surfaces of the solution are conical, i.e., cones in the general sense. A conical surface is the surface generated by a straight line (generatrix) which always passes through the vertex and which moves so as to always pass through a fixed curve (directrix) in space. These surfaces include shock surfaces, body surfaces, and contact discontinuities. Any of these surfaces is developable in a plane. On one of these developed surfaces a natural system of coordinates is a system of plane polar coordinates (to be termed (R, ξ) below). The radial coordinate equals the radial coordinate in a three-dimensional spherical coordinate system; the azimuthal coordinate equals distance along the intercept of the conical surface in the unit sphere.

It is convenient to consider the velocity vector decomposed into a radial component (normal to the unit sphere) and a cross component (tangent to the unit sphere). The projection of the velocity vector on the unit sphere gives an apparent two-dimensional cross flow in the unit sphere. This projected cross flow has cross-flow streamlines; such a cross-flow streamline is the intercept or projection of a conical streamsurface of the three-dimensional flow. In the continuity equation written for the cross flow the radial velocity component enters as a (negative) source term. The customary flow pattern restrictions based upon the absence of sources or sinks do not hold for the cross flow. For example, cross streamlines may terminate in the middle of the field.

In Newtonian conical flow the shock and body surfaces are considered identical as long as there is no separation as, for example, with a free layer. The structure is characterized by our m variable, which evaluated in the unit sphere is the dimensionless variable m/R_1. Conical Newtonian flow has been investigated by Gonor [1; 2; 3], Guiraud [4], Laval [1; 2; 3], and Melnik and Scheuing [1]. The present analysis is similar to the analyses of Laval and of Melnik and

Scheuing. Gonor's analysis is quite different in detail; he uses a stream function constant on cross streamlines as independent variable. We treat first the geometry of conical bodies, next the velocities in the shock layer and the trajectory equation, and then the nature of conical shock layer structure and loci of entering streamlines.

The basic coordinate system chosen is a system of spherical polar coordinates (R, ϑ, ϕ), with ϑ the polar angle from the undisturbed flow direction. The body is described by an equation of the form

$$\vartheta = \vartheta_0(\phi). \tag{3.7.1}$$

The subscript 0 will be dropped here where no confusion would ensue. The shape of the body is represented by a curve on a sphere, which we identify with the unit sphere $R = 1$ in physical space. One of the principal axes of curvature is in the radial direction, the other in the direction lying in an origin-centered sphere. Only the curvature in the second direction is zero, with the curvature K given by

$$K = \frac{1}{R} K_0(\phi). \tag{3.7.2}$$

The quantity K_0 is the curvature in the unit sphere of the curve describing the body shape.

We are now forced to use results of classical analytic geometry, but in the surface of a sphere instead of on a plane. Since this subject is not treated in standard texts, we derive the results we need in a circular cone tangent to the unit sphere at the point of interest (see Fig. 3–15). Local conical coordinates in this developed cone are depicted in Fig. 3–16. The angle α is the inclination of the body surface to a latitude line of constant ϑ, and is given by

$$\tan \alpha = \frac{1}{\sin \vartheta} \frac{d\vartheta}{d\phi}. \tag{3.7.3}$$

A coordinate on the body ξ is defined as the polar angle on the developed body surface. Since $d\xi$ is differential distance on the unit sphere we may write

$$d\xi = \frac{\sin \vartheta}{\cos \alpha}\, d\phi. \tag{3.7.4}$$

For the curvature K_0 of the body in the unit sphere we may borrow from planar polar coordinates. In polar coordinates (r, θ) the curvature is

$$K_{\text{pol}} = \frac{\dfrac{1}{r}\dfrac{d^2 r}{d\theta^2} - 1 - \dfrac{2}{r^2}\left(\dfrac{dr}{d\theta}\right)^2}{r\left(1 + \dfrac{1}{r^2}\left(\dfrac{dr}{d\theta}\right)^2\right)^{3/2}}. \tag{3.7.5}$$

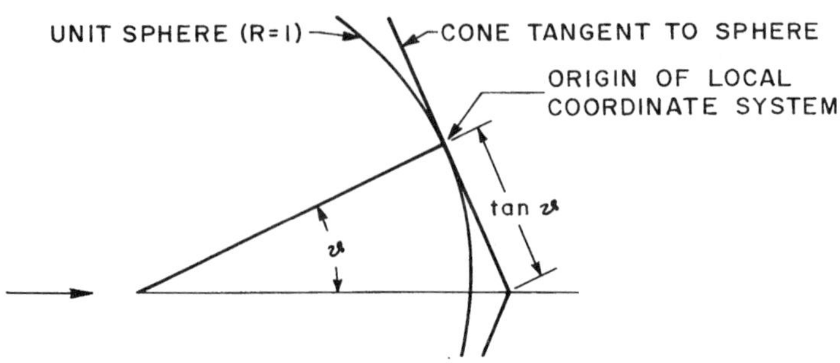

FIG. 3–15. Tangent cone for local coordinates.

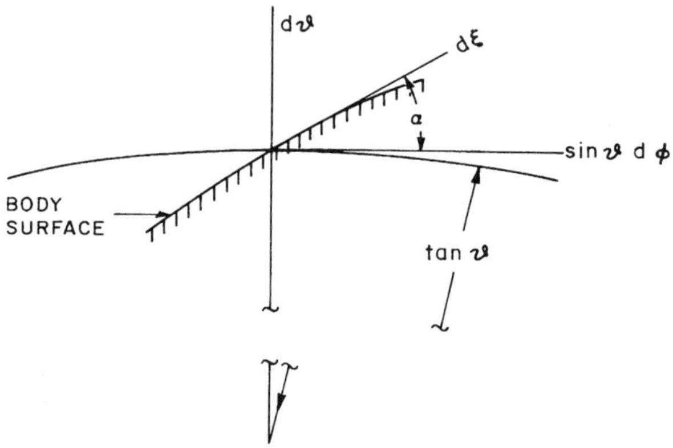

FIG. 3–16. Local conical coordinates.

We replace $r\,d\theta$ by $\sin\vartheta\,d\phi$, dr by $d\vartheta$, $dr/r\,d\theta$ by $\tan\alpha$, and the r left over after the other substitutions by $\tan\vartheta$. The curvature as we have defined it gives $K_0 = -K_{\text{pol}}$, and we obtain

$$K_0 = \cos^3\alpha\left[\cot\vartheta(1 + 2\tan^2\alpha) - \frac{1}{\sin^2\vartheta}\frac{d^2\vartheta}{d\phi^2}\right]. \tag{3.7.6}$$

The entering velocity has a radial component $U\cos\vartheta$ in the radial direction and a cross component $U\sin\vartheta$ in the unit sphere. This latter component has a component $-U\sin\vartheta\sin\alpha$ in the ξ direction and a component $U\sin\vartheta\cos\alpha$ normal to the body. In Fig. 3–17 we have shown these components, with the local coordinate system shown as a projection in the radial direction. Since the component normal to the body is $U\sin\sigma$ we obtain

$$\sin\sigma = \sin\vartheta\cos\alpha. \tag{3.7.7}$$

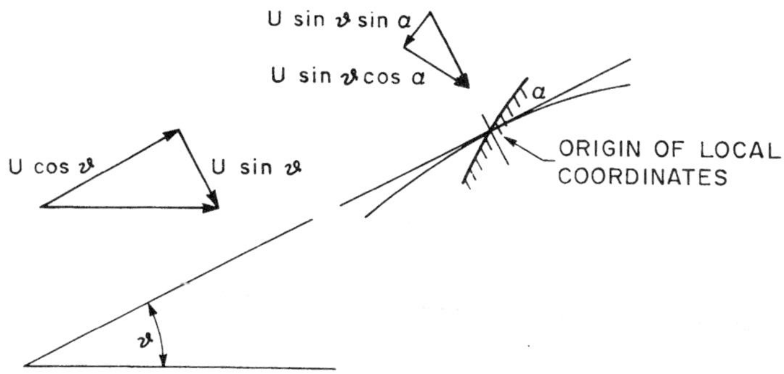

FIG. 3–17. Entering velocity components.

The velocity component in the body surface is given by

$$q^2 = \mathbf{q} \cdot \mathbf{q} = \cos^2 \vartheta + \sin^2 \vartheta \sin^2 \alpha = \cos^2 \sigma. \tag{3.7.8}$$

The initial direction of the velocity vector is described by the angle ψ of the trajectory with respect to the radial direction. This angle is given by

$$\tan \psi = R \frac{d\xi}{dR} \tag{3.7.9}$$

for the trajectory on the developed body surface, and by

$$\tan \psi = -\tan \vartheta \sin \alpha \tag{3.7.10}$$

from the ratio of the two velocity components in the body surface (see Fig. 3–18). From (3.7.7) and (3.7.10) we obtain also

$$\cos \sigma = q = \frac{\cos \vartheta}{\cos \psi}. \tag{3.7.11}$$

An instructive relation may be obtained from the above equations, by differentiating (3.7.7) and (3.7.3). This relation is

$$\cos \sigma \frac{d\sigma}{d\xi} = \frac{\cos \alpha \cos \sigma}{\sin \vartheta} \frac{d\sigma}{d\phi} = \sin \vartheta \sin \alpha \, K_0 . \tag{3.7.12}$$

This formula is checked by the observation that if σ is locally constant, the body is either locally a plane, with $K_0 = 0$, or is a right circular cone, with $\alpha = 0$.

We next consider the body surface developed in a plane, with polar coordinates (R, ξ). A trajectory entering at (R, ξ) and inclined at an angle ψ with respect

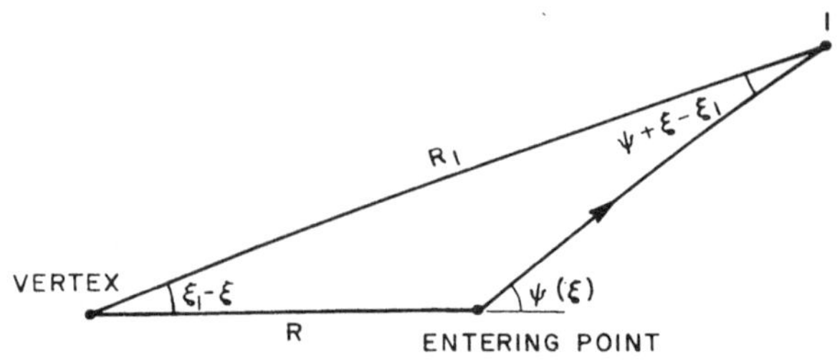

FIG. 3–18. Trajectory on developed cone surface.

to the radial direction will follow a straight line (geodesic) as shown in Fig. 3–18. Applying the law of sines to the triangle in the figure yields

$$R \sin \psi = R_1 \sin (\psi + \xi - \xi_1) \tag{3.7.13}$$

for the trajectory equation.

Applying now (3.6.2) to calculate the structure, we choose point 1′ at the same value of $\xi = \xi_1$, with R_1 increased by δR_1. The differential area is $dS = R\,\delta R\,d\xi$, δ_1 is simply δR_1, and (3.7.13) holds with R and R_1 replaced by δR and δR_1. Equation (3.6.2) becomes

$$dm = \frac{\sin \sigma\, R\, \delta R\, d\xi}{\dfrac{\cos \vartheta}{\cos \psi}\, \delta R \sin \psi}. \tag{3.7.14}$$

With the relations connecting σ, ψ, α, and ϑ given above we may obtain

$$d\,\frac{m}{R_1} = -\left(\frac{R}{R_1}\right)\frac{d\xi}{\tan \alpha}. \tag{3.7.15}$$

Taking $\mathbf{i}$ as a unit vector in the ξ direction and $\mathbf{j}$ as a unit vector in the R direction we may write

$$\mathbf{q} = \mathbf{i}q \sin (\psi + \xi - \xi_1) + \mathbf{j}q \cos (\psi + \xi - \xi_1). \tag{3.7.16}$$

The components of $\mathbf{M}$ and $\mathfrak{P}$ can be obtained from (3.5.12) and (3.5.13) as quadratures over ξ or ϕ, with integration limits to be discussed later.

We are particularly interested in $\mathfrak{P}_{11}$, the only component of $\mathfrak{P}$ which contributes to the pressure. This quantity depends upon

$$q_1 = q \sin (\psi + \xi - \xi_1) = -\left(\frac{R}{R_1}\right) \sin \vartheta \sin \alpha. \tag{3.7.17}$$

The component $\mathfrak{P}_{11}$ may be expressed through

$$\frac{\mathfrak{P}_{11}}{R_1} = -\int \left(\frac{R}{R_1}\right)^3 \sin^2 \vartheta \sin \alpha \cos \alpha \, d\xi. \tag{3.7.18}$$

To indicate the domain of integration it is convenient to introduce a plot of $\psi + \xi$ versus ξ. The function $\psi(\xi)$ is presumed to have been obtained for the body by the preceding analysis, and gives a curve on this plot, termed the trajectory curve. Such a curve is exemplified in Fig. 3–19a. For a given value of ξ_1, the lines $\psi + \xi = \xi_1$ and $\xi = \xi_1$ divide the plot into four quadrants. We impose the condition

$$\vartheta < \frac{\pi}{2}, \tag{3.7.19}$$

and conclude, since then $|\psi| < \pi/2$, that

$$0 \leqslant \frac{R}{R_1} \leqslant 1. \tag{3.7.20}$$

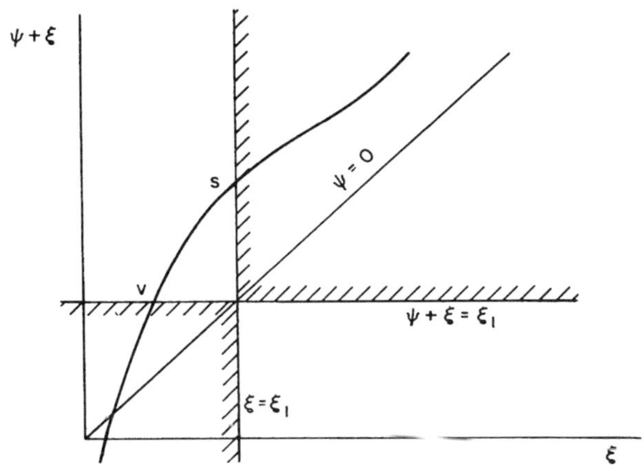

FIG. 3–19a. Plot of trajectory equation

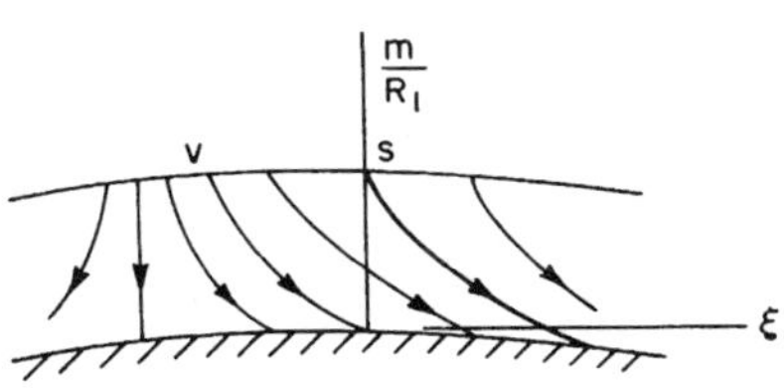

FIG. 3–19b. Cross streamline pattern.

The inequality (3.7.20) leads to the rejection of the upper right and lower left quadrants and the retention of the lower right and upper left quadrants. The quadratures for m, $\mathbf{M}$, and $\mathfrak{P}$ at a given value of ξ_1 are taken over the portions of the trajectory curve within these two quadrants. In Fig. 3–19b is shown the cross streamline pattern corresponding to Fig. 3–19a.

The point s at $\xi = \xi_1$ in Figs. 3–19 corresponds to the point at the shock in the shock layer structure. Since $\psi + \xi$ is a single-valued function of ξ, there can be at most one such point. The point v at $\xi + \psi = \xi_1$ corresponds to $R/R_1 = 0$ and thus to a ray streamline which starts at the vertex. We term such a point a vertex point. In general, a cross streamline enters the shock layer through the shock and moves along the shock layer to a vertex point, with $\xi_1 = \psi(\xi) + \xi$ there. At this point the cross streamline terminates. Corresponding three-dimensional streamlines are asymptotic in direction to the vertex streamline. It may be shown, in general, that a cross streamline comes in to its termination point parabolically, with m/R_1 proportional to $(\Delta\xi)^2$. These termination points generally lie on the body surface. In a gas of finite density the actual cross streamlines do not terminate in the manner predicted by Newtonian theory. Instead there is a thin layer which lies along the locus of the termination points in which the cross streamlines have a significantly different

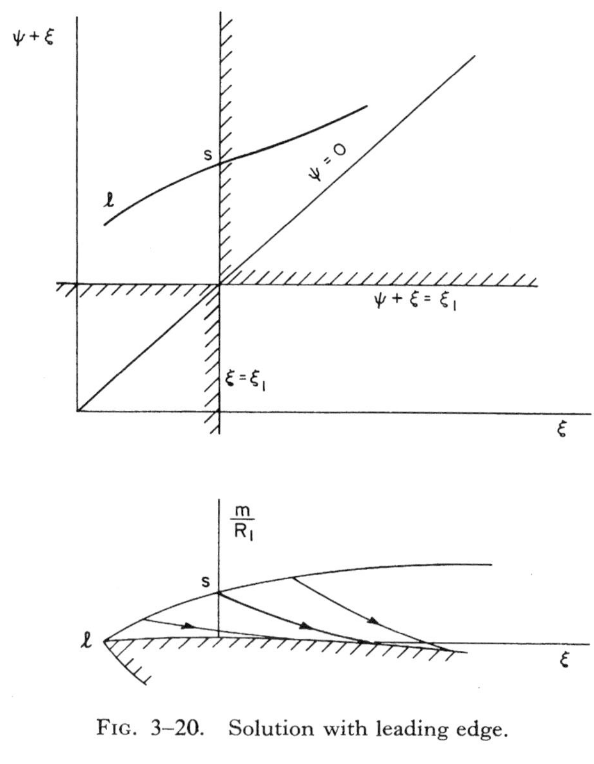

FIG. 3–20. Solution with leading edge.

behavior. This layer is termed a vortical layer and will bė discussed in Section 5.7.

If there is a sharp leading edge in the flow field, the trajectory curve begins at a point at the corresponding value of ξ. The cross streamline from the leading edge passes along the body surface until it reaches its termination point. This case is depicted in Fig. 3–20, with the leading edge labeled l.

If the trajectory curve is monotonic, with $\psi'(\xi) + 1 \geqslant 0$ everywhere, the analysis is completely straightforward. The domain of integration for a given value of ξ_1 is from a leading edge l or from the vertex v along the trajectory curve to the point s. The cases in which the curve is not monotonic yield interesting anomalies. In certain cases a Newtonian shock line must appear, as in the drain trough problem of Section 3.5. One such case is that in which the trajectory curve crosses the line $\psi = 0$ with negative slope. Here we shall consider a case in which a discontinuity surface appears, with no Newtonian shock line or other severe anomaly.

In Fig. 3–21a is sketched a trajectory curve with a single maximum b and a single minimum t, with $\psi > 0$ for both points. Also shown is the resultant cross flow streamline pattern in a plot of m/R_1 versus ξ. The characteristic feature in the flow field is a discontinuity surface, which is a double locus or fold in the

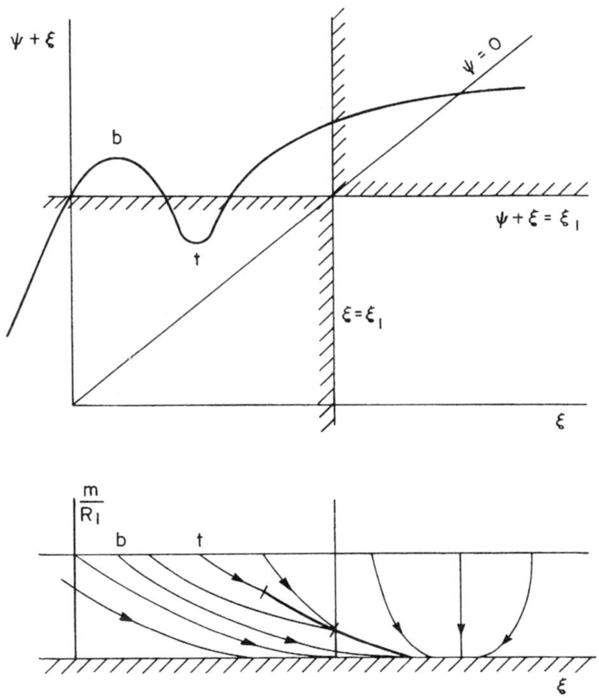

FIG. 3–21a. Solution with discontinuity surface.

locus of terminal points. The streamline from b goes into the point at the base of the discontinuity surface, and that from t into the tip of the surface. For a value of ξ_1 for which this surface is present the m scale has two parts. These parts correspond to two branches of the locus of entering streamlines, as shown in Fig. 3–21b. One branch ending at $\xi = \xi_1$ is conventional, starting at a vertex

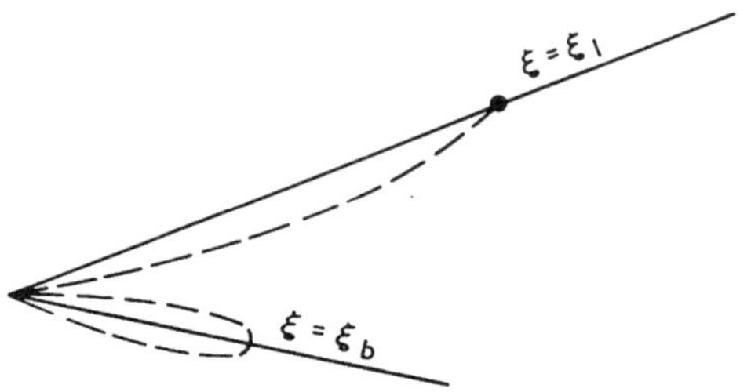

FIG. 3–21b. Locus of entering streamlines.

or termination point. The other branch, which includes the point b, begins and ends with vertex points. In this case, although the locus of entering streamlines has two branches, a 1–1 correspondence with the m scale can be established which is continuous with respect to changes in ξ_1. We do not count the smooth disappearance or appearance of a branch or the coalescence of two branches as a disallowed discontinuity in the 1–1 correspondence. Thus the basic requirement discussed in Section 3.5 is not violated, and a Newtonian structure is possible with no anomaly worse than the appearance of the discontinuity surface. Further discussion of the discontinuity surface was given by Melnik and Scheuing [1] who first noticed its existence.

Once the structure is established the quantity $\mathfrak{P}_{11}$ at $m = m_1$ can be calculated and the pressure obtained from (3.5.20). The formula is

$$\tfrac{1}{2}C_{p_b} = \sin^2\sigma_1 - K_{0_1}\frac{\mathfrak{P}_{11}(m_1)}{R_1}\,. \tag{3.7.21}$$

This pressure can become zero, and a conical free layer can be formed. The shape of such a free layer cannot be expressed generally in a simple form, as the shape depends upon the particular shape of the forebody creating it.

The procedure for calculating the pressure on a conical body may be presented as an algorithm. The procedure is:

(a) Establish the $\xi(\phi)$ scale for the body, and express the geometric variables ϑ, α, σ, ψ, and K_0 as functions of ξ.

(b) Plot $\psi(\xi) + \xi$ as a function of ξ.

(c) For each value of ξ_1 of interest calculate R/R_1 from (3.7.13) and $\mathfrak{P}_{11}/R_1$ from (3.7.18).

(d) Calculate $\tfrac{1}{2}C_{p_b}$ from (3.7.21), and m_1/R_1 from (3.7.15) if desired.

As an example we consider briefly the conical straight axis of symmetry problem already treated in the last section. We have for this case, from (3.6.19),

$$K_0 = \frac{\kappa}{\tan \sigma_0}. \tag{3.7.22}$$

We can derive, with $\xi = x/y$,

$$\sin \vartheta = \sin \sigma_0 - \frac{\cos^2 \sigma_0 \, \kappa(\kappa - 1)}{2 \sin \sigma_0} \xi^2, \tag{3.7.23}$$

$$\tan \alpha = \alpha = -(\kappa - 1) \cot \sigma_0 \, \xi, \tag{3.7.24}$$

and

$$\tan \psi = \psi = (\kappa - 1)\xi. \tag{3.7.25}$$

The ratio R/R_1 is given by

$$\frac{R}{R_1} = \frac{\kappa\xi - \xi_1}{(\kappa - 1)\xi}. \tag{3.7.26}$$

The quadrature for m_1 is

$$\frac{m_1}{R_1} = \frac{\tan \sigma_0}{(\kappa - 1)^2} \int_{\xi_1/\kappa}^{\xi_1} \frac{\kappa\xi - \xi_1}{\xi^2} \, d\xi, \tag{3.7.27}$$

and leads to

$$\frac{m_1}{R_1} = \frac{\tan \sigma_0}{(\kappa - 1)^2} [\kappa \ln \kappa - (\kappa - 1)] \tag{3.7.28}$$

in agreement with (3.6.23a). This result was obtained independently by Gonor [3] in his calculation of the shock angle for a right circular cone at angle of attack. Our κ is his $1 + t$. The pressure behind the shock is obtained by calculating $\sin^2 \sigma$ from (3.7.7) with (3.7.23) and (3.7.24), and agrees with (3.6.22). The quantity $\mathfrak{P}_{11}$ may be computed, and is found to be

$$\frac{\mathfrak{P}_{11}}{R_1} = \frac{\xi_1^2 \sin \sigma_0 \cos \sigma_0}{(\kappa - 1)^2} [\tfrac{1}{2}\kappa(\kappa^2 - 1) - 3\kappa(\kappa - 1) + 3\kappa \ln \kappa - (\kappa - 1)]. \tag{3.7.29}$$

If now (3.6.22) for $\sin^2 \sigma$, (3.7.22), and (3.7.29) are used to evaluate (3.7.21), the result is

$$\tfrac{1}{2} C_{p_b} = \sin^2 \sigma_0 - \xi_1^2 \cos^2 \sigma_0 \, F(\kappa), \tag{3.7.30}$$

where

$$F(\kappa) = \frac{3\kappa^2}{2(\kappa - 1)^2}\,[(\kappa - 1)(\kappa - 3) + 2\ln\kappa]. \tag{3.7.31}$$

This result agrees with (3.6.24).

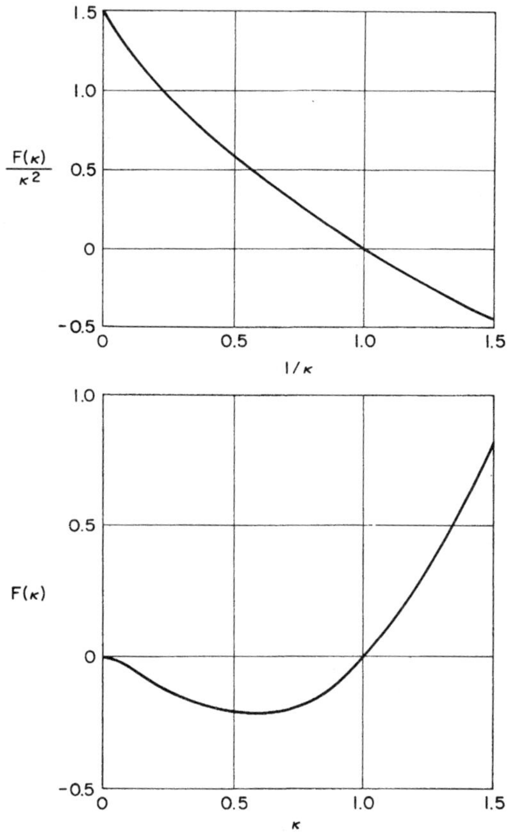

FIG. 3–22. Pressure function for conical axis of symmetry.

The function $F(\kappa)$ is given in Fig. 3–22. For $\kappa \gg 1$, $F(\kappa) \approx \frac{3}{2}\kappa^2$, in agreement with the results for a circular cylinder. For $|\kappa - 1| \ll 1$, $F(\kappa) \approx (\kappa - 1)$, corresponding to negligible centrifugal correction on a body close to a right circular cone. For $0 < \kappa \ll 1$, $F(\kappa) \approx \kappa^2(3\ln\kappa + \frac{9}{2})$. The case $\kappa = 0$ corresponds to an inclined flat plate.

For $\kappa > 1$ the pressure is a maximum on the axis (at $\xi_1 = 0$). For $0 < \kappa < 1$, the pressure is a minimum on the axis. For $\kappa < 0$, although the pressure behind the shock is defined and would be a maximum, the pressure on the body is undefined. The case $\kappa < 0$ is a case for which the trajectory curve crosses the line $\psi = 0$ with negative slope. Thus a regular Newtonian structure does not exist. The range $0 < \kappa < 1$ is one in which the cross streamlines are conically convergent (with respect to ξ), but the physical streamlines on the developed surface diverge. If $\kappa < 0$ the physical streamlines on the developed surface converge.

Other examples may be found treated in the literature. Gonor [1] and Laval [1] have obtained the solution for a circular cone at incidence, and Gonor [2] and Laval [1] have obtained the solution for an elliptic cone. Other cases have been studied by Laval [2; 3].

8. Bodies of revolution at small incidence

The theory of Newtonian flow on bodies of revolution at very small angle of attack was given by Cole [2] for slender bodies. Here we develop the theory for nonslender bodies, and obtain Cole's results in the limit of small thickness ratio. Cylindrical coordinates (r, ϕ, x) are used and are fixed relative to the body, and the body is specified by a known function $r(x)$. At zero angle of attack α, the local incidence angle is σ_0, with

$$\sin \sigma_0 = \frac{r'}{\sqrt{1 + r'^2}}, \tag{3.8.1a}$$

$$\cos \sigma_0 = \frac{1}{\sqrt{1 + r'^2}}. \tag{3.8.1b}$$

To obtain the geodesics, we minimize the integral

$$I = \int \sqrt{r^2 \left(\frac{d\phi}{dx}\right)^2 + 1 + r'^2}\, dx, \tag{3.8.2}$$

representing the distance between two points on the body surface along a given path $\phi = \phi(x)$. The first Euler integral for this variational problem is

$$r^2(r^2 - r_0^2) \left(\frac{d\phi}{dx}\right)^2 = r_0^2(1 + r'^2), \tag{3.8.3}$$

where r_0 is the constant of integration. The geodesic itself is obtained by a quadrature of (3.8.3). The result is

$$\phi = r_0 \int \frac{\sqrt{1 + r'^2}\, dx}{r \sqrt{r^2 - r_0^2}} \tag{3.8.4}$$

in the form of a definite integral.

To interpret the first integral for the geodesics, note first that

$$\tan \psi = \frac{r}{\sqrt{1 + r'^2}} \frac{d\phi}{dx}, \tag{3.8.5}$$

where ψ is the angle between the path in question and a longitude line ($\phi = constant$) on the surface of the body of revolution. Then (3.8.3) may be written

$$r \sin \psi = r_0. \tag{3.8.6}$$

A geodesic may be considered dynamically as the path of a particle moving at constant speed q constrained to the surface. Multiplying (3.8.6) by q times the mass, the equation becomes the classical equation for conservation of angular momentum of the particle about the axis of the body of revolution.

We turn next to the velocity components at entry. The free stream velocity is divided into a component $U \cos \alpha$ in the x direction and a component $U \sin \alpha$ in the r direction at $\phi = \pi$. The total component of velocity normal to the body is $\sin \sigma$, given by

$$\sin \sigma = \cos \alpha \sin \sigma_0 + \sin \alpha \cos \phi \cos \sigma_0 \,. \tag{3.8.7}$$

The tangential velocity component along a longitude line is

$$q_1 = \cos \alpha \cos \sigma_0 - \sin \alpha \cos \phi \sin \sigma_0 \,, \tag{3.8.8}$$

while the azimuthal tangential component is

$$q_2 = \sin \alpha \sin \phi. \tag{3.8.9}$$

The tangential components satisfy

$$q = \sqrt{q_1^2 + q_2^2} = \cos \sigma. \tag{3.8.10}$$

At entry, the relation

$$\tan \psi = \frac{r}{\sqrt{1 + r'^2}} \frac{d\phi}{dx} = \frac{q_2}{q_1} = \frac{r_0}{\sqrt{r^2 - r_0^2}}, \tag{3.8.11a}$$

or

$$\sin \psi = \frac{q_2}{q} = \frac{r_0}{r} \tag{3.8.11b}$$

holds, and may be used to evaluate r_0. With this evaluation of r_0, the geodesic equation (3.8.4) becomes the trajectory equation.

At this point we restrict ourselves to small incidence, to small values of α. We replace sin α by α, cos α by 1, and drop all terms in α^2. Inasmuch as q_2 and ψ are of order α, the centrifugal pressure term in $\mathfrak{P}_{22}$ will be of order α^2 and thus negligible. What we obtain is a theory for bodies of revolution at small incidence. If τ is the thickness ratio of the body, it is the parameter α/τ which must be small compared with one.

The quantity r_0 is of order α, and accordingly we replace $\sqrt{r^2 - r_0^2}$ in (3.8.4) by r. From (3.8.11) we obtain for r_0

$$r_0 = \alpha \frac{\sin \phi}{\cos \sigma_0} r = \alpha \sin \phi \sqrt{1 + r'^2}\, r. \tag{3.8.12}$$

The trajectory equation may now be written

$$\phi_1 - \phi = \alpha \sin \phi \sqrt{1 + r'^2}\, r \int^{x_1} \frac{\sqrt{1 + r'^2}\, dx}{r^2}, \tag{3.8.13}$$

with the lower limit in the integral corresponding to the point of entry. To calculate the entering mass flow we set up the neighboring point to the point of interest at the same value of x_1, with ϕ_1 increased by $\delta\phi_1$. See Fig. 3–23.

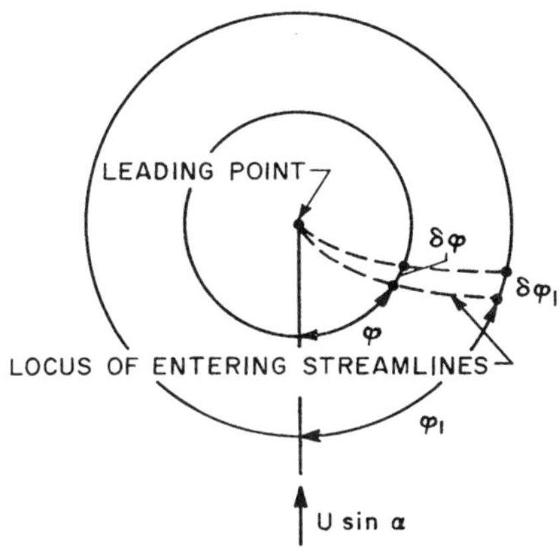

FIG. 3–23. Front view of body of revolution.

The differential area between the locus of entering streamlines and its neighbor is $r \sqrt{1 + r'^2}\, dx\, \delta\phi$. Since both q_2 and ψ are of order α we can simply take q_1 as the longitudinal velocity at values of x greater than that at entry.

We obtain for dm, applying (3.6.2),

$$dm = \frac{\sqrt{1 + r'^2} \sin \sigma}{q_1} \frac{\delta\phi}{\delta\phi_1} \frac{r}{r_1} dx. \tag{3.8.14}$$

The factors in this expression are evaluated,

(3.8.15a) $$\frac{\delta\phi}{\delta\phi_1} = 1 - \alpha\cos\phi\, r\int^{x_1}\frac{\sqrt{1+r'^2}\,dx}{r^2},$$

(3.8.15b) $$\sin\sigma = \sin\sigma_0\,(1+\alpha\cos\phi\cot\sigma_0),$$

(3.8.15c) $$\frac{1}{q_1} = \frac{1}{\cos\sigma_0}(1+\alpha\cos\phi\tan\sigma_0),$$

from (3.8.7), (3.8.8), and (3.8.13), and the integral for m is expressed

(3.8.16) $$m_1 = \frac{1}{r_1}\int_0^1 rr'\sqrt{1+r'^2}\left\{1+\alpha\cos\phi\left(\frac{1+r'^2}{r'} - \sqrt{1+r'^2}\,r\int_x^{x_1}\frac{\sqrt{1+r'^2}\,d\xi}{r^2}\right)\right\}dx.$$

Correspondingly, we may compute

(3.8.17) $$\mathfrak{P}_{11} = \frac{1}{r_1}\int_0^1 \frac{rr'}{\sqrt{1+r'^2}}\left\{1+\alpha\cos\phi\left(\frac{1-r'^2}{r'} - \sqrt{1+r'^2}\,r\int_x^{x_1}\frac{\sqrt{1+r'^2}\,d\xi}{r^2}\right)\right\}dx.$$

If the body has a sharp leading point or tip all loci of entering streamlines will emanate from it, and the lower limit in the integrals above is the value of x at the tip. We designate this point as $x = 0$. If the body is blunt the stagnation point is slightly shifted at angle of attack. The neglect of r_0 in $\sqrt{r^2 - r_0^2}$ fails locally, and a local adjustment for the failure of the assumptions should be made. However, we may expect that except in the immediate neighborhood of the stagnation point only a negligible error will arise from the simplifying assumptions if we use $x = 0$ as the lower limit in the blunt-nose case also. We must require that the body not be flat nosed; thus we have $r(0) = 0$.

One of the terms in (3.8.17) may be integrated by parts, to obtain

(3.8.18) $$-\frac{\alpha\cos\phi}{r_1}\int_0^{x_1} r^2r'\int_x^{x_1}\frac{\sqrt{1+r'^2}\,d\xi}{r^2}\,dx = -\frac{\alpha\cos\phi}{3r_1}\int_0^{x_1} r\sqrt{1+r'^2}\,dr.$$

The local pressure coefficient is

(3.8.19) $$\tfrac{1}{2}C_{p_b} = \sin^2\sigma_0 + \alpha\cos\phi\sin 2\sigma_0 + \cos^3\sigma_0\, r_1''\mathfrak{P}_{11}.$$

This expression may be used to calculate forces on the body.

To obtain the results of Cole [2], let us set the chord of the body equal to one, and let

$$r(x) = \tau F(x), \tag{3.8.20}$$

with τ small. We require that

$$F(0) = 0, \qquad F(1) = 1. \tag{3.8.21}$$

In place of (3.8.16) and (3.8.17) we obtain, neglecting terms of order τ^2,

$$m_1 = \mathfrak{P}_{11} = \tfrac{1}{2}\tau F(x_1) + \tfrac{2}{3}\alpha \cos\phi \frac{1}{F}\int_0^{x_1} F\, dx. \tag{3.8.22}$$

In place of (3.8.19) we have

$$\tfrac{1}{2}C_{p_b} = \tau^2(F'^2 + \tfrac{1}{2}FF'') + \alpha \cos\phi\, \tau \left(2F' + \frac{2}{3}\frac{F''}{F}\int F\, dx\right). \tag{3.8.23}$$

Compare (3.4.38). For the lift coefficient we first multiply by $\cos\phi\, d\phi$ and integrate from 0 to 2π. Then we multiply by $\tau F\, dx$ and integrate from 0 to 1. Dividing by the reference base area $\pi\tau^2$ gives us $\frac{1}{2}$ times the normal force coefficient. In transforming to axes aligned relative to the free stream the normal force coefficient differs from the lift coefficient by a factor $1 + O(\tau^2)$. In addition, the axial force coefficient differs from the drag coefficient by a factor $1 + O(\alpha^2/\tau^2)$. Since both τ and α/τ are required to be small, we can drop these factors.

For the lift coefficient we obtain, from (3.8.23),

$$\tfrac{1}{2}C_L = \alpha\left[1 + \tfrac{2}{3}\int_0^1 F''\int_0^x F\, d\xi\, dx\right] \tag{3.8.24}$$

$$= \tfrac{2}{3}\alpha\left[1 + F'(1)\int_0^1 F\, dx\right].$$

This is the result obtained by Cole [2], using a somewhat different approach. The moment coefficient about the leading tip given by Cole may be obtained similarly, by inserting a factor x in the integration, and is

$$\tfrac{1}{2}C_M = \tfrac{2}{3}\alpha\left[1 + (F'(1) - 1)\int_0^1 F\, dx\right]. \tag{3.8.25}$$

A general remark on the right circular cone at small incidence is in order. With α^2 terms dropped, centrifugal pressure effects do not appear, and the pressure on the body is approximated by the pressure behind the shock. The shock layer does have a structure with cross flow, but the principal effect of this cross flow is the reduction by the factor $\frac{2}{3}$ appearing in (3.8.22) of the change in shock layer thickness resulting from the angle of attack. On the axis of symmetry (3.8.22) agrees with the result (3.6.23a). There we replace β by τ and σ_0 by $\tau + \alpha$, and obtain $\kappa - 1 = \alpha/\tau$. Expansion of (3.6.23a) in α/τ yields the check.

9. Unsteady flow

Our previous analysis of Newtonian flow has been entirely limited to steady flow, and we have made full use of the simplifications which result from this limitation. We next turn to the problem of developing a basic theory for unsteady Newtonian flow. The introduction of the time as a basic independent variable produces about the same additional complication as does an additional space variable; in other words, we may expect two-dimensional unsteady flow to be about as difficult from a conceptual and calculational point of view as is three-dimensional steady flow. And just as we found anomalies of various kinds in three-dimensional flows we may expect to find them in a thorough study of unsteady flow, undoubtedly in greater variety. Here we shall not make any particular attempt to look for anomalous behavior, but shall content ourselves with a general development in which suitably proper behavior of the solution is implicitly assumed.

Our use of vector analysis will necessarily be somewhat involved. In our analysis of three-dimensional steady flow we dealt mostly with two-dimensional vectors (and dyadics or tensors) in the curved two-dimensional space of the shock surface. Since in unsteady flow the shock surface is a moving one we must now deal with velocity vectors and other vectors which are three-dimensional. Part of our problem will be to resolve these vectors into normal and tangential components with respect to the shock surface, of which the tangential components will be two-dimensional vectors in the surface analogous to those of the last two sections. And we shall have to adequately specify the motion of the shock surface itself. The method of treatment is similar to that used by Hayes [5] for a different problem. In unsteady flow the particle trajectories lie in a three-dimensional curved space consisting of the shock surface with the addition of the coordinate time, and our concept of the locus of entering streamlines will have to be modified accordingly. Zadoff [1] has given a method for investigating unsteady Newtonian flows with cylindrical or two-dimensional symmetry, using a Lagrangian formulation. In some fields the Newtonian approximation applied to unsteady problems in one space dimension is known under the term "snowplow" approximation.

We assume that the body has a surface whose motion is known, and that the shock layer follows the body surface without separation. As a first step, we consider the motion of a single particle which is constrained to move on the surface but on which the only force exerted is directed normal to the surface. All velocities are reduced in our notation by being divided by a constant reference velocity U; this velocity, which cannot be defined unambiguously in all cases of unsteady flow, is the one which appears in the definition of the pressure coefficient C_p. In most cases U is to be identified as the average flight speed of a body. The dimensionless velocity of the particle is $\mathbf{q}$, and that of a point on the body $\mathbf{q}_b$. Note that this is a body velocity and not the velocity of a fluid particle at the body boundary. The unit vector normal to the surface directed into the body is denoted $\mathbf{n}$; this convention is consistent with our definition of the curvature as positive on a convex body The velocity vector $\mathbf{q}$ is decomposed into a normal vector and a tangential vector,

$$\mathbf{q} = \mathbf{n}q_n + \mathbf{q}_t\,, \tag{3.9.1}$$

with

$$q_n = \mathbf{q} \cdot \mathbf{n}, \tag{3.9.2}$$

and

$$\mathbf{q}_t = \mathbf{n} \times (\mathbf{q} \times \mathbf{n}). \tag{3.9.3}$$

A tangential vector, one which is normal to $\mathbf{n}$, is a two-dimensional vector in the surface of the type we have considered earlier. A normal vector, one which is parallel to $\mathbf{n}$, may be considered as a scalar with respect to the two-dimensional space of the shock surface.

For the boundary condition on the body, note first that the body can be represented by an equation of the form (2.1.5). The boundary condition on the body may be expressed in the form $\partial f/U\,\partial t + \mathbf{q} \cdot \nabla f = 0$. The same expression holds for a point on the body, with $\mathbf{q}$ replaced by $\mathbf{q}_b$. The relative velocity vector $\mathbf{q}_r$ is defined as

$$\mathbf{q}_r = \mathbf{q} - \mathbf{q}_b\,, \tag{3.9.4}$$

with $\mathbf{q}_b$ the body velocity at the point on the body where the fluid particle of interest is. The difference between the two conditions on $\mathbf{q}$ and $\mathbf{q}_b$ yields $\mathbf{q}_r \cdot \nabla f = 0$. Since ∇f has the direction of $\mathbf{n}$, the boundary condition may be written

$$\mathbf{q}_r \cdot \mathbf{n} = 0. \tag{3.9.5}$$

This condition is then simply that $\mathbf{q}_r$ be a tangential vector.

The body velocity $\mathbf{q}_b$ is a function only of position on the body surface. It may be differentiated in the surface, using the two-dimensional Nabla operator ∇ in the surface. The derivative $\nabla\mathbf{q}_b$ thus has six components, as $\mathbf{q}_b$ is a three-dimensional vector. The quantity $(\nabla\mathbf{q}_b)\cdot\mathbf{n}$ is a tangential vector which can be identified as a vector at right angles to $(\nabla\times\mathbf{q}_b)_t$. It may also be identified as $-d\mathbf{n}/U\,dt$, with d/dt the time derivative for an observer moving with a body point. We define the angular velocity $U\boldsymbol{\omega}_b$ of the body surface for this observer as

$$\boldsymbol{\omega}_b = (\nabla\times\mathbf{q}_b)_t = \mathbf{n}\times\frac{d\mathbf{n}}{U\,dt} = [(\nabla\mathbf{q}_b)\cdot\mathbf{n}]\times\mathbf{n}. \tag{3.9.6a}$$

We may express this relation as

$$\mathbf{n}\times\boldsymbol{\omega}_b = -\frac{d\mathbf{n}}{U\,dt} = (\nabla\mathbf{q}_b)\cdot\mathbf{n} = \nabla(\mathbf{q}_b\cdot\mathbf{n}) + \mathbf{q}_b\cdot\mathfrak{K}, \tag{3.9.6b}$$

where

$$\mathfrak{K} = -\nabla\mathbf{n} \tag{3.9.7}$$

is the curvature dyadic or tensor of the surface. The fact that $\mathfrak{K}$ is symmetric may be readily shown, as for example by showing that $\mathbf{n}$ may be expressed locally in the three-dimensional space as the gradient of a scalar. In defining $\boldsymbol{\omega}_b$ we have again eliminated the dimension time by dividing by U. The angular velocity of the surface as measured by an observer moving with the fluid particle is

$$\boldsymbol{\omega} = \boldsymbol{\omega}_b + \mathbf{q}_r\cdot\mathfrak{K}\times\mathbf{n}, \tag{3.9.8}$$

also defined as a tangential vector. This angular velocity appears in the expression for the time derivative of $\mathbf{n}$ with respect to an observer moving with the particle,

$$\frac{D\mathbf{n}}{Dt} = U\boldsymbol{\omega}\times\mathbf{n}. \tag{3.9.9}$$

We have here used capital D in the derivative to indicate a material derivative. The normal component of the acceleration of a point on the body is denoted

$$a_b = U^{-1}\frac{d\mathbf{q}_b}{dt}\cdot\mathbf{n} \tag{3.9.10}$$

with the dimension time eliminated, and we shall treat it as a scalar. The tangential component of this body acceleration will not enter into our analysis.

The reader may have noticed that the description of the motion of the body surface is arbitrary to some extent. An alternative equally valid description of the surface motion may be obtained by adding an arbitrary tangential vector $\delta\mathbf{q}_b$ to the body velocity $\mathbf{q}_b$. While we could make the description unique by requiring $\mathbf{q}_b$ to be a normal vector, such a procedure would greatly complicate most calculations; for example, the curvature at a body point on a rigid body would become a function of time. It is definitely preferable for us to retain this arbitrariness and to make sure our results are independent of the choice of the description of the body motion. This independence is checked by checking the invariance of our basic equations under the transformation

$$\mathbf{q}_b = \mathbf{q}_b' + \delta\mathbf{q}_b; \qquad \delta\mathbf{q}_b \cdot \mathbf{n} = 0, \tag{3.9.11a}$$

$$\boldsymbol{\omega}_b = \boldsymbol{\omega}_b' + \delta\mathbf{q}_b \cdot \mathfrak{R} \times \mathbf{n}, \tag{3.9.11b}$$

$$a_b = a_b' + 2\boldsymbol{\omega}_b' \times \delta\mathbf{q}_b \cdot \mathbf{n} + \delta\mathbf{q}_b \cdot \mathfrak{R} \cdot \delta\mathbf{q}_b\,. \tag{3.9.11c}$$

Equation (3.9.11b) may be obtained from (3.9.6a) and (3.9.11a), noting that $d/U\,dt = d'/U\,dt + \delta\mathbf{q}_b \cdot \nabla$. Equation (3.9.11c) may be obtained by taking the time derivative of $\delta\mathbf{q}_b \cdot \mathbf{n} = 0$ and using the result with (3.9.6b) in an expansion of (3.9.10). The relative velocity $\mathbf{q}_r$ is changed in this transformation, while the quantities $\mathbf{q}$ and $\boldsymbol{\omega}$ are invariant.

We next take the total time derivative of $\mathbf{q}_t$ as expressed in (3.9.3). Equation (3.9.9) is used, and we obtain directly

$$U^{-1}\frac{D\mathbf{q}_t}{Dt} = \mathbf{n}(\boldsymbol{\omega} \times \mathbf{q} \cdot \mathbf{n}) + U^{-1}\mathbf{n} \times \left(\frac{D\mathbf{q}}{Dt} \times \mathbf{n}\right) + \mathbf{n} \times \boldsymbol{\omega}(\mathbf{q} \cdot \mathbf{n}). \tag{3.9.12}$$

The first term on the right-hand side is the contribution of the change in direction of $\mathbf{q}_t$ normal to the surface. The second term is the component of the particle acceleration tangential to the surface. The third term represents an interaction between angular velocity and normal velocity component.

Our analysis thus far has been purely kinematic, and it is at this point that we turn to dynamical considerations. We take our frame of reference to be unaccelerated, so that the acceleration of a particle is proportional to the force on it. We assume that the motion of the particle on the surface is frictionless, so that the force exerted by the surface on the particle is in the direction normal to the surface. With these assumptions the tangential component of the particle acceleration, i.e. the second term of (3.9.12), is identically zero. From (3.9.12) we obtain

$$\left(\frac{D\mathbf{q}_t}{Dt}\right)_t = U\mathbf{n} \times \boldsymbol{\omega}(\mathbf{q} \cdot \mathbf{n}). \tag{3.9.13a}$$

and

$$\left(\frac{D\mathbf{q}_t}{Dt}\right)_n = U\boldsymbol{\omega} \times \mathbf{q} \cdot \mathbf{n}. \tag{3.9.13b}$$

The first of these equations is the primary equation needed in the computation of trajectories. The second equation is a representation of the kinematic condition that the particle is constrained to the body (or shock) surface.

The force on the particle divided by its mass is equal to its acceleration, and may be calculated with respect to an observer who moves with a body point, with the velocity $\mathbf{q}_b$. The result is

$$\frac{D\mathbf{q}}{Dt} = U\mathbf{n}(a_b + 2\boldsymbol{\omega}_b \times \mathbf{q}_r \cdot \mathbf{n} + \mathbf{q}_r \cdot \mathfrak{K} \cdot \mathbf{q}_r), \tag{3.9.14}$$

in which the second term will be recognized as a Coriolis term. This equation may be obtained by calculating $\mathbf{n} \cdot [D(\mathbf{q}_r + \mathbf{q}_b)/Dt]$, using (3.9.6), (3.9.7), and derivatives of (3.9.5). The analysis is similar to that leading to (3.9.11c), and the similarity between the right-hand side of (3.9.14) and of (3.9.11c) may be noted. This equation, together with the definition (3.9.4) for $\mathbf{q}_r$, will serve as the basis for our calculation of the centrifugal pressure correction in unsteady Newtonian flow, while (3.9.13) is needed for the calculation of the shock layer structure. Note that if $\mathbf{q}_r$ had any normal component it would be annihilated in (3.9.14). Hence we may substitute $\mathbf{q}_t - \mathbf{q}_b$ in place of $\mathbf{q}_r$ in (3.9.14) with no other change, and this we shall do.

We now consider the Newtonian impact process in the unsteady case. The particle before impact has the velocity $U\mathbf{U}$, with $\mathbf{U}$ defined as dimensionless consistent with our definition of other velocities. If the atmosphere in which the body is moving is motionless, we define U as the velocity of our frame of reference, and $\mathbf{U}$ is simply a constant unit vector. The normal component of momentum relative to the body is proportional to $(\mathbf{U} - \mathbf{q}_b) \cdot \mathbf{n}$, and this quantity is transferred in Newtonian impact to the body. The tangential momentum relative to the body is conserved, and we have

$$\mathbf{n} \times [(\mathbf{U} - \mathbf{q}_b) \times \mathbf{n}] = \mathbf{n} \times (\mathbf{q}_r \times \mathbf{n}) = \mathbf{q}_r. \tag{3.9.15}$$

From (3.9.15) and (3.9.4) an expression for the initial value of $\mathbf{q}_t$ immediately after impact is obtained,

$$\mathbf{q}_{t_1} = \mathbf{n} \times (\mathbf{U} \times \mathbf{n}). \tag{3.9.16}$$

Here we have used the subscript 1 as before to denote conditions immediately after impact, at the top of the shock layer.

We are now ready to write down the equations for the pressure at a point on a body in unsteady Newtonian flow. The external flow is assumed to be a uniform one of constant density ρ_∞ and velocity U, with the vector $\mathbf{U}$ a constant unit vector. The mass per unit time per unit area impinging on the body is

$\rho_\infty U$ times $(\mathbf{U} - \mathbf{q}_b) \cdot \mathbf{n}$. The pressure immediately behind the shock is then given by the relation

$$\tag{3.9.17} \begin{aligned} \tfrac{1}{2}C_{p_s} &= [(\mathbf{U} - \mathbf{q}_b) \cdot \mathbf{n}]^2 \\ &= (\mathbf{U} \cdot \mathbf{n})^2 - 2(\mathbf{U} \cdot \mathbf{n})(\mathbf{q}_b \cdot \mathbf{n}) + (\mathbf{q}_b \cdot \mathbf{n})^2. \end{aligned}$$

The quantity $(\mathbf{U} \cdot \mathbf{n})$ is the $\sin \sigma$ of our steady flow analysis, and (3.9.17) appears as a generalization of (3.2.6).

Assuming that the shock layer structure is known in terms of $\mathbf{q}_t$ as a function of m, we form the quantities

$$\tag{3.9.18a} m_1 = \int_0^{m_1} dm,$$

$$\tag{3.9.18b} \mathbf{M}_1 = \int_0^{m_1} \mathbf{q}_t \, dm,$$

$$\tag{3.9.18c} \mathfrak{P}_1 = \int_0^{m_1} \mathbf{q}_t\mathbf{q}_t \, dm,$$

which are equivalent to those defined in (3.5.11) to (3.5.13). If the integrals in (3.9.18) are indefinite ones defined to be zero at the bottom of the layer, we would omit the subscripts 1 and could use the unsubscripted quantities to express the pressure in the layer in an equation analogous to (3.5.19). Here we only obtain the centrifugal pressure correction on the body, using Newton's third law and (3.9.14). This pressure correction is

$$\tag{3.9.19} \begin{aligned} \tfrac{1}{2}C_{p_s} - \tfrac{1}{2}C_{p_b} &= \mathfrak{K} : \mathfrak{P}_1 - 2\mathbf{q}_b \cdot \mathfrak{K} \cdot \mathbf{M}_1 + \mathbf{q}_b \cdot \mathfrak{K} \cdot \mathbf{q}_b m_1 \\ &\quad + 2\boldsymbol{\omega}_b \times \mathbf{M}_1 \cdot \mathbf{n} - 2\boldsymbol{\omega}_b \times \mathbf{q}_b \cdot \mathbf{n} m_1 + a_b m_1 , \end{aligned}$$

and includes a number of terms besides the one giving the generalized Busemann centrifugal correction.

To calculate the velocities in the unsteady shock layer of a general body we must obtain an analogue to the locus of entering streamlines. To do this we must consider the three-dimensional space consisting of the body surface with time as an additional coordinate. A particle which enters the shock layer at a given point and at a given instant follows a trajectory in this three-dimensional space which is to be calculated with the help of (3.9.13) and (3.9.16). These trajectories form a three-parameter family. The trajectories which pass through a given point at a given instant form a one-parameter family, and the locus in the three-dimensional space of the entering points of the trajectories of this family is our unsteady form of the locus of entering streamlines. From such a calculation we would obtain the sequence of velocities $\mathbf{q}_t$ in the shock layer at a given point at a given instant. As in the steady case there will be a trajectory equation of the form $g(\xi, \eta, t; \xi_1, \eta_1, t_1) = 0$, in analogy with (3.6.1). Holding the coordinates (ξ, η, t) of an entering point fixed this equation gives the corresponding trajectory. Holding the coordinates (ξ_1, η_1, t_1) of the point of interest fixed this equation gives the corresponding locus of entering streamlines.

To complete the picture of the shock layer by obtaining the scale variable m we again need an analysis based on continuity. We choose two neighboring points to the point investigated and form a parallelogram with a known differential area on the body at the instant investigated. The locus of the loci of entering streamlines through points on the periphery of the differential area is a tube in the three-dimensional space. The quantity $U(\mathbf{U} - \mathbf{q}_b) \cdot \mathbf{n}$ is proportional to the entering mass flow per unit area. A differential element of volume in the tube times this quantity is equal to the differential area element on the surface times dm. We may thus obtain a relation similar to (3.6.2) which connects m with the locus of entering streamlines. The arbitrariness which we had in the choice of neighboring point in the steady case appears in the unsteady case in a different form. An alternative procedure to that given above may be used, in which one of the neighboring points is chosen at the same instant and the other at a different instant but on the normal through the point investigated. This procedure yields an expression for one of the components of $d\mathbf{M}$ rather than for dm, and the two procedures must be consistent. Again, we may choose the procedure which appears simplest, and we may use the arbitrariness to furnish us with a method of checking our results.

The method outlined above for the general case is not one which is at all easy to apply analytically. The principal difficulty is that the three-dimensional space involved is generally curved and incapable of being represented metrically (as opposed to topologically) in a three-dimensional cartesian space. We would be forced in most cases to a cinematographic or stroboscopic representation, with a sequence of instantaneous pictures of the two-dimensional surface at small time intervals. In such a representation, a locus of entering streamlines would consist of a point on each picture. In the two-dimensional unsteady case we can represent the two-dimensional time-distance space as a surface in a three-dimensional cartesian space.

We present here only one very simple example of the calculation of the structure of an unsteady Newtonian shock layer. The problem is that of a flat plate at constant angle of attack σ which oscillates sinusoidally in a direction normal to its surface with angular frequency Uk and half-amplitude a (see Fig. 3–24a). The quantity k here is a wave number and has the dimensions of inverse distance. We take distance along the plate from the leading edge as x and consider the trajectories on a plot of x versus t. From (3.9.13) and (3.9.16) all particles have the same constant value of q_t, equal to $\cos \sigma$. On the (x, t)

plot the trajectories are all straight lines of slope $U \cos \sigma$ (see Fig. 3–24b). Accordingly, the locus of entering streamlines for a point is a straight line of the same slope through the point, and is given by

(3.9.20) $$x_1 - x = U(t_1 - t) \cos \sigma.$$

With $x = x_0 = 0$ at the leading edge, the corresponding value of t at the leading edge is

(3.9.21) $$t_0 = t_1 - \frac{x_1}{U \cos \sigma}.$$

The distance of the flat plate from its central position is $a \sin Ukt$, its normal velocity is

(3.9.22) $$\mathbf{q}_b \cdot \mathbf{n} = -ak \cos Ukt,$$

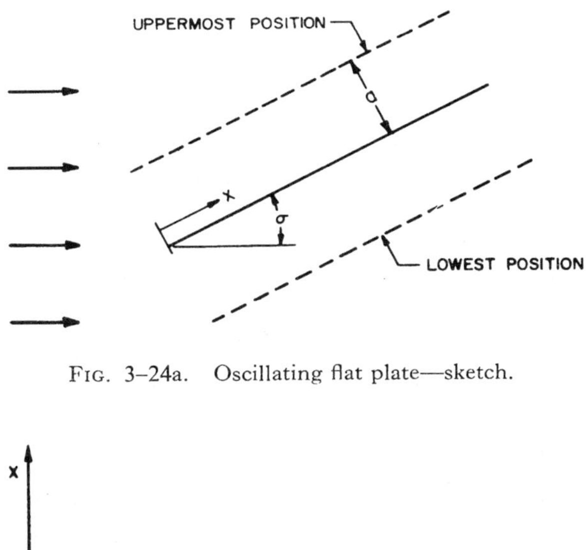

FIG. 3–24a. Oscillating flat plate—sketch.

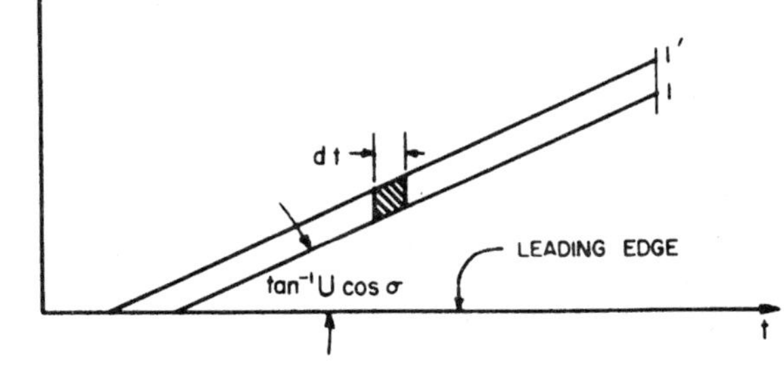

FIG. 3–24b. Oscillating flat plate—calculation of m.

and its acceleration is

$$a_b = ak^2 \sin Ukt. \tag{3.9.23}$$

The entering mass flow per unit area is given by

$$(\mathbf{U} - \mathbf{q}_b) \cdot \mathbf{n} = \sin \sigma + ak \cos Ukt. \tag{3.9.24}$$

We may now calculate m_1 (see Fig. 3–24b), and obtain

$$\begin{aligned} m_1 &= \int_{t_0}^{t_1} (\sin \sigma + ak \cos Ukt) U \, dt \\ &= x_1 \tan \sigma + a \sin Ukt - a \sin (Ukt_1 - kx_1 \sec \sigma). \end{aligned} \tag{3.9.25}$$

The problem chosen is one in which neither curvature nor Coriolis terms appear, and the pressure on the body may be calculated directly from (3.9.17) and (3.9.19),

$$\begin{aligned} \tfrac{1}{2} C_{p_b} &= \sin^2 \sigma + 2ak \sin \sigma \cos Ukt_1 - ak^2 x_1 \tan \sigma \sin Ukt_1 \\ &\quad + a^2k^2 \cos 2Ukt_1 + \tfrac{1}{2}a^2k^2(1 - \cos 2Ukt_1) \cos (kx_1 \sec \sigma) \\ &\quad - \tfrac{1}{2}a^2k^2 \sin 2Ukt_1 \sin (kx_1 \sec \sigma). \end{aligned} \tag{3.9.26}$$

Of the terms appearing in this expression, the first is present in steady flow, the second is a damping term, and the third is an effective-mass term. The remainder of the terms are quadratic in ak, and are small if ak is small.

Certain assumptions are implicit in the foregoing analysis. A principal one is that the mass flow of (3.9.24) is never negative, and that $ak < \sin \sigma$. If the mass flow is negative, incidentally, an intermittent shock layer forms on the reverse side of the plate. A more stringent assumption is that the pressure on the body is never negative. If the pressure as given by the analysis becomes negative, a free layer must form with zero pressure behind it; the appropriate solution with the free layer will be much more complicated than the one given and will involve a reattachment process and perhaps a reimpingement of a more violent nature. A simple necessary condition for the pressure on the body to be never negative in our solution (3.9.26) is that $ak^2c < \sin \sigma \cos \sigma$, where c is the chord of the plate. The condition for nonnegative mass flow is also a necessary condition for nonnegative pressure.

We turn now to the problem of a body for which the deviations from steady flow are small and for which a linearized perturbation theory is adequate. We consider that the quantities $\mathbf{q}_b$, $\boldsymbol{\omega}_b$, and a_b are of the first order of smallness, and express the increment in the pressure from (3.9.17) and (3.9.19). We define an angle change $\delta\boldsymbol{\theta}_b$ for the body such that

$$\delta \mathbf{n} = \delta \boldsymbol{\theta}_b \times \mathbf{n}, \tag{3.9.27}$$

and

$$\frac{d\boldsymbol{\theta}_b}{dt} = \boldsymbol{\omega}_b . \tag{3.9.28}$$

In terms of $\delta\boldsymbol{\theta}_b$ the linearized increment in pressure on the body is given by

$$\begin{aligned} \tfrac{1}{2}\delta C_{p_b} = {} & 2(\mathbf{U}\cdot\mathbf{n})(\mathbf{U}\times\delta\boldsymbol{\theta}_b - \mathbf{q}_b)\cdot\mathbf{n} + 2\mathbf{q}_b\cdot\boldsymbol{\mathfrak{K}}\cdot\mathbf{M}_1 \\ & - 2\boldsymbol{\omega}_b\times\mathbf{M}_1\cdot\mathbf{n} - a_b m_1 - \delta\boldsymbol{\mathfrak{K}} : \boldsymbol{\mathfrak{P}}_1 - \boldsymbol{\mathfrak{K}} : \delta\boldsymbol{\mathfrak{P}}_1 . \end{aligned} \tag{3.9.29}$$

The quantities m_1, $\mathbf{M}_1$, and $\boldsymbol{\mathfrak{P}}_1$ in this equation are defined with respect to the reference steady flow. The effect of the unsteady flow on the structure of the shock layer does appear in this linearized pressure expression, although only in the last term involving $\delta\boldsymbol{\mathfrak{P}}_1$. Hence, an analysis of unsteady shock layer structure is needed even for the linearized problem, but the aditional analysis over that already necessary for the steady flow problem is appreciably simpler than that for the general case. We shall not go into the details of such an incremental analysis here.

If the body is a rigid body we may define the body variables in terms of their values at a given point in the body which is chosen as origin of our coordinate system. We express the linearized relations

$$\delta\boldsymbol{\theta}_b = \boldsymbol{\theta}_c \tag{3.9.30a}$$

$$\boldsymbol{\omega}_b = \boldsymbol{\omega}_c = \frac{d\boldsymbol{\theta}_c}{dt}, \tag{3.9.30b}$$

$$\mathbf{q}_b = \mathbf{q}_c + \boldsymbol{\omega}_c \times \mathbf{r}_b , \tag{3.9.30c}$$

$$a_b = (\mathbf{a}_c + \boldsymbol{\alpha}_c \times \mathbf{r}_b)\cdot\mathbf{n}, \tag{3.9.30d}$$

in which the quadratic terms have been dropped, and quantities with subscript c are for the chosen center of the body. The quantity $\boldsymbol{\alpha}_c$ is the angular acceleration of the body, and $\mathbf{r}_b$ is the distance vector from the center. Combining (3.9.29) with (3.9.30), with $\delta\boldsymbol{\mathfrak{K}} = 0$, there results

$$\begin{aligned} \tfrac{1}{2}\delta C_{p_b} = {} & 2\boldsymbol{\theta}_c\cdot[(\mathbf{U}\cdot\mathbf{n})\mathbf{n}\times\mathbf{U}] \\ & + 2\mathbf{q}_c\cdot[\boldsymbol{\mathfrak{K}}\cdot\mathbf{M}_1 - \mathbf{n}(\mathbf{U}\cdot\mathbf{n})] \\ & + 2\boldsymbol{\omega}_c\cdot[(\mathbf{n}\mathbf{U}\cdot\mathbf{n} - \boldsymbol{\mathfrak{K}}\cdot\mathbf{M}_1)\times\mathbf{r}_b + \mathbf{n}\times\mathbf{M}_1] \\ & - \mathbf{a}_c\cdot\mathbf{n}m_1 + \boldsymbol{\alpha}_c\cdot\mathbf{n}\times\mathbf{r}_b m_1 - \delta\boldsymbol{\mathfrak{P}}_1 : \boldsymbol{\mathfrak{K}}. \end{aligned} \tag{3.9.31}$$

This pressure times $\mathbf{n}\, dS_b$ is integrated to obtain the total force on the body, where dS_b is the differential element of area of the body surface.

If the flow is oscillatory, with the perturbations in body shape and orientation varying sinusoidally in time, the basic parameter is the reduced frequency.

The reduced frequency is the frequency times the body scale divided by the free stream velocity. In the example above it is kc, where c is the chord of the flat plate. This reduced frequency is generally small in applications, and an ordering of terms in powers of the reduced frequency is of interest. The terms in $\delta\boldsymbol{\theta}_b$, $\delta\mathfrak{K}$, and $\delta\mathfrak{P}_1$ in (3.9.29) are of zeroth order, the terms in $\mathbf{q}_b$ and $\boldsymbol{\omega}_b$ are of first order, and the term in a_b is of second order in the reduced frequency. Note that the zeroth order terms are those which would be obtained by perturbing the steady flow result (3.5.20). In (3.9.31) the terms in $\boldsymbol{\theta}_c$ and $\delta\mathfrak{P}_1$ are of zeroth order, those in $\mathbf{q}_c$ and $\boldsymbol{\omega}_c$ are of first order, and the remaining terms are of second order.

Let us close our treatment of unsteady Newtonian flow with a listing of some of the subjects which we have not investigated. The question of anomalies has been mentioned. It should be possible to obtain a threefold condensation in unsteady Newtonian flow, i.e. a "Newtonian shock point". The reimpingement of a free layer presents new aspects, as for example local pressures which instantaneously correspond to an infinite force on the body; such a phenomenon may be termed a "slap". The linearized calculation of $\delta\mathfrak{P}_1$ in (3.9.29) has not been developed; the calculation of complete force coefficients for quasi-steady and oscillatory motions needs such a development. The pseudosteady problem of a body flying at constant velocity on a curved path or with a constant rate of roll has not been investigated. Unsteady Newtonian free layers and sails have not been considered. It is clear that there is ample room for exploration in this subject and that, following the philosophy we have expressed earlier, such exploration should help us develop insight into general unsteady hypersonic flows.

CHAPTER IV

CONSTANT-DENSITY SOLUTIONS

1. The wedge

A fairly common misconception held by students of fluid mechanics is that "incompressible" and "constant density" are synonymous. Any aerodynamicist knows that, while low speed air flows are essentially at constant density, air is a highly compressible fluid. Any oceanographer knows that, while sea water is essentially incompressible, it is far from being a constant-density fluid. The essential point here is that the density in a compressible fluid can be essentially constant if the changes in pressure experienced by the fluid in the flow are small. We shall consider constant-density hypersonic flows, though we should never consider the fluid in a hypersonic flow as incompressible.

In hypersonic flow the approximation that the flow is at a constant density is a useful one in certain cases, and is exact in the case of steady flow on a wedge. The approximation is limited to those cases in which the pressure changes in the flow field are small. Except for the case of a wedge, this limitation usually implies the restriction that the shock layer be thin, and this limitation in turn implies a relationship with Newtonian theory. With but few exceptional cases, then, constant-density theory may be considered a part of thin shock layer theory. The distinction is only one of approach. The principal solutions of this chapter can be obtained by the methods of Chapter V. Conversely, the assumption of constant density is inherent in some of the approximate solutions of Chapter V. In this chapter we treat cases for which the assumption at the outset of constant density is a productive one. The applicable cases are in two categories, those for which the shock inclination angle σ is essentially constant, and those for which the region of interest is near a stagnation point.

There are two principal reasons for our studying constant-density solutions. The first is that we obtain from them new valid information beyond that available from Newtonian theory and with an analysis of minimum complexity. And we obtain thereby some insight into how Newtonian theory should be applied to practical flows. The second reason is that the constant-density solutions are needed as first approximations to more general solutions based upon the assumption that the shock layer is moderately thin.

Our first step in investigating these constant-density solutions will be to obtain as general a solution as is feasible, without regard to applicability. We then consider the case in which the shock layer is thin, and compare these results with those available from Newtonian theory. Finally, we shall discuss the limitations on the validity of our results, and shall estimate the effects of variations in the density.

The basic equation needed for the wedge has already been obtained (1.5.4), and is repeated here for convenience

$$\tan(\sigma - \theta_s) = \epsilon \tan \sigma \tag{4.1.1}$$

(see Fig. 4–1). Variants of this equation which are useful are

$$\frac{\epsilon}{1-\epsilon} = \frac{\cos^2 \sigma (\tan \sigma - \tan \theta_s)}{\tan \theta_s} = \frac{\cos \sigma \sin(\sigma - \theta_s)}{\sin \theta_s}, \tag{4.1.2}$$

$$\sin \theta_s = (1 - \epsilon) \sin \sigma \cos(\sigma - \theta_s). \tag{4.1.3}$$

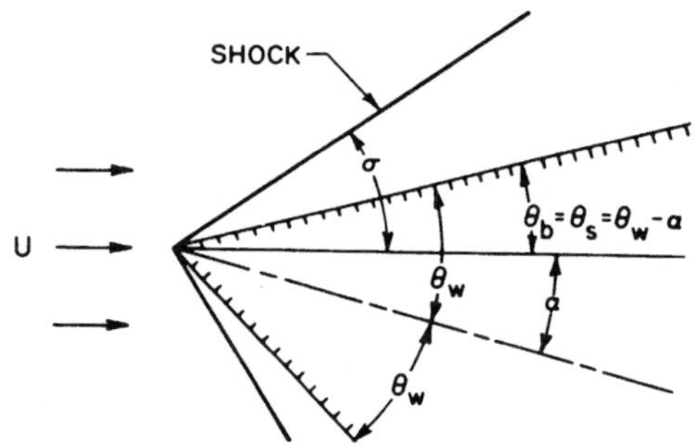

FIG. 4–1. Flow on a wedge.

From (4.1.1) we may also obtain the equation

$$\tan \theta_s \tan^2(\sigma - \theta_s) - (1 - \epsilon) \tan(\sigma - \theta_s) + \epsilon \tan \theta_s = 0, \tag{4.1.4}$$

which we may solve for $\tan(\sigma - \theta_s)$. The result is

$$\tan(\sigma - \theta_s) = \epsilon^{1/2} f(\eta_w) \tag{4.1.5a}$$

with

$$\eta_w = \epsilon^{1/2}(1 - \epsilon)^{-1} \tan \theta_s, \tag{4.1.5b}$$

where

$$f(\eta) = \frac{1}{2\eta} \mp \sqrt{\frac{1}{4\eta^2} - 1} = \frac{2\eta}{1 \pm \sqrt{1 - 4\eta^2}}. \tag{4.1.6}$$

The function $f(\eta)$ is termed the "detachment function", and is plotted in Fig. 4–2. Combining (4.1.1) and (4.1.5) gives us also

$$\tan \sigma = \epsilon^{-1/2} f(\eta_w). \tag{4.1.7}$$

This detachment function is basic for the solution on a wedge, and appears again in the solution for the cone. The detachment function does not give us

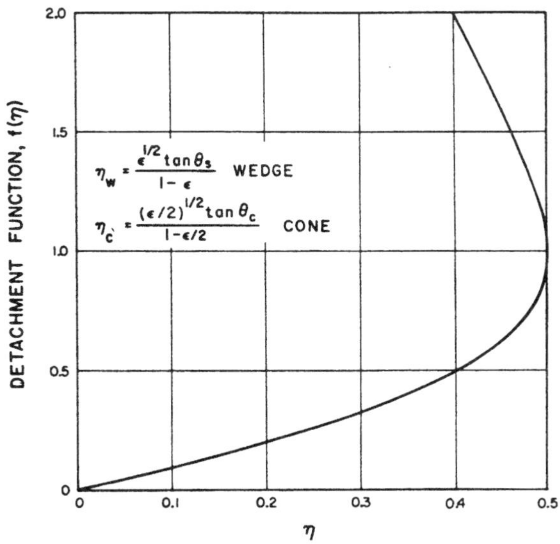

FIG. 4–2. Detachment function.

an immediate solution if θ_s is given, as then ϵ is an unknown. It does, however, give us the solution if σ is known, provided we have determined ϵ as a function of σ. With given free stream conditions, ϵ is a function of σ alone.

The relations just given are in terms of the flow inclination angle θ_s immediately behind the shock, and are valid for any oblique shock. In the wedge solution we are interested in, the shock is straight and the flow behind it is uniform. We have then, for the wedge, the relation

$$\theta_b = \theta_s , \tag{4.1.8}$$

and our equations are all interpretable with the inclination angle of the body surface θ_b in place of θ_s. Note that for a symmetric wedge of half-angle θ_w and angle of attack α the angle θ_b is equal to $\theta_w \pm \alpha$. The importance of the function $f(\eta)$ is that if ϵ is presumed known it gives the wedge solution explicitly in terms of the body angle. The upper sign in (4.1.6) corresponds to the weak solution relative to the detachment point, and the lower sign to the strong

quadratic, which is valid if ϵ_n is small. If $\tan\theta_s$ is of the order of $\epsilon_n^{1/2}$ or less we may drop the first term in (4.1.18). If $\tan\theta_s$ is of order $\epsilon_n^{-1/2}$ (i.e., of the order at detachment) we may drop the last term in (4.1.18). If $\tan\theta_s$ is of the order of 1, we can find the approximate solution

$$\tan\sigma = \left(1 + \epsilon_n + \frac{2}{\gamma+1}\frac{1}{M_\infty^2}\right)\tan\theta_s + \epsilon_n \tan^3\theta_s + \frac{2}{\gamma+1}\frac{1}{M_\infty^2\tan\theta_s}, \tag{4.1.19}$$

by successive approximation based upon the smallness of the first and last term in (4.1.18). This result may also be obtained from our analysis of the detachment function using (4.1.9). It may be noted that $\sigma - \theta_s$ has a minimum between $\theta_s = 0$ and the detachment point.

Solutions for star-shaped conical bodies using wedge solutions have been given by Maikapar [2] and by Gonor [5]. For Maikapar's solution we consider a shock shape in the form of a right pyramid based upon a regular polygon. The streamlines from the edges of the pyramid form a star-shaped conical body. The cross section of the body and the shock system is shown in Fig. 4–3.

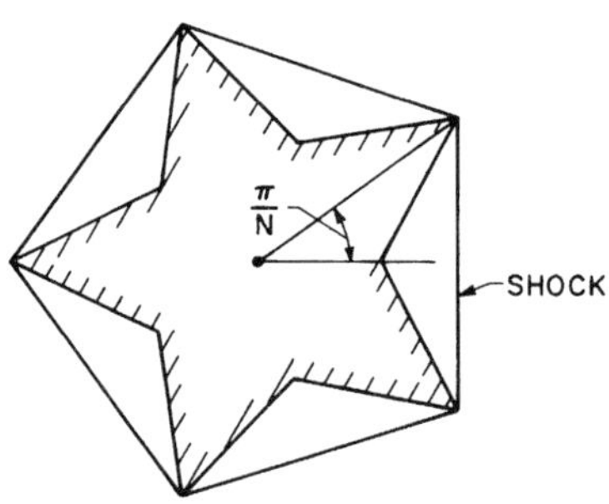

FIG. 4–3. Maikapar's solution.

Such a body may have much less drag than a circular cone of the same cross-sectional area. Delta wings of inverted V or W shape which are essentially the same as pieces of Maikapar's star-shaped conical bodies have been suggested by Nonweiler [1] (see also Peckham [1]).

The shock waves in Maikapar's solution may be either strong or weak. In case it is the strong solution which applies, the solution is not realistic. Gonor's solution is one which permits sharp points to the star with the weak shock solution. The number of points in the star and the inclination angle of the leading edge of the star is selected. A straight shock attached to the leading edge is chosen, and this determines the angle of the body at the tip of the star. The shock wave is reflected from a plane of symmetry whose azimuth angle is π/N different from that of the tip of the star, where N is the number of points. This reflection must exist and be regular. The reflected shock returns and strikes the arm of the star. A suitable change in the angle of the arm of the star permits the reflected shock to terminate at the body without further reflection. The cross section of the body and the shock system is shown in Fig. 4–4. Such a body may also have

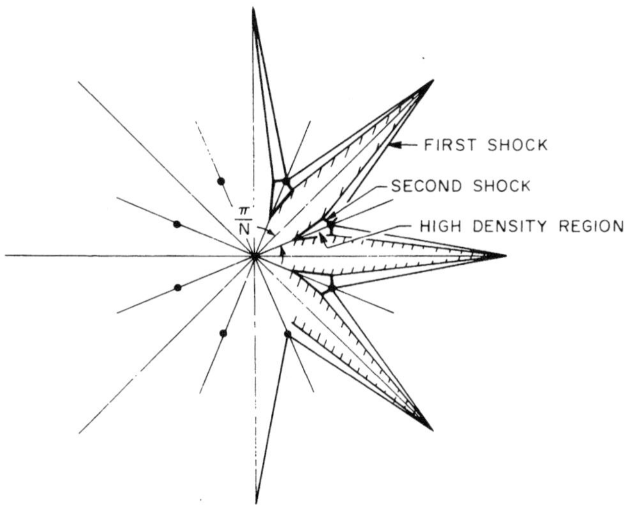

FIG. 4–4. Gonor's solution.

a drag which is far less than that of a circular cone of the same cross-sectional area.

An interesting feature of Gonor's solution with ϵ small is its relation to Newtonian theory. Newtonian theory on a star-shaped body would predict a Newtonian shock line in the trough of the star. The high density flow behind the second shock in Gonor's solution is the vestige of the shock line at finite density ratio.

We close with a look at the case of an almost-normal shock with ϵ small, for which $\tan \sigma$ is approximately $\sec \sigma$ and $(1 - \epsilon)$ is approximately 1. The basic equation for this case may be obtained directly from (4.1.2), and is

$$\cos \sigma = \epsilon \tan \theta_s . \tag{4.1.20}$$

Note here that $\tan \theta_s$ may be of any order of magnitude with ϵ and $\cos \sigma$ both small. The Newtonian approach is valueless in this case even with ϵ very small, as the shock layer is not thin.

2. The cone

For the case of the right circular cone we introduce spherical coordinates, and seek a velocity potential with axial symmetry. With the shock wave at constant inclination the flow is isentropic, and therefore irrotational; thus we know a velocity potential exists. The coordinates are the spherical radius R and the polar angle measured from the downstream axis ϑ (see Fig. 4–5). In

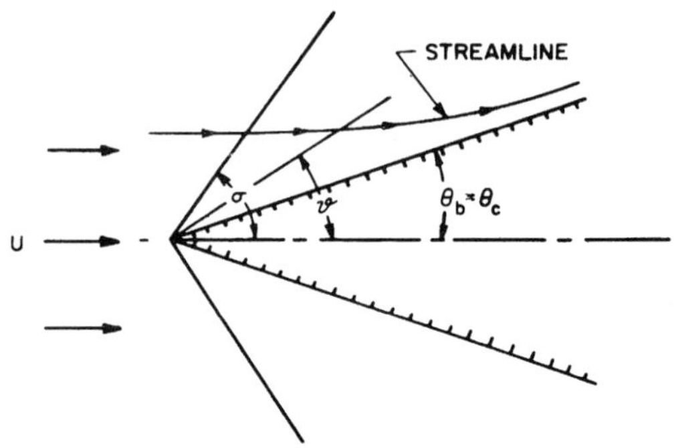

FIG. 4–5. Flow on a cone.

place of ϑ we shall usually use in this section the quantity μ, defined by

(4.2.1) $$\mu = \cos \vartheta.$$

The velocity potential is designated $U\phi$, and with the density constant and the flow axisymmetric ϕ satisfies the equation

(4.2.2) $$((1 - \mu^2)\phi_\mu)_\mu + (R^2\phi_R)_R = 0.$$

The subscripts μ and R here indicate partial differentiation, and the radial and polar velocity components are $U\phi_R$ and $-UR^{-1}\sqrt{1 - \mu^2}\,\phi_\mu$, respectively. A solution which is conical must have ϕ proportional to R, and ϕ must be of the general form

(4.2.3) $$\phi = R[AP_1(\mu) + BQ_1(\mu)],$$

where P_1 and Q_1 are Legendre functions. These functions are defined

(4.2.4a) $$P_1(\mu) = \mu,$$

(4.2.4b) $$Q_1(\mu) = \tfrac{1}{2}\mu \ln \frac{1 + \mu}{1 - \mu} - 1.$$

The function Q_1 may be related to its derivative and to the function Q_0 by the relation

$$Q_1(\mu) = \mu Q_1'(\mu) - \frac{1}{1-\mu^2} = \mu Q_0 - 1, \tag{4.2.5}$$

where

$$Q_0(\mu) = \tfrac{1}{2}\ln\frac{1+\mu}{1-\mu}. \tag{4.2.6}$$

The radial component of velocity is constant across the shock, and equals $U\mu_s$ there, where $\mu_s = \cos\sigma$. The polar component of velocity is normal to the shock, and equals $-\epsilon U\sqrt{1-\mu_s^2}$ immediately behind the shock. These two conditions expressed in terms of the solution (4.2.3) give

$$A + B\mu_s^{-1}Q_1(\mu_s) = 1, \tag{4.2.7a}$$

$$A + BQ_1'(\mu_s) = \epsilon. \tag{4.2.7b}$$

We may evaluate the constants A and B to be

$$A = \epsilon + (1-\epsilon)\mu_s(1-\mu_s^2)Q_1'(\mu_s), \tag{4.2.8a}$$

$$B = -(1-\epsilon)\mu_s(1-\mu_s^2). \tag{4.2.8b}$$

The boundary condition on the body is that the polar component of velocity is zero, and gives us

$$\frac{\epsilon}{1-\epsilon} = \mu_s(1-\mu_s^2)[Q_1'(\mu_c) - Q_1'(\mu_s)], \tag{4.2.9}$$

where $\mu_c = \cos\theta_c$. We use here the subscript c in place of b to indicate our results are for the right circular cone.

The pressure coefficient is a function of μ alone, and may be evaluated by a straightforward application of the form of Bernoulli's equation applicable to constant-density flows. In evaluating the difference in pressure from the pressure immediately behind the shock, we find no contribution from the part of ϕ involving A alone, or no contribution in A^2. The terms in AB and B^2 may be simplified through the use of (4.2.5), so that only the function Q_0 appears. The result may be expressed

$$\begin{aligned}\tfrac{1}{2}C_p - \tfrac{1}{2}C_{p_s} = {} & -\epsilon^{-1}AB[Q_0(\mu) - Q_0(\mu_s)] \\ & + \tfrac{1}{2}\epsilon^{-1}B^2\left[\frac{1}{1-\mu_s^2} - \frac{1}{1-\mu^2} + Q_0^2(\mu_s) - Q_0^2(\mu)\right],\end{aligned} \tag{4.2.10}$$

where

$$\tfrac{1}{2}C_{p_s} = (1-\epsilon)\sin^2\sigma = (1-\epsilon)(1-\mu_s^2) \tag{4.2.11}$$

is the pressure coefficient behind the shock. Of perhaps more interest to us is the pressure gradient, which may be expressed

$$\frac{d\frac{1}{2}C_p}{d\mu} = -\frac{\epsilon^{-1}B}{1-\mu^2}[A + BQ_1'(\mu)] \tag{4.2.12}$$

or with relation (4.2.9) and the values of A and B from (4.2.8),

$$\frac{d\frac{1}{2}C_p}{d\mu} = \frac{(1-\epsilon)^2\mu_s^2(1-\mu_s^2)^2}{\epsilon(1-\mu^2)}[Q_1'(\mu_c) - Q_1'(\mu)]. \tag{4.2.13}$$

It may be noted that the pressure gradient vanishes on the body.

The general analysis just given is based upon the assumption that the density in the flow field behind the shock is constant. This assumption is approximately valid only if the density ratio ϵ is sufficiently small, in which case the shock layer is thin unless the shock is almost normal. If the shock layer is thin we may expand the difference between the two values of Q_1' which appears in (4.2.9) and (4.2.13) in a Taylor series about $\mu = \mu_s$. The result may be expressed

$$\mu_s(1-\mu_s^2)[Q_1'(\mu) - Q_1'(\mu_s)] = 2\bar{\xi} + 4\bar{\xi}^2 + O(\bar{\xi}^3) \tag{4.2.14a}$$

$$= 2\xi + \xi^2 + O(\xi^3), \tag{4.2.14b}$$

where

$$\bar{\xi} = \mu_s(1-\mu_s^2)^{-1}(\mu - \mu_s), \tag{4.2.15}$$

and

$$\xi = \frac{\cos^2\sigma\,(\tan\sigma - \tan\vartheta)}{\tan\vartheta} = \frac{\cos\sigma\sin(\sigma-\vartheta)}{\sin\vartheta}. \tag{4.2.16}$$

On the body the variable ξ has the same form as that appearing in (4.1.2). The limitations of the constant-density approximation require that ξ remain small and restrict us to only one valid term in the series which appear in (4.2.14), provided μ_s is not small. Allowing an error in the second term of an expansion in ξ_c we may satisfy the boundary condition (4.2.9) with the approximate expression

$$\frac{\epsilon}{1-\epsilon} = \frac{2\xi_c}{1-\xi_c}, \tag{4.2.17}$$

which is equivalent to

$$\frac{\frac{1}{2}\epsilon}{1-\frac{1}{2}\epsilon} = \xi_c. \tag{4.2.18}$$

Comparison of this result with (4.1.2) shows that, within the approximation used, the geometric relation between shock angle and body angle in the solution on a cone is the same as that on a wedge with ϵ reduced by the factor $\frac{1}{2}$. To be more accurate we should replace ϵ by $\frac{1}{2}\epsilon(1 + \frac{1}{4}\epsilon)$, but we may not justifiably keep the higher-order term within the limits of our approximation. A similar factor of $\frac{1}{2}$ connects the corrections to the tangent-wedge and tangent-cone pressure results if ϵ is small; cf. (7.3.7) and (7.3.9). Thus we may write in place of (4.1.7), for example,

$$\tan \sigma = (\tfrac{1}{2}\epsilon)^{-1/2} f(\eta_c) \tag{4.2.19a}$$

with

$$\eta_c = (\tfrac{1}{2}\epsilon)^{1/2}(1 - \tfrac{1}{2}\epsilon)^{-1} \tan \theta_c . \tag{4.2.19b}$$

The structure of the solution is different for the two cases, of course. In place of (4.1.8) we have the approximate relation

$$\sigma - \theta_c = \tfrac{1}{2}(\sigma - \theta_s) = \tfrac{1}{2}\epsilon \tan \sigma. \tag{4.2.20}$$

If we re-express (4.2.13) in terms of the polar angle ϑ, we obtain

$$\frac{d\frac{1}{2}C_p}{d\vartheta} = -\frac{(1 - \epsilon)^2 \cos \sigma \sin^2 \sigma}{\epsilon \sin \vartheta} [2(\xi_c - \xi) + O(\xi_c^2 - \xi^2)], \tag{4.2.21}$$

or to the accuracy of our constant-density theory,

$$\frac{d\frac{1}{2}C_p}{d\vartheta} = -2\epsilon^{-1} \cos^2 \sigma \, (\vartheta - \theta_c). \tag{4.2.22}$$

This expression we may integrate to obtain

$$\tfrac{1}{2}C_p = (1 - \epsilon) \sin^2 \sigma + \epsilon^{-1} \cos^2 \sigma \, (\sigma - \vartheta)(\sigma + \vartheta - 2\theta_c). \tag{4.2.23}$$

On the body we may use (4.2.20) to express the pressure

$$\tfrac{1}{2}C_{p_b} = (1 - \tfrac{3}{4}\epsilon) \sin^2 \sigma. \tag{4.2.24}$$

The formula for the cone which is analogous to (4.1.3) is

$$\sin \theta_c = (1 - \tfrac{1}{2}\epsilon) \sin \sigma \cos (\sigma - \theta_c), \tag{4.2.25}$$

and the use of this permits us to rewrite (4.2.24) as

$$\tfrac{1}{2}C_{p_b} = \frac{\sin^2 \theta_c}{(1 - \tfrac{1}{4}\epsilon) \cos^2 (\sigma - \theta_c)} \tag{4.2.26}$$

within the accuracy of our approximation. The result may be compared with (4.1.13) for the wedge, and we note again that the factor in $\cos(\sigma - \theta_c)$ may be dropped if $\epsilon^{1/2} \tan \theta_c$ is small.

Following an alternative approach used earlier by Hayes [3], we could start with the assumption that the shock layer on the cone is thin, obtain an approximate picture of the structure of the layer on the basis of continuity arguments, and derive the equivalents of the approximate relations (4.2.19), (4.2.20), (4.2.24), and (4.2.26). This analysis is much simpler than that given in terms of Legendre functions, and leads to results of equal validity except for the isolated case of the almost-normal shock considered below. We have followed the approach with Legendre functions in order to include this special case and in order to follow the philosophy expressed in the fourth paragraph of Section 4.1 of exhibiting the most general constant-density solution.

Either the analysis we have given or the alternative approach just mentioned leads to a simple geometric picture of the streamline pattern. Analogous to (4.2.20) we may obtain the approximate relation

$$\theta_c - \theta = \vartheta - \theta_c \,, \tag{4.2.27}$$

where θ is the local streamline inclination angle. This relation states that the slope of a streamline with respect to the body surface is equal to minus the distance from the body surface divided by the distance from the vertex. Such streamlines are hyperbolas with the body surface as asymptote. The velocity component parallel to the body surface is approximately constant, while that normal to the surface varies approximately linearly with distance from the surface. The linear variation of the pressure gradient in (4.2.22) is consistent with this picture. As in the case of the wedge, the pressure relations may be obtained on the basis of a mass and momentum balance.

Feldman [3] has independently developed a constant-density theory for the cone which uses Legendre functions and which is essentially equivalent to that given here. He compares results of his theory with calculated results according to the exact theory of Taylor and Maccoll [1] for a number of cases of cones at high flight speeds at an altitude of 100,000 feet. He obtains excellent agreement for those cases in which ϵ is about $\frac{1}{10}$ or less. He does not make any approximation beyond that of assuming constant density; thus with given shock angle he uses his equivalent to (4.2.9) to calculate the cone angle θ_c, and his equivalent of (4.2.10) to calculate the pressure coefficient on the body. As we have pointed out, the validity of the constant-density approach implies the validity of our additional assumptions. With the additional assumptions and the attendant simplification we may use the simpler formulas (4.2.20) and (4.2.24) in place of their counterparts (4.2.9) and (4.2.10) of the full constant-density theory. The errors are generally of the same order of magnitude. And with θ_c given and

ϵ known approximately (4.2.19) and (4.2.26) are available for the calculation of shock angle and pressure coefficient on the cone.

Hord [1] and Zienkiewicz [1] have given theories for the cone in a perfect gas of constant γ, based upon the assumptions of constant density in addition to other assumptions. Hord obtains an approximate formula for the relation between shock angle and cone angle, and compares the predictions of his approximate theory with various approximations and the exact results of Kopal [1]. Our relation (4.2.20) for small ϵ may be deduced from Hord's approximate result, but his approximation breaks down in the detachment region and may not be used to predict detachment. Hord also makes the remark that the complete constant-density solution of the form (4.2.3) had been recognized independently by several persons as being a good approximation. But he gives no indication that he or any of these several unnamed persons has exploited this approach.

Zienkiewicz [1] takes the constant-density form of the Taylor-Maccoll equation (essentially our 4.2.2), and drops one of the terms. His approximation is valid as long as ϵ is small and the shock layer is thin, and represents an intermediate point between the approximation of the complete constant-density theory and the simplest equally valid approximation. Within the limits of validity of constant-density theory a number of almost equivalent approximations of slightly different form are equally valid except in the case of the almost-normal shock. Zienkiewicz derives no geometrical or pressure relationships based upon his approximation, but compares a few numerically calculated examples with the results of Kopal [1]. Results for cones in air at high temperatures obtained from the exact Taylor-Maccoll equations have been reported by Romig [1]. A careful study of thin shock layer theory on a cone has been made by Berger [1].

The result from Newtonian flow analogous to (4.1.14) is

$$m_1 = \tfrac{1}{2}R \tan \sigma \tag{4.2.28}$$

for the cone. Following the same approach that we used for the wedge we obtain the conical analogue of (4.1.1) with $\frac{1}{2}\epsilon$ in place of ϵ and θ_c in place of θ_s. For the geometry of the cone solution with a thin shock layer the Newtonian theory is again adequate, provided that the shock angle is used for the shock layer angle. However, the Newtonian theory cannot give us any information about the structure of the shock layer in terms of the polar velocity component or of the pressure difference between the shock and the body.

We next look at the one special case for which the expansion of (4.2.14) and the approximation of (4.2.18) may not be applied even if ϵ is very small. This is the case of the almost-normal shock, with $\sin \sigma$ very close to one. As long as ϵ is small the constant-density assumption is still valid in this case even though the shock layer may be anything but thin. This is true because the Mach number of the subsonic flow behind the almost-normal shock is of order $\epsilon^{1/2}$ and its square is small. From (4.2.9) we obtain the relation

$$\cos \sigma = \frac{\epsilon}{Q_1'(\mu_c)}, \tag{4.2.29}$$

which is analogous to (4.1.20). Here we have a realistic physical problem for which our complete theory involving Legendre functions is essential.

We turn now to an estimation of the effects of the variations in the density upon our approximate results. In so doing we shall seek a correction to the thickness of the shock layer, and shall base a formula for this correction upon a simple argument. The actual solution is assumed to be at almost constant density and at constant ϵ, and is considered to be almost the same as the constant-density solution with the same shock shape and the same value of ϵ. The two solutions have the same pressures immediately behind the shock and approximately the same pressure distributions at corresponding points in the flow field. Corresponding points are defined as points at the same distance along the shock layer which have the same entering streamline. With these pressure distributions extremely close the mass flow relations in the two compared solutions must be extremely close, and we should be able to approximately set the Howarth-Dorodnitsyn variable $\int_0 \rho \, dy$ (y is here distance from the body) equal at corresponding points in the two flows.

With the assumption that the Howarth-Dorodnitsyn variable, or its equivalent $\rho_\infty m$ used in Newtonian theory, is unchanged, we may write an expression for the correction to the shock layer thickness

$$\frac{\Delta}{\Delta_{CD}} = 1 - \chi = \frac{1}{\Delta_{CD}} \int_0^{\Delta_{CD}} \frac{\rho_s \, dy}{\rho}. \tag{4.2.30}$$

Here the subscript CD indicates the constant-density result, and the density ρ_s is the assumed constant density $\epsilon^{-1}\rho_\infty$ of the constant-density solution. The integral is taken over the constant-density solution. The quantity χ defined in (4.2.30) is a measure of the decrease in thickness of the shock layer due to an increase in density from compressibility. We define γ_s as the value of γ_e defined in (1.4.22) immediately behind the shock, and assume that the speed of sound a may be taken to be constant in the shock layer. The change in density $\rho - \rho_s$ from compressibility is a^{-2} times the change in pressure along a streamline $p - p_s$. Thus, with $\gamma_s = \rho_s a^2/p_s$, we may replace the integrand ρ_s/ρ of (4.2.30) by

$$\frac{\rho_s}{\rho} = 1 - \frac{\rho - \rho_s}{\rho_s} = 1 - \frac{p - p_s}{\gamma_s p_s} \tag{4.2.31}$$

within the accuracy of our approximation. We obtain for χ the expression

$$\chi = \frac{1}{\gamma_s} \int_0^1 \left(\frac{p - p_s}{p_s}\right) d\left(\frac{y}{\Delta_{CD}}\right), \tag{4.2.32}$$

and shall use this expression in this and the two succeeding sections. This gives a correction on the constant-density solution of the same ϵ and the same shock shape; hence the correction gives us a change in the body shape.

We must keep in mind that $p - p_s$ is defined for a streamline, so that p_s is not the pressure behind the shock at the point of interest but the pressure at the point of entry for the streamline of interest. This distinction does not appear in the case of the cone.

In the case of the cone, we may use (4.2.23) to obtain an expression

$$\frac{p - p_s}{p_s} = \frac{\epsilon}{4}\left[1 - \left(\frac{y}{\Delta_{CD}}\right)^2\right], \tag{4.2.33}$$

whence we obtain immediately

$$\chi = \frac{\epsilon}{6\gamma_s}\,. \tag{4.2.34}$$

Using $\frac{1}{2}\epsilon(1 + \frac{1}{4}\epsilon)$ in place of ϵ in the equivalent of (4.1.1) for the cone, we obtain with the density correction

$$\sigma - \theta_c = \tfrac{1}{2}\epsilon[1 + \tfrac{1}{4}\epsilon - \chi + O(\epsilon^2)] \tan \sigma. \tag{4.2.35}$$

in place of (4.2.20). If we may take $\gamma_s - 1$ to be of order ϵ or less we obtain the somewhat simpler result

$$\sigma - \theta_c = \tfrac{1}{2}\epsilon[1 + \tfrac{1}{12}\epsilon + O(\epsilon^2)] \tan \sigma. \tag{4.2.36}$$

For a perfect gas $\gamma_s - 1$ is of order ϵ. We may note that if $0 < \gamma_s^{-1} < \frac{3}{2}$ the result of (4.2.35) lies between our approximate result (4.2.20) and that of the more complete constant-density theory, and that if $\gamma_s < \frac{4}{3}$, our approximate result (4.2.20) should be more accurate than that of the more complete constant-density theory.

Berger [1] has studied the hypersonic flow of a calorically perfect gas on circular cones, using the Taylor-Maccoll equation and a double expansion in ϵ and ϵ_{lim}. His result for $\sigma - \theta_c$ is

$$\sigma - \theta_c = \tfrac{1}{2}\epsilon[1 + \tfrac{1}{12}\epsilon + \tfrac{1}{2}\epsilon\{\tfrac{1}{20}\epsilon(3 - 5\tan^2\sigma) + \tfrac{1}{3}\epsilon_{\text{lim}}\} + O(\epsilon^3)] \tan \sigma. \tag{4.2.37}$$

This result justifies our assumption as to the invariance of the Howarth-Dorodnitsyn variable within the approximation given, as well as extending the result of (4.2.36). Berger also gives an improvement of (4.2.27) and results for the pressure to higher order.

3. Circular cylinder

Our general solution with constant density for a circular cylinder was suggested by the analogous solution of Lighthill for a sphere, which will be presented later. For the cylinder we take two-dimensional polar coordinates R and ϑ, with the axis $\vartheta = 0$ directed upstream. We assume that the shock wave is in the shape of a circular cylinder of radius R_s, that the density ratio ϵ is constant, and that the flow behind the shock is at constant density (see Fig. 4–6). We

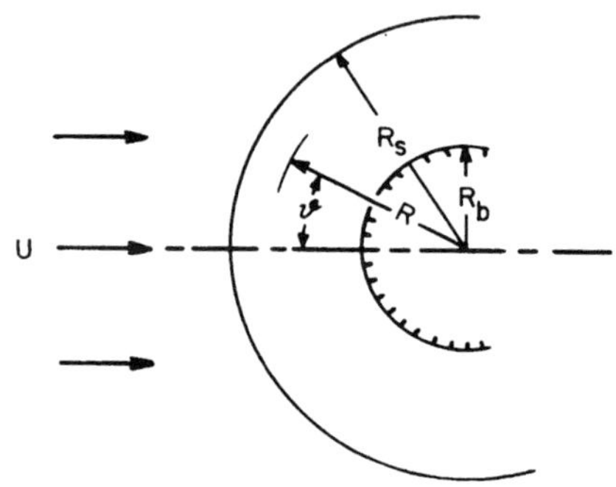

FIG. 4–6. Constant-density flow on a cylinder or sphere.

define ψ as before as the stream function for the flow, with the properties that the outward radial component of the mass flow may be expressed

$$(\rho U v) = -R^{-1}\psi_\vartheta \tag{4.3.1}$$

and the angular or tangential component expressed

$$(\rho U u) = \psi_R , \tag{4.3.2}$$

with the subscripts R and ϑ indicating partial differentiation. The stream function immediately behind the shock equals that immediately in front of it, and we have the condition that

$$\psi_s = \rho_\infty U R_s \sin\vartheta. \tag{4.3.3}$$

The vorticity immediately behind the shock may be obtained from (1.5.12), and is

$$\zeta_s = -\frac{U(1-\epsilon)^2 \sin\vartheta}{\epsilon R_s}. \tag{4.3.4}$$

Since in a steady two-dimensional flow with constant density the vorticity is a function of the stream function alone, we may write

$$\zeta = -\frac{(1-\epsilon)^2}{\epsilon\rho_\infty R_s^2}\psi, \tag{4.3.5}$$

which is a relation obtained from (4.3.3) and (4.3.4) on the shock and which is valid over the entire field. Note that the vorticity ζ must be zero if ψ is zero, and will necessarily be zero at the surface of the body.

The stream function ψ satisfies the equation

$$\psi_{RR} + \frac{1}{R}\psi_R + \frac{1}{R^2}\psi_{\vartheta\vartheta} = -\epsilon^{-1}\rho_\infty\zeta = \frac{(1-\epsilon)^2}{\epsilon^2 R_s^2}\psi \tag{4.3.6}$$

(see Milne-Thomson [1, §4.40]). We may see from (4.3.3) that ψ should be of the form of $\sin\vartheta$ times a function of R. A general solution to (4.3.6) which is of this form does exist and may be expressed

$$\psi = \epsilon^{-1}\rho_\infty U R_s \sin\vartheta\,[AI_1(\kappa R) + BK_1(\kappa R)], \tag{4.3.7}$$

where the parameter κ is defined

$$\kappa = \frac{1-\epsilon}{\epsilon R_s}, \tag{4.3.8}$$

and I_1 and K_1 are modified Bessel functions of order one. The boundary condition on the velocity component normal to the shock is equivalent to (4.3.3), and the condition on the tangential component (4.3.2) is

$$\psi_{R_s} = \epsilon^{-1}\rho_\infty U \sin\vartheta. \tag{4.3.9}$$

The two boundary conditions (4.3.3) and (4.3.9) are expressed

$$AI_1(\kappa R_s) + BK_1(\kappa R_s) = \epsilon, \tag{4.3.10a}$$

$$AI_1'(\kappa R_s) + BK_1'(\kappa R_s) = \frac{\epsilon}{1-\epsilon}, \tag{4.3.10b}$$

from which the constants A and B may be evaluated. From the theory of Bessel functions we take the evaluation of the Wronskian of the modified Bessel functions and write the identity

$$K_1(\kappa R)I_1'(\kappa R) - I_1(\kappa R)K_1'(\kappa R) = \frac{1}{\kappa R}. \tag{4.3.11}$$

The constants A and B are calculated from (4.3.10) and (4.3.11) as

$$A = K_1\left(\frac{1-\epsilon}{\epsilon}\right) - (1-\epsilon)K_1'\left(\frac{1-\epsilon}{\epsilon}\right), \tag{4.3.12a}$$

$$B = (1-\epsilon)I_1'\left(\frac{1-\epsilon}{\epsilon}\right) - I_1\left(\frac{1-\epsilon}{\epsilon}\right). \tag{4.3.12b}$$

The boundary condition from which the body shape is determined is that $\psi = 0$ on the body. From (4.3.7) we see that $\psi = 0$ on the splitting streamline and on a cylindrical body of radius R_b, where

$$AI_1(\kappa R_b) + BK_1(\kappa R_b) = 0. \tag{4.3.13}$$

The general solution of this equation for R_b cannot be obtained in closed form; a numerical solution is given in Fig. 6–29. An approximate solution with ϵ small obtainable from the asymptotic expansions of the Bessel functions is

$$\Delta = R_s - R_b = \tfrac{1}{2}\epsilon R_s \left[\ln \frac{4}{3\epsilon} + \epsilon \ln \frac{4}{3\epsilon} + O\left(\epsilon^2 \ln \frac{1}{\epsilon}\right)\right], \tag{4.3.14}$$

which is in agreement to the lowest-order term with the results of Chester [1] and Freeman [1], and to two terms with Muggia [1]. Chester's result is for a parabolic shock in a perfect gas of constant γ at finite Mach number, while Muggia's is for constant-density flow on a cylinder of quasi-circular cross section. A comparison of this equation with the numerical solution of (4.3.13) is given in Fig. 6–29. The two results are found to be quite close for $\epsilon < 0.3$. The subscript 0 here indicates a result taken on the axis. We shall discuss the result (4.3.14) further when we consider more exact solutions.

The pressure coefficient in our constant-density solution may be obtained by the application of Bernoulli's equation, and is

$$\tfrac{1}{2}C_p = (1 - \tfrac{1}{2}\epsilon) + \frac{(1-\epsilon)^2}{2\epsilon\rho_\infty^2 U^2 R_s^2}\psi^2 - \frac{\epsilon}{2\rho_\infty^2 U^2}\left(\psi_R^2 + \frac{1}{R^2}\psi_\vartheta^2\right). \tag{4.3.15}$$

In this equation the term in ψ^2 is the variable part of the Bernoulli "constant", and is constant along a streamline. On the body we may re-express (4.3.15)

$$\tfrac{1}{2}C_{p_b} = (1 - \tfrac{1}{2}\epsilon) - \frac{(1-\epsilon)^2}{2\epsilon^3}[AI_1'(\kappa R_b) + BK_1'(\kappa R_b)]^2 \sin^2\vartheta. \tag{4.3.16}$$

With $\sin\vartheta$ sufficiently large, the pressure coefficient is negative on and near the body, regardless of the value of ϵ. For small ϵ, W. B. Bush (private communication) and N. H. Kemp (private communication) have found an approximate expression for (4.3.16) from the Bessel function asymptotic expansion

$$\tfrac{1}{2}C_{p_b} = (1 - \tfrac{1}{2}\epsilon) - \tfrac{3}{2}\sin^2\vartheta\left[1 + \tfrac{1}{2}\epsilon\left(\ln\frac{4}{3\epsilon} - 1\right) + O\left(\epsilon^2 \ln\frac{1}{\epsilon}\right)\right]. \tag{4.3.17}$$

Whitham [1] has obtained essentially the same constant-density solution we have presented above from (4.3.1) to (4.3.15), expressed in somewhat different terms. His method is different in one respect, in that he uses the vorticity equation obtained by taking the curl of the vector momentum equation instead of using the known law of dependence of the vorticity on ψ (and on $R \sin \vartheta$ for the Lighthill solution of the next section). Our solution was obtained independently of Whitham's and at about the same time.

A solution of the form of (4.3.7) to the constant-density equations of motion was found and used by Hida [1] in 1953. Hida's application of the solution was as an approximation for the stagnation region on a cylinder in supersonic flow, without considering ϵ as a small parameter. The authors learned of Hida's much earlier results during the preparation of this second edition.

As with the cone, the foregoing general analysis is not valid for a hypersonic flow unless ϵ is very small. Even with ϵ small, the validity is limited by the large pressure changes to a local region in the vicinity of the stagnation point. However, the analysis is instructive and provides a valuable check on results obtained by other methods. But it must be strongly emphasized that only the lowest-order term on the right-hand side of (4.3.14) may be justifiably retained.

We turn now to a quite different approach to the problem, in which only a local solution near the stagnation point is sought. No real restriction is imposed, however, as only such a local solution is valid anyway with the assumption of constant density. Within the limits of validity obtainable from a constant-density theory we have a great deal of freedom in making approximations in this analysis. We choose the coordinate x as distance along the layer from the stagnation point, and approximate $\sin \sigma$ by one within the region of interest except in calculating the pressure (see Fig. 4–7). The quantity x is thus identical with

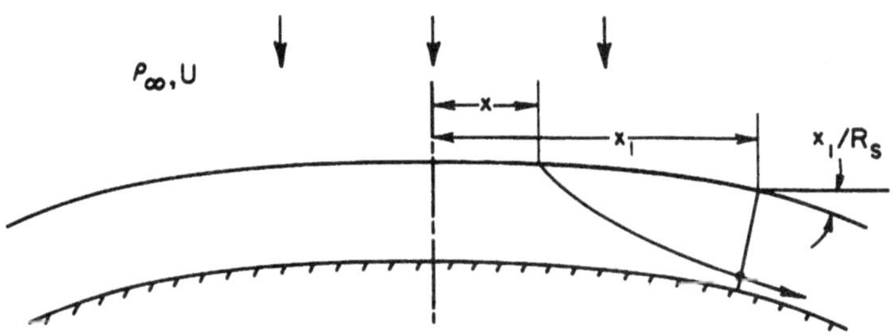

FIG. 4–7. Coordinate system for stagnation region.

the stream function ψ of the free stream flow divided by $\rho_\infty U$. We first neglect the fact that the normal component of velocity is nonzero within the layer, and calculate a correction to the Newtonian velocity distribution using the constant density $\epsilon^{-1}\rho_\infty$ in Bernoulli's equation and the Newtonian pressure distribution. We next calculate on the basis of continuity the corrected vertical scale corresponding to the Newtonian m scale. Finally we calculate the distributions of both velocity components with respect to this scale and obtain an improved pressure distribution. Our reasons for carrying out this alternative analysis are two, first to show how we may obtain results of the same validity as those of the Bessel function analysis starting from our Newtonian results, and second to develop an approach applicable to a more general problem, such as that treated in Section 4.5.

The Newtonian pressure is obtained from the variant of (3.2.7) analogous to (3.5.19), with the aid of (3.3.3) with $n = 2$ and (3.5.6). The result in terms of the pressure coefficient is

$$\tfrac{1}{2}C_p = 1 - \frac{x_1^2}{R_s^2} - \frac{1}{2R_s^2}(x_1^2 - x^2), \tag{4.3.18}$$

with ψ replaced by $\rho_\infty Ux$. The quantity x_1 is the x coordinate of the point being investigated, and the quantity x is the coordinate at entry into the shock layer of the streamline passing through the point being investigated. The velocity given by (3.5.7) is an initial velocity upon entry, assumed within the Newtonian approximation to be purely tangential with respect to the shock layer. We denote this velocity as Uu (essentially the same u as in (4.3.2)) and neglect, for the moment, the normal component of velocity. With the assumed constant density $\epsilon^{-1}\rho_\infty$, and with the initial value of u equal to x/R_s, we may apply Bernoulli's equation to calculate u. We obtain first the equation

$$\frac{1}{2\epsilon}\left(u^2 - \frac{x^2}{R_s^2}\right) = \frac{3}{2R_s^2}(x_1^2 - x^2), \tag{4.3.19}$$

and may then calculate u

$$u = \frac{1}{R_s}\sqrt{x^2 + 3\epsilon(x_1^2 - x^2)}\,. \tag{4.3.20}$$

We now use this corrected velocity distribution to obtain the distance y measured normal to the body. The method we use is based on continuity, analogous to that used to obtain m in Newtonian theory. The quantity y is equivalent to ϵm, and is

$$y = \epsilon\int_0 \frac{dx}{u} = \frac{\epsilon R_s}{\sqrt{1 - 3\epsilon}}\sinh^{-1}\left(\sqrt{\frac{1 - 3\epsilon}{3\epsilon}}\frac{x}{x_1}\right). \tag{4.3.21}$$

We obtain also the result

$$\Delta_0 = \frac{\epsilon R_s}{\sqrt{1 - 3\epsilon}}\sinh^{-1}\sqrt{\frac{1 - 3\epsilon}{3\epsilon}} = \frac{\epsilon R_s}{\sqrt{1 - 3\epsilon}}\cosh^{-1}\frac{1}{\sqrt{3\epsilon}} \tag{4.3.22}$$

for the stand-off or stagnation point detachment distance. By expanding the function $\cosh^{-1}$ this result may be also expressed

$$\Delta_0 = \tfrac{1}{2}\epsilon R_s \left[\ln \frac{4}{3\epsilon} + \tfrac{3}{2}\epsilon \left(\ln \frac{4}{3\epsilon} - 1\right) + O\left(\epsilon^2 \ln \frac{1}{\epsilon}\right)\right], \tag{4.3.23}$$

in agreement with (4.3.14) to the term of lowest order, which again is the only term which may justifiably be retained. The second term, the one in $R_s\epsilon^2 \ln (1/\epsilon)$, is affected by the additional assumptions we have made in our approximate analysis, and is in error for the constant-density theory.

The distribution of the tangential component of the velocity Uu is given by

$$u = \frac{\sqrt{3\epsilon}\, x_1}{R_s} \cosh \left(\frac{\sqrt{1 - 3\epsilon}\, y}{\epsilon R_s}\right), \tag{4.3.24}$$

in terms of y, and the corresponding distribution of outward normal velocity Uv (essentially the same v as in (4.3.1)) calculated to satisfy the cartesian continuity equation in x and y is

$$v = -\epsilon \sqrt{\frac{3\epsilon}{1 - 3\epsilon}} \sinh \left(\frac{\sqrt{1 - 3\epsilon}\, y}{\epsilon R_s}\right) = -\epsilon \frac{x}{x_1}. \tag{4.3.25}$$

From (4.3.24), the velocity gradient on the body is simply $\sqrt{3\epsilon}\, R_s^{-1} U$. The pressure distribution from Bernoulli's equation corresponding to this velocity distribution is

$$\tfrac{1}{2}C_p = 1 - \tfrac{1}{2}\epsilon \left(1 + \frac{x^2}{x_1^2}\right) - \frac{x_1^2}{2R_s^2}\left(3 - \frac{x^2}{x_1^2}\right), \tag{4.3.26}$$

with a term in $\epsilon x^2/R_s^2$ dropped, where

$$\frac{x^2}{x_1^2} = \frac{3\epsilon}{1 - 3\epsilon} \sinh^2 \frac{\sqrt{1 - 3\epsilon}\, y}{\epsilon R_s}. \tag{4.3.27}$$

It may be shown that the pressure given by (4.3.26) satisfies the dynamic equations approximately, with best agreement near the shock wave. On the body (4.3.26) gives simply the Newtonian pressure plus an appropriate correction to give the correct stagnation pressure, and is consistent with (4.3.17).

In the approximate shock layer structure given above we can distinguish two layers, an outer one in which the Newtonian theory holds approximately and an inner one in which the Bernoulli effect evident in (4.3.20) is important but centrifugal effects are small. To divide these layers we select a constant of the order of one and take x^2/x_1^2 equal to 3ϵ times that constant at the point of division. From (4.3.21), the thickness of the inner layer is of order ϵR_s, that of the outer layer is of order $\epsilon R_s \ln \epsilon^{-1}$. This distinction is important in thin shock layer theory and will reappear in Chapter V as well as in the remaining sections of this chapter. In the cylindrical case we may not consider the inner layer as negligibly thin compared with the outer layer. But we shall note that it is the existence of the inner layer which makes the total thickness finite.

A simplified analysis similar to that above was first given by Hayes [3], neglecting the centrifugal correction to the Newtonian pressure distribution. In this analysis the number 2 appeared wherever 3 appears in equations (4.3.19) through (4.3.25). Hayes' results were rederived by Li and Geiger [1] using a different approach. Li and Geiger assumed local similarity of the type evident in (4.3.24) and (4.3.25) and calculated the appropriate stream function in a coordinate system similar to ours. The centrifugal corrections to the pressure have an effect on the results which may not be neglected within the degree of approximation desired, and these earlier analyses which neglect this effect must be considered to be in error on this point. The correct factor 3 appeared first in the results of Chester [1] and Freeman [1].

The stand-off distance Δ as given by (4.3.22) or (4.3.23) is independent of x_1 and thus constant along the body. This would correspond to a radius of curvature R_b for the body simply equal to $R_s - \Delta_0$. However, the theory cannot yield a value for the second derivative of Δ with respect to x_1 accurately, and the body radius R_b is actually undetermined in this approximation. Here we have another example of our principle that Newtonian theory should be used with the shock shape rather than with the body shape. In our present problem the shock shape is the prime determiner of the principal features of the solution, of the velocity distributions and of the pressure field.

At this point we list the principal effects in a hypersonic flow on a blunt body which are not taken into account in the basic Newtonian theory, but need to be considered in thin shock layer theory. In Newtonian theory a shock layer structure exists in most cases, but not in the case of this section. A constant-density analysis such as that of this section takes into account as an improvement over Newtonian the:

(a) Bernoulli effect. This is the effect of change of velocity along a streamline according to the Bernoulli equation.

Not included in a constant-density analysis are the following:

(b) Compressibility effect. This is the effect of the change of density along a streamline resulting from pressure changes.

(c) Nonconstant ϵ effect. This is the effect of the variation of ϵ or ρ_s along the shock.

(d) Noncircularity effect. This is the effect of the deviation of the shock shape from a circular shape.

In our local analysis there are errors arising from the geometrical assumptions made in the coordinate system. These errors do not correspond to any basic flow effects. In this chapter we generally treat the shock shape as given, as though the problem were an inverse problem. Here we consider R_s as specified and the body curvature R_b^{-1} as an unknown to be obtained.

The basic parameter governing the Bernoulli effect is $\rho_\infty/\rho_b = \epsilon\rho_s/\rho_b$. The basic parameter governing the compressibility effect is γ_s^{-1}, where γ_s is the γ_e of (1.4.22) taken immediately behind the shock. The basic parameter governing the nonconstant ϵ effect is the quantity μ defined by

$$\frac{\mu}{1+\mu} = -\frac{p_s - p_\infty}{1-\epsilon}\frac{d\epsilon}{dp_s}, \tag{4.3.28}$$

with the differential taken along the Hugoniot curve with p_∞ and ρ_∞ given. This quantity will reappear in (5.3.14) in Section 5.3. For a calorically perfect gas

$$\mu = M_\infty^{-2} + O(\epsilon^2), \tag{4.3.29}$$

Expansion in a Taylor series about the point corresponding to a normal shock leads to

$$\epsilon = \epsilon_0 + \mu\frac{x^2}{R_s^2} + O\left(\frac{x^4}{R_s^4}\right), \tag{4.3.30}$$

where ϵ_0 and μ are the values of ϵ and μ on the axis. This equation is similar to (5.3.21) and has a similar derivation.

The basic parameter governing the noncircularity effect is

$$\beta = R_s R_s'', \tag{4.3.31}$$

which is the same as the β_2 introduced by Muggia [1]. The differentiation is with respect to distance along the shock. This quantity is 3 for a parabola, > 3 for a hyperbola, and < 3 for an ellipse, all symmetrically oriented. Of course, $\beta = 0$ for a circle.

If the shock and body are concentric at the axis, so that $R_s = R_b + \Delta_0$, the shock layer thickness defined normal to the shock shape is locally constant. As shown in Fig. 4–8, the quantity $\Delta - \Delta_0$ may be defined as the distance between the locally circular body and a circle of radius $R_s - \Delta_0$ concentric with the shock. We thus have

$$\Delta - \Delta_0 = \tfrac{1}{2}x_1^2\left(\frac{1}{R_b} - \frac{1}{R_s - \Delta_0}\right), \tag{4.3.32}$$

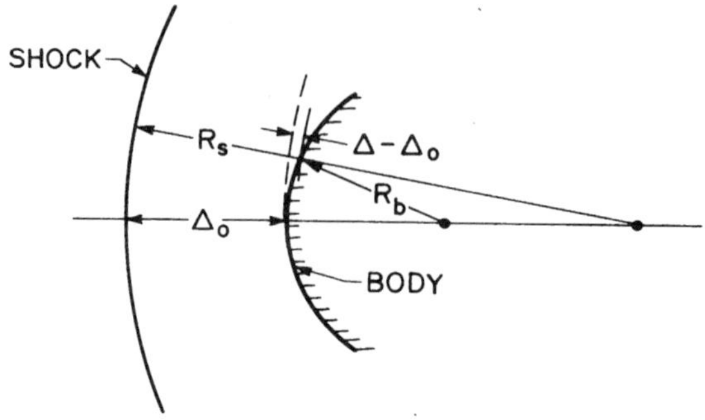

FIG. 4–8. Geometry of nonconcentric shock and body.

or with Δ_0/R_s small

$$R_s\Delta'' = \frac{R_s}{R_b} - 1 - \frac{\Delta_0}{R_s}. \tag{4.3.33}$$

This nonconcentricity we consider as a result of the solution rather than as a basic effect.

Two quantities required to be calculated from the solution are the density correction to the stand-off distance from compressibility and from nonconstant ϵ. The density factor analogous to (4.2.31) may be approximated

$$\begin{aligned}\frac{\rho - \rho_{s_0}}{\rho_{s_0}} &= \frac{\rho - \rho_s}{\rho_{s_0}} + \frac{\rho_s - \rho_{s_0}}{\rho_{s_0}} \\ &= \frac{C_p - C_{p_s}}{\gamma_s C_{p_s}} - \frac{\mu}{\epsilon}\frac{x^2}{R_s^2} \\ &= \tfrac{1}{2}\gamma_s^{-1}\left(1 - \frac{x^2}{x_1^2}\right)\left[\epsilon - \frac{3x_1^2}{R_s^2}\right] - \frac{\mu}{\epsilon}\frac{x_1^2}{R_s^2}\left(\frac{x^2}{x_1^2}\right),\end{aligned} \tag{4.3.34}$$

where ρ_{s_0} is the value of ρ_s on the axis. The quantity χ of (4.2.30) is obtained from the integral (4.2.32) plus a term in μ. The integral may be expressed

$$\chi\Delta_{CD} = \int_0^{\Delta_{CD}} \left[\frac{C_p - C_{p_s}}{\gamma_s C_{p_s}} - \frac{\mu}{\epsilon}\frac{x^2}{R_s^2}\right] dy. \tag{4.3.35}$$

It must be kept in mind that ρ_s and C_{p_s} are taken at the point of entry of the streamline of interest and are different from ρ_{s_1} and $C_{p_{s_1}}$ behind the shock at the point of interest.

In the integral of (4.3.35) we have $dy = \epsilon R_s\, dx/\sqrt{x^2 + 3\epsilon x_1^2}$ within our limits of accuracy.

Carrying out the integration gives

$$\chi\Delta_{CD} = \tfrac{1}{4}\gamma_s^{-1}\epsilon R_s\left[\epsilon - \frac{3x_1^2}{R_s^2}\right]\left[\ln\frac{4}{3\epsilon} - 1\right] - \tfrac{1}{2}\mu R_s\frac{x_1^2}{R_s^2}, \tag{4.3.36}$$

with terms of higher order in ϵ and (x_1/R_s) dropped. This quantity may be subtracted from the constant-density result to obtain a new Δ corresponding to the assumption that the Howarth-Dorodnitsyn variable is unchanged. This is the use we made of χ in Section 4.2. We may also use it to add to a complete result for Δ with the various effects included to obtain an equivalent constant-density or Howarth-Dorodnitsyn thickness Δ_{H-D}.

We next take the solutions of Chester [1] and Muggia [1], and show that they can be combined. We then interpret them through the use of the Howarth-Dorodnitsyn thickness. This interpretation duplicates that given by Hayes [8].

The pressure on the axis at shock and body are given to $O(\epsilon)$ by the same expression in all the solutions considered. These expressions to $O(\epsilon^2)$ are

$$\tfrac{1}{2}C_{p_s} = 1 - \epsilon, \tag{4.3.37a}$$

$$\tfrac{1}{2}C_{p_b} = 1 - \tfrac{1}{2}\epsilon + \tfrac{1}{8}\epsilon^2\gamma_s^{-1} + O(\epsilon^3). \tag{4.3.37b}$$

The density on the axis at the body from (4.3.34) with $x = 0$ is given by

$$\begin{aligned}\frac{\rho_b}{\rho_s} &= 1 + \gamma_s^{-1}(C_{p_b} - C_{p_s}) + O(\epsilon^2) \\ &= 1 + \tfrac{1}{2}\epsilon\gamma_s^{-1} + O(\epsilon^2).\end{aligned} \tag{4.3.38}$$

This last result is needed in relating pressure gradient with velocity gradient on the body, in calculating the Bernoulli effect.

The solution of Muggia [1], corresponding to $\gamma_s^{-1} = 0$ and $\mu = 0$, may be expressed for Δ_{CD} as

$$\frac{\Delta_{CD}}{R_s} = \tfrac{1}{2}\epsilon\ln\frac{4}{3\epsilon} + \tfrac{1}{2}\epsilon^2\left(\ln\frac{4}{3\epsilon} - \frac{1}{12}\beta\right) + \frac{1}{2}\frac{x_1^2}{R_s^2}\left(\frac{7}{12}\epsilon\beta\right) \tag{4.3.39}$$

with terms of higher order in ϵ and x_1/R_s dropped. Chester's result, corresponding to $\gamma_s^{-1} = 1$ and $\beta = 3$, is

$$\begin{aligned}\frac{\Delta}{R_s} = \tfrac{1}{2}\epsilon\ln\frac{4}{3\epsilon} &+ \tfrac{1}{4}\epsilon^2\left(\ln\frac{4}{3\epsilon} - \frac{3}{2} - \frac{2\mu}{\epsilon}\right) \\ &+ \frac{1}{2}\frac{x_1^2}{R_s^2}\left(\frac{3}{2}\epsilon\ln\frac{4}{3\epsilon} - \frac{1}{2}\epsilon + \mu\right).\end{aligned} \tag{4.3.40}$$

These two expressions may be combined into a single formula which fits both cases. This is

$$\frac{\Delta}{R_s} = \tfrac{1}{2}\epsilon \ln \frac{4}{3\epsilon} + \tfrac{1}{4}\epsilon^2 \left[(2 - \gamma_s^{-1}) \ln \frac{4}{3\epsilon} - \gamma_s^{-1} - \frac{\beta}{6} - \frac{2\mu}{\epsilon}\right] + \frac{1}{2}\frac{x_1^2}{R_s^2}\left(\frac{3}{2}\gamma_s^{-1}\epsilon \ln \frac{4}{3\epsilon} - \frac{9}{4}\gamma_s^{-1}\epsilon + \frac{7}{12}\epsilon\beta + \mu\right). \tag{4.3.41}$$

A comparison of this result with exact numerical calculations for the stand-off distance at the axis of a circular cylinder is presented in Fig. 6–29. There we have set $\beta = 0$, $\mu = 0$, and $\gamma_s = 1$. The agreement is found to be excellent.

We next add the correction term (4.3.36) to obtain the expression for the Howarth-Dorodnitsyn thickness

$$\frac{\Delta_{\mathrm{H-D}}}{R_s} = \tfrac{1}{2}\epsilon \ln \frac{4}{3\epsilon} + \tfrac{1}{2}\epsilon^2 \left(\ln \frac{4}{3\epsilon} - \gamma_s^{-1} - \frac{1}{12}\beta - \frac{\mu}{\epsilon}\right) + \frac{1}{2}\frac{x_1^2}{R_s^2}\left(-\frac{3}{4}\gamma_s^{-1}\epsilon + \frac{7}{12}\epsilon\beta\right). \tag{4.3.42}$$

Chester and Muggia both give results for pressure or velocity on the body. From these results we may obtain the composite expression fitting both

$$(R_s u_b')^2 = 3\epsilon\left(1 + \frac{3}{2}\epsilon \ln \frac{4}{3\epsilon} - \frac{1}{2}\epsilon + \gamma_s^{-1}\epsilon + \frac{1}{12}\epsilon\beta + \mu\right). \tag{4.3.43}$$

With this expression we may rewrite (4.3.42) for $\Delta_{\mathrm{H-D}}$ on the axis to the same order of accuracy as

$$\frac{\Delta_{\mathrm{H-D}_0}}{R_s} = \epsilon \ln\left(\frac{2}{R_s u_b'}\right) + \tfrac{1}{4}\epsilon^2\left(5 \ln \frac{4}{3\epsilon} - 1\right). \tag{4.3.44}$$

To interpret (4.3.44) we note first that it is the velocity gradient near the body which results from the Bernoulli effect which permits a solution for a quasi-circular cylinder. The meaning of (4.3.44) is that the stand-off distance at the axis is fully explained by the constant-density result for the circular cylinder provided the interpretation is made in terms of Howarth-Dorodnitsyn variables and thicknesses and provided the actual velocity gradient is used for the Bernoulli effect. This is evidenced by the fact that none of the parameters which measure the basic effects γ_s, μ, or β appear. Of the terms in x_1^2/R_s^2, only that in μ has been eliminated in (4.3.42).

All our previous considerations for quasi-circular cylinders have been made for flows with lateral symmetry. Muggia [1] includes in his analysis the effect of weak asymmetry with constant density. Here we discuss the local solution and relate it to Muggia's solution.

The local solution is the solution we have given above in (4.3.24) and (4.3.25), but with an asymmetric perturbation. For the form of this perturbation the reader may consult Hayes [7, Sect. 5], in which the perturbation is that due to the function G. The unperturbed analytic solution there with $\beta_0' = \gamma_0' = 0$ corresponds to our symmetric solution above. Here we simply present the perturbed solution. That it satisfies the constant-density equations of motion may be readily checked. A dimensionless parameter ν which measures the magnitude of the asymmetry is introduced, and appears in the coefficient of the perturbation. With terms of $O(\epsilon)$ dropped, the perturbed form of (4.3.24) is

$$u = \frac{\sqrt{3\epsilon}\, x}{R_s} \cosh\left(\frac{y}{\epsilon R_s}\right) - \nu\epsilon\sqrt{3\epsilon} \sinh\left(\frac{y}{\epsilon R_s}\right). \tag{4.3.45}$$

Equation (4.3.25) is reexpressed as

$$v = -\epsilon\sqrt{3\epsilon} \sinh\left(\frac{y}{\epsilon R_s}\right). \tag{4.3.46}$$

The shape of the stagnation streamline is obtained by integrating $dx/u = dy/v$. The result is

$$x = \nu\epsilon R_s \tanh\left(\frac{y}{2\epsilon R_s}\right). \tag{4.3.47}$$

The stagnation streamline is asymptotic to a vertical line at $x = x_n$ (see Fig. 4–9), with

$$\frac{x_n}{R_s} = \nu\epsilon. \tag{4.3.48}$$

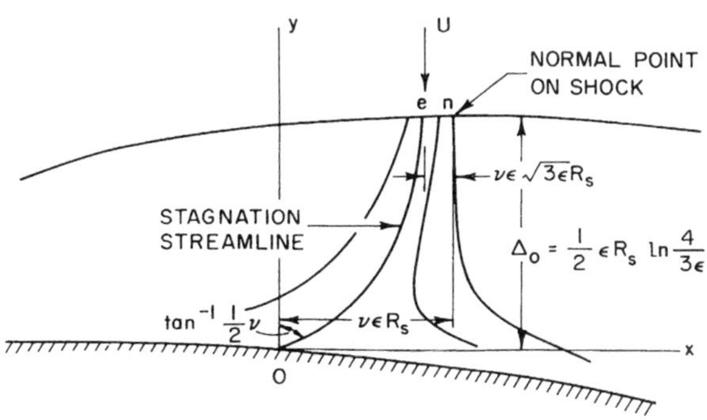

FIG. 4–9. Asymmetric two-dimensional stagnation region.

This value of x_n may be interpreted as corresponding to the normal point on the shock. The streamline which lies closest to $x = x_n$ at large y is given by

$$x = x_n \coth\left(\frac{y}{\epsilon R_s}\right) \tag{4.3.49}$$

and can be shown to have zero vorticity.

The location of the shock is given approximately by $y = \Delta_0$, using (4.3.22) or the leading term of (4.3.23). At the shock the value x_e of x at the entry point of the stagnation streamline is obtained by putting $y = \Delta_0$ in (4.3.48). The result is

$$\frac{x_e}{R_s} = \nu\epsilon(1 - \sqrt{3\epsilon}) = x_n(1 - \sqrt{3\epsilon}). \tag{4.3.50}$$

The geometry of this perturbed solution is shown in Fig. 4–9.

The vorticity immediately behind the shock is expressed

$$\frac{\zeta_s}{U} = (-u_y)_s = \frac{x_n - x}{\epsilon R_s^2}. \tag{4.3.51}$$

Accordingly, from (4.3.50) the vorticity on the stagnation streamline is

$$\frac{\zeta_{\text{st}}}{U} = \frac{\sqrt{3\epsilon}\, x_n}{\epsilon R_s^2} = \frac{\nu\sqrt{3\epsilon}}{R_s}. \tag{4.3.52}$$

The stagnation streamline does not enter the stagnation point normal to the body. Instead, from (4.3.47) we find it is inclined to the normal with an angle whose tangent is $\frac{1}{2}\nu$.

Muggia's solution incorporates the one we have given into a more global analysis. He introduces a parameter $\beta_1 = R_s'$ which is related to our ν by

$$\nu = R_s'/18, \tag{4.3.53}$$

where as before the prime denotes differentiation along the shock. The normal point on the body shown in Fig. 4–10 is at $x = x_b$, with

$$x_b = -18\nu\epsilon R_s = -18x_n. \tag{4.3.54}$$

The slope of the shock is greater than the slope of the body by $(x_n - x_b)/R_s$. This difference in slope makes Δ increase with increasing x, and we may write

$$\Delta' = \frac{x_n - x_b}{R_s} = \frac{19x_n}{R_s} = 19\nu\epsilon. \tag{4.3.55}$$

It may be noted from Muggia's solution that the stagnation point lies on the side of the normal point on the shock for which the body or shock curvature

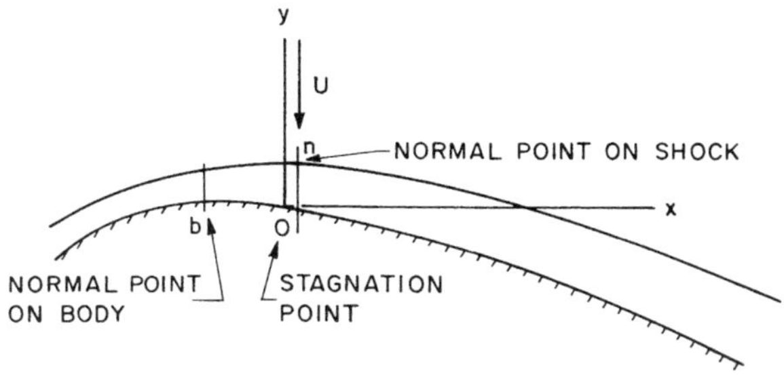

FIG. 4–10. Two-dimensional asymmetric body.

is greater. The streamline of maximum entropy through the normal point on the shock turns in the direction of decreasing body or shock curvature.

It is possible to check some of the results of Muggia by a constant-density analysis following the methods of this section. We choose a shock shape

$$y_s = y_{s_0} - \frac{x^2}{2R_s} + R_s' \frac{x^3}{6R_s^2} \,. \tag{4.3.56}$$

The Newtonian pressure in the layer is calculated as an extension of (4.3.18), and an extension of (4.3.20) is obtained. Manipulation of the quadrature (4.3.21) defining y leads to (4.3.55) in the form

$$\Delta' = 19\epsilon R_s'/18. \tag{4.3.57}$$

This is left as an exercise for the reader, one that is direct but not without difficulty. The matching of this outer solution with the local solution cannot be obtained by this method, which allows for no lateral shift in the point of maximum pressure between shock and body.

The methods of this section may be extended by incorporating a more accurate calculation of the pressure change across the outer layer near the stagnation point (see Hayes [9]). This approach, in many aspects equivalent to that of Section 5.4, yields the basic results (4.3.53) and (4.3.55) of Muggia.

4. The sphere

As with the analogous solution for a cylinder, the analytical solution for the constant-density flow on a sphere was first found by Hida [1]. It was rediscovered by Lighthill [3] and applied to the case of hypersonic flow. This solution is closely connected with Hill's classical solution for a spherical vortex (see Milne-Thomson [1, § 18.50 and § 18.51]). We use spherical coordinates analogous to those we used for the circular cylinder, with the polar angle from the upstream axis, and R the spherical radius. We use the Stokes stream function ψ defined as before, with the outward radial component of the mass flow

$$\rho U v = -(R^2 \sin \vartheta)^{-1} \psi_\vartheta , \tag{4.4.1}$$

and the polar or tangential component

$$\rho U u = (R \sin \vartheta)^{-1} \psi_R ; \tag{4.4.2}$$

the subscripts again denote partial differentiation. The stream function immediately behind the shock equals that immediately in front of it, and we have the condition that

$$\psi_s = \tfrac{1}{2} \rho_\infty U R_s^2 \sin^2 \vartheta, \tag{4.4.3}$$

which may be compared with the two-dimensional value given in (4.3.3). The vorticity immediately behind the shock is given again by (4.3.4), but in axisymmetric constant-density flow the vorticity is proportional to the cylindrical radius $R \sin \vartheta$. In place of (4.3.5) we have

$$\zeta = -\frac{U(1-\epsilon)^2}{\epsilon R_s^2} R \sin \vartheta \tag{4.4.4}$$

for the vorticity in the flow field, independent of ψ. Note that the vorticity ζ is never zero except on the axis $\vartheta = 0$, and hence is nonzero at the surface of the body. The stream function satisfies the equation

$$\psi_{RR} + \frac{\sin \vartheta}{R^2} \left(\frac{1}{\sin \vartheta} \psi_\vartheta \right)_\vartheta = \frac{(1-\epsilon)^2 \rho_\infty U}{\epsilon^2 R_s^2} R^2 \sin^2 \vartheta. \tag{4.4.5}$$

From (4.4.3) we note that ψ should be of the form of $\sin^2 \vartheta$ times a function of R, and we find that the general solution of (4.4.5) of this form exists and may be expressed

$$\psi = \frac{\rho_\infty U R_s^2 \sin^2 \vartheta}{30 \epsilon^2} \left[3(1-\epsilon)^2 \left(\frac{R}{R_s} \right)^4 + A \left(\frac{R}{R_s} \right)^2 + B \left(\frac{R_s}{R} \right) \right]. \tag{4.4.6}$$

Evaluation of the constants A and B using (4.4.3) and the condition that the tangential component of velocity $U \sin \vartheta$ is continuous across the shock gives

$$A = -5(1 - 4\epsilon), \tag{4.4.7a}$$

$$B = 2(1-\epsilon)(1-6\epsilon). \tag{4.4.7b}$$

The boundary condition which determines the location of the body is that $\psi = 0$, and this yields a spherical body of radius R_b, with

$$(4.4.8)\quad 3(1-\epsilon)^2\left(\frac{R_b}{R_s}\right)^4 - 5(1-4\epsilon)\left(\frac{R_b}{R_s}\right)^2 + 2(1-\epsilon)(1-6\epsilon)\left(\frac{R_s}{R_b}\right) = 0.$$

This equation is a quintic in R_b/R_s for which a closed form solution is not available; a numerical solution is given in Figs. 6–27 and 6–28. Equation (4.4.8) is a quadratic equation for ϵ, and so may be solved explicitly for ϵ in terms of R_b/R_s. An approximate solution with ϵ small is

$$(4.4.9)\quad \Delta = R_s - R_b = \epsilon R_s\left[1 - \sqrt{\frac{8\epsilon}{3}} + 3\epsilon - \frac{25}{8}\epsilon\sqrt{\frac{8\epsilon}{3}} + O(\epsilon^2)\right],$$

which agrees with the result of Chester [2] and Muggia [2] to two terms. Chester's result is for a paraboloidal shock in a calorically perfect gas at finite Mach number, while Muggia's is for a quasi-circular cross section in a calorically perfect gas at infinite Mach number. W. J. McCroskey (private communication) has applied Muggia's approach to the constant-density case and obtained agreement with (4.4.9) to four terms in the case of a circular shock. As with the cylindrical case, we shall compare (4.4.9) with the more complete solutions later. The Newtonian solution (3.5.3) gives us the first term in (4.4.9), but no more. The pressure may be expressed in an equation analogous to (4.3.15) as

$$(4.4.10)\quad \tfrac{1}{2}C_p = (1 - \tfrac{1}{2}\epsilon) + \frac{(1-\epsilon)^2}{\epsilon\rho_\infty U R_s^2}\psi - \frac{\epsilon}{2\rho_\infty^2 U^2 R^2 \sin^2\vartheta}\left(\psi_R^2 + \frac{1}{R^2}\psi_\vartheta^2\right).$$

On the body W. B. Bush (private communication) and N. H. Kemp (private communication) obtain the approximate result

$$(4.4.11)\quad \tfrac{1}{2}C_{p_b} = (1 - \tfrac{1}{2}\epsilon) - \tfrac{4}{3}\sin^2\vartheta\left[1 + \frac{1}{4}\epsilon - \frac{2}{3}\sqrt{\frac{8}{3}}\epsilon^{3/2} + O(\epsilon^2)\right]$$

by a suitable but laborious expansion. Regardless of the value of ϵ, with $\sin\vartheta$ sufficiently large, the pressure coefficient is negative on and near the body.

As with the analogous solution for the circular cylinder, the solution of Lighthill is only valid with ϵ small and for a local region in the vicinity of the stagnation point. Only the first two terms (to order $R_s\epsilon^{3/2}$) in (4.4.9) may be justifiably retained. As with the cylinder, the solution is instructive and has intrinsic interest. We turn now to the local solution of the type obtained for the circular cylinder. The same coordinate system is used, except that now the Stokes stream function is replaced by the quantity $\frac{1}{2}\rho_\infty U x^2$. The quantity x is essentially the same as $R_s \sin\vartheta$ of (4.4.3).

The Newtonian pressure is again obtained from the variant of (3.2.7) analogous to (3.5.19), with the aid of (3.3.3) and (3.5.1), and is

$$\tfrac{1}{2}C_p = 1 - \frac{x_1^2}{R_s^2} - \frac{1}{3x_1R_s^2}(x_1^3 - x^3). \tag{4.4.12}$$

This equation may be compared with (4.3.18). Again, x_1 is the coordinate of the point being investigated and x is the coordinate at entry of the streamline of interest. It can be shown that the term x^3 in (4.4.12) may be replaced by x_1x^2 or may be dropped without affecting the principal results and with but a small error in terms of higher order. We choose to replace x^3 by x_1x^2 to simplify the form of the expressions for u and v, and obtain in place of (4.4.12)

$$\tfrac{1}{2}C_p = 1 - \frac{x_1^2}{R_s^2} - \frac{1}{3R_s^2}(x_1^2 - x^2) \tag{4.4.13}$$

in a form similar to (4.3.18). Application of Bernoulli's equation as before gives us for the tangential velocity component Uu (essentially the same u as in (4.4.2)) the result

$$u = \frac{1}{R_s}\sqrt{x^2 + \frac{8\epsilon}{3}(x_1^2 - x^2)} \tag{4.4.14}$$

in place of (4.3.20). The vertical scale may now be calculated, with the result

$$y = \epsilon\int_0 \frac{x\,dx}{x_1u} = \frac{\epsilon R_s}{x_1\left(1 - \frac{8\epsilon}{3}\right)}\left[\sqrt{x^2 + \frac{8\epsilon}{3}(x_1^2 - x^2)} - \sqrt{\frac{8\epsilon}{3}}\,x_1\right], \tag{4.4.15}$$

analogous to (4.3.21). The stand-off distance is the value of y at $x = x_1$, or

$$\Delta_0 = \frac{\epsilon R_s}{1 + \sqrt{\dfrac{8\epsilon}{3}}}, \tag{4.4.16}$$

analogous to (4.3.22). This simple result turns out to fit experimental results and exact numerical calculations very well (see Fig. 6–28 and discussion), and may be considered to have empirical support for values of $\epsilon^{1/2}$ which are not small. If we expand Δ_0 in a power series in $\epsilon^{1/2}$, we obtain

$$\Delta_0 = \epsilon R_s\left[1 - \sqrt{\frac{8\epsilon}{3}} + \frac{8}{3}\epsilon + O(\epsilon^{3/2})\right], \tag{4.4.17}$$

which may be compared with (4.4.9) and is analogous to (4.3.23). The third term, the one in $R_s\epsilon^2$, is in error for two reasons. First, it is in error because of the approximations involved in our simplified coordinate system, and second, because of the approximation made in getting (4.4.13). This term cannot be justifiably retained anyway in a constant-density analysis.

The distribution of u is given by

$$u = \sqrt{\frac{8\epsilon}{3}}\,\frac{x_1}{R_s}\left[\frac{\kappa y}{\epsilon R_s} + 1\right], \tag{4.4.18}$$

where

$$\kappa = \frac{1 - \dfrac{8\epsilon}{3}}{\sqrt{\dfrac{8\epsilon}{3}}}, \tag{4.4.19}$$

and the velocity gradient on the body is $\sqrt{8\epsilon/3}\,R_s^{-1}U$. The corresponding distribution of the normal velocity component Uv (essentially the same v as in (4 4.1)) is obtained through a cylindrical continuity equation in x and y, and is

$$v = -\sqrt{\frac{8\epsilon}{3}}\,\frac{y}{R_s}\left[\frac{\kappa y}{\epsilon R_s} + 2\right] = -\epsilon\,\frac{x^2}{x_1^2}. \tag{4.4.20}$$

The pressure distribution is

$$\tfrac{1}{2}C_p = 1 - \tfrac{1}{2}\epsilon\left(1 + \frac{x^4}{x_1^4}\right) - \frac{x_1^2}{3R_s^2}\left(4 - \frac{x^3}{x_1^3}\right), \tag{4.4.21}$$

with x/x_1 obtainable as a function of y from (4.4.20). Again, (4.4.21) satisfies the dynamic equations approximately, and is consistent with (4.4.11).

As in the last section, we can divide the shock layer into an inner and an outer layer, setting x^2/x_1^2 equal to $8\epsilon/3$ times a constant as a dividing criterion. The outer layer, in which the Bernoulli effect may be neglected, is of order ϵR_s in thickness. The inner layer, in which the centrifugal pressure gradient may be neglected, is of order $\epsilon^{3/2}R_s$ in thickness. The thickness of the inner layer is of order $\epsilon^{1/2}$ times that of the outer layer, and can be neglected if $\epsilon^{1/2}$ is small enough. Such an approximation is inherent in the constant-streamtube-area approximation of Section 5.3. We can consider the inner layer as one within the shock layer which has a negative displacement thickness. The presence of the inner layer has the effect of making the layer thinner.

In general, the same comments that we made on the extent of validity of the results for the circular cylinder apply with equal force for the sphere. However, we should point out an essential difference between the two-dimensional and the axisymmetric blunt body problems. In the two-dimensional case the Newtonian solution was divergent, and we obtained a reasonable solution for shock layer structure and stand-off distance only in our constant-density solution in which the fluid acceleration is taken into account. In the axisymmetric case the Newtonian solution gives a first approximation to the structure of the shock layer, and a stand-off distance corresponding to the first term in (4.4.9) or (4.4.17). This corresponds to neglecting the existence of the inner layer.

An approximate analysis similar to that given above was given by Hayes [2], neglecting the centrifugal correction to the Newtonian pressure distribution, and the results of Hayes were rederived by Li and Geiger [1] using a different approach, the one mentioned in the last section using a stream function. In these analyses the number 4 appeared wherever 3 does in equations (4.4.14) through (4.4.20). As with the circular cylinder, the centrifugal corrections to the pressure should not be neglected and these earlier analyses are in error on this point. The correct number 3 appeared first in the results of Chester [2].

We next obtain the density correction corresponding to (4.3.36). In place of (4.3.34) we have

$$\text{(4.4.22)} \qquad \frac{\rho - \rho_{s0}}{\rho_{s_0}} = \tfrac{1}{2}\gamma_s^{-1}\epsilon\left(1 - \frac{x^4}{x_1^4}\right) - \frac{x_1^2}{R_s^2}\left[\tfrac{1}{3}\gamma_s^{-1}\left(4 - 3\frac{x^2}{x_1^2} - \frac{x^3}{x_1^3}\right) + \frac{\mu}{\epsilon}\left(\frac{x^2}{x_1^2}\right)\right].$$

We have, approximately, $dy = \epsilon R_s x\, dx/x_1\sqrt{x^2 + 8\epsilon x^2/3}$, from (4.4.15). The integration (4.2.30) or (4.3.35) for χ yields

$$\text{(4.4.23)} \qquad \chi\Delta_{CD} = \gamma_s^{-1}\epsilon^2 R_s\left[\frac{2}{5} - \frac{1}{2}\sqrt{\frac{8\epsilon}{3}}\right] - \epsilon R_s\frac{x_1^2}{R_s^2}\left[\frac{1}{3}\gamma_s^{-1}\left(\frac{11}{4} - 4\sqrt{\frac{8\epsilon}{3}}\right) + \frac{\mu}{3\epsilon}\right],$$

again with terms of higher order dropped. As in the last section, we shall use (4.4.23) to add to a complete result for Δ to obtain an equivalent Howarth-Dorodnitsyn thickness Δ_{H-D}.

We next take the solutions of Chester [2], Muggia [2], W. J McCroskey (private communication), and Lighthill [1], and show they can be combined. We then interpret them through the Howarth-Dorodnitsyn thickness. This interpretation duplicates that of Hayes [8].

The pressure and density on the axis at shock and body are given by (4.3.37) and (4.3.38). The stand-off distance may be given in general by an equation analogous to (4.3.41), by

$$(4.4.24)\qquad \frac{\Delta}{R_s} = \epsilon - \epsilon\sqrt{\frac{8\epsilon}{3}} + \left(3 - \frac{2}{5}\gamma_s^{-1}\right)\epsilon^2 - \left(\frac{25}{8} - \frac{19}{42}\gamma_s^{-1} + \frac{\beta}{36} + \frac{5}{12}\frac{\mu}{\epsilon}\right)\epsilon^2\sqrt{\frac{8\epsilon}{3}} + \frac{1}{2}\frac{x_1^2}{R_s^2}\left[\left(\frac{11}{6}\gamma_s^{-1} + \frac{\beta}{9} + \frac{2}{3}\frac{\mu}{\epsilon}\right)\epsilon - \left(\frac{10}{3}\gamma_s^{-1} - \frac{2\beta}{5}\right)\epsilon\sqrt{\frac{8\epsilon}{3}}\right],$$

with terms of higher order in ϵ and x_1/R_s dropped. In (4.4.24), γ_s^{-1}, β, and μ are defined as in the last section, as in (4.3.28) and (4.3.31). To obtain McCroskey's result for Δ_{CD}/R_s we set $\gamma_s^{-1} = 0$ and $\mu = 0$ in (4.4.24); for Lighthill's result we set also $\beta = 0$. To obtain Muggia's result we set $\gamma_s^{-1} = 1$ and $\mu = 0$. To obtain Chester's result we set $\gamma_s^{-1} = 1$, $\beta = 3$, and $\mu = M_\infty^{-2}$.

We next add the correction term (4.4.23) to obtain the Howarth-Dorodnitsyn thickness

$$(4.4.25)\qquad \frac{\Delta_{\mathrm{H-D}}}{R_s} = \epsilon - \epsilon\sqrt{\frac{8\epsilon}{3}} + 3\epsilon^2 - \left(\frac{25}{8} + \frac{1}{21}\gamma_s^{-1} + \frac{\beta}{36} + \frac{5}{12}\frac{\mu}{\epsilon}\right)\epsilon^2\sqrt{\frac{8\epsilon}{3}} + \frac{1}{2}\frac{x_1^2}{R_s^2}\left[\frac{\beta}{9}\epsilon - \left(\frac{2}{3}\gamma_s^{-1} - \frac{2\beta}{5}\right)\epsilon\sqrt{\frac{8\epsilon}{3}}\right].$$

Chester and Muggia both give results for pressure or velocity on the body. In Muggia's expression for V/V_∞ there is clearly an error; the number 121 appearing should be 107 for this expression to be consistent with the other expressions given there. With this change there still is a discrepancy between the results of Muggia and Chester. If there were a change in Muggia's α_{11} such that the 107 above became 93, Muggia would agree with Chester. The authors have not been able to determine which is correct.

Here we use the value of the coefficient in question according to the result given by Chester. The velocity gradient expression analogous to (4.3.43) is

$$(4.4.26)\qquad (R_s u_b')^2 = \frac{8\epsilon}{3}\left[1 + \left(\frac{9}{4} + \frac{2}{21}\gamma_s^{-1} + \frac{\beta}{18} + \frac{5}{6}\frac{\mu}{\epsilon}\right)\epsilon - \frac{8}{3}\epsilon\sqrt{\frac{8\epsilon}{3}}\right]$$

with terms of $O(\epsilon^3)$ dropped. As in the last section, we can re-express (4.4.25) on the axis using the velocity gradient from (4.4.26) as

$$(4.4.27)\qquad \frac{\Delta_{\mathrm{H-D}_0}}{R_s} = \epsilon - \epsilon(R_s u_b') + 3\epsilon^2 - 2\epsilon^2\sqrt{\frac{8\epsilon}{3}}.$$

Analogous to (4.4.16) we can re-express (4.4.27) to the same order of accuracy as

$$\frac{\Delta_{\text{H-D}_0}}{R_s} = \frac{\epsilon}{1 + (R_s u_b')} + \tfrac{1}{3}\epsilon^2 + \tfrac{2}{3}\epsilon^2\sqrt{\frac{8\epsilon}{3}} \tag{4.4.28}$$

in a form better adapted for computation.

As before, we interpret (4.4.27) or (4.4.28) as indicating that to the given order of accuracy the stand-off distance at the axis is fully explained by the constant-density result for the sphere provided the interpretation is made in terms of Howarth-Dorodnitsyn variables and thicknesses and provided the actual velocity gradient is used for the Bernoulli effect. This interpretation would be weakened if Muggia's coefficient were correct and Chester's wrong in the disputed term. Of the terms in x_1^2/R_s^2, both that in $\gamma_s^{-1}\epsilon$ and in μ have been eliminated in (4.4.25).

For the asymmetric case, we present the local perturbed asymmetric solution in close analogy to that of Section 4.3. Again, the solution is taken from Hayes [7, Sect. 3], in which the unperturbed solution is that with $\alpha_0' = \beta_0' = \gamma_0' = 0$. In our notation we replace (4.4.18) by the perturbed form

$$u = \frac{x}{R_s}\left[\frac{y}{\epsilon R_s} + \sqrt{\frac{8\epsilon}{3}}\right] - \nu\epsilon\left[\frac{y^2}{\epsilon^2 R_s^2} + 2\sqrt{\frac{8\epsilon}{3}}\frac{y}{\epsilon R_s}\right]^{1/2}. \tag{4.4.29}$$

We interpret u as U^{-1} times the cartesian velocity component rather than the radial cylindrical component. For the other cartesian velocity component in the z direction parallel to the body we rewrite (4.4.18) in unperturbed form

$$w = \frac{z}{R_s}\left[\frac{y}{\epsilon R_s} + \sqrt{\frac{8\epsilon}{3}}\right]. \tag{4.4.30}$$

The normal velocity is given again by (4.4.20), slightly simplified

$$v = -\epsilon\left[\frac{y^2}{\epsilon^2 R_s^2} + 2\sqrt{\frac{8\epsilon}{3}}\frac{y}{\epsilon R_s}\right]. \tag{4.4.31}$$

As in Section 4.4, it may be checked that this velocity distribution satisfies the constant-density equations of motion.

As before, the equation for the stagnation streamline can be obtained

$$x = \frac{\nu\epsilon R_s}{\left[1 + 2\sqrt{\frac{8\epsilon}{3}}\frac{\epsilon R_s}{y}\right]^{1/2}}, \qquad z = 0. \tag{4.4.32}$$

This streamline comes in to the stagnation point tangent to the body. For large y this curve is asymptotic to $x = x_n$, $z = 0$, where again

$$\frac{x_n}{R_s} = \nu\epsilon. \tag{4.4.33}$$

The streamline which lies closest to this straight line is

$$x = x_n \frac{1 + \sqrt{\dfrac{8\epsilon}{3}}\dfrac{\epsilon R_s}{y}}{\left[1 + 2\sqrt{\dfrac{8\epsilon}{3}}\dfrac{\epsilon R_s}{y}\right]^{1/2}}, \tag{4.4.34}$$

and can be shown to carry zero vorticity. Again, we identify x_n with the normal point on the shock.

The location of the shock is given approximately by $y = \Delta_0 = \epsilon R_s$, and the value x_e of x at the point of entry of the stagnation streamline is obtained from (4.4.32) as

$$\frac{x_e}{R_s} = \nu\epsilon\left(1 - \sqrt{\frac{8\epsilon}{3}}\right) = \frac{x_n}{R_s}\left(1 - \sqrt{\frac{8\epsilon}{3}}\right). \tag{4.4.35}$$

The geometry of this solution is shown in Fig. 4–11.

The vorticity immediately behind the shock on the plane $z = 0$ is expressed

$$\frac{\zeta_s}{U} = (-u_y)_s = \frac{x_n - x}{\epsilon R_s^2}, \tag{4.4.36}$$

as before. The vorticity on the stagnation streamline at the point of entry is

$$\frac{\zeta_e}{U} = \sqrt{\frac{8\epsilon}{3}}\frac{x_n}{\epsilon R_s^2} = \sqrt{\frac{8\epsilon}{3}}\frac{\nu}{R_s}. \tag{4.4.37}$$

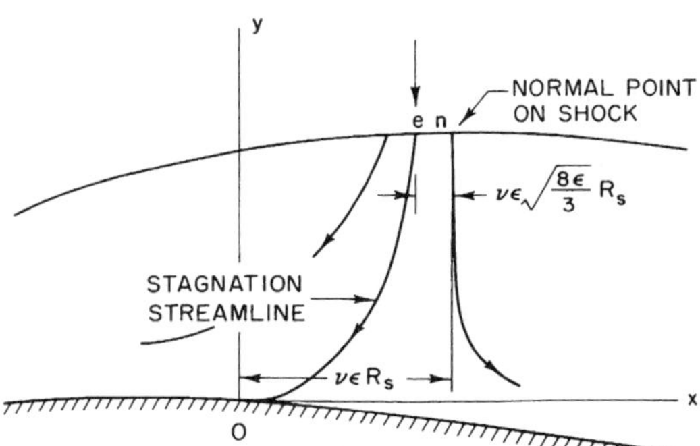

FIG. 4–11. Asymmetric quasi-spherical stagnation region.

The vorticity at the stagnation point is infinite, as it is everywhere on the body surface.

As in the cylindrical case, it is possible to obtain an approximate outer solution with asymmetry, in this case by Newtonian theory without the Bernoulli effect of the constant-density theory. The quasi-spherical shock shape

$$y_s = y_{s_0} - \frac{x^2}{2R_s} - \frac{z^2}{2R_s} + R_s' \frac{x^3}{6R_s^2} \tag{4.4.38}$$

analogous to (4.3.56) is chosen. A Newtonian calculation similar to that for the general stagnation point in Section 3.6 leads to the result

$$m_1 = R_s\left(1 + R_s' \frac{x_1}{6R_s}\right), \tag{4.4.39}$$

with terms of higher order in $R_s' x_1$ dropped. This gives

$$\Delta' = \epsilon R_s'/6 \tag{4.4.40}$$

in analogy with (4.3.57). This is left as an exercise for the reader. For purposes of checking, the pressure on the body in this Newtonian example is

$$\tfrac{1}{2}C_{p_b} = 1 - \frac{x_1^2}{R_s^2}\left(\frac{4}{3} - \frac{89R_s' x_1}{60R_s}\right) - \frac{y_1^2}{R_s^2}\left(\frac{4}{3} + \frac{R_s' x_1}{20R_s}\right). \tag{4.4.41}$$

As in the cylindrical case, the matching of this outer solution with the local solution cannot be obtained by this simple method.

As in the cylindrical case, however, the method may be extended by calculating more accurately the pressure change across the outer layer near the stagnation point (see Hayes [9]), supplementing the three-dimensional Newtonian calculation. The result (4.4.40) is obtained, of course, and also the result

$$x_n = \epsilon R_s' R_s/40; \qquad \nu = R_s'/40. \tag{4.4.42}$$

No global asymmetric perturbation analysis analogous to that of Muggia [1] but based upon the axisymmetric solution (4.4.6) appears to be available.

5. Solutions with cross flow

Within the limitations inherent in the constant-density approximation, we may seek for constant-density solutions for problems with cross flow. The approach we need is very like the approach we used for Newtonian solutions with cross flow. However, we must now include the influence of the pressure field on the trajectory as well as on the velocity of a fluid particle. Following the alternative approximate constant-density approach used in the last two sections, we use the pressure distribution from the Newtonian theory for this purpose. The additional complexity prevents us from finding analytical solutions in closed form except in trivial cases, but the procedures involved are straightforward and we can readily obtain a number of solutions with numerical quadratures.

To exemplify the method we present here the constant-density solution for a general stagnation point flow, for which we have already obtained the Newtonian solution. We shall take advantage of the fact that the Newtonian pressure we shall use, (3.6.34) and (3.6.35), is the sum of a function of x and a function of y. Since the value of $u = q_x$ upon entry from (3.6.26) and the pressure gradient in the x direction are independent of y, u will remain independent of y, and we can calculate u from a one-dimensional Bernoulli equation. The result obtained, analogous to (4.3.20) and (4.4.14), is

$$u = K_1 \sqrt{x^2 + 2\epsilon(1 + \pi_1)(x_1^2 - x^2)}, \tag{4.5.1}$$

with x_1 the coordinate of the point investigated and x the coordinate at entry of a streamline. In a similar way we may obtain an expression for the other dimensionless tangential velocity component $v = q_y$,

$$v = K_2 \sqrt{y^2 + 2\epsilon(1 + \pi_2)(y_1^2 - y^2)}. \tag{4.5.2}$$

The quantities π_1 and π_2 are pressure correction terms from the Newtonian analysis. The accuracy of these expressions is about the same as that of (4.4.14) for our approximate solution for a blunt body of revolution. The quantities K_1 and K_2 are the principal curvatures of the shock, and are not generally the same as those of the body.

The particle trajectories are the integral curves of the equation

$$\frac{dx_1}{u} = \frac{dy_1}{v}, \tag{4.5.3}$$

with the coordinates x and y of the point of entry kept fixed. The result of this integration is the equation for the particle trajectories,

$$\begin{aligned} \frac{1}{K_1 \sqrt{2\epsilon(1 + \pi_1)}} \left[\sinh^{-1} \frac{\alpha_1 x_1}{x} - \sinh^{-1} \alpha_1\right] \\ = \frac{1}{K_2 \sqrt{2\epsilon(1 + \pi_2)}} \left[\sinh^{-1} \frac{\alpha_2 y_1}{y} - \sinh^{-1} \alpha_2\right], \end{aligned} \tag{4.5.4}$$

where the parameters α_1 and α_2 are defined as

$$\alpha_1 = \sqrt{\frac{2\epsilon(1+\pi_1)}{1-2\epsilon(1+\pi_1)}}, \tag{4.5.5a}$$

$$\alpha_2 = \sqrt{\frac{2\epsilon(1+\pi_2)}{1-2\epsilon(1+\pi_2)}}. \tag{4.5.5b}$$

As in the Newtonian theory the locus of entering streamlines for a point is given by (4.5.4) with x_1 and y_1 fixed and x and y considered the variables.

The quantity m of Newtonian theory is replaced by $\epsilon^{-1}z$ in our constant-density theory, with z distance normal to the body in this case. Using the same reasoning as for the Newtonian analysis and in complete analogy with (4.3.21) and (4.4.15), we obtain

$$\epsilon^{-1}z = \int_0 \frac{y\,dx}{y_1 u} = \int_0 \frac{x\,dy}{x_1 v}. \tag{4.5.6}$$

The derivative of this expression may be looked upon as a differential equation for the locus of entering streamlines, and (4.5.4) can be obtained with this alternative approach. To calculate the stand-off distance, we must carry out the integral (4.5.6) over the entire range of integration, for example from $x = 0$ to $x = x_1$ in the first integral. The result is

$$\Delta_0 = \frac{\epsilon\alpha_2}{K_1\sqrt{1-2\epsilon(1+\pi_1)}} \int_{x=0}^{x=x_1} \frac{d\sinh^{-1}\left(\dfrac{x}{\alpha_1 x_1}\right)}{\sinh\left(\sinh^{-1}\alpha_2 + \mathscr{K}\sinh^{-1}\dfrac{\alpha_1 x_1}{x} - \mathscr{K}\sinh^{-1}\alpha_1\right)} \tag{4.5.7}$$

expressed as a Stieltjes integral, where the parameter $\mathscr{K}$

$$\mathscr{K} = \frac{K_2\sqrt{1+\pi_2}}{K_1\sqrt{1+\pi_1}} = \frac{\sqrt{1+\pi_2}}{\kappa\sqrt{1+\pi_1}} \tag{4.5.8}$$

is a function of the parameter κ of the Newtonian analysis alone. A transformation of variables changes the integral of (4.5.7) to

$$\Delta_0 = \frac{\epsilon\alpha_2}{K_1\sqrt{1-2\epsilon(1+\pi_1)}} \int_{\sinh^{-1}\alpha_1}^{\infty} \frac{d\xi}{\sinh\xi\,\sinh(\sinh^{-1}\alpha_2 + \mathscr{K}\xi - \mathscr{K}\sinh^{-1}\alpha_1)}. \tag{4.5.9}$$

To simplify (4.5.9), let us first note that we may drop the factors of the form $\sqrt{1-2\epsilon(1+\pi)}$ in (4.5.5) and (4.5.9). Then $\mathscr{K} = \alpha_2/\alpha_1\kappa$. We replace

$\sinh^{-1} \alpha_{1,2}$ by $\alpha_{1,2}$. The second sinh in the denominator of the integrand is expressed approximately through the addition law as

$$\sinh [\alpha_2(1 - \kappa^{-1}) + \mathscr{K}\xi] = \sinh \mathscr{K}\xi + \alpha_2(1 - \kappa^{-1}) \cosh \mathscr{K}\xi, \tag{4.5.10}$$

for the range $\kappa \geqslant 1$, $\mathscr{K} \leqslant 1$. We then separate the integrand of (4.5.9) with α_2 included into five parts:

$$\begin{aligned}\frac{\alpha_2}{\sinh \xi \sinh [\alpha_2(1 - \kappa^{-1}) + \mathscr{K}\xi]} &= \frac{\alpha_1 \kappa}{\xi[\xi + \alpha_1(\kappa - 1)]} \\ &- \left(\frac{1}{\xi} - \frac{1}{\sinh \xi}\right) \frac{\alpha_1 \kappa}{\xi + \alpha_1(\kappa - 1)} - \alpha_1 \kappa \left(\frac{1}{\xi^2} - \frac{\mathscr{K}}{\sinh \xi \sinh \mathscr{K}\xi}\right) \\ &+ \alpha_1 \kappa \left(\frac{1}{\xi} - \frac{1}{\sinh \xi}\right) \frac{1}{\xi} + \alpha_1 \kappa \left\{\frac{\alpha_1(\kappa - 1)}{\xi[\xi + \alpha_1(\kappa - 1)]}\right. \\ &\left. - \frac{\mathscr{K}\alpha_2(1 - \kappa^{-1}) \cosh \mathscr{K}\xi}{\sinh \mathscr{K}\xi\, [\sinh \mathscr{K}\xi + \alpha_2(1 - \kappa^{-1}) \cosh \mathscr{K}\xi]}\right\} \frac{1}{\sinh \xi}.\end{aligned} \tag{4.5.11}$$

To obtain Δ_0 correct with an error no larger than $O(\alpha^2 \ln \alpha^{-1})$ we integrate the first part from α_1 to ∞ and the next three parts from 0 to ∞. The integral of the fifth part is dropped.

We define the following two functions by

$$F(\mathscr{K}) = \int_0^\infty \left(\frac{1}{\xi^2} - \frac{\mathscr{K}}{\sinh \xi \sinh \mathscr{K}\xi}\right) d\xi, \tag{4.5.12}$$

and

$$G(\lambda) = \int_0^\infty \left(\frac{1}{\xi} - \frac{1}{\sinh \xi}\right) \frac{d\xi}{\xi + \lambda}. \tag{4.5.13}$$

In addition we note the following two definite integrals,

$$\int_0^\infty \left(\frac{1}{\xi} - \frac{1}{\sinh \xi}\right) \frac{d\xi}{\xi} = F(0) = G(0) = \ln 2 = 0.6931 \tag{4.5.14}$$

and

$$\int_0^\infty \frac{\xi\, d\xi}{\sinh \xi} = \frac{\pi^2}{4}. \tag{4.5.15}$$

The second of these will be used later.

We may now express Δ_0 in the form

$$\frac{\Delta_0}{\epsilon R_1} = \frac{\kappa \ln \kappa}{\kappa - 1} - \alpha_1 \kappa \{F(\mathscr{K}) + G[\alpha_1(\kappa - 1)] - \ln 2\}, \tag{4.5.16}$$

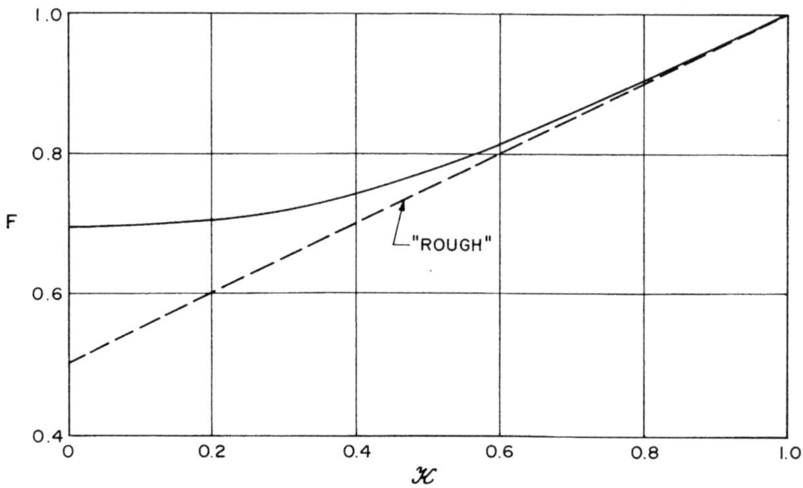

FIG. 4–12. The function $F(\mathscr{K})$.

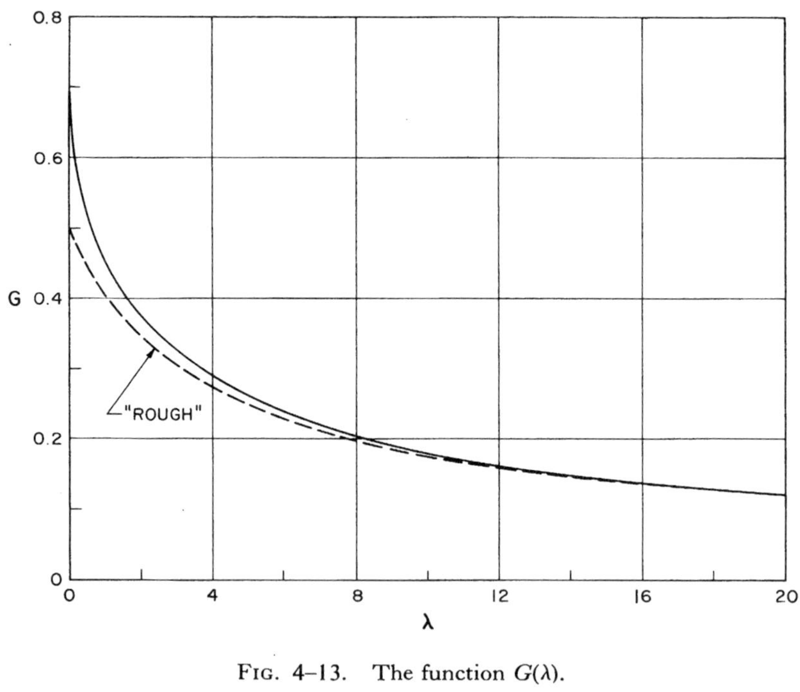

FIG. 4–13. The function $G(\lambda)$.

in which the first term gives the Newtonian result of (3.6.31). The functions $F(\mathscr{K})$ and $G(\lambda)$ are shown plotted in Figs. 4–12 and 4–13. To assist in the computation of the α's and of $\mathscr{K}$ the quantities

$$\sqrt{1+\pi_1}, \qquad \sqrt{1+\pi_2}, \qquad \text{and} \qquad \sqrt{(1+\pi_2)/(1+\pi_1)} = \alpha_2/\alpha_1 = \kappa\mathscr{K}$$

are plotted in Fig. 4–14.

The function F satisfies the functional relation

$$F(\mathscr{K}^{-1}) = \mathscr{K}^{-1}F(\mathscr{K}). \tag{4.5.17}$$

For $\mathscr{K}$ small we have, using (4.5.15),

$$F(\mathscr{K}) = \ln 2 + \frac{\pi^2}{24}\mathscr{K}^2 + O(\mathscr{K}^4). \tag{4.5.18}$$

For $|\mathscr{K} - 1|$ small we have

$$F(\mathscr{K}) = 1 + \tfrac{1}{2}(\mathscr{K} - 1) + O[(\mathscr{K} - 1)^2]. \tag{4.5.19}$$

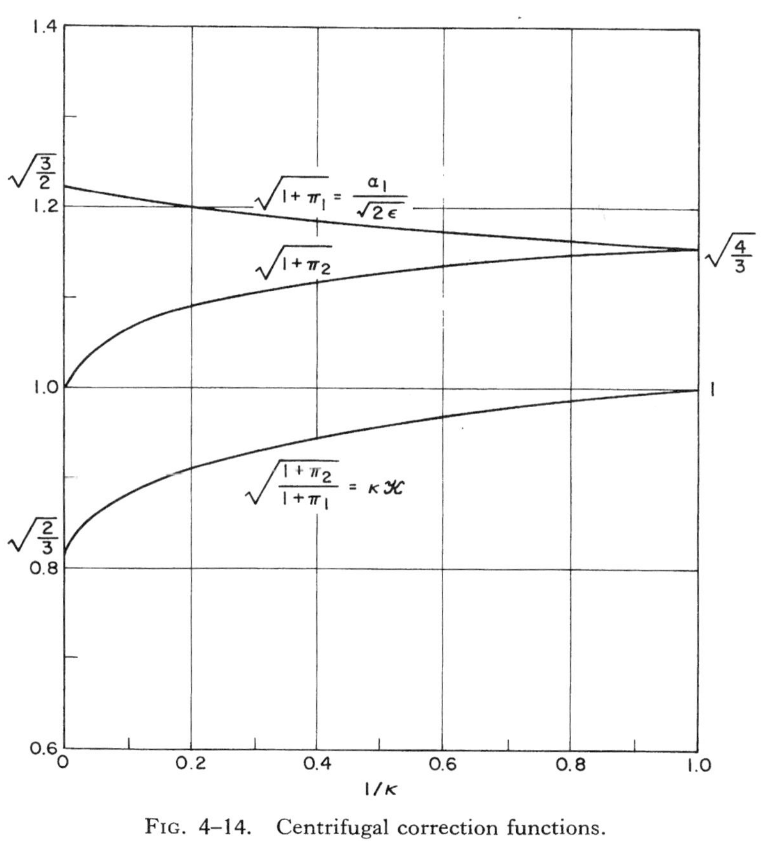

FIG. 4–14. Centrifugal correction functions.

For the function G, when λ is small we have

$$G(\lambda) = \ln 2 - \frac{\lambda}{6}\ln\frac{1}{\lambda} - \lambda\int_0^\infty\left(\frac{1}{\xi^2} - \frac{1}{\xi\sinh\xi} - \frac{1}{6(1+\xi^2)}\right)\frac{d\xi}{\xi} + O(\lambda^2). \tag{4.5.20}$$

When λ is large we have, using (4.5.15),

$$G(\lambda) = \frac{1}{\lambda}\ln\frac{\lambda}{2} + \frac{\pi^2}{4\lambda^2} + O\left(\frac{1}{\lambda^3}\right). \tag{4.5.21}$$

In the central range of the parameters for which κ and $\mathscr{K}$ are neither large nor small, we may set $G = \ln 2$ in (4.5.16). This follows from the fact that we are neglecting terms of $O(\alpha^2 \ln \alpha^{-1})$ and $G - \ln 2$ is of that order. This result we may express in the form

$$\frac{\Delta_0}{\epsilon R_1} = \frac{\kappa}{\kappa - 1}[\ln\kappa - \alpha_1(\kappa - 1)F(\mathscr{K})]. \tag{4.5.22}$$

To the order of α_1^3, this expression may in turn be put in the form

$$\frac{\Delta_0}{\epsilon R_1} = \frac{\kappa}{\kappa - 1}\ln\kappa\,\frac{1 - \frac{1}{2}\alpha_1(\kappa - 1)F(\mathscr{K})}{1 + \frac{1}{2}\alpha_1(\kappa - 1)F(\mathscr{K})}. \tag{4.5.23}$$

In either form the result is invariant with respect to a transformation in which the subscripts 1 and 2 are interchanged and κ and $\mathscr{K}$ replaced by their inverses. In showing this (4.5.17) is used.

In the transition range of the parameters for which $\mathscr{K}$ and κ^{-1} are of no larger order than $\alpha^{1/4}$ (or $\epsilon^{1/8}$) and of no smaller order than $\alpha^{1/2}$ (or $\epsilon^{1/4}$), we may make the replacement

$$F + G - \ln 2 = \ln 2 + \frac{\pi^2}{24}\mathscr{K}^2 - \frac{\alpha_1(\kappa - 1)}{6}\ln\frac{A^6}{\alpha_1(\kappa - 1)} \tag{4.5.24}$$

in (4.5.16). Here we have set

$$\ln A = \int_0^\infty\left(\frac{1}{\xi^2} - \frac{1}{\xi\sinh\xi} - \frac{1}{6(1+\xi^2)}\right)\frac{d\xi}{\xi}. \tag{4.5.25}$$

In the quasi-cylindrical range of the parameters with $\mathscr{K}$ and κ^{-1} of the order of $\alpha^{1/2}$ (or $\epsilon^{1/4}$) or smaller, we may set $F = \ln 2$ in (4.5.16). In the cylindrical range of the parameters with $\mathscr{K}$ and κ^{-1} of the order of α^2 (or ϵ) or smaller we can put (4.5.16) into the form

$$\frac{\Delta_0}{\epsilon R_1} = \ln \frac{2}{\sqrt{3\epsilon} + \dfrac{\pi^2}{4\kappa}} . \tag{4.5.26}$$

In the cylindrical range we have set $\alpha_1 = \sqrt{3\epsilon}$. The approach to the first term of (4.3.23) is evident. If κ^{-1} is of the order of α^3 or $\epsilon^{3/2}$ we may drop the $\pi^2/4\kappa$ term in (4.5.26).

The foregoing treatment has distinguished the various ranges of the parameters. A "rough" approximation over the entire range, one that gives the right order of magnitude to the Bernoulli effect correction, is

$$\frac{\Delta_0}{\epsilon R_1} = \frac{\kappa}{\kappa - 1} \ln \frac{\kappa}{1 + \frac{1}{2}(\kappa - 1)(\alpha_1 + \alpha_2\kappa^{-1})}, \tag{4.5.27}$$

for the range $\kappa \geqslant 1$. This result would fit in the form (4.5.16) with

$$F(\mathscr{K}) = \tfrac{1}{2}(1 + \mathscr{K}) \tag{4.5.28a}$$

$$G(\lambda) = \frac{1}{\lambda} \ln (1 + \tfrac{1}{2}\lambda) \tag{4.5.28b}$$

$$F(0) = G(0) = \tfrac{1}{2}, \tag{4.5.28c}$$

with ln 2 replaced by $\frac{1}{2}$. The corresponding approximate curves are included dotted in Figs. 4–12 and 4–13. In this rough and much simpler approximation the distinction between the various ranges of the parameters does not appear. This form was found in an early attempt to integrate (4.5.9) and is not fully rational. It has the virtue of being simple while estimating the order of magnitude correctly.

The pressure on the body is governed by the Newtonian pressure distribution, within the approximations of our theory. The velocity gradients on the body are obtained by setting $x = 0$ and $y = 0$ in (4.5.1) and (4.5.2). The results are

$$R_1 u_{b_x} = \sqrt{2\epsilon(1 + \pi_1)} \tag{4.5.29a}$$

$$u_{b_y} = 0, \tag{4.5.29b}$$

and

$$v_{b_x} = 0, \tag{4.5.30a}$$

$$R_2 v_{b_y} = \sqrt{2\epsilon(1 + \pi_2)}. \tag{4.5.30b}$$

The evaluation of the corresponding indefinite integral is necessary for the establishment of a vertical scale. For example, we may integrate (4.5.7) from $x = 0$ only to x to obtain the quantity $z(x, x_1)$. This function inverted gives $x(z, x_1)$. The velocity u of (4.5.1) may then be expressed as $u(z, x_1)$, as a velocity profile. We may do the same with $v(z, y_1)$, calculate $\partial w/\partial z$ from the equation of continuity, and obtain the normal velocity profile $w(z)$ by a quadrature.

In general, the comments we have made for the corresponding two-dimensional and axisymmetric constant-density solutions for stagnation point regions apply for the solution we have just obtained. The second derivatives of Δ are undetermined and the body curvature is unspecified within this approximate theory. In fact, unless the body is suitably symmetric, we have no assurance that even the principal axes of curvature are the same for the body surface as for the shock. The most important limitation in the accuracy of the results probably lies in the effect of compressibility.

Before a calculation of the density correction would be worthwhile, it would be necessary for us to improve our constant-density analysis to take account of the errors inherent in our approximate analysis. The sources of error include the approximations made in the simplified coordinate system and the approximations made in establishing the equations for the velocity components (4.5.1) and (4.5.2).

As an additional simpler example of a solution with cross flow we consider the finite yawed cylindrical body for which the Newtonian results are given in (3.6.16) to (3.6.18). We must keep in mind here that it is the shock which we consider to have the cylindrical shape, not the body (see Fig. 4–15a). A cross section through the axis of symmetry would show a flow pattern as that in Fig. 4–15b. Distance along the body is designated R instead of y_1 as in Section 3.6.

This problem is particularly simple because the analysis is essentially precisely

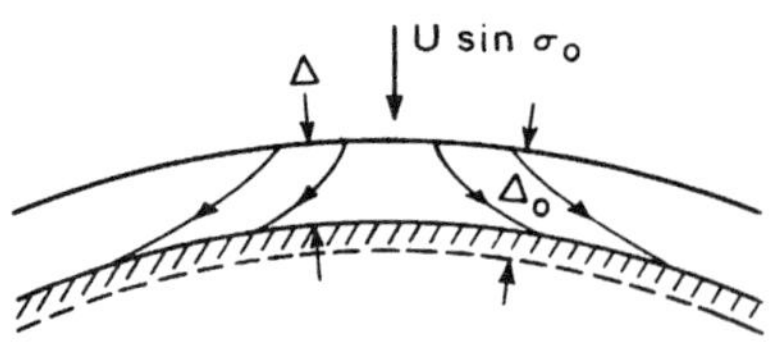

FIG. 4–15a. Transverse section of finite cylindrical shock.

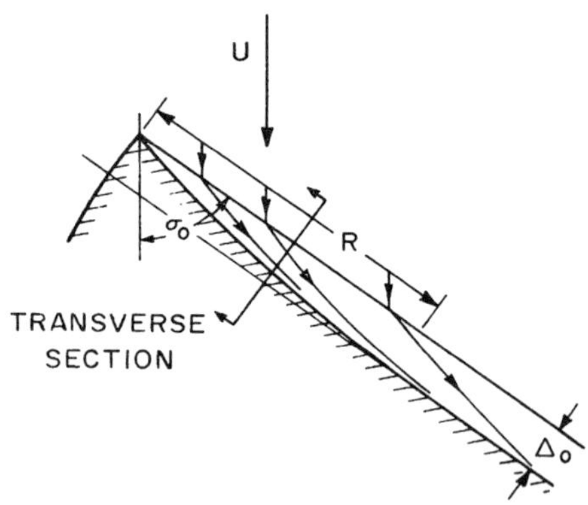

FIG. 4–15b. Axial cross section of finite cylindrical shock.

the same as that of Section 4.3. We use the point of view of cylindrical flow theory, and consider the distance R divided by the velocity component $U \cos \sigma_0$ along the axis of symmetry as the age of the shock layer. The structure of the shock layer is that of Section 4.3 with normal velocity $U \sin \sigma_0$, and fills out from the top ($y = \Delta_0$) with time. The age of the layer with respect to the shock layer structure is $1/U \sin \sigma_0$ times $\int_s dy/v$, where v is given by (4.3.25).

Equating these two times gives

$$\frac{R}{\cos \sigma_0} = \frac{R_s}{\sqrt{3\epsilon} \sin \sigma_0} \int_{\Delta_0 - \Delta}^{\Delta_0} \frac{d(y/\epsilon R_s)}{\sinh(y/\epsilon R_s)}. \tag{4.5.31}$$

The indefinite integral here is $\log \tanh(y/2\epsilon R_s)$. In manipulating the definite integral to calculate Δ we use the result

$$\tanh \frac{\Delta_0}{2\epsilon R_s} = 1 - \sqrt{3\epsilon} + \tfrac{3}{2}\epsilon + O(\epsilon^{3/2}), \tag{4.5.32}$$

derivable from $\Delta_0/\epsilon R_s = \ln 2/\sqrt{3\epsilon}$ (see (4.3.23)). To carry out the calculation we first calculate the quantity $\tanh(\sqrt{3\epsilon}\, R \tan \sigma_0/2R_s)$ as a function of Δ. We can then solve for Δ to obtain

$$\frac{\Delta}{\epsilon R_s} = \ln \frac{1 + \dfrac{2}{\sqrt{3\epsilon}} \tanh \dfrac{\sqrt{3\epsilon}\, R \tan \sigma_0}{2R_s}}{1 + \dfrac{\sqrt{3\epsilon}}{2} \tanh \dfrac{\sqrt{3\epsilon}\, R \tan \sigma_0}{2R_s}}. \tag{4.5.33}$$

To check this result we may first verify that $\Delta \to \Delta_0$ as $R \to \infty$ with ϵ fixed. If we identify Δ/ϵ as m_1 and let $\sqrt{3\epsilon} \to 0$ with R fixed, (4.5.33) approaches the Newtonian result (3.6.16). For large R with ϵ fixed we may deduce

$$\frac{\Delta}{\epsilon R_s} = \tfrac{1}{2} \ln \frac{4}{3\epsilon} - \frac{2}{1 + \frac{1}{2}\sqrt{3\epsilon}} \exp\left(-\frac{\sqrt{3\epsilon}\, R \tan \sigma_0}{R_s}\right) + O\left[\exp\left(-\frac{2\sqrt{3\epsilon}\, R \tan \sigma_0}{R_s}\right)\right]. \tag{4.5.34}$$

This result shows that the full cylindrical thickness is approached exponentially, with a characteristic decay distance of $R_s/\sqrt{3\epsilon} \tan \sigma_0$.

CHAPTER V

THE THEORY OF THIN SHOCK LAYERS

1. Basic concepts

The methods given in the last two chapters are generally inadequate for the calculation of a sufficiently good approximation to the flow field in the shock layer on a blunt body. Although the Newtonian theory gives us a valuable conceptual picture of phenomena in a thin shock layer, its estimate of velocities and pressures is too rough to be satisfactory in engineering practice, and its anomalies are unrealistic and also unacceptable in practice. The improvement afforded by the constant-density solutions is limited in scope and validity, and an improvement of Newtonian theory adequate for most purposes cannot be based upon the assumption of constant density.

In order to utilize the information gained from the other methods, however, we must assume that the shock layer is thin and take advantage of this assumption in our analysis. With the shock layer thin the shape of the shock wave cannot be greatly different from the shape of the body; and the pressure on the body is determined primarily by the shock shape. In order that the shock layer be thin we must be able to make assumption D, that the density ratio ϵ across the shock be small compared with one. This assumption on the behavior of the fluid is appreciably weaker than the assumption required for Newtonian theory, but still puts some limitation on the validity of the resulting development.

Some investigators who have based their work on a shock layer which is thin have used the assumption that the fluid is a perfect gas with constant ratio of specific heats γ which is close to one. In practice reasonably small values of the density ratio ϵ are only obtainable with gases which depart significantly from the perfect gas laws, and it is preferable to postulate a fluid obeying a general equation of state. The greater generality that we gain thereby in our development is at little or no expense in conceptual or analytical simplicity. In addition, it is not necessary to make the assumption that the fluid is in thermal equilibrium. The principal approaches and concepts apply with nonequilibrium fluids. We shall consider the effects of nonequilibrium in the final section of this chapter.

The concepts of the theory of thin shock layers apply to asymmetric flows, flows with cross flow, and unsteady flows, through a direct extension of the methods of the last two chapters. We shall be primarily concerned with steady axisymmetric and two-dimensional flows with lateral symmetry, and with the concepts which arise in these simpler flows. However, we shall also consider certain properties of three-dimensional flows, and of conical flows in particular. And we shall present certain similitudes for flat-faced bodies with small density ratio. We shall consider integral methods based upon the basic hydrodynamic conservation laws. And we shall consider in more detail some of the problems connected with Newtonian separation, i.e. the creation of a free layer.

The basic problem considered may be one of several types, of which we shall list three. Our attention will be focused principally on the first two types, termed the "inverse problem" and the "direct problem". But we include the "pressure problem" primarily to make it clear that the problems of interest are not necessarily limited to those of the first two types. The problem types considered are:

Inverse Problem (*Shock Given*). The shape of the shock wave is given, and it is required to find the corresponding shape of the body and the pressures on this body.

Direct Problem (*Body Given*). The shape of the body is given, and it is required to find the shape of the shock wave and the pressures on the body.

Pressure Problem (*Pressure Given*). The pressure on the body is given as a function of distance along the body and it is required to find the corresponding body shape and shock shape.

In any of these problems information about the structure of the shock layer (pressures, densities, and velocities) may or may not be required, but is normally available as part of the results of the problem solution. The direct problem may be considered the problem normally met with in practice, while the pressure problem may arise when it is desirable to control the behavior of the viscous boundary layer in some manner. The inverse problem does not appear directly as a design problem, but has a physical simplicity which the others do not have. If the shock shape is given, so also are all components of the velocity and the thermodynamic state of the fluid behind the shock. In fact, certain mathematical difficulties appear in the inverse problem because the quantities behind the shock are so completely specified (see Section 6.5).

With the shock layer very thin, there is a close correspondence between the shock shape and the body shape, and in the Newtonian limit the two shapes coincide in an unseparated flow. With the shock layer not so thin, we find that a relatively large change of small wavelength in the body shape corresponds to a small change in the shock shape. This insensitiveness of the shock shape to changes in the body shape, discussed in more detail in Section 6.5, is related to the fact that the shock shape in front of a body in a hypersonic (or supersonic) flow may not be chosen arbitrarily, but must satisfy certain restrictions.

Qualitatively speaking, the shock shape must be smooth. It is probably necessary that the shock shape be analytic except at special singular points, if the pertinent equations of state of the fluid are analytic.

The coordinate system used in an investigation of the hypersonic flow around a blunt body may be cartesian or cylindrical. Various authors have used these straightforward systems with success. Our principle of taking advantage of the thinness of the shock layer leads naturally to a different type of coordinate system, generally designated as of the boundary layer type. In a coordinate system of this type, the two coordinates are chosen to correspond to distances along the layer and across the layer, respectively. Figure 5–1 depicts the coordinate system we shall use.

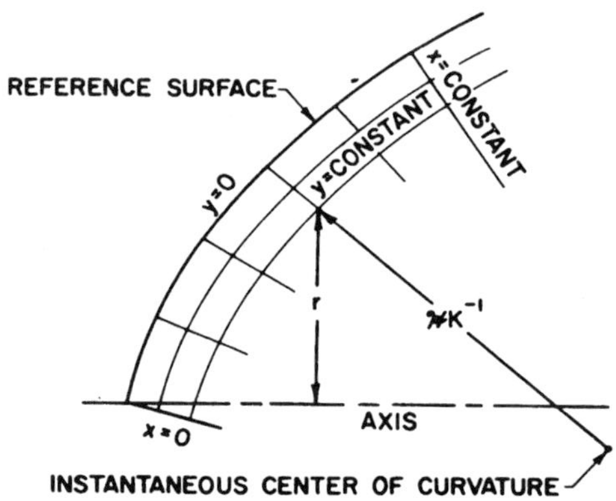

FIG. 5–1. Coordinate system of the boundary layer type.

To serve as a basis for our coordinate system a reference surface is chosen, generally either the shock surface or the body surface. This surface must be smooth, by which we mean that the surface must have a finite curvature at every point except perhaps at a leading edge, and that this curvature must be continuous. The coordinate x is defined on the surface as distance on the surface from the leading edge or stagnation point for the portion of the shock layer investigated. For the coordinate system as a whole we set x constant on the straight lines normal to the surface. The coordinate y is distance from the surface along one of these normals, defined as positive in the direction toward the body from the shock (the purpose of this choice of direction is to make the corresponding velocity components positive at the shock). The surfaces of constant y form with the reference surface a family of parallel surfaces (or geodesic parallels) as defined in standard texts on differential geometry. The curvature of the reference surface is denoted $K(x)$, defined as positive when the surface is concave on the side of positive y.

The curvature of any of the other surfaces of constant y is $\mathscr{H}^{-1}K$, where

$$\mathscr{H} = 1 - Ky. \tag{5.1.1}$$

The domain within which we may use the coordinate system is limited by the requirement that $\mathscr{H}$ be always positive (see Fig. 5–1).

We now have a coordinate system which is orthogonal and generally left-handed, and for which the metric is

$$ds^2 = \mathscr{H}^2\, dx^2 + dy^2. \tag{5.1.2}$$

Thus x corresponds to distance only at the reference surface, for which $\mathscr{H} = 1$. For axisymmetric problems we need also the distance from the axis $r(x, y)$. If the angle of incidence of the surface is $\sigma(x)$ (if the reference surface is the shock surface), the quantity r has the form

$$r(x, y) = \int \sin\sigma\, dx - y\cos\sigma. \tag{5.1.3}$$

The curvature in this case is given by

$$K(x) = -\frac{d\sigma}{dx}, \tag{5.1.4}$$

and r from (5.1.3) satisfies the equation

$$dr = \sin\sigma\, \mathscr{H}\, dx - \cos\sigma\, dy. \tag{5.1.5}$$

If the surface is the body surface, (5.1.3) to (5.1.5) are still valid with θ_b in place of σ.

The component of the velocity in the x direction is denoted u, and the component in the y direction is denoted v. The equations of motion in our coordinate system are

$$\frac{\partial r^j \rho u}{\partial x} + \frac{\partial \mathscr{H} r^j \rho v}{\partial y} = 0, \tag{5.1.6a}$$

$$u\frac{\partial u}{\partial x} + \mathscr{H} v\frac{\partial u}{\partial y} - Kuv + \frac{1}{\rho}\frac{\partial p}{\partial x} = 0, \tag{5.1.6b}$$

$$u\frac{\partial v}{\partial x} + \mathscr{H} v\frac{\partial v}{\partial y} + Ku^2 + \frac{\mathscr{H}}{\rho}\frac{\partial p}{\partial y} = 0, \tag{5.1.6c}$$

$$u\frac{\partial S}{\partial x} + \mathscr{H} v\frac{\partial S}{\partial y} = 0, \tag{5.1.7}$$

in which the entropy equation (5.1.7) is used in place of the energy equation.

A transformation which is convenient for many purposes replaces the normal coordinate y by a stream function ψ but does not change the dependent variables. Such a transformation is called a von Mises transformation after the well known transformation of boundary layer theory. We define ψ in terms of its partial derivatives

$$\psi_x = \mathscr{H} r^j \rho v, \tag{5.1.8a}$$

$$\psi_y = -r^j \rho u; \tag{5.1.8b}$$

the existence of such a quantity is ensured by the continuity equation (5.1.6a). In changing to the new coordinate system (x, ψ) there is no possibility of confusion in partial derivatives with respect to y or ψ, as they are both taken with x constant. In order to distinguish clearly between partial derivatives with respect to x taken with y constant and with ψ constant, we shall reserve the usual notation for the former and introduce the notation

$$\frac{D}{Dx} = \left(\frac{\partial}{\partial x}\right)_\psi = \frac{\partial}{\partial x} + \mathscr{H} \frac{v}{u} \frac{\partial}{\partial y}, \tag{5.1.9a}$$

for the latter, with

$$\frac{\partial}{\partial x} = \frac{D}{Dx} + \mathscr{H} r^j v \frac{\partial}{\partial \psi}. \tag{5.1.9b}$$

In the new coordinate system with the variable y eliminated, (5.1.7) and a suitable combination of (5.1.6b) and (5.1.6c) become (assuming u not equal to zero)

$$\frac{DS}{Dx} = 0, \tag{5.1.10}$$

$$\frac{D}{Dx}\left[\tfrac{1}{2}(u^2 + v^2)\right] + \frac{1}{\rho}\frac{Dp}{Dx} = \frac{D}{Dx}\left(h + \tfrac{1}{2}q^2\right) = 0. \tag{5.1.11}$$

These two equations are immediately recognizable as stating that the entropy and total enthalpy are constant along streamlines. Since the total enthalpy is constant across the shock and thus across the entire flow field we shall use that fact in place of (5.1.11). And (5.1.10) will be automatically taken care of by requiring the entropy to be a function $S(\psi)$ of the stream function alone.

Equation (5.1.6c) becomes

$$\mathscr{H} r^j \frac{\partial p}{\partial \psi} = \frac{Dv}{Dx} + Ku, \tag{5.1.12}$$

which gives pressure changes across streamlines. Equation (5.1.6a) may be put into alternative forms

$$\mathscr{H}\frac{\partial\left(\dfrac{v}{u}\right)}{\partial\psi}+\frac{D\left(\dfrac{1}{r^j\rho u}\right)}{Dx}+\left(\frac{1}{r^j\rho u}\right)\left(\frac{v}{u}\right)K=0, \tag{5.1.13a}$$

$$\frac{\partial\left(\mathscr{H}r^j\dfrac{v}{u}\right)}{\partial\psi}+\frac{D\left(\dfrac{1}{\rho u}\right)}{Dx}-j\mathscr{H}\sin\sigma\left(\frac{1}{r}\right)\left(\frac{1}{\rho u}\right)=0, \tag{5.1.13b}$$

in which the σ in the third term of (5.1.13b) indicates that the reference surface is the shock, and is to be replaced by θ_b if the body surface is used. The quantity $(r^j\rho u)^{-1}$ is proportional to the width of a streamtube measured on an $x = constant$ section, while v/u is the tangent of the inclination angle of a streamline relative to the coordinate system. Equations (5.1.13) express a geometric relation between these quantities.

The equation governing u may be expressed

$$u\frac{Du}{Dx}-Kuv+\frac{1}{\rho}\frac{Dp}{Dx}+\mathscr{H}r^jv\frac{\partial p}{\partial\psi}=0, \tag{5.1.14}$$

as an equivalent of (5.1.6b). The variable y is a dependent variable after the von Mises transformation, and obeys the equations

$$\frac{Dy}{Dx}=\mathscr{H}\frac{v}{u}, \tag{5.1.15a}$$

$$\frac{\partial y}{\partial\psi}=-\frac{1}{r^j\rho u}. \tag{5.1.15b}$$

The latter of these is equivalent to (5.1.8b). The cross derivative relation for (5.1.15) yields (5.1.13).

Provided the shock layer is very thin and the reference surface is suitably chosen we may expect that the velocity component v should be much smaller than u over most of the range of the solution. However, in certain regions this is not true, particularly in the immediate neighborhood of a stagnation point. Where it is true, we may note that Dv/Dx in (5.1.12) is of smaller order of magnitude than is Ku, and the quantity v/u in (5.1.13) is small compared with one. This result, that certain quantities are small, is one of the reasons for choosing a coordinate system of the boundary layer type.

So far, we have said little about the choice of reference surface. The only choices considered will be the two obvious ones, the shock surface and the body surface. A coordinate system of the boundary layer type with the reference surface on the shock wave is termed a "shock-oriented" coordinate system.

Analogously a system with the reference surface on the body surface is termed a "body-oriented" coordinate system.

It is clear that there is a natural relation between the problem set the investigator and the choice of coordinate system. With the inverse problem (shock given), for example, the shock-oriented coordinate system does not depend upon the solution to the problem. With the pressure problem (pressure given), of course, neither coordinate system may be fixed beforehand.

There is an undisputable advantage in being able to fix the coordinate system for a problem before solving the problem. However, it is not necessary to do so, and we may attack the inverse problem with a body-oriented coordinate system or the direct problem with a shock-oriented coordinate system. If we put this factor aside, we may note that, intrinsically, a shock-oriented system has certain advantages over a body-oriented system. Probably the principal one of these is that the shock surface is smoother than is the body surface, as we have noted above in discussing the insensitiveness of the shock shape to small changes in the body shape. In addition, certain of the equations involved may appear in simpler form with shock orientation. In the case of the direct problem, these intrinsic advantages of shock orientation may outweigh the one of being able to fix the coordinate system in advance. Alternatively, we could choose a system which roughly approximates the shock shape, but is fixed. Thereby, we would have a fixed coordinate system with some of the intrinsic advantages of a shock-oriented system.

In attacking the inverse problem a direct "marching-ahead" method is available; this approach and its intrinsic difficulties are considered in Section 6.5. The general approach for the problems of interest which we shall consider here is that of successive approximations. Successive approximation schemes may be of many types, ranging from relaxation methods applied to the differential equations of motion to integral methods employing integral relations analogous to those well known in boundary layer theory or employing direct quadratures as part of the scheme. Successive approximation schemes are considered in the following section, and other related methods are reported in Chapter VI.

We close this section with a brief discussion of the physical nature of the direct problem, with the body shape given. The shock surface, the location of which is unknown, is a free surface in a sense closely analogous to that of classical water-wave or channel-flow theory. The existence of such a free surface suggests an analogy between our hypersonic blunt-body problem and the general hydraulic problem of an inviscid constant-density fluid under gravity with a free surface. We note that with the density ratio very small, the density outside the free surface in the hypersonic case is negligible compared to that inside the surface, as is the case in the usual hydraulic problem. And in both problems there is a continuous supply of unidirectional momentum to the fluid, from the free stream in the hypersonic case and from gravity in the hydraulic case.

The correctness of this analogy stops with the similarities mentioned above. The momentum flux per unit area in the hydraulic case is proportional to the depth of the layer rather than constant, and is distributed to the fluid uniformly through the layer instead of being concentrated at a shock surface. Also, the hypersonic flow is rotational and is not strictly at constant density. Nevertheless, an analogy does exist which has a limited qualitative correctness, and which is useful conceptually.

In hydraulic shallow-water theory there is a critical velocity equal to $\sqrt{gh}$, where h is the depth of the fluid and g the acceleration due to gravity, which corresponds to the propagation velocity of weak disturbances. In the elementary theory given in Section 5.3 for blunt bodies of revolution there is a critical point at which the (steady) velocity of the flow in the thin shock layer immediately behind the shock is $\sqrt{\epsilon}U$. It may be noted that in both cases the square of the critical velocity times the fluid density is equal to the momentum flux per unit area. In either problem, the critical velocity disappears as a strict concept in a more accurate theory for thin fluid layers. This does not mean that the critical phenomenon disappears, but only that it is "smoothed over". The phenomena associated with the critical velocity are essential consequences of a thin layer of which one boundary is free, i.e. not determined in advance by the body shape.

In a calorically perfect gas a coincidence arises. The sonic point immediately behind the shock coincides in thin-shock-layer theory with the critical point. This has led some investigators to work with concepts such as Mach number and Mach angle when dealing with critical phenomena. Inasmuch as the theories of these investigators are essentially constant-density theories, compressibility and sound speed obviously play no role in them.

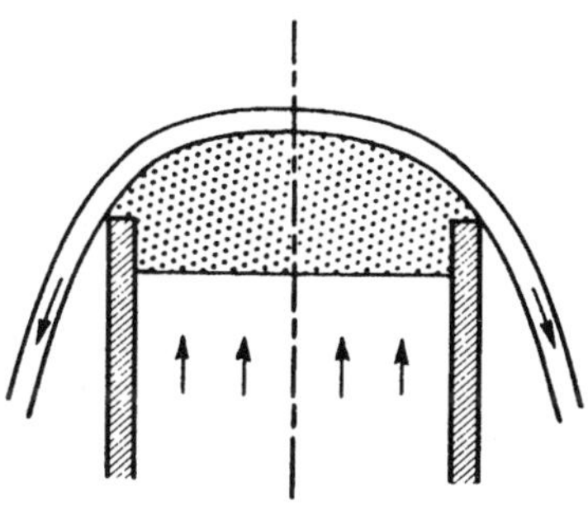

FIG. 5–2. Artesian-well analogy.

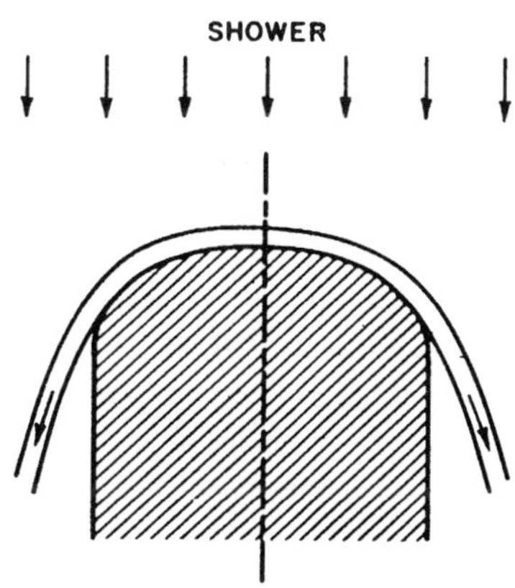

FIG. 5–3. Shower-bath analogy.

Two figures are given to illustrate the analogy. Figure 5–2 depicts the "Artesian-well" analogy, in which the water flow under gravity over a blunt shape is likened to a hypersonic flow. The water is considered to come from a pipe through a porous block of appropriate shape. Figure 5–3 depicts the "Shower-bath" analogy, which differs from the analogy discussed before in several respects. The water is supplied from a unidirectional flow of small droplets from a shower head, and the principal part of the momentum can come from this flow. This second analogy would avoid several of the weaknesses of the direct hydraulic or artesian-well analogy. Certain practical difficulties may arise in setting up laboratory models for these analogies, principally that of finding a suitable porous material and that of avoiding excessive splashing in the shower bath.

2. Successive approximation schemes

The small-density-ratio assumption (D) is not essential for the application of a successive approximation scheme to the calculation of a hypersonic flow (see Sections 6.2 to 6.4). However, the Newtonian theory gives in most cases a solution for zero density ratio which may be used as a zeroth approximation, and an attack based upon improvement of the Newtonian results appears logical and attractive if the density ratio is small. A number of authors independently introduced approximation methods amounting to at least the first approximation in a successive approximation scheme, including Hayes [3; 4], Chester [1; 2], Freeman [1], Fraenkel [1], Chernyi [1; 3], and Cole [1]. Of the later work, that of Muggia [1; 2] reported in Sections 4.3 and 4.4 is particularly significant. One quite standard method is to expand the solution in powers of the density ratio ϵ. This approach must be used with caution, primarily because of the nonanalyticities already evident in the treatment in Chapter IV.

All these authors considered both two-dimensional and axisymmetric flow, and all except Chester considered the direct problem (body given) with a general body shape. The shapes considered by Fraenkel and Chernyi are restricted to those with shock waves attached on a leading edge, while Cole's analysis is devoted to slender bodies for which the small-disturbance theory discussed in Chapter II is available. Chester considered a special case of the inverse problem with a shock of parabolic shape, and carried out the analytic solution to a high order of approximation. All the authors except Hayes and Freeman limited their investigations to a perfect gas. Chester and Cole used cartesian and cylindrical coordinate systems, while the others used systems of the boundary layer type. Freeman, Fraenkel, and Chernyi used body orientation, while Hayes considered both types of orientation. Chester, Freeman, Fraenkel, and Chernyi used the von Mises transformation in their analyses.

We shall not attempt to present individually the methods of these authors, or of others such as that of Maslen and Moeckel [1] (Integral Method). Instead we shall outline a general successive approximation scheme which combines various features of all these contributions, which is a representative composite emphasizing the philosophy of the general approach. The assumption of thermal equilibrium is made here. An analogous approach can be made with a simple nonequilibrium fluid. Five basic relations besides the shock relations are needed. One of these is the constancy of total enthalpy (5.1.11), restated as

$$h + \tfrac{1}{2}q^2 = h_\infty + \tfrac{1}{2}U^2 = H. \tag{5.2.1}$$

The second is the constancy of entropy along a streamline (5.1.10) indicated by its functional form

$$S = S(\psi). \tag{5.2.2}$$

The third is an equation for the pressure gradient across streamlines, for which we may use (5.1.12). The fourth is an equation derived from the continuity equation which relates the streamline inclination with a measure of the mass flow. Such an equation is either of (5.1.13), with v/u describing the streamline inclination and ρu the mass flow. The fifth is the equation of state of the fluid material, which we need in the form

$$h = h(p, S). \tag{5.2.3}$$

The density of the fluid is also needed, and may be obtained from (5.2.3) by the relation

$$\frac{1}{\rho} = \left(\frac{\partial h}{\partial p}\right)_S \tag{5.2.4}$$

A number of subsidiary relations are needed, including (5.1.1), (5.1.4), and (5.1.8), and, with axisymmetric problems, (5.1.3) and (5.1.5). The pressure, entropy, and streamline inclination as a function of a shock angle σ must be available, and are obtainable from the shock relations given in Chapter I and the equation of state.

We shall divide our successive approximation scheme into two principal stages, with the first stage starting with a given shock shape. In the inverse problem (shock given) this shock shape is the correct one and with its associated shock-oriented coordinate system is the same for all approximations. In the direct problem (body given) the shock shape is an approximate one based on the last approximation and, with the coordinate system if shock orientation is used, is different for each approximation as long as a correct solution has not been reached. With this distinction between the two problem types in mind, we may consider them together.

For the first stage, the quantities shock wave inclination $\sigma(x)$ and pressure difference from pressure behind the shock $p(x, \psi) - p_s(x)$ are given; these have been estimated from an earlier approximation. On the shock wave $\psi(x)$ is known, in fact is simply

$$\psi_s = \frac{\rho_\infty U r^{1+j}}{1+j}, \tag{5.2.5}$$

where r is the lateral cartesian coordinate in the two-dimensional case. From $\sigma(x)$ and the shock relations $p_s(x)$ and $S(\psi)$ are obtained. The quantity p_s is added to the pressure difference $p(x, \psi) - p_s(x)$ to give the pressure $p(x, \psi)$. From the equation of state (5.2.3) are obtained $h(x, \psi)$ and $\rho(x, \psi)$. Finally $q^2(x, \psi) = u^2 + v^2$ is obtained from (5.2.1). This completes the first stage. A block diagram of this stage is shown in Fig. 5–4, in which the steps just described are shown as directed lines.

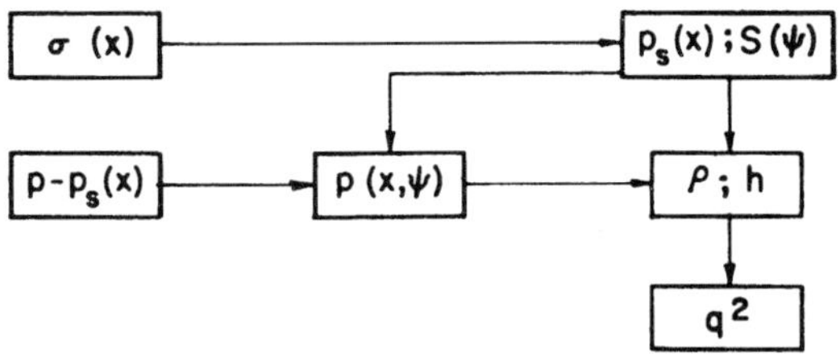

FIG. 5–4. First stage of successive approximation scheme.

The second stage starts with the approximate functions $\rho(x, \psi)$ and $q^2(x, \psi)$ from the first stage, an approximation for $v(x, \psi)$ or v/u, and for axisymmetric flow an approximation for $r(x, \psi)$. The values of v or v/u are used to obtain $u(x, \psi)$ by the relation

$$u = \sqrt{q^2 - v^2} = q/\sqrt{1 + (v/u)^2}\,. \tag{5.2.6}$$

An equation for streamline inclination (5.1.13), which uses the quantities ρ, u, and r^j, is used to check the approximation given for v or v/u or to give new approximate values for the next approximation.

The quadrature of (5.1.8b)

$$y = \int^{\psi_s} \frac{d\psi}{r^j \rho u} \tag{5.2.7}$$

on an $x = constant$ line then gives new approximate values for $r(x, \psi)$ in the axisymmetric case, and carried out across the whole layer gives the shock layer thickness

$$\Delta = \int_0^{\psi_s} \frac{d\psi}{r^j \rho u} = y(x, 0). \tag{5.2.8}$$

This second stage has the property of generating the quantities needed for starting it, apart from ρ and q^2. It may thus be repeated without an intervening application of the first stage in an attempt to obtain the flow geometry consistent with the given functions $\rho(x, \psi)$ and $q^2(x, \psi)$. Finally, (5.1.12) is used to obtain a new approximation for the pressure difference $p(x, \psi) - p_s(x)$. This second stage is diagrammed in Fig. 5–5. Note that v/u and y are connected by the relation

$$\frac{v}{u} = \frac{1}{\mathscr{H}}\left(\frac{\partial y}{\partial x}\right)_{\psi} = \frac{1}{\mathscr{H}}\frac{Dy}{Dx}. \tag{5.2.9}$$

The dashed lines in Fig. 5–5 indicate steps needed only in axisymmetric problems ($j = 1$).

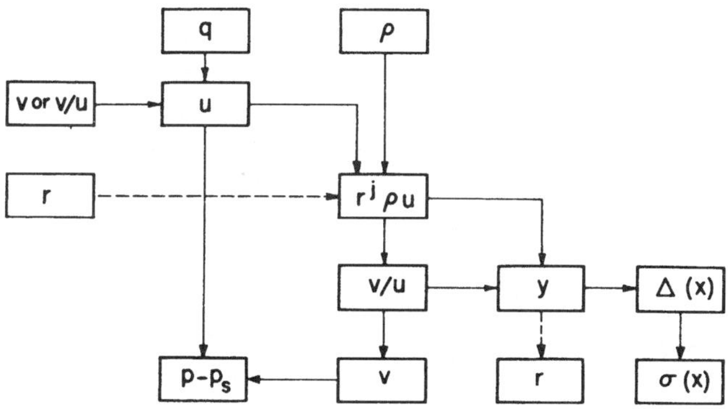

FIG. 5–5. Second stage of successive approximation scheme.

In the inverse problem the resulting shock layer thickness $\Delta(x)$ gives the desired shape of the body. In the direct problem, with the body shape known, $\Delta(x)$ gives the shape of the shock, which in turn yields the function $\sigma(x)$ to be used in the next application of the first stage. The objective of the second stage is to determine the geometry corresponding to a given distribution of ρ and q^2 as functions of x and ψ. The interplay between the two stages has the objective of determining the pressure field and, in the direct problem, the shock shape.

If the body is blunt so that the shock is detached, there is a stagnation point on the body. In the immediate vicinity of this stagnation point lies a region of the shock layer within which our assumptions of orders of magnitude do not hold. In particular, not only is v/u not much smaller than one here, but may be much larger than one in part of this stagnation region. We have no positive assurance that a successive approximation scheme of the type outlined above converges, and in the stagnation region the scheme as stated above undoubtedly diverges. A modification of the scheme is needed to take care of this region. Of the authors mentioned who considered problems with stagnation regions, Hayes depended on the constant-density solutions and on anticipated improvements on them for this region; Chester chose as dependent variables quantities for which his Newtonian zeroth approximation was a uniformly valid one as $\psi \to 0$; and Freeman obtained a first approximation without considering the stagnation region, then checked that his solution was consistent within this region.

The root of the difficulty lies in the problem character which is exploited by the boundary layer approach. This approach assumes that, with the solution established for smaller values of x, the solution may be continued along the layer to larger values of x. Away from the stagnation region the problem does indeed have such a quasi-parabolic character if the shock layer is thin. But in the stagnation region the problem is fundamentally elliptic in character and not amenable to boundary layer treatment. With the shock shape given (the inverse problem), the solution may be obtained directly by the "marching-ahead" process discussed in Section 6.5. Thus Chester [1; 2] has no difficulty, in principle, in computing higher approximations for the inverse problems which he solves. A minor difficulty, existent but unimportant in Chester's problems, is that the coordinate system after the von Mises transformation is singular along the dividing streamline; this difficulty may be readily circumvented if necessary by using a different coordinate system locally. In asymmetric problems, the von Mises transformation has a difficulty more serious than that of local singular behavior. A range generally appears in which the transformation is not one-to-one, so that two physical points correspond to the same values of (x, ψ). Here the use of a different local coordinate system may be necessary.

Muggia [1; 2] treats inverse problems in coordinate systems of polar type. He considers quasi-circular shocks and bodies, whose shapes vary by small amounts from coordinate surfaces. In [2] he uses the fact that in a steady axisymmetric flow of constant enthalpy the vorticity is proportional to $r^j \rho T$.

With the direct problem (body given), the difficulty is more fundamental. The most obvious attack is to guess the shock shape, calculate by a "marching-ahead" process, find the corresponding body shape, and use the error in body shape as a guide in guessing a new shock shape. Another approach is to set up a scheme similar to that described above but based upon different assumptions as to what quantities are small. Account must be taken of the essential elliptic nature of the problem, with both components of the pressure gradient of equivalent importance, and with upstream influence in the shock layer important.

An additional difficulty arises in the vicinity of the Newtonian separation point, the point at which the pressure on the body according to the Newtonian theory drops to zero. As this point is approached in a shock layer of finite thickness the layer thickens markedly and the boundary layer approach becomes invalid. This difficulty was noted and discussed by Freeman [1], who found that his successive approximation procedure diverged strongly near this singular point. In his first approximation the expression for the shock layer thickness approached infinity as this point was approached. This difficulty was studied further by Freeman [5], who obtained a uniformly valid solution in the separation region. We consider this solution later in this chapter, together with a treatment of Freeman [6] on the asymptotic flow far downstream of a free layer.

The flow in the shock layer of a blunt body is actually transonic, with a sonic line dividing subsonic and supersonic domains. Of the upstream-running characteristics leaving the sonic line there is a last one, termed the limiting characteristic. This divides the supersonic hyperbolic region into a region which is properly hyperbolic and one which is "transonic" or pseudo-elliptic, in which a disturbance has an influence in the elliptic region of the flow (see also Section 6.1). All the classical difficulties associated with rotational transonic flow thus appear here in principle. However, if the shock layer is very thin the influence of the free surface in a direct problem is predominant and more important than the influence of the sonic line. It is likely that the early approximations following a successive approximation scheme of the boundary layer type are not affected by any peculiar transonic behavior, but difficulty on this point may appear in higher approximations. It is not known in general whether such a scheme may be expected to converge or not in a transonic region. A successive approximation scheme for determining the sonic line shape in a direct problem has been described by Bloor [1]. As discussed below the scheme starts with the Newtonian solution with the shock shape the same as the body shape. To this lowest-order approximation the sonic line is given by the streamline through the sonic point on the shock. The shock shape and pressure distribution are corrected and a modified sonic line obtained. Results are presented for a sphere and cylinder. Comparison with numerical calculations for a sphere shows reasonable agreement.

So far we have said little about how to start our successive approximation scheme, about what we should use as our lowest-order approximation in a direct problem. The obvious choice with a thin shock layer is the Newtonian solution, with the shock shape the same as the body shape. The Newtonian theory gives $p(x, \psi)$ directly, with $v = 0$. This choice of a starting approximation is the one used by most authors. There are certain body shapes, however, for which the Newtonian solution is inappropriate for a lowest-order approximation. One such is a flat-faced strip or disc, for which no steady solution is obtainable from Newtonian theory. Another is a shape with a blunt wedge or conical nose such that the actual solution involves a detached shock wave. In such cases the approximate theories of the next two sections may be used to obtain a lowest-order approximation, or, alternatively, we may depend upon an "educated guess". Of course, the procedure of the next two sections may be used to obtain a starting approximation for a successive approximation procedure when the Newtonian approximation could have been used.

An alternative start in an inverse problem might be obtained from the method of Maslen [2]. Maslen's method does not correspond to a rational theory and overestimates the centrifugal effect on convex bodies with thin shock layers. However, it has shown very good agreement with experiment and with more detailed calculations. In essence, his method amounts to estimating the pressure as a function of stream function from Newtonian theory, but using $P = \psi \cos \sigma_1$, where σ_1 is the local shock angle, in place of the $\int \cos \sigma \, d\psi$ of (3.2.4). With $\rho(p, S)$ known and u used in place of q in Bernoulli's equation, the quadratures (5.2.7) and (5.2.8) yield the solution and body shape.

With the pressure problem (pressure given), the procedure is not much different from that used with the direct problem (body given). The second stage described above is used to obtain the quantity

$$p_b - p_s = p(x, 0) - p_s(x), \tag{5.2.10}$$

which with the given value of p_b gives a new value for $p_s(x)$. The relation between shock angle and pressure is applied backwards to obtain the new $\sigma(x)$ and, by integration, the shock shape if desired. The information needed to recommence the first stage is then available.

To illustrate the successive approximation approach let us consider the flow of a perfect gas with $\gamma - 1$ small on a body with a pointed nose at $M_\infty = \infty$. With the pointed nose we avoid the difficulties of the stagnation region, and with $M_\infty = \infty$ the quantity ϵ is constant. This example is a simplification of those treated by Freeman [1], Fraenkel [1], and Chernyi [1]. With this approach based on small ϵ the shock and body shapes are treated as identical for the first approximation, and no distinction between the direct and indirect problems appears until the second approximation is considered. We assume also that we do not have a shape which would lead to a detached free layer in the Newtonian theory within the region of interest.

The shock shape is assumed to be known (approximately the same as the body shape in the direct problem), and is specified in terms of the quantity $\cos \sigma$ as a function of the stream function ψ. In the lowest-order approximation ϵ is considered to be zero and the density in the shock layer infinite. Consistent with the assumption of infinite density the enthalpy h is considered a function of the entropy S alone, and hence q is constant along each streamline. With these considerations the first stage is complete in this lowest-order approximation.

With infinite density the shock layer is infinitesimally thin and v is zero everywhere. Thus we obtain the result $q = u$. The quantity Δ from (5.2.8) is zero because of the assumed infinite density. The pressure difference $p - p_s$ is given directly by the Newtonian theory. With this result the second stage is complete in the lowest-order approximation.

We thus start what we shall call the first approximation with the results of the Newtonian theory at our disposal. In calculating successive approximations we have a choice in procedure. We may at each stage use the complete equation of state in calculating ρ and h, or we may use approximate expressions consistent with the level of approximation involved. The principal danger in the latter course lies in the stagnation or Newtonian separation regions, in which too coarse an approximation may preclude a convergent result. In addition, too coarse an approximation for the equation of state may unnecessarily slow down the convergence of the successive approximation scheme.

The pressure immediately behind the shock is expressed

$$p_s = \rho_\infty U^2 \sin^2 \sigma, \tag{5.2.11}$$

with the factor $1 - \epsilon$ dropped. The pressure within the shock layer is given by the Newton-Busemann pressure law (3.2.7) as

$$p = \rho_\infty U^2 \left[\sin^2 \sigma_1 - \left(\frac{d \cos \sigma}{d\psi}\right)_1 \int^{\psi_1} \cos \sigma \, d\psi\right], \tag{5.2.12}$$

with the integral a function of its lower limit ψ and with the subscript 1 used as in Chapters III and IV.

The density immediately behind the shock is $\rho_s = \rho_\infty/\epsilon$, where ϵ is given by (1.4.18) and is constant. The density within the layer is determined by the condition of constant entropy along a streamline, and may be expressed in terms of p_s and ρ_s for the same value of ψ as

$$\rho = \rho_s \left(\frac{p}{p_s}\right)^{\frac{1}{\gamma}} \approx \frac{\rho_s}{p_s} p \left(1 - \frac{\gamma - 1}{\gamma} \ln \frac{p}{p_s}\right) \approx \frac{\rho_s}{p_s} p. \tag{5.2.13}$$

These approximate expressions are based on the assumption that $\gamma - 1$ is small. Which of the expressions is used depends upon the choice of the person carrying out the approximation.

The velocity u is considered equal to q and is determined by the constancy of total enthalpy. Thus we have

$$u^2 = q^2 = u_s^2 + \frac{2\gamma}{\gamma - 1} \frac{p_s}{\rho_s} \left[1 - \left(\frac{p}{p_s}\right)^{\frac{\gamma-1}{\gamma}}\right] \approx u_s^2 - 2 \frac{p_s}{\rho_s} \ln \frac{p}{p_s} \approx u_s^2, \tag{5.2.14}$$

in terms of conditions at the point of entry of the streamline being considered. Again, there is a freedom of choice in the expression used.

With shock orientation we may calculate y from (5.2.7) using (5.2.12) and the simplest expressions of (5.2.13) and (5.2.14). The result is the same as that of Freeman [1, p. 376]

$$y(x, \psi) = \frac{\epsilon}{\rho_\infty U r_1^j} \int^{\psi_1} \frac{\sin^2 \sigma \sec \sigma \, d\psi}{\sin^2 \sigma_1 - \left(\dfrac{d \cos \sigma}{d\psi}\right)_1 \displaystyle\int^{\psi_1} \cos \sigma \, d\psi}, \tag{5.2.15}$$

with the integrals again functions of their lower limits ψ. The shock layer thickness is given by (5.2.8). The result (5.2.15) may be improved without repeating the entire procedure by using the more complete expressions for ρ and u from (5.2.13) and (5.2.14). Of particular importance is the improvement afforded by using a more accurate expression for the velocity from (5.2.14) instead of simply $u = u_s$. Freeman obtained convergence in the stagnation region of a two-dimensional blunt body only with this improvement, obtaining the first term of (4.3.14) or (4.3.23). Thus an appreciably better result may be obtained by replacing the term $\sec \sigma$ representing U/u in (5.2.15) by the quantity U/u obtained from

$$\frac{u^2}{U^2} = \cos^2 \sigma - 2\epsilon \sin^2 \sigma \ln \frac{p}{p_s}, \tag{5.2.16}$$

with p and p_s given by (5.2.12) and (5.2.11).

For a second approximation the pressure distribution $p - p_s$ should be recomputed on the basis of the first approximation, the shock shape should be corrected if the problem is a direct one, and the quantities $p_s(x)$, $p(x, \psi)$, $\rho(x, \psi)$, and $q^2(x, \psi)$ should be obtained as before to complete the first stage. The streamline slope according to (5.2.9) may be calculated from (5.2.15) and used to compute u. Finally ρ and u are used in (5.2.7) to recalculate y. Examples of calculations for the second approximation may be found in the cited references.

3. Constant-streamtube-area approximation

In the last three sections of Chapter IV we have considered solutions based upon the assumption of constant density and also perturbations of these solutions. There we used generally the approach of considering inverse problems, with the shock shape given. Only to the extent that body shape was predicted could these theories be used for direct problems.

One approach to thin layer theory for the direct problem on very blunt or flat-faced bodies would be to apply the constant-density theory directly. Because of the mathematical difficulties involved, this approach has not been fruitful. What has been accomplished with success are treatments based upon constant density with additional simplifying assumptions. Here the inner and outer layers which appeared in the last three sections of Chapter IV are distinguished in the theory and treated differently. In the outer layer the Bernoulli effect is neglected, and the Newtonian velocity and pressure distributions hold as a function of stream function. In the inner layer the pressure gradient across the layer is neglected. Where the inner layer is treated, it takes its outer boundary conditions from the outer layer and must fit it with respect to pressure and velocity.

The simplest theories of this type treat the outer layer only. Here the appropriate assumptions of constant density and constant (Newtonian) velocity along streamlines lead to constant ρq for a streamtube, hence to constant streamtube area. Because of the convergence difficulty discussed following (3.5.7), such a theory is unavailable for two-dimensional bodies with detached shock waves. Several approaches lead to essentially equivalent theories. The approaches which have been used are those of using (a) integral equations for the shock shape, (b) equations governing the total shock layer thickness, and (c) approximate equations of motion.

In approach (a), the shock shape is specified in terms of an unknown function. The thickness at any point is obtained as an integral over this function. This thickness is related to the known body shape, and the resulting relation yields an integral equation for the shock shape. This method was used by Hayes [3] in the solution for a circular flat-faced body. The later report of Hida [2] uses essentially this same approach. The integral equation obtained may be transformed into a differential equation.

Approach (b), taken from Hayes [4], will be followed in the present section. In this method the total layer thickness is expressed as an integral over differential streamtube thickness, and the relation between shock and body boundary conditions to changes in this total thickness gives a differential equation for shock shape. In consonance with the first edition of this book, we continue to refer to this approach as the "constant-streamtube-area approximation". In this section we treat primarily axisymmetric bodies.

In approach (c), the equations of motion are set down in an approximate form. These are based either upon the first terms in a suitable expansion in powers of $\epsilon^{1/2}$, as by Freeman [3] or by J. D. Cole and his associates, or upon simplifying assumptions applied to the full equations, as by Serbin [1] or by Maslen and Moeckel [1] (Integral Method). These equations are constant-density ones, with the pressure changes along streamlines neglected.

One feature of these methods is that, although compressibility and Bernoulli effects are neglected, the effect of nonconstant ϵ need not be neglected. The method of this section encompasses this effect, through the parameter μ. In most cases, however, the influence of this effect is properly neglected in the theory. Only for fluids obeying equations of state that are somewhat special is the parameter μ of $O(1)$ in ϵ and thereby properly included.

The analysis of this section, taken from Hayes [4], is for the direct problem of the hypersonic flow on a given blunt body of revolution. The analysis is based on the assumption that $\epsilon^{1/2}$ is small (assumption D-strong), and thus has a limited validity. The results have less accuracy than do the results of the constant-density theory, as is indicated by the fact that in the analogue of (4.4.16) the factor $1 + \sqrt{8\epsilon/3}$ is not obtained. What this theory does do is to permit a straightforward solution of the direct problem in which the body shape is given and the shock shape is unknown, to give this shape on the blunt part of the body by a simple and direct procedure. The theory thus provides an improvement in obtaining shock shapes over Newtonian theory which is of a different type than the improvement offered by the constant-density solutions; the constant-density solutions give us no information on the relation between shock and body shape. In particular, solutions are obtained for problems for which there is no Newtonian solution. Although the part of the shock layer to which the theory may apply is limited, its extent is larger by the order of $\epsilon^{-1/2}$ than that of the stagnation region discussed in the previous section. The constant-density solutions of Section 4.4 are restricted to shock and body shapes of circular or quasi-circular cross section.

The primary purpose of this theory is to obtain the shock shape as a function of the distance x along the body. For the blunt shapes we are considering the shock angle σ is close to $\pi/2$ over the range of interest and is equal to $\pi/2$ on the axis of symmetry if the shock is detached. We shall use the tangent of $\pi/2 - \sigma$ as the principal independent variable describing the shock shape, or $\cot \sigma$. Similarly, we shall use $\cot \theta_b$ considered as a known function of x as the quantity describing the body shape. This choice of variables is partly dictated by the fact that we would like our formulation to be available for a wider range of σ. We shall not use it in this wider range here.

The theory leads to a differential equation for $\cot \sigma$ considered as a function of x, once the body shape is given. The solution to the problem of determining the shock shape is a solution to this differential equation. Thus we shall be primarily concerned with the nature of the solutions to this differential equation, with a study of its singular points, and with interpretations of its mathematical behavior in terms of physical solutions.

Certain simplifying assumptions are now introduced for the purposes of this section. We assume that the thickness of the layer Δ is small enough so that $K\Delta$ may be neglected, and hence that $\mathscr{H}$ defined in (5.1.1) may be set equal to one. Similarly, we neglect the dependence of r on y in (5.1.3) and take r to be a function of x alone. And we assume that for a given value of x, the quantity $\sigma - \theta_b$ is small even though it may be the quantity of direct interest to us. With these assumptions we may integrate (5.1.13b) to obtain

$$\left(\frac{v}{u}\right)_s - \left(\frac{v}{u}\right)_b = \frac{j \sin \sigma}{r} \Delta + \mathscr{D}, \tag{5.3.1}$$

where

$$\mathscr{D} = \int_0^{\Delta} \frac{1}{\rho u} \frac{D\rho u}{Dx}\, dy. \tag{5.3.2}$$

In these equations (5.1.8b) has been used to replace the stream function ψ by y. Shock orientation has been assumed here; with body orientation $\sin \theta_b$ would appear in (5.3.1) in place of $\sin \sigma$, and the limits in the integral of (5.3.2) would be $-\Delta$ and 0 in place of 0 and Δ. Regardless of the choice of orientation we may write

$$\frac{d\Delta}{dx} = \tan(\sigma - \theta_b) = \frac{\cot \theta_b - \cot \sigma}{1 + \cot \sigma \cot \theta_b}. \tag{5.3.3}$$

With our assumption that $\sigma - \theta_b$ is small, we may replace $\tan(\sigma - \theta_b)$ in (5.3.3) by $\sin(\sigma - \theta_b)$ and obtain

$$\Delta = \Delta_0 - \int_0 \sin \theta_b \sin \sigma (\cot \sigma - \cot \theta_b)\, dx. \tag{5.3.4}$$

With shock orientation we may evaluate the left side of (5.3.1) with the aid of (1.5.4) to give

$$\begin{aligned} \left(\frac{v}{u}\right)_s - \left(\frac{v}{u}\right)_b &= \tan(\sigma - \theta_s) - \tan(\sigma - \theta_b) \\ &= \epsilon \tan \sigma - \tan(\sigma - \theta_b). \end{aligned} \tag{5.3.5}$$

Combining this result with (5.3.1) we obtain

$$\frac{\cot \theta_b - \cot \sigma}{1 + \cot \sigma \cot \theta_b} = \frac{\epsilon}{\cot \sigma} - \frac{j \sin \sigma}{r} \Delta - \mathscr{D}, \tag{5.3.6}$$

which with $\sigma - \theta_b$ small and $1 - \epsilon$ replaced with 1 may be rewritten

$$\cot \sigma + \frac{\epsilon}{\cot \sigma} = \cot \theta_b + \frac{j\Delta}{r \sin \theta_b} + \frac{\mathscr{D}}{\sin^2 \sigma}. \tag{5.3.7}$$

With body orientation we have, in place of (5.3.5),

$$\begin{aligned} \left(\frac{v}{u}\right)_s - \left(\frac{v}{u}\right)_b = \left(\frac{v}{u}\right)_s &= \tan(\theta_b - \theta_s) \\ &= \frac{\epsilon \tan \sigma - \tan(\sigma - \theta_b)}{1 + \epsilon \tan \sigma \tan(\sigma - \theta_b)}. \end{aligned} \tag{5.3.8}$$

For the class of problems we are considering, the quantity $\tan \sigma \tan(\sigma - \theta_b)$ remains of the order of one or smaller; this fact may be checked later by examining extreme examples. Thus, the denominator in the right side of (5.3.8) is $1 + O(\epsilon)$ and may be dropped as a factor. We conclude that (5.3.8) is the same as (5.3.5) within our order of approximation, and thus that there is no important difference between shock orientation and body orientation. Equations (5.3.4) and (5.3.7) are the basic ones for this section.

We now turn to a somewhat cruder approximation applicable only to blunt bodies of revolution with $\epsilon^{1/2} \ll 1$ and limited to a region on the body within which $\cot \sigma$ is also very small. With the restriction that $\cot \sigma$ is small we set $j = 1$, replace $\sin \theta_b$ and $\sin \sigma$ in (5.3.4) and (5.3.7) by one, and replace r by x. On the right side of (5.3.7) are left the terms Δ/x and $\mathscr{D}$.

We may interpret (5.3.7) as an equation expressing the effects which tend to make the shock layer become thicker or thinner with increasing x. The mass flow fed into the shock layer across the shock tends to make it thicker, an increase in the mass flow per unit area ρu tends to make it thinner (the influence of the term in $\mathscr{D}$), and in an axisymmetric flow the radial geometrical effect tends to make it thinner (the influence of the term in $j\Delta/r$). With our approximations for a blunt body of revolution the quantity $\mathscr{D}$ is smaller than Δ/x by a factor which is of the order of $\epsilon^{1/2}$, so that with $\epsilon^{1/2}$ small the quantity Δ/x predominates. This conclusion is based on estimates of the relative order of magnitude of the two terms.

The approximation of this section is based on dropping this $\mathscr{D}$ term, on assuming that we may set

$$\mathscr{D} = 0 \tag{5.3.9}$$

approximately. Such an approximation is not generally available to us in two-dimensional flow. Combining (5.3.4) and (5.3.7) we obtain finally

$$x\left(\cot \sigma + \frac{\epsilon}{\cot \sigma}\right) = x \cot \theta_b + \Delta_0 - \int_0 (\cot \sigma - \cot \theta_b)\, dx. \tag{5.3.10}$$

Since the quantity $\mathscr{D}$ corresponds to changes in the cross-sectional areas of the streamtubes in the flow, this approximation is termed the constant-streamtube-area approximation. One result may be noted immediately, that

$$\Delta_0 = \lim_{x\to\infty} \left(\frac{\epsilon x}{\cot \sigma}\right) = \epsilon_0 R_s \tag{5.3.11}$$

in agreement with the lowest-order approximation of (4.4.9) or (4.4.17). This result is obtainable from the Newtonian theory, in which, however, there is no fundamental distinction between shock and body radii.

Equation (5.3.10) is an integral equation for the unknown function $\cot \sigma$ determining the shock shape in terms of the known function $\cot \theta_b$ describing the body shape. Although for some purposes we might wish to use the integral equation proper, a differential equation is generally preferable in this problem. This is obtained directly by differentiating (5.3.10), and is

(5.3.12)
$$\left(1 - \frac{\epsilon}{\cot^2 \sigma} + \frac{1}{\cot \sigma}\frac{d\epsilon}{d \cot \sigma}\right) x \frac{d \cot \sigma}{dx} + 2 \cot \sigma + \frac{\epsilon}{\cot \sigma} = f(x),$$

where $f(x)$ is given by

$$f(x) = 2 \cot \theta_b + x \frac{d \cot \theta_b}{dx} = \frac{d\, x^2 \cot \theta_b}{x\, dx}. \tag{5.3.13}$$

The term involving the derivative of ϵ appearing in (5.3.12) may be evaluated in terms of quantities evaluated from the Hugoniot curve for the fluid. With ρ^{-1} the specific volume of the fluid, we may obtain the dimensionless quantity μ defined by

$$\frac{\mu}{1 + \mu} = - \frac{p_s - p_\infty}{\rho_\infty^{-1} - \rho_s^{-1}} \frac{d\rho_s^{-1}}{dp_s} = - \frac{p_s - p_\infty}{1 - \epsilon} \frac{d\epsilon}{dp_s} \tag{5.3.14}$$

from the Hugoniot curve. We have already defined this quantity in (4.3.28). Differentiation of the relation (1.4.5b) for p_s in terms of ϵ and σ yields

$$dp_s = 2\rho_\infty U^2(1 - \epsilon) \sin \sigma \cos \sigma \, d\sigma - \rho_\infty U^2 \sin^2 \sigma \, d\epsilon. \tag{5.3.15}$$

A combination of the last two equations gives the expression

$$\frac{1}{\cot \sigma}\frac{d\epsilon}{d \cot \sigma} = \frac{2(1 - \epsilon)\mu}{1 + \cot^2 \sigma}. \tag{5.3.16}$$

Within the accuracy of our approximations, we may replace (5.3.16) by

$$\frac{1}{\cot \sigma}\frac{d\epsilon}{d \cot \sigma} = 2\mu. \tag{5.3.17}$$

The quantity μ is generally small if ϵ is small, and is positive and of smaller order than ϵ for a perfect gas. However, it could be large in particular cases of a general fluid, and may be of either sign.

The basic differential equation (5.3.12) has a singular point where $\cot \sigma + \epsilon/\cot \sigma$ has a minimum value, where the parenthesis of (5.3.12) vanishes. This point is termed the "critical" point, and occurs for a shock angle given by

$$\cot^2 \sigma_{cr} = \frac{\epsilon_{cr}}{1 + 2\mu_{cr}}. \tag{5.3.18}$$

In order that a solution of (5.3.12) pass through the critical point, it is necessary that $f(x) - 2 \cot \sigma - \epsilon/\cot \sigma$ be zero there, or that

$$f(x_{cr}) = (3 + 2\mu_{cr}) \cot \sigma_{cr} = \frac{\sqrt{\epsilon_{cr}}\,(3 + 2\mu_{cr})}{\sqrt{1 + 2\mu_{cr}}}. \tag{5.3.19}$$

If ϵ is constant so that μ is zero, (5.3.18) and (5.3.19) become accordingly simpler in form. In any case, the quantity x_{cr} is independent of the solution for the shock shape. The stand-off or detachment distance at the critical point is given from (5.3.10) and (5.3.4) as

$$\Delta_{cr} = x_{cr}[2(1 + \mu_{cr}) \cot \sigma_{cr} - \cot \theta_{b_{cr}}]. \tag{5.3.20}$$

In order to exhibit some sample solutions we could take ϵ to be constant. The case in which μ is taken to be constant (instead of zero) involves but little more complexity; we shall consider this case as an example which includes effects of variations in ϵ. The case $\mu = 0$ is simpler in form, of course. With μ constant and equal to μ_0 we may integrate (5.3.17) to obtain

$$\epsilon = \epsilon_0 + \mu_0 \cot^2 \sigma. \tag{5.3.21}$$

We may now rewrite the basic differential equation (5.3.12) as

$$\left(1 + \mu_0 - \frac{\epsilon_0}{\cot^2 \sigma}\right) x \frac{d \cot \sigma}{dx} + (2 + \mu_0) \cot \sigma + \frac{\epsilon_0}{\cot \sigma} = f(x). \tag{5.3.22}$$

We next introduce a new variable z defined by

$$z = \epsilon_0^{-1/2} \sqrt{1 + \mu_0} \cot \sigma, \tag{5.3.23}$$

in terms of which (5.3.22) may be rewritten

$$\left(1 - \frac{1}{z^2}\right) \frac{x}{z} \frac{dz}{dx} + 1 + \frac{1}{z^2} = \frac{f(x)}{\epsilon_0^{1/2}(1 + \mu_0)^{1/2} z} - \frac{1}{1 + \mu_0}. \tag{5.3.24}$$

The quantity z is equal to one at the critical point.

A particular family of solutions to (5.3.24) which are of interest appears if the body shape function $f(x)$ is so chosen that a constant times $f(x)$ is a solution for z. Solutions for the sphere and for the axisymmetric flat-nosed body are included in this family. To investigate these solutions we set the right side of (5.3.24) equal to a constant, say to $1 - a$. In this case the solution of interest may be expressed in the form

$$x = \epsilon_0^{1/2}(1 + \mu_0)^{-1/2} R_s z (1 + a z^2)^{-\frac{1+a}{2a}}, \tag{5.3.25}$$

in which the arbitrary multiplicative constant has been evaluated in terms of the shock radius at the nose R_s. For this solution $f(x)$ must satisfy the relation

$$f(x) = \cot \sigma[(1 + \mu_0)(1 - a) + 1]. \tag{5.3.26}$$

It will be observed that (5.3.26) is apparently clearly inconsistent with (5.3.19) except in the case $a = -1$. In this special case (5.3.25) tells us that $\cot \sigma$ is proportional to x, corresponding to a spherical or parabolic nose shape. If $\mu_0 \neq 0$ we obtain a result for the body radius of curvature different from that of the shock. With the definition (5.3.13) for $f(x)$ and the relation (5.3.26) with $a = -1$ we obtain the result

$$\frac{R_s}{R_b} = 1 + \tfrac{2}{3}\mu_0 \tag{5.3.27}$$

for the ratio of these radii of curvature at the nose. The term in μ_0 may justifiably be retained only if μ_0 is of larger order of magnitude than ϵ_0. With $\mu_0 = 0$ the shock and body radii are the same for this particular solution.

For the other solutions with $a \neq -1$, we first must impose the condition $a > -1$. The apparent inconsistency between (5.3.19) and (5.3.26) is resolved by the observation that $f(x)$ may be a discontinuous function, with θ_b continuous and $d\theta_b/dx$ discontinuous. The critical point may then be determined as the point at which $f(x)$ jumps across the value given by (5.3.19). In such a case the critical point on the body is determined by the presence of a shoulder at which a choking phenomenon occurs. This shoulder is at a given value of x at which $f(x)$ jumps from a value consistent with (5.3.26) to a value greater than that given by (5.3.19). Thus the solution is determined by the given value of x at the shoulder set equal to x_{cr}, which from (5.3.25) must be related to R_s by the relation

$$x_{cr} = \epsilon_0^{1/2}(1 + \mu_0)^{-1/2} R_s (1 + a)^{-\frac{1+a}{2a}}. \tag{5.3.28}$$

Of particular interest is the case of a flat-nosed body, with $f(x) = 0$ for all $x < x_{cr}$. In this case $a = (2 + \mu_0)/(1 + \mu_0)$ or, if $\mu_0 = 0$, $a = 2$. With $\mu_0 = 0$ and ϵ constant, the stand-off or detachment distance on the axis is given by

$$\Delta_0 = \epsilon_0 R_s = 3^{3/4}\, \epsilon^{1/2} x_{cr} \tag{5.3.29}$$

from (5.3.28). The detachment distance at the shoulder is

$$\Delta_{cr} = 2\epsilon^{1/2} x_{cr} \tag{5.3.30}$$

from (5.3.20). These particular results were obtained by an alternative method by Hayes [3], in which the integral equation approach (a) was used.

The same shock shape as that corresponding to the solution (5.3.29) and (5.3.30) was obtained by Serbin [1], who also obtained the corresponding solution for a sphere and noted agreement with the results of Hayes. Serbin did not use equations such as (5.3.11) or (5.3.20) to determine the shock layer thickness, and so obtains only $\Delta_0 - \Delta_{cr}$ in agreement with our results. Instead, he estimates Δ_{cr} by the condition of uniform sonic flow at the critical point, corresponding to the useful semiempirical condition introduced by Moeckel [1]. The use of this condition gives Serbin good agreement with experiment, but destroys the internal consistency of this theory. It can be argued that Serbin's choking condition may be justified as corresponding to the approximation that $r\mathscr{D}$ is an unknown constant, to be determined by some means outside the theory. Such an approximation is not rational. As we can learn from the following section, the quantity $r\mathscr{D}$ is not constant but is much larger near the shoulder than at the axis. It should be noted that for values of ϵ of the order of 0.1, equation (5.3.29) overestimates the detachment distance by roughly a factor of 2.

Within the approximation of this section there is a definite relation between the deviation of the shock shape from spherical and the relative radii of curvature of the shock and the body. With the body shape given locally by

$$\cot \theta_b = \frac{x}{R_b} + O(x^3) \tag{5.3.31}$$

the shock shape must have the form

$$\cot \sigma = \frac{x}{R_s} + \frac{3x^3}{2\epsilon R_s^2}\left(\frac{1 + \frac{2}{3}\mu}{R_s} - \frac{1}{R_b}\right) + O(x^5), \tag{5.3.32a}$$

in which ϵ and μ are evaluated at $x = 0$. This result may be obtained directly from (5.3.22) by substituting and equating like powers of x, and is important because it shows that the shock radius is not determined by the body radius on the axis alone.

Another equivalent expression which generalizes (5.3.27) and is related to (4.4.24) is

$$\frac{R_s}{R_b} = 1 + \frac{2}{3}\mu + \frac{\epsilon\beta}{9}, \tag{5.3.32b}$$

where $\beta = R_s R_s''$ as defined in (4.3.56). This may be obtained from (5.3.32a) by noting that the second term on the right-hand side may be identified as $-\beta x^3/6R_s^3$.

A distinctive feature of the constant-streamtube-area approximation is the possibility of a jump in the value of $\cot \sigma$. This possibility is most readily seen from (5.3.10), in which the quantity $\cot \sigma + \epsilon/\cot \sigma$ appears. A value of

$\cot \sigma + \epsilon/\cot \sigma$ above its minimum or critical value corresponds to two possible values of $\cot \sigma$ (barring an exceptional case in which more than two might be possible), of which one is greater than $\cot \sigma_{cr}$ and the other less. As none of the other terms appearing in (5.3.10) are affected, a jump from one of the two corresponding values of $\cot \sigma$ to the other is permitted. Such a jump is analogous to the hydraulic jump of classical free-surface hydraulic theory, and gives a jump in the pressure. The apparently anomalous fact that there is no jump in the thickness of the shock layer is consistent with our basic assumption $\mathscr{D} = 0$, whereby all changes in streamtube area are neglected. We may expect that such a jump should be considered possible only in one direction, in the direction of decreasing $\cot \sigma$ from a supercritical value ($\cot \sigma > \cot \sigma_{cr}$) to a subcritical value ($\cot \sigma < \cot \sigma_{cr}$). Although we cannot prove this is the case without going to a more refined approximation, this restriction appears physically reasonable and we shall adopt it. This jump does not generally appear in problems for which $f(x)$ is monotonic increasing, as the solution of interest with supercritical downstream conditions never passes from the supercritical to the subcritical domain.

A plot of $\cot \sigma$ versus x is instructive in the study of solutions of (5.3.12). Such a plot is shown for a typical blunt body in Fig. 5–6. The mathematical solutions having shock angle σ equal to $\pi/2$ at $x = 0$ form a one-parameter family of integral curves, each of which passes through the origin on the

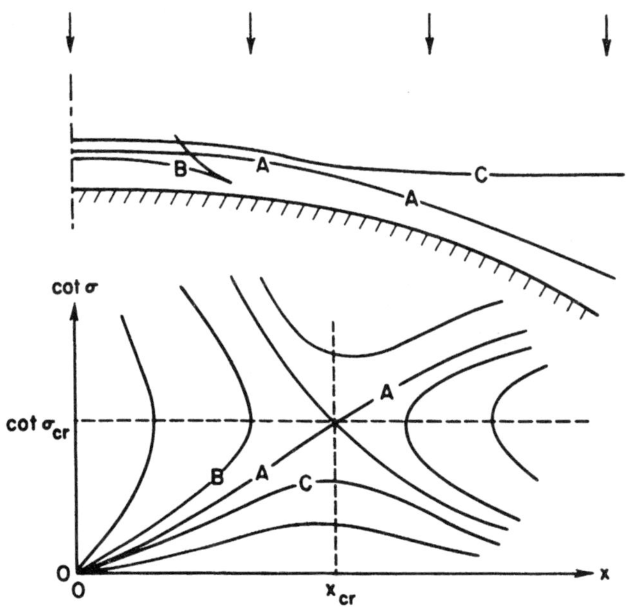

FIG. 5–6. Constant-streamtube-area solution for a blunt axisymmetric body.

$(x, \cot \sigma)$ plot. The slope of each of these curves corresponds to the shock curvature. The critical point $(x_{cr}, \cot \sigma_{cr})$ is a saddle point, and the solution A which passes through this point is the one of physical interest. If the initial shock curvature had been chosen too large, as in solution B, the solution reverses onto a different Riemann sheet for the variable x and is not physically interpretable through this reversal point. If the initial shock curvature had been chosen too small, as in solution C, the solution remains subcritical and approaches the solution $\cot \sigma = 0$ for a normal shock. In this case the solution does have a direct physical interpretation and is rejected only because it is not the solution desired. In general, we have a one-parameter family of initially possible solutions; of these solutions only a single one passes from the subcritical region into the supercritical region and is the desired solution. In addition, there is a one-parameter family of solutions following solution A through the critical point into the supercritical region, jumping discontinuously to a subcritical solution, and approaching a solution for a normal shock. As with solution C, a solution of this type is rejected as not representing the solution desired.

So far we have implicitly assumed that $f(x)$ is a monotonic increasing function (not necessarily a continuous one) and that the body is initially blunt with a finite curvature. In order to generalize our study of the basic equation (5.3.12) we may take one of a number of directions. A fruitful and instructive direction is toward the study of attached shock waves on blunt ogival bodies, for which the initial value of $\cot \theta_b$ is not zero. On such bodies we find an analogue of the Crocco point, a detachment point, and detached as well as attached solutions.

The shape of an ogival body of revolution may be expressed in terms of the Taylor series for the function $f(x)$ under the assumption that this shape is analytic. Thus we have

$$f(x) = f_0 + f_0' x + \tfrac{1}{2} f_0'' x^2 + \cdots . \tag{5.3.33}$$

In order to simplify our analysis we shall assume generally that ϵ is constant and hence that μ is zero. A similar analysis would hold with μ constant or with μ analytic in $\cot^2 \sigma$, but no significant additional features appear in such a more general analysis. The coefficients appearing in (5.3.33) may be evaluated in terms of $\cot \theta_b$ as

$$f_0 = 2\,(\cot \theta_b)_0 , \tag{5.3.34a}$$

$$f_0' = 3\left(\frac{d \cot \theta_b}{dx}\right)_0 , \tag{5.3.34b}$$

$$f_0'' = 4\left(\frac{d^2 \cot \theta_b}{dx^2}\right)_0 . \tag{5.3.34c}$$

We are interested primarily in attached solutions, for which we may assume $\cot\sigma$ finite and $d\cot\sigma/dx$ of smaller order than x^{-1} near the origin. From (5.3.12) and (5.3.34a) we obtain

$$(5.3.35)\qquad (\cot\theta_b)_0 = \cot\sigma_0 + \frac{\epsilon_0}{2\cot\sigma_0}.$$

In order that an attached solution may exist, it is necessary that $(\cot\theta_b)_0$ be greater than a minimum value determined by the minimum of the right side of (5.3.35), its value at the detachment point (cf. Section 4.2). At this detachment point for which $(\cot\theta_b)_0$ takes on its minimum value from (5.3.35), the shock angle is determined by the condition

$$(5.3.36)\qquad \cot^2\sigma_{\text{det}} = \frac{\epsilon_{\text{det}}}{2(1+\mu_{\text{det}})},$$

and the initial body slope has the value

$$(5.3.37)\qquad (\cot\theta_b)_{\text{det}} = (2+\mu_{\text{det}})\cot\sigma_{\text{det}}.$$

We assume in the treatment below that an attached solution exists, i.e. that $(\cot\theta_b)_0$ is greater than $(\cot\theta_b)_{\text{det}}$. The possible solutions in the vicinity of the origin are divided into three principal classes, into detached solutions, "strong" attached solutions for which $\cot\sigma < \cot\sigma_{\text{det}}$, and "weak" attached solutions for which $\cot\sigma > \cot\sigma_{\text{det}}$. At least one solution of each class exists. A weak attached solution is generally of one of three types, determined by whether $(\cot\theta_b)_0$ is greater or less than two reference values corresponding to the Crocco point (defined below; cf. Crocco [1]) and to the critical point. A weak solution may thus be classed as a weak sub-Crocco solution, as a subcritical super-Crocco solution, or as a supercritical solution. It should be emphasized that this Crocco point is quite different in nature from the original one of Crocco [1]. Ours appears in a constant-streamtube-area analysis in an axisymmetric flow. The classic Crocco point appears in a compressible two-dimensional flow.

The Crocco point is an intermediate point between the critical point and the detachment point. It is defined as a point for which the second term obtained by expanding the left side of (5.3.12) in a Taylor's series in x vanishes. This second term is $x(d\cot\sigma/dx)_0$ times $3+4\mu-2\epsilon/\cot^2\sigma$ evaluated at $x=0$, and the Crocco point is accordingly defined by the relation

$$(5.3.38)\qquad \cot^2\sigma_{\text{crocco}} = \frac{2\epsilon_{\text{crocco}}}{3+4\mu_{\text{crocco}}}.$$

From (5.3.35) we see that this corresponds to an initial body slope of

$$(5.3.39)\qquad (\cot\theta_b)_{\text{crocco}} = (\tfrac{7}{4}+\mu_{\text{crocco}})\cot\sigma_{\text{crocco}}.$$

The critical point is defined by (5.3.18) applied at $x = 0$ and the corresponding initial body slope obtained from (5.3.19) or (5.3.35) is

$$(\cot \theta_b)_{cr} = (\tfrac{3}{2} + \mu_{cr}) \cot \sigma_{cr} \, . \tag{5.3.40}$$

We now set $\mu = 0$ and $\cot \sigma = \cot \sigma_0 + \delta \cot \sigma$, so that $\delta \cot \sigma$ is the change in $\cot \sigma$ from its value at $x = 0$. Equation (5.3.12) is now expanded with $\delta \cot \sigma$ as the dependent variable, giving

$$\begin{aligned} &\left(1 - \frac{\epsilon}{\cot^2 \sigma_0} + \frac{2\epsilon \, \delta \cot \sigma}{\cot^3 \sigma_0}\right) x \frac{d\, \delta \cot \sigma}{dx} \\ &\qquad + \left(2 - \frac{\epsilon}{\cot^2 \sigma_0} + \frac{\epsilon \, \delta \cot \sigma}{\cot^3 \sigma_0}\right) \delta \cot \sigma = f_0' x + \tfrac{1}{2} f_0'' x^2 + \cdots \end{aligned} \tag{5.3.41}$$

with terms of order $(\delta \cot \sigma)^3$ dropped. With terms of order $(\delta \cot \sigma)^2$ dropped, (5.3.41) is linear and has the general solution

$$\delta \cot \sigma = A x^{\alpha} - \frac{f_0' x}{\dfrac{2\epsilon}{\cot^2 \sigma_0} - 3} - \frac{\tfrac{1}{2} f_0'' x^2}{\dfrac{3\epsilon}{\cot^2 \sigma_0} - 4} - \cdots , \tag{5.3.42}$$

where the exponent α is given by

$$\alpha = \frac{2 - \dfrac{\epsilon}{\cot^2 \sigma_0}}{\dfrac{\epsilon}{\cot^2 \sigma_0} - 1} \tag{5.3.43}$$

and A is an arbitrary constant. This solution is invalid at the critical point, for which $\cot^2 \sigma_0 = \epsilon$, at the detachment point, for which $\cot^2 \sigma_0 = \frac{1}{2}\epsilon$, at the Crocco point, for which $\cot^2 \sigma_0 = \frac{2}{3}\epsilon$, or at any point for which $\cot^2 \sigma_0 = n\epsilon/(n + 1)$ with n a positive integer. With n larger than 2 the solution (5.3.42) is still correct in its leading terms.

The behavior of the solution (5.3.42) depends critically on the value of α. For a strong solution, one for which $\cot^2 \sigma_0 < \frac{1}{2}\epsilon$, we have $0 > \alpha > -1$. With α thus negative the local singularity is a saddle point and only a single integral curve is possible with $(\delta \cot \sigma)_0 = 0$, that for which the constant A is zero. Since in general we need a one-parameter family of solutions from which to choose the one which can go from a subcritical state to a supercritical state, we must reject such a strong solution as not being of general physical interest. The remaining solutions are either weak or detached.

If the weak solution is supercritical, with $\cot^2 \sigma_0 > \epsilon$, we have $-2 > \alpha > -\infty$. Again α is negative, the singularity is a saddle point, and only a single integral curve is possible. In this case, however, the solution is initially supercritical and the multiplicity of initial solutions is not needed. If $f(x)$ is monotonic increasing from its initial value we are assured that the entire solution will remain supercritical.

If the initial solution is weak and subcritical, so that $\frac{1}{2}\epsilon < \cot^2 \sigma_0 < \epsilon$, then $0 < \alpha < \infty$ and we always have a one-parameter family of solutions, with the singularity a node. The Crocco point divides this range into a range for which $\frac{2}{3}\epsilon < \cot^2 \sigma_0 < \epsilon$ and $1 < \alpha < \infty$, designated as super-Crocco, and a range for which $\frac{1}{2}\epsilon < \cot^2 \sigma_0 < \frac{2}{3}\epsilon$ and $0 < \alpha < 1$, designated as sub-Crocco. In the super-Crocco range with $\alpha > 1$ the leading term is the one in f_0' and all integral curves have the same finite slope at the origin (shock curvature) given by

$$\left(\frac{d \cot \sigma}{dx}\right)_0 = \frac{f_0' \cot^2 \sigma_0}{3 \cot^2 \sigma_0 - 2\epsilon}. \tag{5.3.44}$$

In the sub-Crocco range with $\alpha < 1$ the leading term is the one in x^α, and the only integral curve of finite slope corresponding to (5.3.44) is the one with $A = 0$. All the other integral curves have infinite slope at the origin, corresponding to infinite shock curvatures. It is interesting to note that these results correspond completely with the analogous results for supersonic flow on a two-dimensional ogive. See, for example, Ferri [2, Art. H, 10].

In order to illustrate the solutions of the type just discussed, we present in Fig. 5–7 the integral curves for an ogive in the sub-Crocco range for which the shock is attached. The solution of interest is solution A and passes through the critical point into the supercritical region. Solution B reverses itself and is to be rejected. Solutions C, D, and E approach solutions for normal shocks. All the attached weak solutions have infinite shock curvature at the nose except the single solution C. There is a single strong solution D. And there is a one-parameter family of detached solutions such as solution E for which $(\cot \sigma)_0 = 0$.

The solution (5.3.42) is invalid at the various special points, and special solutions are needed for these cases. If the initial point is the Crocco point the solution may be expressed

$$\delta \cot \sigma = Ax - 2f_0' x \ln x - f_0'' x^2 + \cdots. \tag{5.3.45}$$

For the other special points other than the detachment or critical points (5.3.42) needs only a modification involving a logarithmic term in $x^\alpha \ln x$. If the initial point is the detachment point the term in $\delta \cot \sigma$ in the second parenthesis of (5.3.41) must be retained, with the approximate solution

$$\delta \cot \sigma = \frac{\cot \sigma_{\text{det}}}{2(A - \ln x)} - f_0' x + \cdots. \tag{5.3.46}$$

The solutions (5.3.46) all start at the detachment point and form a one-parameter family.

The case in which the initial point is the critical point is a little more involved. The term in $\delta \cot \sigma$ in the first parenthesis of (5.3.41) must necessarily be

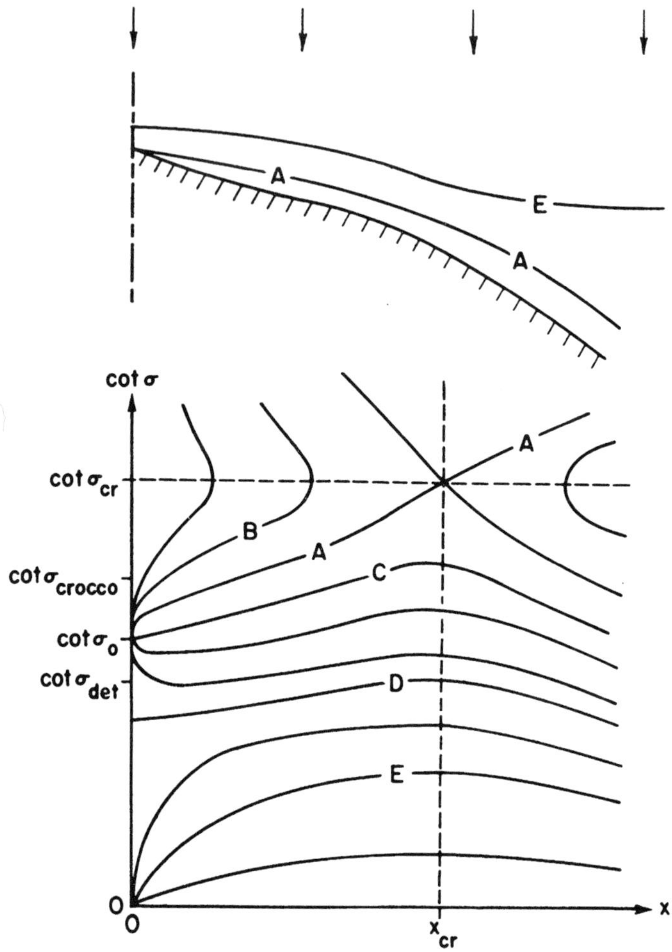

FIG. 5–7. Solution for an ogive of revolution in the sub-Crocco range.

retained, and the resulting nonlinear equation has no general approximate solution. If $f_0' \neq 0$ and $\delta \cot \sigma - f_0'x$ is small compared with $f_0'x$ we can obtain the approximate solution for small x

$$\delta \cot \sigma \approx f_0'x + \tfrac{1}{2}f_0''x^2 - \frac{3f_0'^2x^2}{\cot \sigma_{cr}} + A \exp\left(\frac{\cot \sigma_{cr}}{2f_0'x}\right) + O(x^3), \tag{5.3.47}$$

with A an arbitrary constant. If $f_0' \neq 0$ and $\delta \cot \sigma$ is small compared with $f_0'x$ we can obtain

$$\delta \cot \sigma \approx \pm\sqrt{f_0' \cot \sigma_{cr}(x - a)}\,, \tag{5.3.48}$$

where a is an arbitrary constant. A study of these two approximate solutions indicates that if $f_0' > 0$ there is but a single initially critical solution and that this solution is supercritical for $x > 0$. If $f_0' < 0$ there is a one-parameter

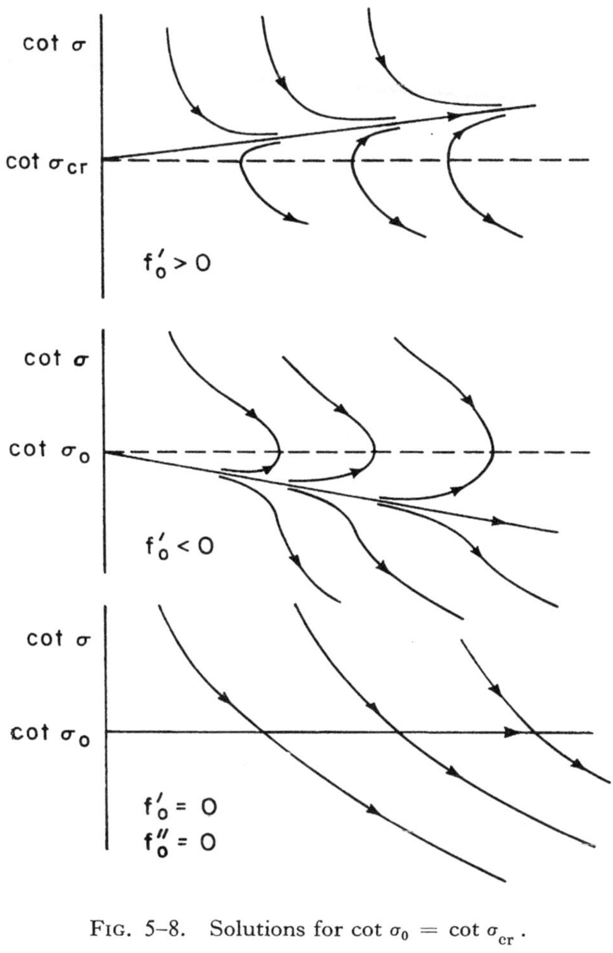

FIG. 5–8. Solutions for $\cot \sigma_0 = \cot \sigma_{cr}$.

family of solutions which are initially critical, all of which are subcritical for x small enough but nonzero. Thus we have here another example of the principle that the solutions of interest should form a one-parameter family in the subcritical region but should not have this multiplicity in the supercritical region. See Fig. 5–8 for sketches of the solution curves in this case. The approach to the limit $f_0' = 0$ is nonuniform. If $f \equiv f_0$ in this case, so that f_0' and other derivatives are zero, the solutions have a discontinuous behavior. A solution may come from the supercritical region to the critical line, pass along the critical line to an arbitrary point on it, then pass into the subcritical region.

As one final example, we show in Fig. 5–9 a solution with the initial angle supercritical, but for which $f(x)$ decreases to a subcritical value and subsequently increases and returns into the supercritical region. The singularity at a critical point may be shown to be always a saddle point if $f'(x_{cr}) > 0$ and to be either a node or a focus (spiral) if $f'(x_{cr}) < 0$. In Fig. 5–9 the first critical point, with $f(x)$ decreasing, is shown as a focus. The initial part of the solution lies on a single curve in the supercritical region. The locus of points in the subcritical region to which points on this integral curve may jump is termed the "image" locus and is shown as a dotted curve in Fig. 5–9. If the value of $\cot \sigma$ on this image locus is denoted $\cot \sigma_*$, the jump condition reads

$$\cot \sigma + \frac{\epsilon}{\cot \sigma} = \cot \sigma_* + \frac{\epsilon_*}{\cot \sigma_*}. \tag{5.3.49}$$

It may be shown that $\cot \sigma_*$ satisfies a differential equation of exactly the same form as (5.3.12), but with $f(x) - (\cot \sigma - \cot \sigma_*)$ on the right-hand side in place of $f(x)$. Since if $\cot \sigma_*$ is in the subcritical regime $\cot \sigma - \cot \sigma_*$ is positive and the coefficient of $d \cot \sigma_*/dx$ is negative, we may conclude that $d \cot \sigma_*/dx$ is always greater than the slope of the local integral curves passing through the same point. In the case of Fig. 5–9, this ensures that there is a one-parameter family of subcritical integral curves leaving the region bounded by the image locus and the critical line in the direction of increasing x. With the possibility of a jump from supercritical to subcritical permitted, the single supercritical integral curve jumps to a one-parameter family of subcritical integral curves. The one of these which is of physical interest is the one which traverses the saddle point at the second critical point. Thus, for the case shown in Fig. 5–9, the presence of a supercritical-to-subcritical jump has given us a single solution which is unique.

Note that a discontinuity in shock slope in Newtonian theory would give us a concentrated force. This illustrates an important difference between Newtonian theory and our constant-streamtube-area theory. In the latter, the results of Newtonian theory apply in terms of the shock angle for velocity and pressure immediately behind the shock, and pressure is discontinuous at a slope discontinuity. The results of Newtonian theory do *not* apply for the centrifugal correction. In certain cases the centrifugal correction may be obtained from Newtonian theory by using the body curvature, but this is not true in general. The analysis of the next section will yield the pressure within the shock layer.

In assessing the constant-streamtube-area approximation we have presented, we must emphasize the limitations inherent in the assumptions. The theory should be valid in an asymptotic sense as $\epsilon \to 0$, in the same sense in which the Newtonian theory is valid. What the theory does accomplish is that it does

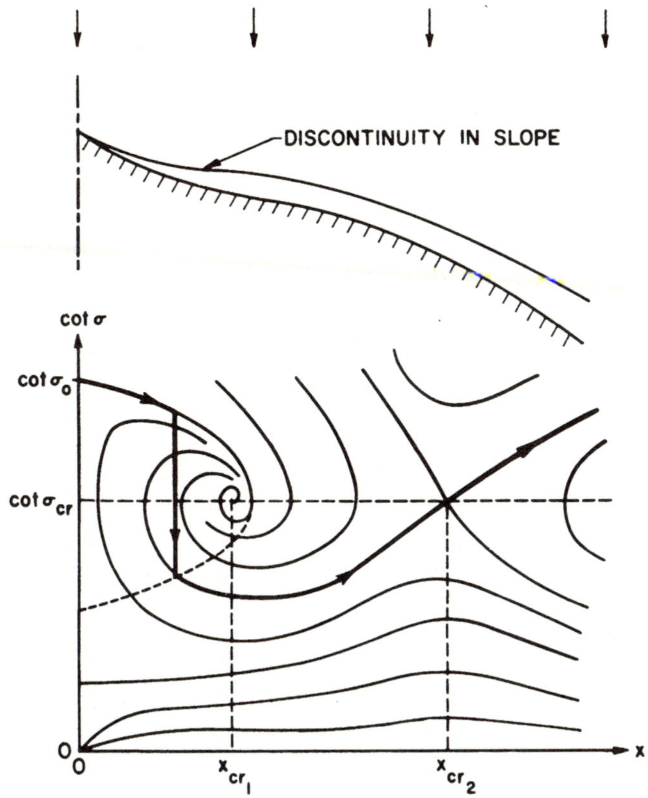

FIG. 5–9. Solution for an ogive of revolution with a jump in the shock slope.

give nontrivial solutions to the direct problem, in which the shock shape is initially unknown and in which the initial approximation that the shock and body shapes are the same may be fundamentally incorrect. The existence of such an elementary theory in the axisymmetric case and not in the two-dimensional case points out again the important differences between these two cases. It should be noted that the concept of the speed of sound or of the compressibility of the gas has no place in the constant-streamtube-area approximation.

We now look at the two-dimensional case, with both $j = 0$ and $\mathscr{D} = 0$ in (5.3.7). This governing equation then becomes

$$\cot \sigma + \frac{\epsilon}{\cot \sigma} = \cot \theta_b \, . \tag{5.3.50}$$

We note here that with $\tan \sigma \gg 1$ this equation may be derived from (4.1.1) for wedge flow.

Solutions exist only if

$$\cot\theta_b > 2\cot\sigma_{cr}\,(1+\mu_{cr}), \tag{5.3.51}$$

where $\cot\sigma_{cr}$ is that defined in (5.3.18). Note that this minimum value of $\cot\theta_b$ is not the axisymmetric critical value of (5.3.40).

We require, therefore, that (5.3.51) hold everywhere for our two-dimensional body. On a plot of $\cot\sigma$ versus x there are two curves, one corresponding to the strong solution of (5.3.50), the other to the weak solution. See Fig. 5–10. The strong curve is precisely the image locus of the weak curve, inasmuch as (5.3.49) holds for the two solutions. A physical solution may follow the weak curve, then jump to the strong curve and follow it. Such a solution is indicated in Fig. 5–10.

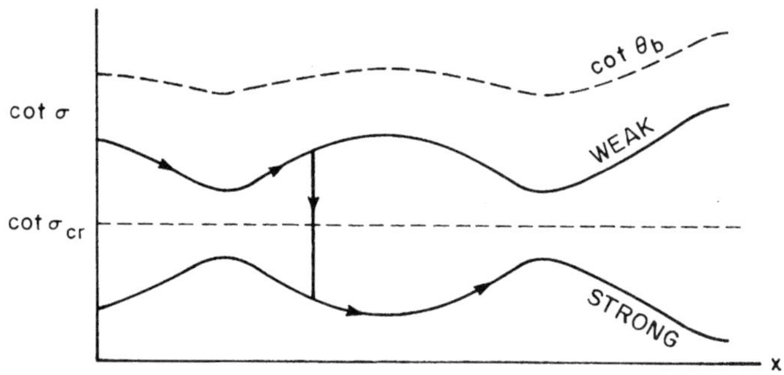

FIG. 5–10. Two-dimensional constant-streamtube-area solution.

This type of solution is a "tangent-wedge" type of solution, somewhat related to the tangent-wedge method discussed in Section 7.3. As we shall see in the next section, the pressure includes a centrifugal correction based upon body curvature.

4. Two-dimensional and axisymmetric blunt-faced bodies

In the first edition of the book, Section 5.4 was entitled "variable-streamtube-area approximations". It contained an attempt to extend the approach of Section 5.3 by estimating the quantity $\mathscr{D}$ of (5.3.2). An essential part of that approach was the estimation of the pressure gradient or velocity gradient at the body surface. This estimation is needed for the Bernoulli effect in the inner layer.

This earlier attempt has turned out not to be rationally sound, because the estimate of the pressure and velocity gradients near the axis was incorrect. The Newtonian formulas for pressure on the body near a stagnation point, leading to the factors $\sqrt{3\epsilon}$ $(j=0)$ and $\sqrt{8\epsilon/3}$ $(j=1)$ in expressions for the velocity gradient, are valid with a thin shock layer only if the shock and body have approximately the same curvature. This validity requires that the β appearing in Sections 4.3 and 4.4 be of the order of 1, and that the parameter $R_s \Delta''$ appearing there be no larger than of order $\epsilon \ln \epsilon^{-1}$ or ϵ. With bodies that have very blunt faces these requirements are not met, and the Newtonian pressure distributions do not hold. A method is needed in which the pressure distribution near the body is given correctly as a result of the theory.

To the lowest order of magnitude such a theory has been given by Cole and Brainerd [1] for the two-dimensional case. The key to the difficulties lies in treating the inner and outer layers separately. We present here the theory of Cole and Brainerd, in a somewhat altered form to fit the other material in this chapter. The extension of this theory to the axisymmetric case is taken from Antonov and Hayes [1].

The body shapes considered vary but slightly from a plane. Such bodies we will term "blunt-faced bodies". In the present section the reference plane is normal to the flow direction, and "slightly" means that the inclination of the body from the plane is $O(\epsilon^{1/2})$. We choose a cartesian coordinate system with the y axis in the free stream direction and with the other coordinate designated z. The coordinate system is of the boundary layer type discussed in Section 5.1 oriented with respect to the reference plane, but with x and u replaced by z and w. See Fig. 5–11. The quantity ϵ is taken to be a constant in the theory, assuming that the effect of nonconstant ϵ is of smaller order.

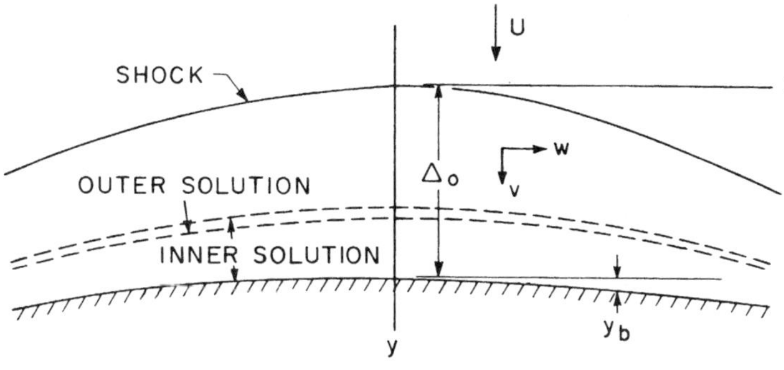

FIG. 5–11. Coordinate system for Cole and Brainerd solution.

A coordinate transformation is made, in which z is unchanged and y is altered by

$$y = \epsilon^{1/2} y^*. \tag{5.4.1}$$

The velocity components are transformed according to

$$v = U\epsilon v^* + O(\epsilon^2), \tag{5.4.2}$$

$$w = U\epsilon^{1/2} w^* + O(\epsilon^{3/2}), \tag{5.4.3}$$

while the pressure is transformed by

$$\frac{p - p_\infty}{\rho_\infty U^2} = 1 + \epsilon(p^* - 1) + O(\epsilon^2). \tag{5.4.4a}$$

It should be noted that our $p^* - 1$ is the quantity denoted p^* by Cole and Brainerd. The density is given by

$$\frac{\rho_\infty}{\rho} = \epsilon + O(\epsilon^2). \tag{5.4.4b}$$

We use the von Mises formulation, with ψ replaced by

$$\frac{\psi}{\rho_\infty U} = \psi^*. \tag{5.4.5}$$

Equations (5.1.12) to (5.1.15) hold, with D/Dx replaced by $\partial/\partial z$ and u replaced by w. Equations (5.1.10) and (5.1.11) do not apply here, but could be used in a more complete theory justifying the constant-density assumption. Equation (5.1.14) becomes, with $K = 0$ in our system,

$$w^* \frac{\partial w^*}{\partial z} + \epsilon\left(\frac{\partial p^*}{\partial z} + z^j v^* \frac{\partial p^*}{\partial \psi^*}\right) = O(\epsilon). \tag{5.4.6}$$

To lowest order the term in ϵ is dropped, and we obtain

$$w^* = w^*(\psi^*). \tag{5.4.7}$$

This corresponds to the Newtonian result of constant velocity along streamlines.

We use (5.4.7) to replace ψ^* by w^* as an independent variable. This procedure may be used only in this approximation, and we should look on w^* as a modified stream function. Equations (5.1.15) become

$$\frac{\partial y^*}{\partial z} = \frac{v^*}{w^*}, \tag{5.4.8}$$

$$\frac{\partial y^*}{\partial \psi^*} = w^{*\prime}(\psi^*)\frac{\partial y^*}{\partial w^*} = -\frac{1}{z^j\, w^*}. \tag{5.4.9}$$

Finally, we have for (5.1.12),

$$z^j \frac{\partial p^*}{\partial \psi^*} = -\frac{\partial p^*}{\partial w^*}\Big/ w^* \frac{\partial y^*}{\partial w^*} = \frac{\partial v^*}{\partial z}. \tag{5.4.10}$$

These are the basic equations for the theory.

Examination of (5.4.9) shows that the quantity $z^j\, \partial y^*/\partial w^*$ is a function of w^* alone. We designate this function as $f''(w^*)$. We can then integrate with respect to w^* to obtain y^*, with an arbitrary function $g(z)$ appearing. The result is

$$y^*(w^*, z) = \frac{1}{z^j} f'(w^*) + g(z). \tag{5.4.11}$$

From this expression (5.4.8) yields

$$v^*(w^*, z) = -\frac{j}{z^{1+j}}\, w^* f'(w^*) + w^* g'(z). \tag{5.4.12}$$

The pressure can now be obtained from (5.4.10),

$$p^* = p_b^*(z) - \frac{g''(z)}{z^j} \int_{w_b^*}^{w^*} t^2 f''(t)\, dt - \frac{j(1+j)}{z^{2+2j}} \int_{w_b^*}^{w^*} t^2 f''(t) f'(t)\, dt, \tag{5.4.13}$$

where $p_b^*(z)$ is an arbitrary function which represents the pressure on the body. The lower limit in the integrals w_b^* is the value of w^* corresponding to conditions on the body. With a detached flow $w_b^* = 0$. We thus have the complete solution expressed in terms of the three functions f, g, and p_b^*.

To interpret this result, we may identify $-tf''\, dt$ as proportional to the differential mass flow $d\psi$, and $-t^2 f''\, dt$ as proportional to the differential momentum flow dP of Newtonian theory. The quantity $g''(z) + j(1+j)f'(w^*)/z^{2+j}$ is equal to $\partial^2 y^*/\partial z^2$, the streamline curvature. The integrals in (5.4.13) taken together thus represent a centrifugal correction, one in which the curvature is variable across the layer if $j \neq 0$.

At the shock wave, we have $\sin\sigma = 1 + O(\epsilon)$, $\cos\sigma = \cot\sigma = (dy/dz)_s$. The shock conditions may be put into the form

$$w_s^* = \left(\frac{dy^*}{dz}\right)_s, \tag{5.4.14}$$

$$v_s^* = 1 + w_s^{*2} = -p_s^* + 1. \tag{5.4.15}$$

Note that w_s^* plays the same role here as does $\cot\sigma/\epsilon^{1/2}$ in Section 5.3.

The shock shape in physical space is given by a function $y_s^*(z)$. Through (5.4.14) it may also be represented as a function

$$y_s^*(w^*) = y^*[w^*, z_s(w^*)], \tag{5.4.16}$$

with z on the shock expressed as

$$z_s = z_s(w^*). \tag{5.4.17}$$

The shock condition (5.4.14) may be expressed as

$$\frac{dy_s^*}{dw^*} = w^* \frac{dz_s}{dw^*}. \tag{5.4.18}$$

Substituting $y^* = y_s^*$ in (5.4.11) yields

$$\left(w^* - g'(z_s) + \frac{j}{z^{1+j}} f'(w^*)\right) \frac{dz_s}{dw^*} = \frac{1}{z_s^j} f''(w^*). \tag{5.4.19}$$

Combining (5.4.12) and (5.4.15) applied at the shock gives

$$g'(z_s) - \frac{j}{z_s^{1+j}} f' = w^* + \frac{1}{w^*}, \tag{5.4.20}$$

and this substituted into (5.4.19) gives

$$\frac{dz_s}{dw^*} = -\frac{w^*}{z_s^j} f''(w^*). \tag{5.4.21}$$

This equation can be integrated to yield

$$\begin{aligned} \frac{1}{1+j} z_s^{1+j} &= f(w^*) - w^* f'(w^*) \\ &= -w^{*2} \frac{d}{dw^*}\left(\frac{f}{w^*}\right) = \psi_s^*. \end{aligned} \tag{5.4.22}$$

(a) Two-dimensional attached and axisymmetric outer solutions. We define the coordinate system so that $y^* = 0$ at the dividing point on the shock $z_s(w_b^*) = 0$, with

$$y_s^*(w_b^*) = \lim_{w^* \to w_b^*} y^*[w^*, z_s(w_b^*)] = 0 \tag{5.4.23}$$

Note that $w_b^* = 0$ for a detached solution, while w_b^* is finite for an attached solution. There is a certain arbitrariness in the definition of f' and g in (5.4.11); a constant k_1 may be added to f' and $k_1 z^j$ subtracted from $g(z)$ with no change. We consider first those cases in which we satisfy (5.4.23) with

$$f'(w_b^*) = 0, \tag{5.4.24a}$$

$$g(0) = \lim_{w^* \to w_b^*} \frac{f'(w^*)}{z_s(w^*)^j}, \tag{5.4.24b}$$

taking advantage of the arbitrariness. With detached shocks in two-dimensional flow this cannot be done. With attached solutions $g(0) = 0$, and this is necessary for (5.4.24) in the two-dimensional case. We then obtain from (5.4.11)

$$y_b^*(z) = g(z). \tag{5.4.25}$$

In the two-dimensional case ($j = 0$), (5.4.20) gives us

$$\frac{dy_b^*(z_s)}{dz_s} = w^* + \frac{1}{w^*}. \tag{5.4.26}$$

This equation requires that

$$\frac{dy_b^*}{dz} > 2 \tag{5.4.27}$$

at all points, corresponding to the condition (5.3.51). The solution obtained is then the "tangent-wedge" solution described at the end of Section 5.3. The new information obtained is with regard to the pressure. The pressure equation may be integrated with the help of (5.4.18), (5.4.21), and (5.4.25), with the integral over t transformed into an integral along the shock from $y_s^*(0) = 0$. The result for the pressure on the body is

$$p_b^* = p_s^*(z) - y_b^{*\prime\prime}(z)\, y_s^*(z). \tag{5.4.28}$$

The term y_s^* may be interpreted as the P of Newtonian theory defined in (3.2.4), and $y_b^{*\prime\prime}$ as the body curvature. Thus in the two-dimensional "tangent-wedge" case the Newtonian pressure law holds with the body curvature used in the centrifugal correction. The discontinuity or jump in the quantity w^*, which is allowed by (5.4.26), has been discussed in the last section, and gives a discontinuity in p_b^*. Note also that a discontinuity in $y_b^{*\prime\prime}$ also gives a discontinuity in p_b^*.

In the axisymmetric case ($j = 1$) we can solve (5.4.20) for f' and substitute into the expressions (5.4.16) and (5.4.11) for y_s^*. We then differentiate y_s^* with respect to z_s and apply (5.4.14). The result, replacing g by y_b^* according to (5.4.25), is

$$\left(1 - \frac{1}{w^{*2}}\right) z_s \frac{dw^*}{dz} + \frac{1}{w^*} + 2w^* = 2\frac{dy_b^*(z_s)}{dz_s} + z_s \frac{d^2 y_b^*}{dz_s^2}. \tag{5.4.29}$$

This equation may be recognized as the same as (5.3.24), so that the theory of this section yields an alternative derivation.

The pressure on the body may be expressed in a form analogous to (5.4.28),

$$p_b^* = p_s^*(z) - \frac{y_b^{*\prime\prime}(z)}{z} \int_0^{y_s^*(z)} \zeta \, dy_s^*(\zeta) + \frac{2}{z^3} \int_0^{y_s^*(z)} (y_b^*(z) - y^*\{z, y_s^*(\zeta)\}) \, \zeta \, dy_s^*(\zeta), \tag{5.4.30}$$

where $y^*\{z, y_s^*\} = y^*(z, w^*)$ for $y_s^* = y_s^*(w^*)$. The first integral in (5.4.30) is proportional to the Newtonian impulse P, and gives a centrifugal term based on body curvature. To interpret the second term, we first identify $2(y_b^* - y^*)/z^2$ with the curvature of streamlines relative to the body surface. The second term may then be interpreted as a centrifugal term arising from this curvature. It is equivalent to the term on the right-hand side of (4.2.22) in the solution for the cone, and is responsible for the change from $1 - \epsilon$ to $1 - \frac{1}{2}\epsilon$ in equations such as (4.3.15) or (4.4.10).

(b) Two-dimensional detached local solution. In the two-dimensional detached case, a local solution near the stagnation point is required. We assume the function $z_s(w^*)$ has the form

$$z_s = Aw^* + Nw^{*2} + Bw^{*3} + O(w^{*4}), \tag{5.4.31}$$

with the coefficient N a measure of the asymmetry of the solution. Using (5.4.18), the shock shape is expressed parametrically by (5.4.31) and

$$y_s^* = \tfrac{1}{2}Aw^{*2} + \tfrac{2}{3}Nw^{*3} + \tfrac{3}{4}Bw^{*4} + O(w^{*5}). \tag{5.4.32}$$

Integrating (5.4.22), and for our convenience choosing the arbitrary constant to be A we have

$$f = Aw^* - Aw^* \ln w^* - Nw^{*2} - \tfrac{1}{2}Bw^{*3} + O(w^{*4}), \tag{5.4.33}$$

and

$$f' = -A \ln w^* - 2Nw^* - \tfrac{3}{2}Bw^{*2} + O(w^{*3}). \tag{5.4.34}$$

We can invert the series (5.4.31) to obtain

$$\frac{1}{w^*} = \frac{A}{z_s} + \frac{N}{A} + \left(\frac{B}{A^2} - \frac{N^2}{A^3}\right) z_s + O(z_s^2). \tag{5.4.35}$$

This result may be used to integrate (5.4.20), to give

$$g(z) = A \ln \frac{z}{A} + \frac{N}{A} z + \frac{1}{2}\left(\frac{A+B}{A^2} - \frac{N^2}{A^3}\right) z^2 + O(z^3). \tag{5.4.36}$$

Here the arbitrary constant has been chosen to have $y_s^* = 0$ at $z = 0$. Equations (5.4.34) and (5.4.35) together determine y^* from (5.4.11), and can be used to obtain v^* from (5.4.12) if desired.

The pressure may be calculated from (5.4.13), and is expressed

$$p^* = p_b^*(z) - \frac{1}{2}\frac{A^2 w^2}{z^2}\left[1 - \left(\frac{A+B}{A} - \frac{N^2}{A^2}\right)\frac{z^2}{A^2} + O(z^3)\right]\left[1 + \frac{4}{3}\frac{N}{A} w + \frac{3}{2}\frac{B}{A} w^2 + O(w^3)\right]. \tag{5.4.37}$$

We must next express z_s^{-2} from (5.4.31) in order to evaluate p_s^*. The result is

$$p_s^* = p_b^*(z_s) - \frac{1}{2}\left[1 - \frac{2}{3}\frac{N}{A} w^* - \left(1 + \frac{3}{2}\frac{B}{A} - \frac{4}{3}\frac{N^2}{A^2}\right) w^{*2} + O(w^{*3})\right]. \tag{5.4.38}$$

Inasmuch as $p_s^* = -w^{*2}$ from (5.4.15), we may express $p_b^*(z)$ as

$$\frac{1}{2} - p_b^*(z) = \frac{1}{3}\frac{N}{A}\frac{z}{A} + \left(\frac{3}{2} + \frac{3}{4}\frac{B}{A} - \frac{N^2}{A^2}\right)\frac{z^2}{A^2} + O(z^3). \tag{5.4.39}$$

An asymptotic relation between w^* and y^* will be needed for small w^* and large y^*. From (5.4.34) and (5.4.11) this relation is

$$w^* \sim \exp\left(-\frac{y^* - g(z)}{A}\right). \tag{5.4.40}$$

The fact that y^* is unbounded as $w^* \to 0$ prevents the satisfaction of body boundary conditions without an inner solution.

(c) Two-dimensional detached inner solution and matching. For the inner solution in the two-dimensional case we may leave the coordinate system unchanged, and merely introduce new velocity variables. In place of (5.4.2) and (5.4.3) we set

$$v = U\epsilon^{3/2}\tilde{v} + O(\epsilon^{5/2}), \tag{5.4.41}$$

$$w = U\epsilon\tilde{w} + O(\epsilon^2). \tag{5.4.42}$$

The orders of error in the outer solution fail as the inner region is approached. Thus the two systems are related by $v^* = \epsilon^{1/2}\tilde{v}$ and $w^* = \epsilon^{1/2}\tilde{w}$.

The equations of motion are written without the von Mises transformation as

$$\frac{\partial \tilde{v}}{\partial y^*} + \frac{\partial \tilde{w}}{\partial z} = 0, \tag{5.4.43}$$

$$\tilde{v}\,\frac{\partial \tilde{w}}{\partial y^*} + \tilde{w}\,\frac{\partial \tilde{w}}{\partial z} + \frac{\partial p^*}{\partial z} = 0, \tag{5.4.44}$$

$$\frac{\partial p^*}{\partial y^*} = 0, \tag{5.4.45}$$

with terms of $O(\epsilon)$ dropped.

A solution of the equations of motion is

$$\tilde{w} = \tilde{w}_b(z)\cosh\left(\frac{y^* - y_b^*(z)}{\tilde{A}}\right) - \tilde{w}_b(0)\sinh\left(\frac{y^* - y_b^*}{\tilde{A}}\right), \tag{5.4.46}$$

$$\tilde{v} = -(\tilde{A}\tilde{w}_b' + \tilde{w}_b(0)y_b^{*\prime})\sinh\left(\frac{y^* - y_b^*}{\tilde{A}}\right) + \tilde{w}_b\, y_b^{*\prime}\cosh\left(\frac{y^* - y_b^*}{\tilde{A}}\right). \tag{5.4.47}$$

This solution satisfies the proper boundary condition $\tilde{v}/\tilde{w} = y_b^{*\prime}$ on the body at $y^* = y_b^*$, and is chosen to have $\tilde{w} \to 0$ at $z = 0$ far from the body. The pressure is given by

$$p^* = p_b^*(z) = \tfrac{1}{2} + \tfrac{1}{2}\tilde{w}_b^2(0) - \tfrac{1}{2}\tilde{w}_b^2, \tag{5.4.48}$$

by the Bernoulli equation obtained from (5.4.44).

In order to match the inner and outer solutions, we first take the asymptotic expansion for (5.4.46) for large *negative* $y^* - y_b^*$. We replace $\tilde{w}$ by $w^*/\epsilon^{1/2}$ and obtain

$$w^* \sim \tfrac{1}{2}\epsilon^{1/2}[\tilde{w}_b - \tilde{w}_b(0)]\exp\left(-\frac{y^* - y_b^*}{\tilde{A}}\right). \tag{5.4.49}$$

The matching is accomplished by requiring that (5.4.40) and (5.4.49) agree. The basic condition is

$$A = \tilde{A}, \tag{5.4.50}$$

and shows why the local outer solution is necessary in the matching. With this condition satisfied we obtain

$$g(z) = A \ln \tfrac{1}{2}\epsilon^{1/2}[\tilde{w}_b - \tilde{w}_b(0)] + y_b^*(z) \tag{5.4.51}$$

as a condition analogous to (5.4.25).

The pressure is also required to match, so that (5.4.39) and (5.4.48) must agree. The maximum pressure corresponds to $\tilde{w}_b(z_{st}) = 0$, and is at a minimum of $-p_b^*(z)$. This stagnation point is at $z = z_{st}$, where

$$\frac{z_{st}}{A} = -\frac{\dfrac{1}{9}\dfrac{N}{A}}{\left(1 + \dfrac{1}{2}\dfrac{B}{A}\right)} \tag{5.4.52}$$

if $N/A = O(\epsilon^{1/2})$. The value of $\tilde{w}_b(0)$ may be found to be

$$\tilde{w}_b(0) = \frac{\sqrt{3}\dfrac{1}{9}\dfrac{N}{A}}{\left(1 + \dfrac{1}{2}\dfrac{B}{A}\right)^{1/2}}, \tag{5.4.53}$$

and we may verify that

$$\tilde{w}_b - \tilde{w}_b(0) = \sqrt{3}\sqrt{1 + \frac{1}{2}\frac{B}{A}}\,\frac{z}{A} + O(z^2). \tag{5.4.54}$$

We substitute (5.4.54) into (5.4.51) and apply (5.4.36). We obtain thereby, by setting $z = 0$, the stand-off distance

$$y_b^*(0) = \frac{\Delta}{\epsilon^{1/2}} = +\frac{1}{2}A \ln \frac{4}{3\epsilon\left(1 + \dfrac{1}{2}\dfrac{B}{A}\right)}. \tag{5.4.55}$$

If N/A is $O(1)$ we cannot evaluate z_{st} and $\tilde{w}_b(0)$ in general because then the higher order terms enter. We can write, from (5.4.39) and (5.4.48),

$$\tilde{w}_b - \tilde{w}_b(0) = \frac{1}{3}\frac{N}{A\tilde{w}_b(0)}\frac{z}{A} + O(z^2) \tag{5.4.56}$$

in place of (5.4.54). In place of (5.4.55) we obtain

$$y_b^*(0) = \frac{\Delta_0}{\epsilon^{1/2}} = +A \ln \frac{6A\tilde{w}_b(0)}{\epsilon^{1/2}N}. \tag{5.4.57}$$

To interpret the parameters involved, we may eliminate w^* between (5.4.31) and (5.4.32) to obtain $y_s^*(z)$. The result is

$$y_s^* = \frac{y_s}{\epsilon^{1/2}} = \frac{z^2}{2A} - \frac{Nz^3}{3A^2} + \frac{z^4}{A^3}\left(\frac{N^2}{2A^2} - \frac{B}{4A}\right) + O(z^4). \tag{5.4.58}$$

We readily interpret

$$A = \epsilon^{1/2} R_{s_0} . \tag{5.4.59}$$

Carrying out the calculation for the parameters β defined in (4.3.31) and R_s' which appears in (4.3.53) gives

$$\frac{N}{A} = \tfrac{1}{2}\epsilon^{1/2} R_s' , \tag{5.4.60}$$

$$\frac{B}{A} = \frac{2}{3}\frac{N^2}{A^2} + \tfrac{1}{6}\epsilon\beta = \tfrac{1}{6}\epsilon(R_s'^2 + \beta). \tag{5.4.61}$$

In these terms (5.4.55) reads, with $N/A = O(\epsilon^{1/2})$,

$$\frac{\Delta}{R_s} = \tfrac{1}{2}\epsilon \ln \frac{4}{3\epsilon(1 + \frac{1}{12}\epsilon\beta)} , \tag{5.4.62}$$

consistent with the β variation in (4.3.39), and

$$\frac{z_{\text{st}}}{R_s} = \frac{-\frac{1}{18}\epsilon R_s'}{1 + \frac{1}{12}\beta\epsilon} , \tag{5.4.63}$$

consistent with (4.3.48) and (4.3.53).

(d) Solution procedures for two-dimensional detached flows. In order to obtain a full solution for a two-dimensional body the local pressure matching above must be extended to a matching of the pressure over the entire body. This is done by relating $\tilde{w}_b$, which depends on the pressure on the body, with w^* at the shock at the same value of z. The result (5.4.51) obtained from the general velocity matching is also used. We will consider the function $z_s(w^*)$ inverted to a function $w^*(z_s)$, and then we shall drop the subscript s. Thus w^* will mean the value of w^* at the shock for the given value of z.

We differentiate (5.4.51) and substitute in (5.4.20) to obtain

$$g'(z) = w^* + \frac{1}{w^*} = \frac{A\tilde{w}_b'}{\tilde{w}_b - \tilde{w}_b(0)} + y_b^{*\prime} . \tag{5.4.64}$$

Equation (5.4.20) is differentiated, to obtain

$$g'' = -\frac{1 - w^{*2}}{w^{*2}}\frac{dw^*}{dz} . \tag{5.4.65}$$

We calculate the quantity $(p_s^* - p_b^*)/g''$ using (5.4.15), (5.4.48), (5.4.65), and (5.4.13). The result is

$$\frac{p_s^* - p_b^*}{g''} = \frac{\frac{1}{2}[1 + 2w^{*2} + \tilde{w}_b^2(0) - \tilde{w}_b^2]}{\dfrac{1 - w^{*2}}{w^{*2}} \dfrac{dw^*}{dz}} \tag{5.4.66}$$

$$= -\int_0^{w^*} t^2 f'' \, dt.$$

We replace tf'' in the integral by the value $-dz/dw^*$ given by (5.4.21), and differentiate. The result is

$$w^* = \frac{1}{2} \frac{d}{dz} \left\{ \frac{w^{*2}[1 + 2w^{*2} + \tilde{w}_b^2(0) - \tilde{w}_b^2]}{(1 - w^{*2})\, dw^*/dz} \right\}. \tag{5.4.67}$$

This equation together with (5.4.64) provides a pair of differential equations for the two unknown functions $w^*(z)$ and $\tilde{w}_b(z)$. The solution of this pair of equations gives the solution to the problem of determining shock shape and body velocity distribution for a given body shape.

The nominal order of the pair of differential equations is three, one from (5.4.64) and two from (5.4.67). In the problem of determining the solution, the effective order is the number of independent parameters which must be determined to obtain the solution desired. We may consider that the solution is represented by a point in some parameter space, and that the effective order is the dimension of that parameter space.

For the purpose of determining effective order we take the point of view that we will start the solution from the "origin" at the point for which $w^* = 0$. Equation (5.4.64) permits solutions for which $\tilde{w}_b - \tilde{w}_b(0)$ is nonanalytic in w^* or z. These solutions are rejected, and the condition of analyticity reduces the nominal order to two. In a general symmetric problem $\tilde{w}_b(0) = 0$, and the effective order is also two, corresponding to the two parameters $w^{*\prime}(0)$ and $\tilde{w}_b'(0)$. These are generally to be determined by the condition that the solution pass through the critical point of (5.4.67) at $w^* = 1$, $\tilde{w}_b = \sqrt{3}$. In a symmetric problem for which $y_b^{*\prime}$ is constant, for which the body is flat-faced or V-shaped, the scale of the problem may be eliminated and the solution made to depend only on the parameter $d\tilde{w}_b/dw$ at $w^* = 0$. The effective order in this case is one.

In an asymmetric problem two additional parameters need to be determined, the location of the origin and the quantity $\tilde{w}_b(0)$. The general asymmetric problem thus has effective order four. The parameters are generally to be determined by the conditions that the solution $(w^*, \tilde{w}_b)$ pass through the two points $\pm (1, \sqrt{3 + \tilde{w}_b^2})$. If $y_b^{*\prime}$ is constant, so that the body is flat-faced but at an angle of incidence, the effective order is reduced to two. The two parameters are $d\tilde{w}_b/dw^*$ at $w^* = 0$ and $\tilde{w}_b(0)$. If the body is V-shaped, with $y_b^{*\prime}$ constant on either side of a vertex, the effective order is three. It is of interest to note that these effective orders are the same as those for the two-strip integral relations method ("reduced" if asymmetric) of Section 6.3; they are discussed further there.

In the case of the symmetric flat-faced body the solution was obtained numerically by Cole and Brainerd [1]. The basic results are

$$A = 1.23\, z_{cr}\,, \tag{5.4.68a}$$

$$\frac{dw_b}{dz} = 0.64\, \epsilon\, z_{cr}^{-1}\,, \qquad \text{at} \quad z = 0, \tag{5.4.68b}$$

$$\Delta_0 = -1.23\, \epsilon^{1/2}\, z_{cr} \ln (0.79\, \epsilon^{1/2}), \tag{5.4.68c}$$

$$\beta\epsilon = 6\,\frac{B}{A} = -2.0. \tag{5.4.68d}$$

It may be noted that the pressure gradient on the body near the axis is increased over that behind the shock by the factor 1.25 instead of by the Newtonian factor $\frac{3}{2}$.

Plots of y_s^*/z_{cr}, $\Delta_0/\epsilon^{1/2} z_{cr}$, and $\tilde{w}_b$ against z/z_{cr} will be found in Figs. 5–12, taken from the results of Cole and Brainerd [1]. A plot of the shock shape

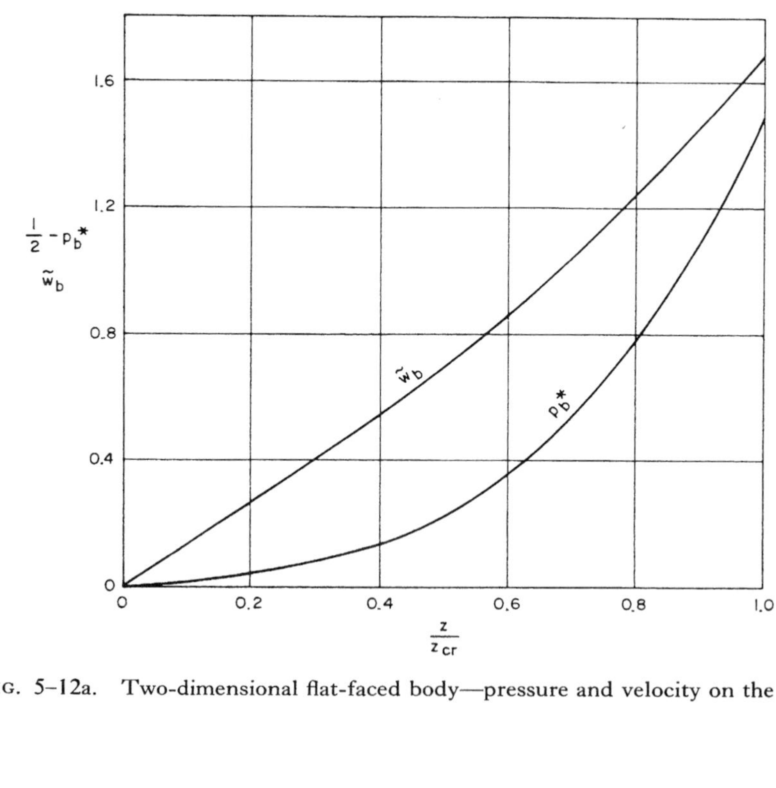

FIG. 5–12a. Two-dimensional flat-faced body—pressure and velocity on the body.

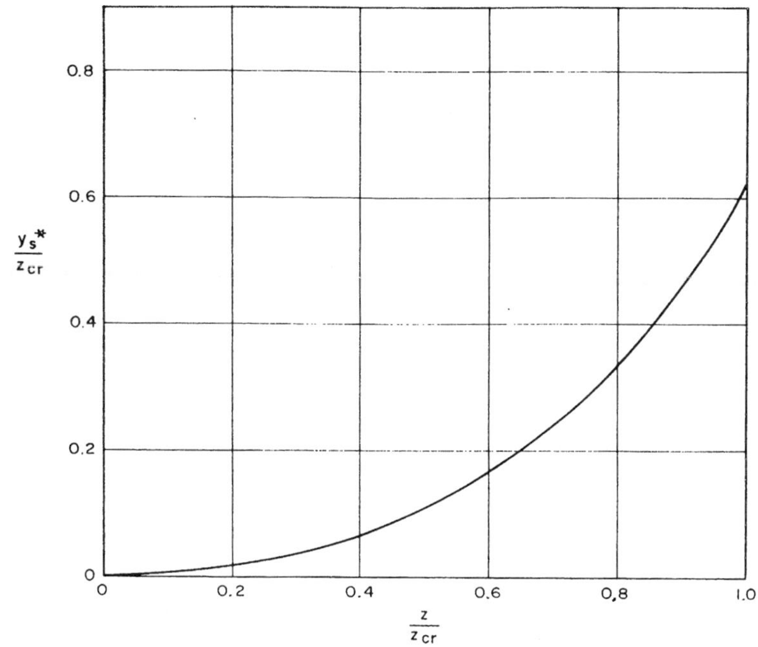

FIG. 5–12b. Two-dimensional flat-faced body—shock shape.

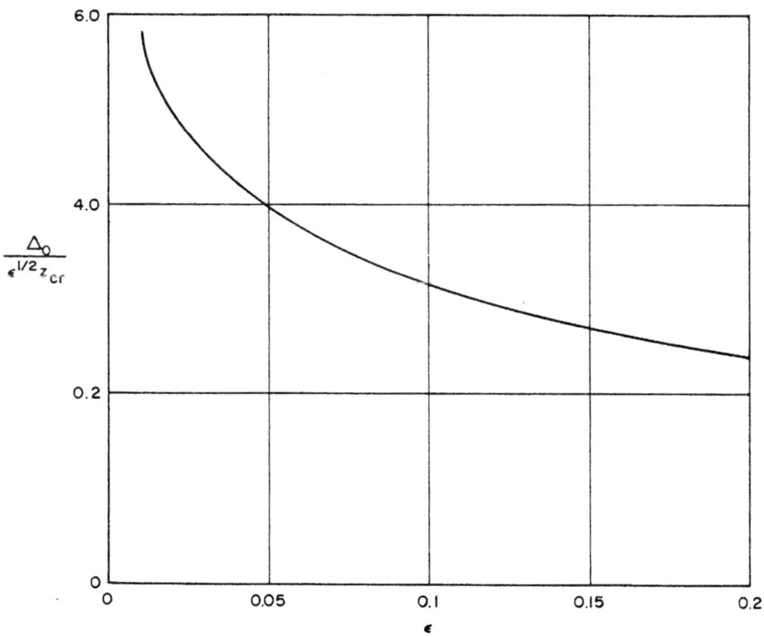

FIG. 5–12c. Two-dimensional flat-faced body—stand-off distance.

for $\epsilon = 0.16$ will be found in Fig. 6–13, and a plot of the pressure for $\epsilon = 0.167$ in Fig. 6–14.

(e) Axisymmetric detached solutions with inner layer. Following the extension of Cole and Brainerd's work by Antonov and Hayes [1], we give the local solution for the axisymmetric case. The approach and details are but slightly different from the two-dimensional case. In place of (5.4.31) we have

$$z_s = Aw^* + Bw^{*3} + O(w^{*5}), \tag{5.4.69}$$

and in place of (5.4.32),

$$y_s^* = \tfrac{1}{2}Aw^{*2} + \tfrac{3}{4}Bw^{*4} + O(w^{*6}). \tag{5.4.70}$$

In place of (5.4.33) and (5.4.34), from (5.4.22), we have

$$f = -\tfrac{1}{2}A^2w^{*2} - \tfrac{1}{3}ABw^{*4} + O(w^{*6}), \tag{5.4.71}$$

and

$$f' = -A^2w^* - \tfrac{4}{3}ABw^{*3} + O(w^{*5}). \tag{5.4.72}$$

where as before the arbitrary constant is chosen for convenience. With $N = 0$, (5.4.35) holds, and from (5.4.20) we obtain

$$g(z) = A + \frac{1}{2A}\left(1 + \frac{2}{3}\frac{B}{A}\right)z^2 + O(z^4). \tag{5.4.73}$$

The arbitrary constant here is chosen to have $y_s^* = 0$ at $z = 0$. The pressure is given by

(5.4.74)
$$p^* = p_b^*(z) + \frac{Aw^{*3}}{3z}\left[1 + \frac{12}{5}\frac{B}{A}w^{*2} + O(w^{*4})\right]\left[1 + \frac{2}{3}\frac{B}{A} + 3\frac{B}{A}\left(\frac{B}{A} - 1\right)\frac{z^2}{A^2} + O(z^4)\right] - \frac{A^4w^{*4}}{2z^4}\left[1 + \frac{32}{9}\frac{B}{A}w^{*2} + O(w^{*4})\right],$$

and the pressure at the shock by

$$p_s^* = p_b^*(z_s) - \frac{1}{2} + \frac{1}{3}\left(1 + \frac{4}{3}\frac{B}{A}\right)w^{*2} + O(w^{*4}). \tag{5.4.75}$$

The function $p_b^*(z)$ can be evaluated, with

$$\frac{1}{2} - p_b^*(z) = \frac{4}{3}\left(1 + \frac{1}{3}\frac{B}{A}\right)\frac{z^2}{A^2} + O(z^4). \tag{5.4.76}$$

The asymptotic relation between w^* and y^* for small w^* is obtained from (5.4.11) and (5.4.72), and is

$$w^* \sim \frac{z}{A^2}(g(z) - y^*). \tag{5.4.77}$$

Setting $w^* = 0$ on the boundary leads to the condition (5.4.25) considered earlier.

The inner layer, as we have seen in Section 4.4, has a thickness which is of the order $\epsilon^{3/2}R_{s_0} \approx \epsilon A \approx \epsilon^{1/2}\Delta_0$. This layer lies on the body, and to properly represent it in a reduced form we should use a curvilinear coordinate system of the boundary layer type oriented with respect to the body. We avoid this approach in order to maintain formal simplicity and to use the same quantities previously defined in dealing with the two-dimensional case. We expect the inner solution to hold only in a region for which $y^* - g$ or $y^* - y_b^*$ is $O(\epsilon^{1/2})$, and we expect $\tilde{v} - g'\tilde{w}$ or $\tilde{v} - y_b^{*}\tilde{w}$ to be of order $\epsilon^{1/2}$. In addition, $\partial\tilde{w}/\partial z$ should be $O(\epsilon^{-1/2})$ if $y_b^{*\prime}$ is not zero. With this understanding, we use unaltered coordinates and the same reduced variables as in the two-dimensional case.

Thus (5.4.41) to (5.4.45) hold, with the exception of the continuity equation (5.4.43). The continuity equation in this case reads

$$\frac{\partial\tilde{v}}{\partial y^*} + \frac{\partial\tilde{w}}{\partial z} + \frac{\tilde{w}}{z} = 0. \tag{5.4.78}$$

The appropriate solution of these equations is

$$\tilde{w} = \tilde{w}_b(z) + \frac{z(y_b^* - y^*)}{\epsilon^{1/2}\tilde{A}^2}, \tag{5.4.79}$$

$$\tilde{v} = \tilde{w}_b y_b^{*\prime} + (y_b^* - y^*)\left(\tilde{w}_b' + \frac{\tilde{w}_b}{z} + \frac{z y_b^{*\prime}}{\epsilon^{1/2}\tilde{A}^2}\right) + \frac{(y_b^* - y)^2}{\epsilon^{1/2}\tilde{A}^2}. \tag{5.4.80}$$

This solution satisfies the boundary condition $\tilde{v}/\tilde{w} = y_b^{*\prime}$ on the body, with the pressure given by

$$p_b^*(z) = \tfrac{1}{2} - \tfrac{1}{2}\tilde{w}_b^2. \tag{5.4.81}$$

The matching is carried out by requiring $\epsilon^{1/2}\,\tilde{w}$ from (5.4.79) to agree with (5.4.77). The basic requirement is again (5.4.50). In addition, we have the relation

$$g(z) = y_b^*(z) + \epsilon^{1/2}A^2\,\frac{\tilde{w}_b(z)}{z}. \tag{5.4.82}$$

Compare (5.4.25), for which the inner layer was neglected. The term $\epsilon^{1/2}A^2 w_b/z$ is a negative displacement thickness resulting from the Bernoulli effect taken into account in the inner layer. See Fig. 5–13.

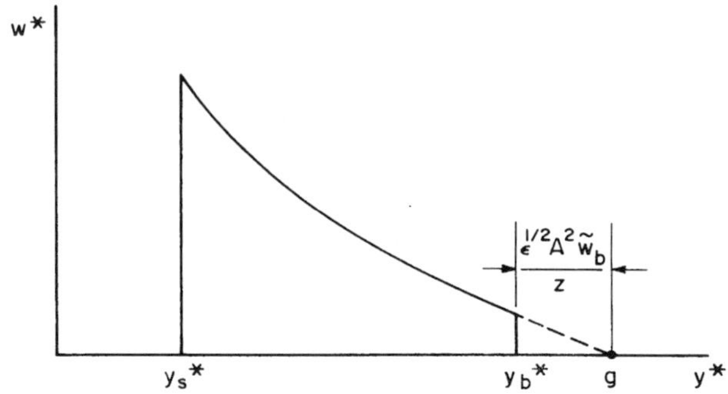

FIG. 5–13. Velocity profile in axisymmetric detached case.

We return now to the local solution to obtain an expression for the stand-off distance. Combining (5.4.76) and (5.4.81) we obtain

$$\tilde{w}_b = \sqrt{\frac{8}{3}\left(1 + \frac{1}{3}\frac{B}{A}\right)}\,\frac{z}{A} + O(z^3). \tag{5.4.83}$$

For the stand-off distance we obtain

$$y_b^*(0) = \frac{\Delta_0}{\epsilon^{1/2}} = A\left[1 - \sqrt{\frac{8\epsilon}{3}\left(1 + \frac{1}{3}\frac{B}{A}\right)}\,\right]. \tag{5.4.84}$$

With $A = \epsilon^{1/2}R_{s_0}$ from (5.4.59) and $B/A = \epsilon\beta/6$ from (5.4.61) we can reexpress this distance as

$$\Delta_0 = \epsilon R_s\left[1 - \sqrt{\frac{8\epsilon}{3}\left(1 + \frac{1}{18}\epsilon\beta\right)}\,\right] \tag{5.4.85}$$

in a form consistent with the β variation in (4.4.24).

For the solution procedure it is possible to proceed as in the two-dimensional case to obtain a set of differential equations to be solved for the complete problem with the inner solution. Because two differentiations are needed to eliminate the two integrals in the pressure expression, the equation system is of high order and quite ungainly. Such an approach fails to take advantage of the fact that the inner layer and the negative displacement thickness $g - y_b^*$ are small.

A much better method is the following: The outer solution is first obtained using the boundary condition (5.4.25) by solving (5.4.29) (or (5.3.25)). The function $p_b^*(z)$ is obtained by carrying out the indefinite integrals in the pressure expression (5.4.13) (with $j = 1$) and matching the pressure along the shock. Then $\tilde{w}_b$ is obtained from (5.4.81) and a new $g(z)$ calculated from (5.4.82). The process is then repeated, using the new $g(z)$ in place of y_b^* in (5.4.29), to obtain the corrected shock shape and $\tilde{w}_b$ distributions. Further iterations are possible, and may be desirable in practice in some cases. However, since this is a perturbation scheme in the small parameter $\epsilon^{1/2}$, further iterations involve corrections of small order in ϵ which cannot be defended rationally.

We look now again at the problem of the flat disc, with $y_b^{*\prime}(z) = 0$. The lowest order solution has already been obtained. From (5.3.25) with $a = 2$ and $\mu_0 = 0$ and (5.3.29) we have

$$g(z) = A = 3^{3/4}\, z_{cr} \tag{5.4.86}$$

$$z_s = \frac{Aw^*}{(1 + 2w^{*2})^{3/4}}, \tag{5.4.87}$$

with (5.4.18) giving

$$y_s^* = A\left[1 - \frac{1 + w^{*2}}{(1 + 2w^{*2})^{3/4}}\right]. \tag{5.4.88}$$

From (5.4.11) we get

$$f'(w^*) = -A^2\frac{w^*(1 + w^{*2})}{(1 + 2w^{*2})^{3/2}}. \tag{5.4.89}$$

The second integral in (5.4.13) may be computed to be

$$\int_0^{w^*} t^2 f'' f'\, dt = \frac{A^4}{32(1 + 2w^{*2})^3}[2w^{*2}(1 + 9w^{*2} + 10w^{*4}) - (1 + 2w^{*2})^3 \ln(1 + 2w^{*2})]. \tag{5.4.90}$$

With this result we may calculate

$$\tilde{w}_b^2 = \frac{(1 + 2w^{*2})^3}{8w^{*4}} \ln(1 + 2w^{*2}) - \frac{(1 + 5w^{*2} + 2w^{*4})}{4w^{*2}}. \tag{5.4.91}$$

From (5.4.91) the negative displacement thickness $g - y_b^* = \epsilon^{1/2}A^2\tilde{w}_b/z$ may be calculated. On the axis we have

$$B/A = -3/2, \tag{5.4.92}$$

and (5.4.84) becomes

$$\frac{\Delta_0}{\epsilon^{1/2}} = A\left[1 - \sqrt{\frac{4\epsilon}{3}}\right]. \tag{5.4.93}$$

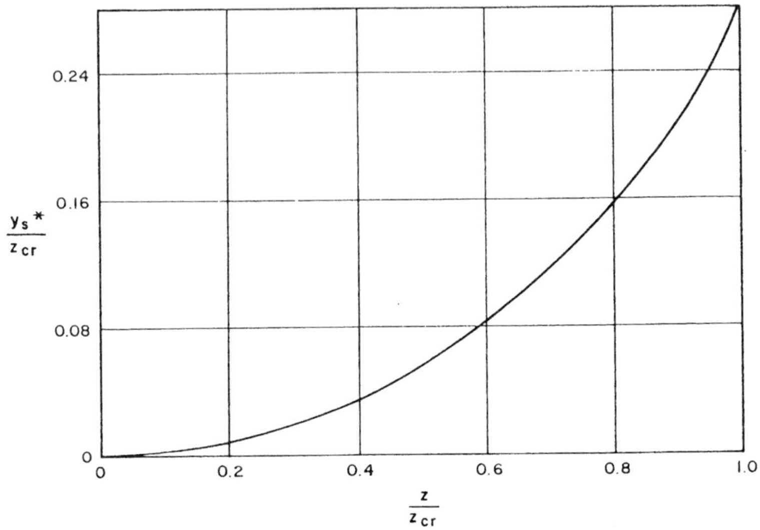

FIG. 5–14a. Flat disc—lowest order shock shape.

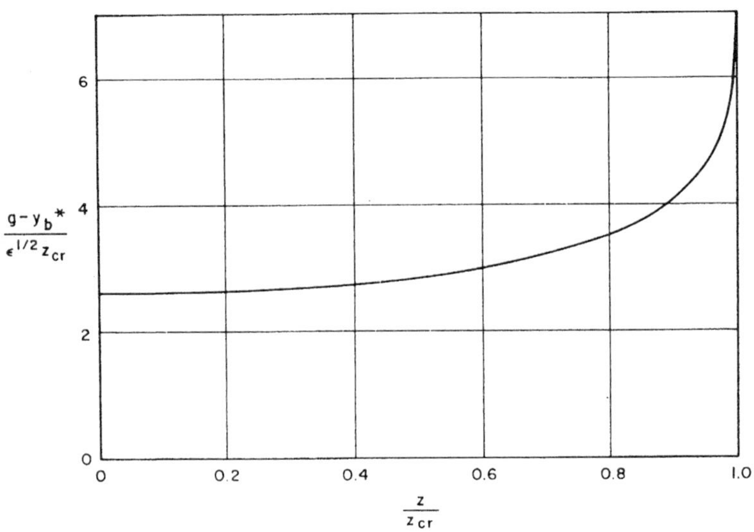

FIG. 5–14b. Flat disc—negative displacement thickness.

We see that instead of an increase of the body pressure gradient near the axis over that behind the shock by a factor 4/3, there is a reduction by the factor 2/3.

Plots of the lowest order shock shape (without correction for the inner layer) and the corresponding function giving the negative displacement thickness are shown in Figs. 5–14. The uncorrected stand-off distance is simply $\epsilon^{1/2}A$. The correction to the stand-off distance has not been calculated, but is expected to be a correction of $O(\epsilon^{1/3})$ instead of $O(\epsilon^{1/2})$, because of a shift in the critical point.

(f) Flows with slope discontinuities. As we have already discussed in this section and the last, discontinuities in the slope of the shock profile must appear in certain flows. Among the related questions that arise we discuss three. What is the nature of two-dimensional flows that start with a weak attached shock and pass to a condition in which (5.4.26) cannot be met? What can we say about the structure of slope discontinuities? What happens if the body slope is discontinuous?

First we must make some remarks about the Bernoulli effect. In cases in which the shock is attached, in either two-dimensional or axisymmetric flow, there may be no identifiable inner layer with low velocities. In these cases the Bernoulli effect will appear to lowest order as a perturbation on the outer flow. The details and orders will differ from those in the inner layer theory, but the qualitative behavior will be the same. Secondly, the Newtonian divergence appears only in those two-dimensional problems for which a normal point on the shock appears. With no normal point present, the concept of a finite displacement thickness from the Bernoulli effect may be applied. It must be applied where (5.4.26) fails. Thirdly, the displacement thickness from the Bernoulli effect is negative and decreasing where the pressure is decreasing in the flow direction along the body. These conditions generally hold with a convex shock shape. If the pressure is increasing in the flow direction, as with a concave shock shape, the displacement thickness is increasing and may be positive.

We consider in Fig. 5–15 a two-dimensional flow in which a slope discontinuity must occur. The dashed line in both the physical plot and the w^* plot shows the "tangent-wedge" solution, which exists up to the point labeled cr. After the slope discontinuity the solution must deviate from the tangent-wedge solution because of the Bernoulli effect. The flow which entered the shock layer upstream of the discontinuity lies in a high-velocity sublayer next to the body. Above this sublayer is an inner layer in which the bulk of the Bernoulli effect appears. The velocities in this layer are not as small as in the inner layer of a detached flow.

For the simplest picture of the structure of a slope discontinuity we neglect the centrifugal effect and assume that the displacement thickness is locally a function of shock pressure alone, one that is monotonically increasing. Thus we assume

$$g - y_b^* = h(w^{*2}), \tag{5.4.94}$$

with $h' > 0$.

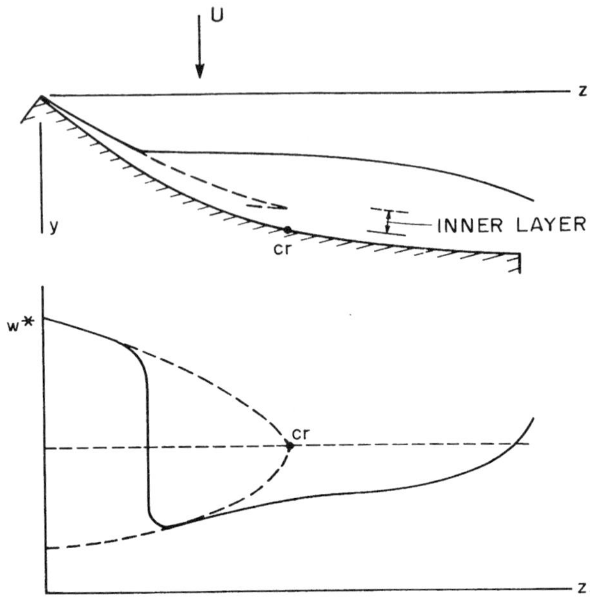

FIG. 5–15. Two-dimensional flow with slope discontinuity.

We restrict ourselves to two-dimensional flow, and apply (5.4.20) with $j = 0$,

$$w^* + \frac{1}{w^*} = y_b^{*\prime} + 2w^*h'(w^{*2})w^{*\prime}. \tag{5.4.95}$$

We next take $y_b^{*\prime}$ to be constant, and denote the weak and strong solutions of (5.4.26) by w_1^* and w_2^*, respectively. We can then rewrite (5.4.95) as

$$-\frac{2w^{*2}h'(w^{*2})\,dw^*}{(w_1^* - w^*)(w^* - w_2^*)} = dz. \tag{5.4.96}$$

The quadrature of this expression gives the structure. If the discontinuity is weak, we may replace w^{*2} in the numerator above by 1, to obtain

$$z - z_0 = \frac{2h'(1)}{w_1^* - w_2^*}\log\frac{w_1^* - w^*}{w^* - w_2^*}. \tag{5.4.97}$$

This result shows the analogy with the classical weak shock wave structure, for which an analogous law holds.

The effect of centrifugal effects complicates the analysis somewhat, raising the order of the differential equation by one if a simple model is used. The results are thus different in details from (5.4.96), but not in qualitative results.

The centrifugal effect acts as a perturbation on the structure, and causes no essential change in its nature.

In any case, the fact that the displacement thickness should increase with increasing pressure leads to the conclusion that slope discontinuities in which w^* and $\cot \sigma$ increase are permitted, while those in which they decrease are forbidden. Our discussion of the role of slope discontinuities in the integral curves of Section 5.3 is thus supported.

If the body slope is discontinuous, the lowest order theory for the outer layer requires a corresponding discontinuity in $w^* + 1/w^*$. The quantity g'' has a delta function behavior at the discontinuity, as then does also the pressure on the body. If $y_b^{*\prime}$ decreases across the discontinuity, there is a concentrated force on the body. The discontinuity shifts the solution nearer to critical in this case. If $y_b^{*\prime}$ increases across the discontinuity, a separation analogous to Newtonian separation should appear.

If we now apply the model of (5.4.95) to this case, w^* is continuous and a discontinuity appears in dw^*/dz. If the flow is supercritical (weak solution) on both sides of the discontinuity, the w^* distribution has a structure upstream which brings w^* to the correct value downstream. No structure appears downstream of the discontinuity. If the flow is subcritical (strong solution), there is no upstream structure and only a downstream structure.

However, in this case we cannot say that the centrifugal effects do not affect the structure in an essential manner. Centrifugal effects must help smooth out the pressure hump on the body and eliminate the discontinuity in dw^*/dz. Thus centrifugal effects as well as Bernoulli effects must be taken into account for a realistic theory for the solution near a body slope discontinuity.

(g) Similitude. In the theory presented above, a similitude is available in those cases for which the inner layer may be neglected. We choose $c\ (= z_{cr})$ as a reference scale in the z direction, and describe the body shape by

$$y_b(z) = \tau c f(z/c). \tag{5.4.98}$$

The quantity τ is a thickness ratio for the body profile. All the equations above which concern solutions involving only an outer layer are invariant, provided the parameter

$$\tau^* = \frac{\tau}{\epsilon^{1/2}} \tag{5.4.99}$$

is held fixed. This yields a similitude for our constant-streamtube-area theory.

Where the inner layer plays an essential role, no similitude of this type holds. An exception is the case of detached two-dimensional flows. In this case all the equations involving reduced quantities, including those for the inner layer, are invariant with τ^* fixed except for those concerning the matching. These fail to be invariant only with respect to the y scale, and thus the total stand-off distance does not follow the similitude. However, shock shape and body pressure and velocity distributions do follow the similitude. And for the stand-off distance, the quantity $\Delta + \epsilon^{1/2}A \ln \epsilon^{1/2} = \Delta + \epsilon R_s \ln \epsilon^{1/2}$ does follow the similitude.

5. Quasi wedges and quasi cones

In this section we present a simple theory for thin shock layers on two-dimensional bodies which closely approximate wedges or bodies of revolution which closely approximate right circular cones. Our results are closely related to those already obtained in Sections 4.1, 4.2, and 5.3. Although the results are perhaps somewhat trivial in the sense of this chapter, they are essential to the establishment of a general similitude picture in thin shock layer theory and are of interest in their own right. The method we use is that labeled (a) at the beginning of Section 5.3, that of using an integral for m to express the shock position. The method is similar to that we shall use in the next section, on conical flow. We could just as well use the method (c) which was used in Section 5.4.

As before, the theory is a constant-density one without Bernoulli correction. The body is a blunt-faced one in that it approximates a reference surface at constant angle of incidence α (Fig. 5–16). This angle of incidence is large enough that we may consider ϵ in a hypersonic flow to be small and to vary but slightly with small changes in incidence. The angle of incidence α is small enough that the flow is definitely in the weak or supercritical regime far from

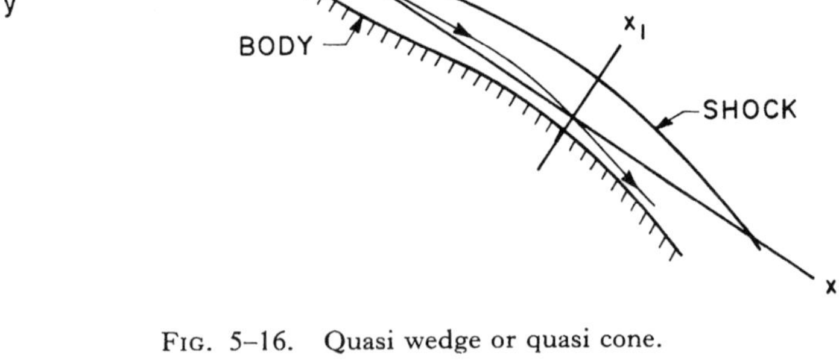

FIG. 5–16. Quasi wedge or quasi cone.

the critical point. This requires that the η of (4.1.5b) be small, or that $\cot \alpha/\epsilon^{1/2}$ be large. Body and shock inclinations relative to the reference surface are restricted to $O(\epsilon \tan \alpha)$, so that the relative pressure variations with respect to $\rho_\infty U^2 \sin^2 \alpha$ are of $O(\epsilon)$. Consequently, with γ_s of the order of one, relative density variations with respect to ρ_∞/ϵ are also of $O(\epsilon)$, and a constant-density theory is applicable in lowest order.

The coordinate system, shown in Fig. 5–16, is of the boundary layer type oriented with respect to the reference surface. The velocity in the x direction is given by

$$u = U \cos \alpha \, (1 + O(\epsilon)); \tag{5.5.1}$$

we shall not investigate variations of u. The mass flow entering the layer between 0 and x is $\rho_\infty U \sin \alpha \, x^{1+j}/(1+j)$. This is equated, at $x = x_1$, to $\rho_\infty U m x_1^j \cos \alpha$, where m is a Howarth-Dorodnitsyn variable of the type used in Chapter III. Thus, conservation of mass flow gives us

$$m(x, x_1) = \frac{1}{1+j} \left(\frac{x}{x_1}\right)^{1+j} x_1 \tan \alpha, \tag{5.5.2}$$

in agreement with the Newtonian result for a wedge or cone. Here x designates the point of entry of a streamline, and thus serves as a stream function. The point of interest is at x_1. Distances normal to the body are given by ϵm, and we may write

$$y_b(x_1) - y(x, x_1) = \epsilon m = \frac{\epsilon}{1+j} \left(\frac{x}{x_1}\right)^{1+j} x_1 \tan \alpha. \tag{5.5.3}$$

At the shock we have

$$y_b(x_1) - y_s(x_1) = \frac{\epsilon}{1+j} x_1 \tan \alpha = \epsilon m_1 . \tag{5.5.4}$$

The slope of a streamline is obtained from (5.5.3), differentiating with respect to x_1, by

$$\frac{\partial y}{\partial x_1} = y' = y_b' + \frac{\epsilon j}{1+j} \left(\frac{x}{x_1}\right)^{1+j} \tan \alpha, \tag{5.5.5}$$

and its curvature by

$$\frac{\partial^2 y}{\partial x_1^2} = y'' = y_b'' - \epsilon j \left(\frac{x}{x_1}\right)^{1+j} \frac{\tan \alpha}{x_1} . \tag{5.5.6}$$

The pressure is obtained by integrating $\rho_\infty U^2 \cos^2 \alpha$ times $y''\, dm$. The result is

$$p = p_b(x_1) + \rho_\infty U^2 \sin\alpha \cos\alpha \frac{x_1}{1+j}\left(\frac{x}{x_1}\right)^{1+j} y_b'' - \tfrac{1}{2}\rho_\infty U^2 \sin^2\alpha \frac{\epsilon j}{1+j}\left(\frac{x}{x_1}\right)^{2(1+j)}. \tag{5.5.7}$$

At the shock, we have, from (5.5.4)

$$\sigma = \alpha - y_s' = \alpha - y_b' + \frac{\epsilon}{1+j}\tan\alpha. \tag{5.5.8}$$

The pressure there is $p_\infty + \rho_\infty U^2(1-\epsilon)\sin^2\sigma$, or

$$p_s = p_\infty + \rho_\infty U^2 \sin^2\alpha - \rho_\infty U^2 \epsilon \sin^2\alpha + 2\rho_\infty U^2 \sin\alpha\cos\alpha\left[-y_b' + \frac{\epsilon}{1+j}\tan\alpha\right]. \tag{5.5.9}$$

Evaluating (5.5.7) at the shock, we obtain

$$p_b - p_\infty - \rho_\infty U^2 \sin^2\alpha = \frac{2-j}{2(1+j)}\epsilon\rho_\infty U^2 \sin^2\alpha - \rho_\infty U^2 \sin\alpha\cos\alpha\left(2y_b' + \frac{x_1 y_b''}{1+j}\right). \tag{5.5.10}$$

These results may be readily interpreted. The thickness of the shock layer is that given in Sections 4.1 and 4.2 for the wedge and right circular cone at the angle α. The pressure on the body is that given by Sections 4.1 and 4.2 for an angle of incidence $\alpha - y_b'$, with the Newtonian centrifugal correction based upon the impulse on a wedge or cone and upon the curvature of the body. This theory gives a tangent-wedge or tangent-cone result for pressure on the body, with a centrifugal correction proportional to body curvature. The theory thus gives a certain rational basis to Section 7.3.

We next make a reduction of the variables. The variable x is unchanged, and we set

$$y = \epsilon \tan\alpha\, y^*, \tag{5.5.11}$$

$$p - p_\infty - \rho_\infty U^2 \sin^2\alpha = \epsilon\rho_\infty U^2 \sin^2\alpha\,(p^* - 1). \tag{5.5.12}$$

We may now rewrite the equations above in terms of y^* and p^*. Equation (5.5.3) becomes

$$y_b^* - y^*(x, x_1) = \frac{x_1}{1+j}\left(\frac{x}{x_1}\right)^{1+j}, \tag{5.5.13}$$

(5.5.7) becomes

$$p^* = p_b^*(x_1) + \frac{x_1}{1+j}\left(\frac{x}{x_1}\right)^{1+j} y_b^{*\prime\prime} - \frac{j}{2(1+j)}\left(\frac{x}{x_1}\right)^{2(1+j)}, \tag{5.5.14}$$

(5.5.9) becomes

$$p_s^* = \frac{2}{1+j} - 2y_b^{*\prime}, \tag{5.5.15}$$

and (5.5.10) becomes

$$p_b^* = \frac{4+j}{2(1+j)} - 2y_b^{*\prime} - \frac{x_1 y_b^{*\prime\prime}}{1+j}. \tag{5.5.16}$$

We now set $y_b(x)$ proportional to a thickness ratio τ, with

$$y_b(x) = \tau c f(x/c), \tag{5.5.17}$$

where c is a reference body length. If we transform x and x_1 in (5.5.13) to (5.5.16) to x/c and x_1/c, the equations are invariant provided the parameter

$$\tau^* = \frac{\tau}{\epsilon \tan \alpha} = \frac{\tau \cot \alpha}{\epsilon} \tag{5.5.18}$$

is an invariant. This result thus yields a similitude.

6. Conical bodies

In conical flow with thin shock layers two questions are suggested by the theory already developed in this chapter. One question asks what the nature of the direct problem is on cones. What parameters control shock detachment from a leading edge, and what parameters control critical phenomena? The second question asks how the Bernoulli effect influences conical flow. Is there an inner layer analogous to those considered in Section 5.4? These questions form the subject matter of this section. We shall find that direct problems on conical bodies with Bernoulli and compressibility effects neglected are very similar to those on the bodies considered in Sections 5.3 and 5.4. But there are important differences in the theories, deriving primarily from the fact that the governing equation is a differential-difference equation instead of a differential equation. As to the Bernoulli effect, we find that it creates a thin layer near the body termed the Newtonian vortical layer, which has a nonanalytic behavior in ϵ. In contrast to the inner layers studied in Section 5.4, this layer has little displacement effect on the outer flow.

For direct problems our interest focuses on bodies whose local angle of incidence is nearly constant. Bodies with strongly variable angle of incidence are readily treated by the approach of Section 5.2 and do not pose any special problems. We consider, then, bodies which closely approximate a reference surface of constant incidence, either a plane or a right circular cone. We term such bodies again as blunt-faced. Our theory for direct problems, as before, is a constant-density theory with Bernoulli effect neglected. Such a theory can be obtained as the lowest-order approximation to a complete theory based upon an expansion in ϵ. We follow the simpler approach of assuming constant density at the outset.

Two theories of this type may be obtained, one with the reference surface a plane and one with the reference surface a circular cone. The theory of Messiter [1] is the first theory. He considers delta wings for which the body itself lies in a plane. So as not to repeat his analysis we present the second theory, for conical bodies that closely approximate a right circular cone at zero angle of attack. The more interesting theoretical problems all appear for bodies of small aspect ratio which are laterally symmetric or have only weak asymmetry. In this case the two theories merge into one. Thus, we shall obtain Messiter's basic equations for delta wings of small aspect ratio.

Hida [3] has extended the theory of Messiter somewhat. In calculating the shock layer thickness at a conical axis of symmetry he reproduces our (3.6.23a). A singular solution which he considers is that for a right circular cone at zero incidence. He presents a simple approximation (two or three terms of a power series expansion) for laterally symmetric wings, one which gives the right order of magnitude for quantities such as the shock layer thickness.

We draw heavily on the results of Section 3.7, and refer repeatedly to that section for various relations. Our reference surface is the right circular cone $\vartheta = \vartheta_0$. We avoid using the symbol α for ϑ_0 in order to avoid confusion with the quantity α used in Section 3.7. We make all our calculations in the unit sphere, and simply set $R_1 = 1$ wherever appropriate.

The theory here is simpler than that of Section 3.7 in some respects, because of the restriction to small perturbations of ϑ from the reference value ϑ_0. Thus the quantity α defined in (3.7.3) is small; $\cos \alpha$ is set equal to 1 except where the pressure behind the shock is calculated with the aid of (3.7.7). The quantity ψ is related to α through (3.7.10) and is also small. The quantity $\xi_1 - \xi$ always lies between 0 and ψ in the theory and is therefore also to be considered small.

The theory of this section is different from that of Section 3.7 in the same way that that of Section 5.4 is different from Newtonian. The shock and body shapes are different. The shock shape is used to calculate the Newtonian velocity field and the pressure behind the shock. The shock layer thickness is given by ϵm_1, where m_1 is the Howarth-Dorodnitsyn variable of Newtonian theory, and this gives the relation between shock and body position. The pressure on the body is obtained by a separate calculation.

(a) Basic theory. The boundary layer coordinate system used is essentially that of Fig. 3–16, with

$$d\xi = \sin \vartheta_0 \, d\phi. \tag{5.6.1}$$

The y axis is in the direction of decreasing ϑ, as shown on Fig. 5–17, and we have (with $R_1 = 1$)

$$y = \vartheta_0 - \vartheta. \tag{5.6.2}$$

The independent variables to be used are ξ and y. The quantity α defined in (3.7.3) is obtained from the shock shape, and is expressed as

$$\alpha = -y_s' . \tag{5.6.3}$$

The streamline deflection angle ψ is given from (3.7.10) as

$$\psi = \tan \vartheta_0 \, y_s' . \tag{5.6.4}$$

The quantity R/R_1 is obtained from (3.7.13) with the sines replaced by their arguments, and the quantity m is given by

$$m(\xi, \xi_1) = \tan \vartheta_0 \int_{\xi_2}^{\xi} \frac{\bar{\psi} + \bar{\xi} - \xi_1}{\bar{\psi}^2} \, d\bar{\xi}. \tag{5.6.5}$$

Here m is the Howarth-Dorodnitsyn variable between the body and the streamline which entered at ξ. The variable $\bar{\xi}$ is a dummy variable, and $\bar{\psi} = \psi(\bar{\xi})$. The relation between these variables and ξ_2 as generally defined is indicated in Fig. 5–17.

The quantity ξ_2 is a function of ξ_1. It is generally defined in terms of the relation

$$\psi(\xi_2) + \xi_2 = \xi_1 , \tag{5.6.6}$$

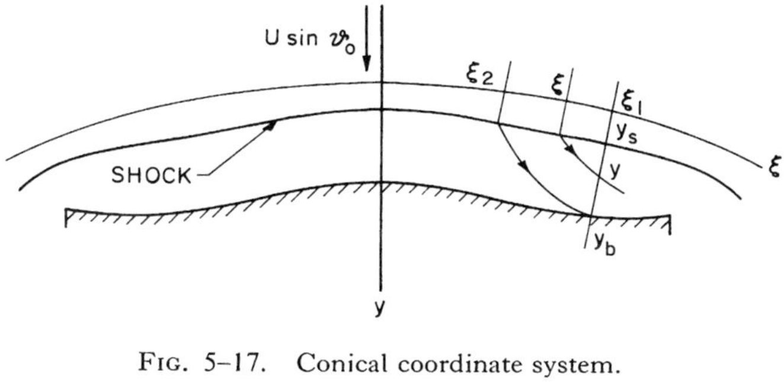

FIG. 5–17. Conical coordinate system.

as the value of ξ at entry of a cross streamline which terminates at ξ_1. If a leading edge at ξ_l exists with $\psi(\xi_l) + \xi_l - \xi_1$ having the same sign as $\xi_1 - \xi_l$, then $\xi_2 = \xi_l$ and is locally a constant. The quantity ξ_2 may also be locally a constant at the value of ξ for which the shock slope has a discontinuity. An example of this case will appear in the example of a delta wing (Ω large) discussed below.

If a discontinuity surface of the type discussed in Section 3.7 is present at ξ_1, the domain of integration of (5.6.5) has two segments, and ξ_2 is multivalued. The second derivative of m with respect to ξ_1 has terms in $d\xi_2/d\xi_1$ (see (5.6.9) below), and in this case more than one term of this type appears.

The shock layer thickness at ξ_1 is ϵm_1, and we have

$$y_b(\xi_1) - y_s(\xi_1) = \epsilon m_1(\xi_1). \tag{5.6.7}$$

This equation is differentiated to obtain

$$y_b' - \frac{\psi_1}{\tan\vartheta_0} = \epsilon \tan\vartheta_0 \left(\frac{1}{\psi_1} - \int_{\xi_2}^{\xi_1} \frac{d\bar{\xi}}{\bar{\psi}^2}\right) \tag{5.6.8}$$

with the help of (5.6.4). The term in the derivative of m_1 from the variation of ξ_2 is zero because relation (5.6.6) is valid if ξ_2 is not constant. We differentiate again to obtain

$$y_b'' + \left(\frac{\epsilon\tan\vartheta_0}{\psi_1^2} - \frac{1}{\tan\vartheta_0}\right)\frac{d\psi_1}{d\xi_1} + \epsilon\tan\vartheta_0\left(\frac{1}{\psi_1^2} - \frac{1}{\psi_2^2}\frac{d\xi_2}{d\xi_1}\right) = 0. \tag{5.6.9}$$

This equation and (5.6.8) are the basic ones for our theory. Equation (5.6.8) is an integral equation of Volterra type for $\psi(\xi)$, while (5.6.9) is a differential-difference equation for $\psi(\xi)$.

Next we calculate the streamline curvature. A streamline at (ξ_1, y) corresponds to a value of m given by

$$\epsilon m(\xi, \xi_1) = y_b(\xi_1) - y(\xi, \xi_1). \tag{5.6.10}$$

We differentiate this with respect to ξ_1, and obtain

$$y_b' - \frac{\partial y}{\partial \xi_1} + \epsilon\tan\vartheta_0 \int_{\xi_2}^{\xi} \frac{d\bar{\xi}}{\bar{\psi}^2} = 0. \tag{5.6.11}$$

A second differentiation gives

$$\frac{\partial^2 y}{\partial \xi_1^2} = y'' = y_b'' - \epsilon\tan\vartheta_0 \frac{1}{\psi_2^2}\frac{d\xi_2}{d\xi_1}. \tag{5.6.12}$$

This quantity is independent of ξ. This means that the streamline curvature is a function only of ξ_1, and constant across the layer at any point. The streamline curvature is not y'', but $y'' + \cot \vartheta_0$, as we may see from (3.7.6).

The total velocity divided by U is

(5.6.13) $$q = \cos \vartheta_0 ,$$

from (3.7.11), and the lateral velocity divided by U is

(5.6.14) $$q_1 = (\psi + \xi - \xi_1) \cos \vartheta_0 .$$

The pressure is then given by the integral of ρq_1^2 times curvature over dm, or by

(5.6.15) $$p(\xi, \xi_1) = p_b(\xi_1) + \rho_\infty U^2 \sin \vartheta_0 \cos \vartheta_0 \left(y_b'' - \epsilon \tan \vartheta_0 \frac{1}{\psi_2^2} \frac{d\xi_2}{d\xi_1}\right) \int_{\xi_2}^{\xi} \frac{(\bar{\psi} + \bar{\xi} - \xi_1)^3}{\bar{\psi}^2} d\bar{\xi}.$$

Since the curvature is constant across the layer, this integral is the same as that in the quantity $\mathfrak{P}_{11}$ of Section 3.7. At the shock wave we expand $\sin \sigma$ from (3.7.7) to obtain, using (5.6.2) and (5.6.3),

(5.6.16) $$\sin \sigma = \sin \vartheta_0 (1 - \tfrac{1}{2} y_s'^2) - \cos \vartheta_0 \, y_s .$$

The pressure behind the shock is then

(5.6.17) $$p_s = p_\infty + \rho_\infty U^2 [\sin^2 \vartheta_0 (1 - \epsilon - y_s'^2) - 2 \sin \vartheta_0 \cos \vartheta_0 \, y_s].$$

(b) Reduced equations and similitude. In order to put these results into a reduced form, we carry out the transformation

(5.6.18a) $$\xi = \epsilon^{1/2} \tan \vartheta_0 \, \xi^*,$$

(5.6.18b) $$\psi = \epsilon^{1/2} \tan \vartheta_0 \, \psi^*,$$

(5.6.19) $$y = \epsilon \tan \vartheta_0 \, y^*,$$

(5.6.20) $$p - p_\infty - \rho_\infty U^2 \sin^2 \vartheta_0 = \rho_\infty U^2 \epsilon \sin^2 \vartheta_0 (p^* - 1).$$

Our $p^* - 1$ is the same as the p^* used by Messiter. We may rewrite the basic equations above in reduced form. These reduced equations are

(5.6.21) $$\psi^* = y_s^{*\prime}, \qquad ((5.6.4))$$

(5.6.22) $$y_b^{*\prime} - \psi_1^* = \left(\frac{1}{\psi_1^*} - \int_{\xi_2^*}^{\xi_1^*} \frac{d\bar{\xi}^*}{\bar{\psi}^{*2}}\right), \qquad ((5.6.8))$$

$$y_b^{*''} + \left(\frac{1}{\psi_1^{*2}} - 1\right)\frac{d\psi_1^*}{d\xi_1^*} + \frac{1}{\psi_1^{*2}} - \frac{1}{\psi_2^{*2}}\frac{d\xi_2^*}{d\xi_1^*} = 0, \tag{5.6.23} ((5.6.9))$$

$$\frac{\partial^2 y^*}{\partial \xi_1^{*2}} = y^{*''} = y_b^{*''} - \frac{1}{\psi_2^{*2}}\frac{d\xi_2^*}{d\xi_1^*}, \tag{5.6.24} ((5.6.12))$$

$$p^* = p_b^*(\xi_1^*) + \left(y_b^{*''} - \frac{1}{\psi_2^{*2}}\frac{d\xi_2^*}{d\xi_1^*}\right)\int_{\xi_2^*}^{\xi^*} \frac{(\bar{\psi}^* + \bar{\xi}^* - \xi_1^*)^3}{\bar{\psi}^{*2}}\, d\bar{\xi}^*, \tag{5.6.25} ((5.6.15))$$

$$p_s^* = -y_s^{*'2} - 2y_s^*, \tag{5.6.26} ((5.6.17))$$

Generally, we shall treat the equations in this reduced form.

The flat delta wing is a portion of a plane. A plane cuts the unit sphere in a great circle. A great circle whose plane is inclined at an angle ϑ_0 to the axis and aligned to be tangent to the cone $\vartheta = \vartheta_0$ at $\phi = 0$ can be shown to satisfy the equation

$$\cos^2\vartheta = \frac{\cos^2\vartheta_0}{1 + \sin^2\vartheta_0 \tan^2\phi}. \tag{5.6.27}$$

With $\tan\phi$ small, using (5.6.2) with $\vartheta = \vartheta_b$, we may express

$$y_b = -\tfrac{1}{2}\cot\vartheta_0\, \xi^2. \tag{5.6.28}$$

Shifting to reduced coordinates, we have

$$y_b^{*'} = -\xi^*, \tag{5.6.29a}$$

$$y_b^{*''} = -1. \tag{5.6.29b}$$

With these functions substituted, (5.6.22) may be recognized as (3.20) of Messiter [1], and (5.6.23) as (3.21).

To obtain the conical similitude of Messiter [1] in somewhat more general form, we first require (with lateral symmetry assumed) that the wing extend from $\xi = -A\!\!R$ to $\xi = +A\!\!R$. In terms of reduced coordinates,

$$\xi^* = \xi_l^* = \Omega \tag{5.6.30a}$$

at one leading edge, and $\xi^* = -\Omega$ at the other. Here we have set

$$\Omega = \frac{A\!\!R \cot\vartheta_0}{\epsilon^{1/2}} \tag{5.6.30b}$$

for Messiter's similarity parameter. The body shape is described by an equation of the form

$$y_b = \tau f(\xi/A\!\!\!R). \tag{5.6.31}$$

The equations governing the solution are invariant if the quantities Ω and

$$\tau^* = \frac{\tau \cot \vartheta_0}{\epsilon} \tag{5.6.32}$$

are invariants, as well as the shape function $f(\xi^*/\Omega)$. Thus Ω and τ^* are basic similarity parameters in this theory. It is clear that lateral symmetry is not required in the similitude.

In case we wished to use the plane as reference surface we would have a choice of two approaches. We could introduce a cartesian coordinate system oriented with respect to the plane, and develop a conical theory therein. This was Messiter's approach. Or we could again take advantage of the conical theory already developed in Section 3.7, with the body slightly perturbed from the planar shape given by (5.6.27). In this case it would be advantageous to use the line $R \cos \xi =$ *constant* instead of the intercept on the unit sphere on which to describe shock and body shapes, and to use $\tan \xi$ instead of ξ as the basic independent variable. We would then arrive at Messiter's equations with additional body shape terms, without the restriction that $A\!\!\!R$ be small.

(c) Flat delta wing. We turn next to the problem of determining the flow field on a narrow flat delta wing at large angle of attack by the theory just developed. A particular case appears, that for which $\Omega \ll 1$. Messiter has found the solution for this case.

Here we do not present the details of his solution, but only note certain features of his results. In particular, it is instructive to compare them with the results of Cole and Brainerd reported in Section 5.4. One result of Messiter is that in the unit sphere the radius of curvature on the axis is given by

$$\epsilon^{1/2} R_{s_0} = 2A\!\!\!R. \tag{5.6.33}$$

For an infinite two-dimensional strip, with $A\!\!\!R$ identified as z_{cr}, (5.4.59) and (5.4.68a) give 1.23 $A\!\!\!R$ in place of 2 $A\!\!\!R$. Thus the two solutions give significantly different shock shapes. For the stand-off distance, Messiter's solution gives

$$\Delta_0 = \epsilon R_{s_0} \ln \frac{1}{2e\Omega}, \tag{5.6.34}$$

where $\ln e = 1$. The solution of Cole and Brainerd is of similar form, but with $0.79\, \epsilon^{1/2}$ in place of $2e\Omega$.

Messiter's theory neglects the effect of the inner layer, and requires that $\Omega \gg \epsilon^{1/2}$. Cole and Brainerd's theory excludes any conical effect, and applied to a conical body requires that $\Omega \ll \epsilon^{1/2}$. As pointed out by Messiter, it is to be expected in a combined theory that $\Omega/\epsilon^{1/2}$ is a basic similarity parameter in determining the shock shape and the flow in the outer layer. We shall explore such a combined theory briefly in Section 5.7.

Before examining the nature of delta wing solutions for arbitrary values of Ω, we discuss critical phenomena and attached shocks. In (5.6.23), it will be noted that the coefficient of the derivative $d\psi_1^*/d\xi_1^*$ is zero if $\psi_1^{*2} = 1$. This occurs at two critical points, one at $\psi_1^* = 1$ for flow in the positive ξ^* direction and one at $\psi_1^* = -1$ for flow in the reverse direction. Physically, these points are equivalent to the points $\cot^2 \sigma = \epsilon/(1 + 2\mu)$ or $w^{*2} = 1$ of Sections 5.3 and 5.4. If $\psi_1^{*2} > 1$ the flow is termed supercritical, if $\psi_1^{*2} < 1$ the flow is termed subcritical. As before, a jump in ψ_1^* is possible in (5.6.22) with $y_b^{*\prime}$ continuous allowed if it goes from supercritical to subcritical in the flow direction. As before, a discontinuity in $y_b^{*\prime}$ corresponding to a concave corner gives a discontinuity in ψ_1^* which shifts it closer to critical. A discontinuity in the integral in (5.6.22) acts in the same manner.

At a leading edge the integral in (5.6.22) is zero if the flow is attached. In this case we have

$$\psi_l^{*2} - y_{b_l}^{*\prime}\psi_l^* + 1 = 0. \tag{5.6.35}$$

If $y_{b_l}^{*\prime 2} < 4$ this equation cannot be satisfied, no attached shock is possible, and the solution must be detached. If $y_{b_l}^{*\prime 2} > 4$ two attached solutions are possible locally, in the direction of decreasing ξ if $y_{b_l}^{*\prime} < 0$ and conversely. One of these solutions is supercritical and is termed weak; the other is subcritical and is termed strong. But in such cases detached solutions are also possible locally. In the flat delta wing case $y_{b_l}^{*\prime} = -\Omega$ and the condition for existence of attached shocks is $\Omega^2 > 4$.

In integrating either (5.6.22) or (5.6.23) we must proceed in the flow direction. At any point on the body, the quantity $\psi^* + \xi^* - \xi_1^*$ always has the same sign across the layer, the same sign as ψ_1^*. This was shown in Section 3.7 as a consequence of the exclusion of Newtonian shock lines. The direction of flow is then defined as the direction of increasing ξ if ψ is positive and of decreasing ξ if ψ is negative. This property is characteristic of Volterra-type integral equations and of differential-difference equations. Another characteristic of equations of this type is the propagation of nonanalytic behavior in the flow direction.

Turning now to the laterally symmetric delta wing, we restrict ourselves to the region $\xi^* \geqq 0$. At $\xi^* = 0$ symmetry dictates that $\psi^* = 0$. If the solution is analytic it locally corresponds to that for a local axis of symmetry treated in Sections 3.6 and 3.7. That solution was characterized by a parameter κ which must be greater than zero. If $d\psi^*/d\xi^* > 0$ at $\xi^* = 0$, then $\kappa > 1$, and the flow is outward. If $d\psi^*/d\xi^* < 0$ there, $0 < \kappa < 1$, and the flow is inward. Three solution ranges are found, corresponding to Ω large, Ω intermediate, and Ω small. Although jumps from supercritical to subcritical may appear in the general case, no such jump appears in this particular case. Other singular behavior does appear in the Ω-large regime, and there is no dearth of slope discontinuities.

With Ω sufficiently large, we expect an attached weak shock on the leading edge. Starting with this solution, there is a one-parameter family of solutions possible. Some of these solutions involve a jump from supercritical to subcritical. In this case these solutions do not turn out to be of interest. To understand the other possible solutions, set $\xi_2^* = \Omega = constant$ and $y_b^{*\prime\prime} = -1$ in (5.6.23). The result is

$$\left(\frac{1}{\psi_1^{*2}} - 1\right)\left(\frac{d\psi_1^*}{d\xi_1^*} + 1\right) = 0. \tag{5.6.36}$$

The attached shock solution satisfies this equation by having $\psi_1^* + \xi_1^*$ constant. Without a jump to subcritical, this solution must continue to the critical line at $\psi_1^* = -1$. Then, as long as $\xi_2(\xi_1)$ is constant, the solution may follow the critical line. At any point on this critical line solution it may shift again to a $\psi_1^* + \xi_1^* = constant$ subcritical solution. Thus there is a one-parameter family of solutions each incorporating a critical segment.

One property of a solution with $\psi_1^* + \xi_1^* = constant$ is that it creates a discontinuity in the integral of (5.6.22) through a discontinuity of the lower limit ξ_2^*. This has the same effect as a concave body slope discontinuity, which in the subcritical regime causes a discontinuous weakening of the shock with a convex discontinuity in the shock slope. At this point there is a delta function in the pressure on the body, as discussed at the end of Section 5.4. The discontinuity in the shock slope has the result that further downstream a region appears with ξ_2^* constant again, for which (5.6.36) holds. Thus we again have a shock segment with $\psi_1^* + \xi_1^* = constant$, which in turn engenders a slope discontinuity. Solutions appear of the type shown in Fig. 5–18, highly nonanalytic and with an infinite sequence of jumps in shock slope. One of these solutions passes through the point $\xi^* = 0$, $\psi^* = 0$ and is the solution desired.

Messiter indicated that there is a certain difficulty in carrying out the solution in this case, in matching the uniform attached shock near the leading edge to a nonuniform solution near the axis. The origin of the difficulty is here evident, and is the highly discontinuous behavior the solution necessarily has. The inclusion of Bernoulli effects may be expected to smooth out the singularities, but their existence would still have to be accounted for in a uniformly valid solution for small ϵ.

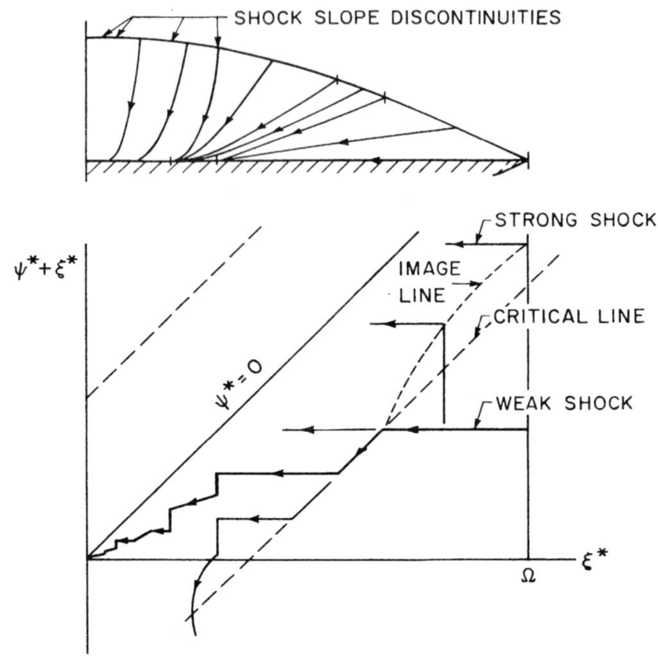

FIG. 5–18. Flat delta wing with $\Omega > 2$.

There is clearly little hope of obtaining a precise analytic solution giving the stand-off distance on the axis. However, we may expect that the nonanalytic solution is probably closely approximated sufficiently near the axis by an analytic one, one corresponding to a conical axis of symmetry solution with $0 < \kappa < 1$.

This regime of large Ω extends down to $\Omega = 2$, through the range of attached shocks. At $\Omega = 2$ the solution begins with the critical segment, and a solution pattern is obtained which is somewhat different from that shown in Fig. 5–18.

Below $\Omega = 2$ the solution regime for Ω intermediate appears. A solution in this regime is sketched in Fig. 5–19. As Ω is made less than 2, a $\psi^* = 0$ point appears near the leading edge. This point divides the flow into two parts. One flows outboard over the leading edge and satisfies the critical condition $\psi^* = 1$ at $\xi^* = \Omega$. The other part flows inboard to the $\psi^* = 0$ point on the axis of symmetry. This solution regime extends down to some value of Ω which we term $\bar{\Omega}$.

It is conceivable that another regime exists between the one just described and the one for small Ω. We assume here, as a reasonable conjecture, that this is not the case. As Ω approaches $\bar{\Omega}$ from above the $\psi_1^* = 0$ dividing point which appeared near the leading edge moves inboard till it coalesces with the

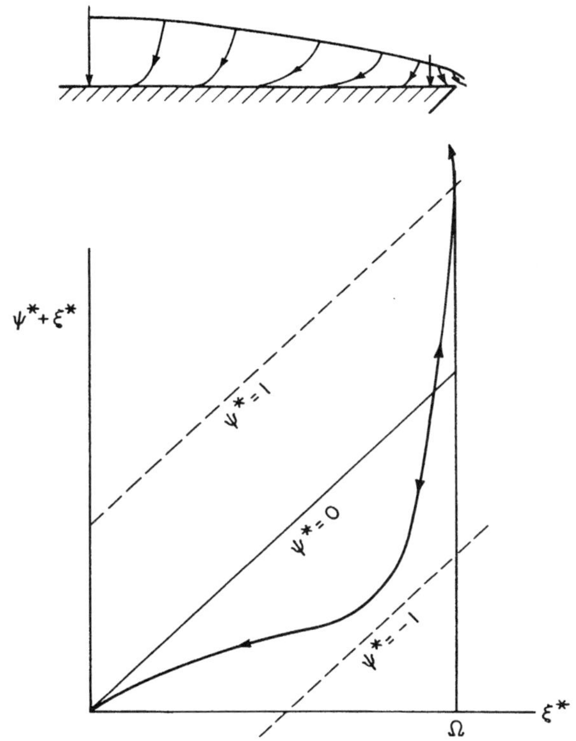

FIG. 5–19. Flat delta wing with $\bar{\Omega} < \Omega < 2$.

point on the axis. At $\Omega = \bar{\Omega}$ the flow is entirely outboard, with the solution near the axis corresponding to $\kappa = 1$. At this point the shock osculates a right circular cone at the axis of symmetry.

For $\Omega < \bar{\Omega}$, the solution is as shown in Fig. 5–20. The flow near the axis corresponds to $\kappa > 1$, and the entire flow is outward. At $\xi_1^* = \Omega$ the solution is critical, with $\psi_1^* = 1$.

As an exercise, the reader may verify that for a V-shape delta wing of appropriate shape solutions appear in this theory which correspond to the Maikapar-Nonweiler solutions discussed in Section 4.1. Further, the Maikapar-Nonweiler solutions with weak shocks on both leading edges form a boundary case between solution regimes of different types. V-shape delta wings with smaller negative dihedral have solutions of the type possessed by flat delta wings with $\Omega > 2$, without supercritical to subcritical jumps. V-shape delta wings with larger negative dihedral correspond, with weak attached shocks, to Newtonian solutions requiring a Newtonian shock line and thus are not covered by the present theory. Physically, these would correspond to solutions with lambda shocks.

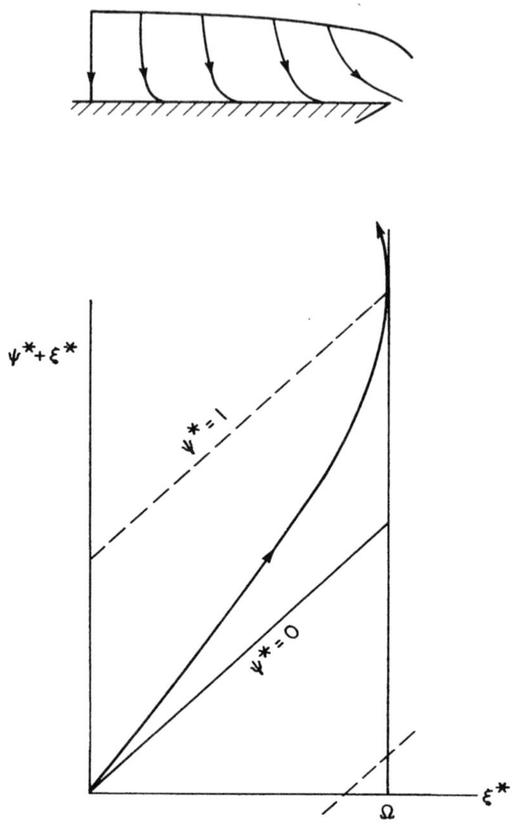

FIG. 5–20. Flat delta wing with $0 < \Omega < \bar{\Omega}$.

(d) Newtonian vortical layer. The solutions we have been studying neglect the Bernoulli effect and involve only an outer layer. In general, the Bernoulli effect with ϵ small is weak in conical flows, because the outer flow has no regions of very low velocity except in certain very special cases. Such a case would appear where locally $\cos \sigma$ in the theory of Section 3.7 is very small. However, there are regions of very low cross velocity in conical flows, in particular near where Newtonian cross streamlines terminate on the body or on a discontinuity surface. The Bernoulli effect is significant in these regions in altering the cross velocity even though ϵ is very small. This effect appears in a thin layer termed a vortical layer, and is important only with respect to the details of the cross-flow structure. The influence of the vortical layer on pressure distributions and shock layer thickness is weak.

There are two quite distinct types of vortical layers in conical flows. The type first discovered by Ferri [1; 3, Art. 15] appears on conical bodies which are slightly perturbed right circular cones, without the condition that ϵ be small. Such a perturbation appears, for example, if the cone is placed at a small angle of attack or is distorted into an elliptic cone of small ellipticity. This type of vortical layer is not characteristic of thin shock layer theory and will be discussed in Section 7.4.

A vortical layer of the type here considered will be termed a Newtonian vortical layer to distinguish it from one of the Ferri type. A combined Newtonian-Ferri vortical layer arises if both ϵ and the deviation from circularity of the cross section are small. Guiraud [4] estimated the thickness of the combined layer. Cheng [2] presented a double expansion in ϵ and angle of attack for the outer flow, and made a separate treatment for the combined vortical layer which results in this case. Sapunkov [1] corrected Cheng's solution for the outer flow, and [2] presented an alternative treatment of the vortical layer based on coordinate distortion by the PLK method.

On the problem of the Newtonian vortical layer on general cones, Melnik and Scheuing [1] have made a comprehensive study. They apply two methods, the classical boundary layer method for the vortical layer structure and the PLK method based on coordinate deformation. Our approach will be close to their boundary layer approach. Melnik [2] has presented an extension of his earlier work with Scheuing, with the aim of finding uniformly valid solutions. Bulakh [2] has treated the problem, and in particular made a careful study of the essential nature and limitations of expansions in ϵ with respect to the Newtonian vortical layer. Laval [3] has treated the problem using expansions in which a reduced stream function is used as an auxiliary independent variable.

Our approach is that of successive approximations, consistent with the approach described in Section 5.2. We assume that a Newtonian solution or constant-density solution without Bernoulli effect (outer layer) is available for our lowest approximation. We use the value of ξ at entry as an independent variable which serves as a stream function. The pressure distribution is then given as a function of ξ and ξ_1, although only to the point $\xi_1 = \psi + \xi$ for each cross streamline. For values of ξ_1 beyond the Newtonian terminal point the pressure on the body (or on a discontinuity surface, if appropriate) is taken. With the pressure field given for each cross streamline its geometry is recalculated with the Bernoulli effect properly taken into account. Continuity is used to recalculate the physical coordinate scale across the layer and the total shock layer thickness. The new thickness gives a shock shape perturbation in a direct problem and a body shape perturbation in an inverse problem. New pressure distributions may be calculated and the process repeated. With ϵ small the structure of the vortical layer may be obtained to lowest order by a simple procedure.

One result will be that the vortical layer may be considered to be divided into two sublayers. One of these, termed the inner layer by Melnik and Scheuing

[1], is a very thin layer next to the wall in which the cross-flow velocity is in a kind of equilibrium with the transverse pressure gradient. The other sublayer, termed the transitional layer by Melnik and Scheuing [1], is a layer in which the transition from the outer layer solution to the inner layer solution occurs. The vortical layer is made up of the transitional layer plus an inner layer.

We first examine the dynamics of the flow entering a thin conical shock layer at a given value of ξ. The simplifications of thin-shock-layer theory are used, so that, for example, the factor $\mathscr{H}$ appearing in Section 5.1 is set equal to one. The pressure is assumed known as a function of ξ_1. A new dependent variable β is introduced, the inclination of the streamline relative to a conical ray. See Fig. 5–21. This variable equals ψ at $\xi_1 = \xi$.

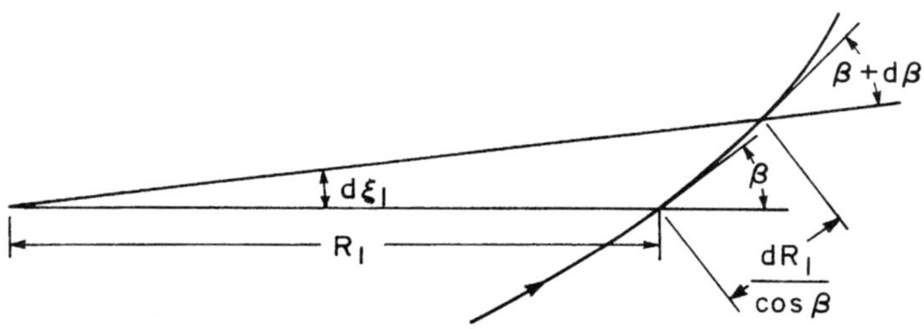

FIG. 5–21. Conical streamline geometry.

The differential equation for the streamline is

$$\frac{R_1\, d\xi_1}{dR_1} = \tan \beta. \tag{5.6.37}$$

The entropy $S(\xi)$ is known from entry conditions, so that the pressure $p(\xi_1, \xi)$ with the equation of state yields the density $\rho(\xi_1, \xi)$. The complete Bernoulli equation also yields the total velocity $q(\xi_1, \xi)$. A function $g(\xi_1, \xi)$ is defined by the relation

$$\epsilon g(\xi_1) = -\frac{1}{\rho q^2 U^2} \frac{dp}{d\xi_1}, \tag{5.6.38}$$

in which ξ is treated as a fixed parameter and ϵ is a fixed reference value of the density ratio. As before qU is the velocity component parallel to the body in the shock layer. The function g is a dimensionless pressure gradient, in which the reduction has been made relative to the dynamic pressure for a particular streamline. For a circular cone this dimensionless pressure gradient g at any point off the plane of symmetry is of the order of the angle of attack, and proportional to it if the angle of attack is much smaller than the cone half-angle. In a general case, as long as $g(\xi_1)$ is a smooth function, it is of the order of ψ or of $\xi_1 - \xi_2$.

The streamline curvature is given by

$$K = -\frac{1}{\rho q^2 U^2}\frac{\partial p}{\partial n} = \frac{\epsilon g(\xi_1)\cos\beta}{R_1}, \tag{5.6.39}$$

where $\partial p/\partial n$ refers to the component of the pressure gradient normal to the streamline. The change in β resulting from a change in ξ_1 may be expressed (see Fig. 5–21) by

$$d\beta = \frac{K\,dR_1}{\cos\beta} - d\xi_1\,. \tag{5.6.40}$$

We may now substitute (5.6.39) and eliminate dR_1 through (5.6.37). The result is

$$\frac{d\beta}{d\xi_1} + 1 = \epsilon g(\xi_1)\cot\beta. \tag{5.6.41}$$

Integration of this equation yields, with quadrature of (5.6.37), the streamline shape.

If ϵ is small we may simplify (5.6.41). If we drop the right-hand side the integral is simply

$$\beta + \xi_1 = \mathit{constant} = \psi + \xi, \tag{5.6.42}$$

the Newtonian result, with an error of $O(\epsilon g)$ in β as long as β is not small. Recall that in this case the range of integration $\xi_1 - \xi_2$ is of the order of g. This approximate solution fails for β small, because of the behavior of $\cot\beta$. For an approximate solution we must approximate $\epsilon g\cot\beta$ accurately where β is small, or where $\psi + \xi - \xi_1$ is small. We replace $g(\xi_1)$ by the constant $g(\psi + \xi)$ and $\cot\beta$ by $1/\beta$. This approximation satisfies our requirements as long as $g(\xi_1)$ is a smooth function. In the outer layer, with β not small, the error in g is then of $O(g'g)$ and the error in β is $O(\epsilon g^2 g')$. Here g' is a measure of the derivative $dg/d\xi_1$, which is also assumed smooth. With this approximation we may integrate (5.6.41) and obtain

$$\beta + \xi_1 + \epsilon g\ln(\beta - \epsilon g) = \psi + \xi + \epsilon g\ln(\psi - \epsilon g), \tag{5.6.43}$$

with the boundary condition $\beta(\xi, \xi) = \psi$ imposed. With β not small this satisfies (5.6.42) with an error of $O(\epsilon g^2 g')$. With $\beta - \epsilon g$ not small but β small the solution is for what Melnik and Scheuing term the transitional layer. With $\beta \approx \epsilon g$ the approximate solution

$$\beta - \epsilon g = (\psi - \epsilon g)\exp\left(\frac{\psi + \xi - \xi_1}{\epsilon g}\right) \tag{5.6.44}$$

holds, in what Melnik and Scheuing term the inner layer. A more refined approximate solution for the inner layer, valid with $g(\xi_1)$ variable and with an error of $O(\epsilon^3)$ times a function of ξ_1, may be shown to be

$$\beta - \epsilon g - \epsilon^2 g g' = \frac{A(\xi)}{g} \exp\left(-\int \frac{d\xi_1}{\epsilon g}\right). \tag{5.6.45}$$

We do not derive this result here, and use it only for the purpose of comparing it with (5.6.44). This equation reduces to (5.6.44) if g is constant, and shows that (5.6.44) is a good approximation if g'/g is $O(1)$.

In treating the transitional layer, it is convenient to introduce a function $t(x)$ defined by

$$x = t + \ln t. \tag{5.6.46}$$

In terms of this function we can solve (5.6.43) explicitly for β, as

$$\beta = \epsilon g + \epsilon g t \left(\frac{\psi + \xi - \xi_1}{\epsilon g} + \ln\left(\frac{\psi}{\epsilon g} - 1\right) - 1\right). \tag{5.6.47}$$

A plot of this transitional function is given in Fig. 5–22.

For an approximate solution for R_1 we replace $g(\xi_1)$ in (5.6.41) by $g(\psi + \xi)/\cos\beta$, and replace $d\xi_1$ by dR_1 through (5.6.37). The resulting solution is

$$R_1(\sin\beta - \epsilon g) = R(\sin\psi - \epsilon g). \tag{5.6.48}$$

Again, this fits the Newtonian result with an error of $O(\epsilon g^2)$ in R/R_1 as long as β is not small. The fact that we have used slightly different approximations for $g(\xi_1)$ in obtaining (5.6.43) and (5.6.48) will not affect our analysis.

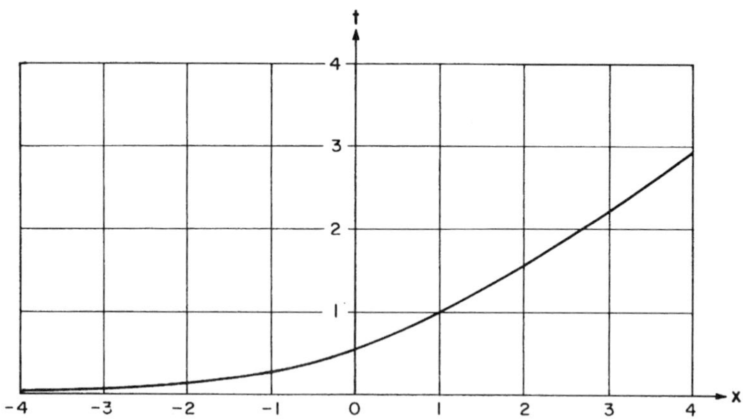

FIG. 5–22. Transitional function.

To obtain the scale we apply the principle of continuity. The mass flow entering the shock layer between ξ and $\xi + d\xi$ and between the vertex and R is $\rho_\infty U \sin\sigma$ times the area $\frac{1}{2}R^2\, d\xi$. At a different point in the shock layer this mass flow equals $\rho q U \sin\beta$ times the area $\frac{1}{2}R_1\, dy$. We replace $\rho\, dy$ by $\rho_\infty\, dm$ and apply (3.7.7) to obtain

$$\frac{dm}{R_1} = \frac{\sin\vartheta\cos\alpha}{q\sin\beta}\left(\frac{R}{R_1}\right)^2 d\xi. \tag{5.6.49}$$

Neglecting errors of $O(\epsilon)$ we may replace q by its value $\cos\vartheta/\cos\psi$ from (3.7.11). Replacing $\tan\vartheta\cos\alpha$ by $\sqrt{\tan^2\vartheta - \tan^2\psi}$ from (3.7.10), we rewrite (5.6.49) as

$$\frac{dm}{R_1} = \frac{\sqrt{\tan^2\vartheta - \tan^2\psi}\,\cos\psi}{\sin\beta}\left(\frac{R}{R_1}\right)^2 d\xi. \tag{5.6.50}$$

If we replace β by its Newtonian value $\psi + \xi - \xi_1$ and R/R_1 by its value from (3.7.13) we obtain agreement with (3.7.15).

In the outer layer the integral for m is the same as in Newtonian flow, with an error of $O(\epsilon)$. In the vortical layer the integral takes the form, at $R_1 = 1$,

$$dm = \frac{\sqrt{\tan^2\vartheta - \tan^2\psi}\,\cos\psi}{(\sin\psi - \epsilon g)^2}\,\frac{(\beta - \epsilon g)^2}{\beta}\, d\xi, \tag{5.6.51}$$

with $\beta(\xi_1, \xi)$ to be obtained from (5.6.43). For the inner layer, we restrict ψ to be small in order to simplify the expression and apply (5.6.44). In place of (5.6.51) we obtain

$$dm = \frac{\tan\vartheta}{\epsilon g}\exp\left(2\,\frac{\psi + \xi - \xi_1}{\epsilon g}\right) d\xi. \tag{5.6.52}$$

In integrating dm to obtain the shock layer structure we hold ξ_1 fixed and vary ξ. We must first establish the integration limits for ξ. In the Newtonian case the integration range is from $\xi = \xi_2$ to $\xi = \xi_1$, where the cross streamline entering at ξ_2 terminates at ξ_1 (excluding leading edges). The cross streamlines flow in the direction of decreasing pressure. In the vortical layer cross streamlines do not terminate until they reach a minimum of pressure. The integration range then is from ξ_{max} to ξ_1, where ξ_{max} is the point upstream in the cross flow for which the pressure is at a maximum. If ξ_2 is at a leading edge no vortical layer is present. The body is always wetted by fluid from a maximum pressure point or from a leading edge.

As we have seen, the range of ξ contributing to the outer layer integration is of the order of g. The integrand in (5.6.50) is of the order of $\tan\vartheta/\sin\beta$ or of the order of $\tan\vartheta/g$. Thus m_1 is of the order of $\tan\vartheta$. The thickness of the layer is of the order of ϵm_1 or of $\epsilon\tan\vartheta$. This estimate is in accord with the results for the circular cone in Section 4.2.

For the transitional layer, the argument of the function t in (5.6.47) has a range of the order of 1, so that ξ has a range of the order of ϵg. Both β and $\beta - \epsilon g$ are of the order of ϵg, while $\sin \psi$ is still of the order of g. The integrand of (5.6.51) is of the order of $\epsilon \tan \vartheta / g$, and the portion of m_1 attributable to the transitional layer is of the order of $\epsilon^2 \tan \vartheta$. Thus the transitional layer has a thickness which is of the order of ϵ^2 times the total shock layer thickness.

The range of ξ contributing to the inner layer is of the order of $\xi_2 - \xi_{max}$, which we take to be of the order of one. The order of magnitude of the integrand in (5.6.52) depends on the criterion used for defining the outer boundary of the inner layer. We choose $\psi + \xi - \xi_1 = -a$, where a is of the order of one and has the same sign as does g. The integral of (5.6.52) is then of the order of $\tan \vartheta \exp(-2a/\epsilon g)$. The thickness of the inner layer is then of the order of $\exp(-2a/\epsilon g)$ times the total shock layer thickness.

These estimates show that the inner layer is much thinner than the transitional layer, which is in turn much thinner than the total shock layer. A sketch of a typical cross-streamline pattern is shown in Fig. 5–23, and shows qualitatively the different layers involved. The dotted lines indicate the streamlines as they would be without the Bernoulli effect in the vortical layer.

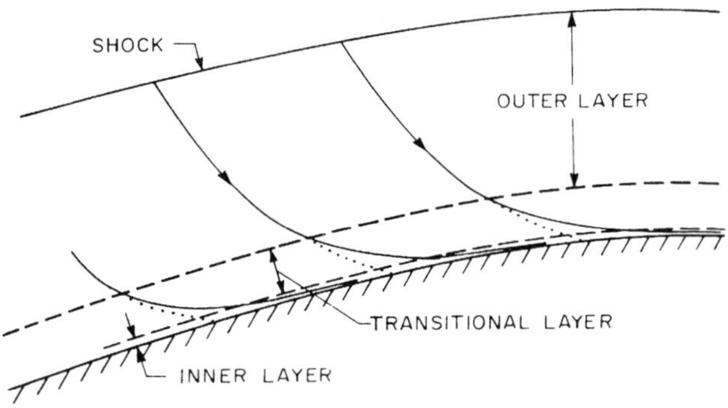

FIG. 5–23. Vortical layer structure.

The change in total shock layer thickness resulting from Bernoulli effects may be of either sign. This change is dominated by the Bernoulli effects in the outer layer, which give a change of $O(\epsilon)$ in m_1 and thus a change of $O(\epsilon^2)$ in shock layer thickness. These effects in the outer layer are generally of the same order of magnitude as compressibility effects, and have been neglected in our study of the vortical layer. The presence of the transitional layer provides a change in shock layer thickness of $O(\epsilon^3)$, and the influence of the inner layer is still weaker.

It is to be emphasized that the vortical layer in conical flows is a far different entity than the inner layers of the type appearing in Sections 4.3 to 4.5 and in Section 5.4. In those cases the Newtonian velocity on the body is zero, and the Bernoulli effect near the body is essential in the two-dimensional case or relatively strong in the axisymmetric case. In those cases the total Bernoulli effect from the inner layer is stronger than from the outer layer. In the conical case the total Bernoulli effect from the vortical layer is weaker than from the outer layer.

Near a maximum or minimum pressure point the analysis leading to (5.6.43) is not valid. That analysis was based on the assumption that g could be approximated by a constant. We define the ξ scale so that $\xi = 0$ at the pressure extremum point in question, and approximate g by

$$g(\xi_1) = g'\xi_1 , \tag{5.6.53}$$

with g' considered constant. An approximate solution to (5.6.41) in this case, with $\cot\beta$ approximated by $1/\beta$, is

$$(\beta + \xi_1)^{1-\epsilon g'}(\beta - \epsilon g'\xi_1)^{\epsilon g'} = (\psi + \xi)^{1-\epsilon g'}(\psi - \epsilon g'\xi)^{\epsilon g'}. \tag{5.6.54}$$

A corresponding approximate solution for R_1 is

$$R_1 \frac{\beta - \epsilon g'\xi_1}{\beta + \xi_1} = R \frac{\psi - \epsilon g'\xi}{\psi + \xi} . \tag{5.6.55}$$

The additional approximation made in obtaining both (5.6.54) and (5.6.55), besides those that g' is constant and β is small, consists of replacing $\sqrt{1 + 4\epsilon g'}$ by $1 + 2\epsilon g'$ or $(1 - 2\epsilon g')^{-1}$ after carrying out the quadrature.

Near a pressure maximum $g' > 0$, and $\psi = (\kappa - 1)\xi$ with $\kappa > 0$. The parameter κ is the parameter for the axis of symmetry of Sections 3.6 and 3.7. The quantity g' has the same magnitude as $\kappa - 1$. In place of (3.7.27) we may write, for $R_1 = 1$,

$$m_1 = \frac{\tan\vartheta_0}{(\kappa - 1 - \epsilon g')^2} \int_0^{\xi_1} \frac{(\beta - \epsilon g'\xi)^2}{\xi^2\beta} \left(\frac{\psi + \xi}{\beta + \xi_1}\right)^2 d\xi. \tag{5.6.56}$$

With $\beta(\xi_1 , \xi)$ taken from (5.6.54) this may be integrated to obtain the Howarth-Dorodnitsyn shock layer thickness. The transitional layer has a thickness of $O(\epsilon^3 \tan\vartheta_0)$. Only a weak inner layer appears in this case. Near a pressure minimum g' and $(\kappa - 1)$ are negative, and β is generally negative if ξ is positive. The same equation (5.6.56) holds, with certain changes. We replace the limits of integration 0 to ξ_1 by ξ_{max} to ξ_1, where ξ_{max} is the value of ξ at the next pressure maximum. The limits may be changed to ξ_1 to ξ_{max} by replacing β

by the positive quantity $-\beta$. The simplifications made in the integral, for example that of putting $\psi = (\kappa - 1)\xi$ in the denominator of the integrand, cannot be made over the entire range of ξ. These simplifications are valid in the transitional layer and the outer layer, and fail only for the inner layer. If the contribution of the inner layer to m_1 is neglected, (5.6.56) with the altered limits holds through the transitional layer. The Newtonian variable $\xi_2 = \xi_1 + (1 - \kappa)\xi_1/\kappa$ in this case. In place of ξ_{max} we should put some value far enough beyond ξ_2 from ξ_1 to encompass the transitional layer. The value $\xi_1 + 2(1 - \kappa)\xi_1/\kappa$ would have this property.

Melnik [1] has investigated the problem of the symmetry axes with the aim of expressing a composite solution valid in all regions. He used a PLK method of distorted coordinates.

At a pressure minimum on the body the point is a vortical singularity of the type discussed by Ferri [1; 3, Art. 15]. This type of singularity will be discussed in Section 7.4 when we consider conical flows in more generality. The more anomalous properties of this point in a thin shock layer come from the inner layer, which is very thin and is made up of material which came from far away. The contribution of the inner layer to the shock layer thickness or to other gross properties of the layer is generally negligible.

In general, the inner layer does not in practice affect either the shock layer thickness or the pressure field. The inner layer can have an essential effect on the boundary layer on a conical body, because the boundary layer grows first where the inner layer should be. The boundary layer has a thickness on a cone proportional to $R^{1/2}$, and so prevents any inner layer from forming near the vertex. If an inner layer is to appear in a conical flow in a real gas, it must form itself around the boundary layer at large distances downstream. In summary, we should not take the inner layer too seriously, even though it is a clear-cut entity in inviscid thin-shock-layer theory. The transitional layer is much more likely to be important.

7. General blunt-faced bodies and related similitudes

In this section we consider the entire class of blunt-faced bodies in thin-shock-layer theory. Here again, a blunt-faced body refers to a body whose shape closely approximates a smooth reference surface which is at a constant angle of incidence α. With the shock layer thin, the shock shape also closely approximates the same reference surface. The body is characterized by a reference length c in the direction of what has been chosen to be the principal direction of flow along the body in the shock layer. The lateral scale of the body is characterized by a reference length Ac. In those special cases for which $A \gg 1$ we will relabel A as A'. The deviation of the body from the reference surface is assumed to be of the order of the shock layer thickness.

Our purpose in this analysis will be to develop a family of related similitudes for flows of the type considered. These similitudes will encompass those mentioned earlier in this chapter. To construct the appropriate theory we use concepts and estimates of orders of magnitude from these previous sections. Then, by means of various limiting processes, we show how various theories considered earlier come out as special cases.

We will first develop the theory with the Bernoulli effect neglected, and deduce similitudes for those cases in which this simplification is appropriate. We then examine the inner layer to determine what modifications are needed when the inner layer must be taken into account.

For convenience in the analysis we treat only the case for which the reference surface is a plane. The extension to other cases is straightforward, but the analysis would be more involved. Again, we treat the flow as a constant-density flow with ϵ constant and compressibility neglected. We could equally well set up an expansion scheme without these simplifying assumptions and derive the constant-density equations as those appropriate for the lowest order approximation.

(a) Basic theory. The coordinate system used is the cartesian system shown in Fig. 5–24. It is of boundary layer type. The x direction (with velocity component u) is the chosen principal direction of flow along the body. The z direction (with velocity component w) is the direction of flow across the body. The y direction (with velocity component v) is the direction normal to the reference surface, positive into the body.

The key to the analysis will be the proper estimation of the orders of magnitude of the various flow variables. This estimation is obtained through an estimate of the shock layer thickness. If the body is narrow, the theories of Sections 5.3, 5.4, and 5.6 tell us Δ is of the order of $\epsilon^{1/2}\mathcal{R}c$. If the body is wide and at

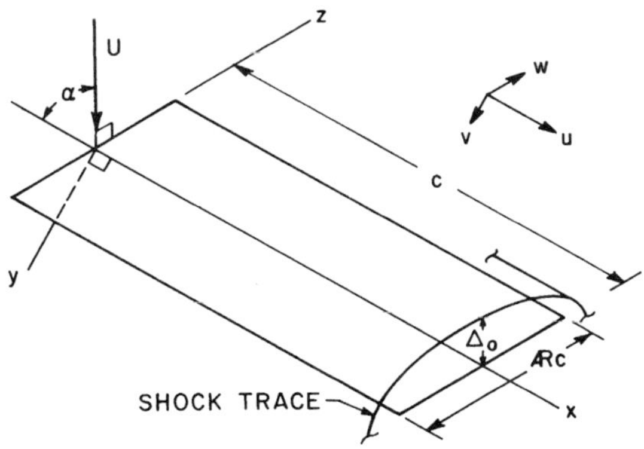

FIG. 5–24. Planar blunt-faced body.

nonnegligible incidence, the theories of Sections 4.1 and 5.5 tell us Δ is of the order of $\epsilon c \tan \alpha$. If the body is wide and at negligible incidence, the theory of Section 5.4 again tells us Δ is of the order of $\epsilon^{1/2}c$. The actual shock layer thickness is of the order of the least of these three quantities. We introduce the Messiter similarity parameter Ω, defined by

$$\Omega = \frac{AR \cot \alpha}{\epsilon^{1/2}}, \tag{5.7.1a}$$

and an additional parameter Λ given by

$$\Lambda = 1 + \Omega + AR. \tag{5.7.1b}$$

The reference value of the shock layer thickness is denoted $\Delta_0 c$, with Δ_0 defined as

$$\Delta_0 = \frac{\epsilon^{1/2} AR}{\Lambda} = \frac{\Omega \epsilon \tan \alpha}{\Lambda} = \frac{AR^2 \cot \alpha}{\Omega \Lambda}. \tag{5.7.2}$$

It may be checked that $\Delta_0 c$ does have the property of being of the order of magnitude of the least of $\epsilon^{1/2} AR c$, $\epsilon c \tan \alpha$, and $\epsilon^{1/2}c$.

We may now estimate the order of magnitude of the velocity components. The component v is of the order of $U\epsilon \sin \alpha$, while w is of the order of $U \sin \alpha$ times the slope Δ_0/AR. The quantity $u - U \cos \alpha$ is of the order of $U \sin \alpha$ times the slope Δ_0. With these estimates are the scale estimates to be used in transforming the coordinate variables to obtain correct orders of magnitudes for gradient terms. The scale estimates are simply $(c, \Delta_0 c, AR c)$.

The pressure perturbation from the full value $p_\infty + \rho_\infty U^2 \sin^2 \alpha$ at $\epsilon = 0$ on the reference surface is governed either by the shock angle perturbation for the principal flow or by the critical point pressure for the lateral flow. The order of magnitude estimate is the same in either case, equal to $\rho_\infty U^2 \epsilon \sin^2 \alpha$.

Based upon the estimates obtained above, we make a transformation of both dependent and independent variables. For the velocity we set

$$u = U \cos \alpha + \Delta_0 U \sin \alpha \, u^*, \tag{5.7.3a}$$

$$v = U\epsilon \sin \alpha \, v^*, \tag{5.7.3b}$$

$$w = \Delta_0 U \sin \alpha \, w^*/AR. \tag{5.7.3c}$$

For the pressure we set

$$p = p_\infty + \rho_\infty U^2 \sin^2 \alpha + \rho_\infty U^2 \epsilon \sin^2 \alpha \, (p^* - 1). \tag{5.7.4}$$

It should be noted that our $p^* - 1$ is the p^* used by J. D. Cole and his associates in similar developments.

The coordinates are transformed according to

$$x = cx^*, \tag{5.7.5a}$$

$$y = \Delta_0 cy^*, \tag{5.7.5b}$$

$$z = Ꝃcz^*. \tag{5.7.5c}$$

The density is the constant ρ_∞/ϵ and requires no transformation.

Equations (5.7.2) to (5.7.5) may be put into the equations of motion for a fluid of constant density ρ_∞/ϵ. The continuity equation takes the form

$$\frac{Ꝃ^2}{\Omega\Lambda}\frac{\partial u^*}{\partial x^*} + \frac{\Lambda}{\Omega}\frac{\partial v^*}{\partial y^*} + \frac{1}{\Omega\Lambda}\frac{\partial w^*}{\partial z^*} = 0. \tag{5.7.6}$$

For the momentum equations, it is convenient to introduce the operator D defined by

$$D = \left(1 + \frac{Ꝃ^2}{\Omega\Lambda}u^*\right)\frac{\partial}{\partial x^*} + \frac{\Lambda}{\Omega}v^*\frac{\partial}{\partial y^*} + \frac{1}{\Omega\Lambda}w^*\frac{\partial}{\partial z^*}. \tag{5.7.7}$$

In terms of this operator the momentum equations read

$$Du^* + \left(\epsilon\frac{\Lambda}{\Omega}\frac{\partial p^*}{\partial x^*}\right) = 0, \tag{5.7.8a}$$

$$Dv^* + \frac{\Lambda}{\Omega}\frac{\partial p^*}{\partial y^*} = 0, \tag{5.7.8b}$$

$$Dw^* + \left(\epsilon\frac{\Lambda}{\Omega}\frac{\partial p^*}{\partial z^*}\right) = 0. \tag{5.7.8c}$$

Where the Bernoulli effect is neglected, the parentheses of (5.7.8) are dropped, and (5.7.8b) is used to obtain the pressure by quadrature.

The body shape is given by an equation of the form

$$y_b = \tau^*\Delta_0 c\, f(x/c,\, z/Ꝃc). \tag{5.7.9}$$

The direction cosines of the normal to the body are given by

$$\{\mathbf{n}_b\} = \{-\tau^*\Delta_0 f_1,\ \sqrt{1 - \tau^{*2}\Delta_0^2 f_1^2 - \tau^{*2}\Delta_0^2 f_2^2/Ꝃ^2},\ -\tau^*\Delta_0 f_2/Ꝃ\}, \tag{5.7.10}$$

where

$$f_1 = \frac{\partial f}{\partial x/c}, \tag{5.7.11a}$$

$$f_2 = \frac{\partial f}{\partial z/Ꝃc}. \tag{5.7.11b}$$

The boundary condition on the body is the condition that the velocity be tangent to the body. The condition may be expressed in the form

$$\frac{\Lambda}{\Omega} v_b^* = \tau^* \left[\left(1 + \frac{Æ^2}{\Omega\Lambda} u_b^* \right) f_1 + \frac{1}{\Omega\Lambda} w_b^* f_2 \right], \tag{5.7.12}$$

with terms of higher order in Δ_0 dropped.

The shock shape is assumed to be given by an equation similar to (5.7.9), by

$$y_s = \Delta_0 c\, g(x/c, z/Æc). \tag{5.7.13}$$

The direction cosines of the normal to the shock surface are given by (5.7.10) with $\mathbf{n}_b$ replaced by $\mathbf{n}_s$, τ^* by 1, and f replaced by g. Analogously, we set

$$g_1 = \frac{\partial g}{\partial x/c}, \tag{5.7.14a}$$

$$g_2 = \frac{\partial g}{\partial z/Æc}. \tag{5.7.14b}$$

The shock conditions for the principal and lateral velocity components appear in the simple form

$$u_s^* = g_1, \tag{5.7.15a}$$

$$w_s^* = g_2. \tag{5.7.15b}$$

The velocity component normal to the reference surface may be shown to satisfy the relation immediately behind the shock

$$\frac{\Lambda}{\Omega} v_s^* = \frac{\Lambda}{\Omega} + g_1 + \frac{Æ^2}{\Omega\Lambda} g_1^2 + \frac{1}{\Omega\Lambda} g_2^2. \tag{5.7.16}$$

For the pressure immediately behind the shock we note first that $\sin\sigma$ is obtainable as the cosine of the angle between the normal to the shock and the flow direction. We obtain with the help of direction cosines

$$\sin\sigma = \sin\alpha \sqrt{1 - \Delta_0^2 g_1^2 - \Delta_0^2 g_2^2/Æ^2} - \cos\alpha\, \Delta_0 g_1. \tag{5.7.17}$$

With this result the quantity p^* behind the shock may be readily shown to be

$$p_s^* = -2\frac{\Omega}{\Lambda} g_1 - \frac{Æ^2}{\Lambda^2} g_1^2 - \frac{1}{\Lambda^2} g_2^2 + O(\epsilon). \tag{5.7.18}$$

(b) Constant-streamtube-area similitudes. We now are in a position to discuss those problems for which the constant-streamtube-area approximation is valid, for which the Bernoulli effect may be neglected. The general question of validity of this approximation we shall look at later. Neglecting the Bernoulli effect in this thin-layer theory amounts simply to dropping the two pressure gradient terms in (5.7.8) which bear the coefficient ϵ. With these terms dropped the theory presented above is independent of ϵ. If the parameters τ^*, Ω, and $Æ$ are fixed, with the body shape function $f(x^*, z^*)$ fixed, the theory above is invariant under a transformation in which ϵ is changed. This yields a general constant-streamtube-area similitude with similarity parameters τ^*, Ω, and $Æ$.

For the direct problem of determining the flow on a given body, we can give no general method now. For the inverse problem in which the shock shape $g(x^*, z^*)$ is given, we can present an algorithm for the solution. With g given the shock conditions yield values for all the dependent variables behind the shock. We obtain u^* and w^* everywhere as functions of the coordinates at entry from (5.7.8a, c). A Newtonian analysis of the type presented in Section 3.6 gives an m scale, which in turn gives the y^* scale at each point. The component v^* may be obtained by a quadrature of the continuity equation (5.7.6), and p^* by a quadrature of (5.7.8b). The body shape $\tau^* f$ is obtained either by integrating the boundary condition (5.7.12) from a leading edge or by locating the body where the y^* scale ends with all three velocity components zero.

We next consider a number of special cases obtained by limiting processes. These correspond to special theories and special similitudes which encompass the blunt-faced body theories and similitudes obtained in earlier sections. Instead of writing all the equations out in full in each case, we exhibit the relative values of the four constant coefficients appearing in the operator D defined in (5.7.7). In the general case these coefficients are

$$\{D\} = \left\{1, \frac{Æ^2}{\Omega\Lambda}, \frac{\Lambda}{\Omega}, \frac{1}{\Omega\Lambda}\right\}. \tag{5.7.19}$$

The last three of these are the coefficients appearing in the continuity equation (5.7.6). All four of the coefficients appear in the body boundary condition (5.7.12), and again in the shock condition for v_s^*, (5.7.16). Thus these coefficients are all that are needed to determine the form of all the equations in the theory.

A table of these coefficients in the special cases considered is presented in Table 5–1. There are three principal special cases, labeled A, B, and C, each giving a similitude. In the table the coefficients which are to be dropped are placed in large parentheses. In any one of the special theories the terms corresponding to these coefficients are to be set equal to zero. They are retained in parentheses in the table because they serve as error estimates and reappear in some cases in the corresponding inner layer analysis.

In order to show the nature of the bodies for which the special cases apply, we present Fig. 5–25. The lines with arrows indicate fall lines on the shock

TABLE 5–1

Constant-streamtube-area similitudes

	Similitude	Condition	Λ	Similarity parameter (besides τ^*)	Relative coefficient in D of: $\frac{\partial}{\partial x^*}$	$u^* \frac{\partial}{\partial x^*}$	$v^* \frac{\partial}{\partial y^*}$	$w^* \frac{\partial}{\partial z^*}$	Section, example
A	Messiter	$\mathit{AR} \ll 1 + \Omega$	$1 + \Omega$	Ω	1	$\left(\frac{\mathit{AR}^2}{\Omega + \Omega^2}\right)$	$\frac{1 + \Omega}{\Omega}$	$\frac{1}{\Omega + \Omega^2}$	5.6, conical
B	Almost normal	$\Omega \ll 1 + \mathit{AR}$	$1 + \mathit{AR}$	AR	(Ω)	$\frac{\mathit{AR}^2}{1 + \mathit{AR}}$	$1 + \mathit{AR}$	$\frac{1}{1 + \mathit{AR}}$	5.3, 5.4, axisymmetric ($\mathit{AR} = 1$)
C	Almost two-dimensional	$1 \ll \Omega + \mathit{AR}'$	$\Omega + \mathit{AR}'$	$\frac{\mathit{AR}'}{\Omega}$	1	$\frac{\mathit{AR}'^2}{\Omega \mathit{AR}' + \Omega^2}$	$\frac{\Omega + \mathit{AR}'}{\Omega}$	$\left(\frac{1}{\Omega \mathit{AR}' + \Omega^2}\right)$	5.3, 5.4, "tangent-wedge" ($\mathit{AR}' = \infty$)
AB	(invalid without inner layer)	$\Omega + \mathit{AR} \ll 1$	1	—	(Ω)	(AR^2)	1	1	—
AC	Quasi wedge	$1 + \mathit{AR} \ll \Omega$	Ω	—	1	$\left(\frac{\mathit{AR}^2}{\Omega^2}\right)$	1	$\left(\frac{1}{\Omega^2}\right)$	5.5, quasi wedge ($\mathit{AR}' = \infty$)
BC	(invalid without inner layer)	$1 + \Omega \ll \mathit{AR}'$	AR'	—	$\left(\frac{\Omega}{\mathit{AR}'}\right)$	1	1	$\left(\frac{1}{\mathit{AR}'^2}\right)$	—

surface, the lines having the direction of (u^*, w^*). These are used to indicate the general location and inclination of the shock surface.

In special case A we assume that the aspect ratio $\mathcal{R}$ is small and that Ω is of the order of one. In the resulting theory the continuity equation is two-dimensional, and x^* appears essentially as a time-like variable. The theory is

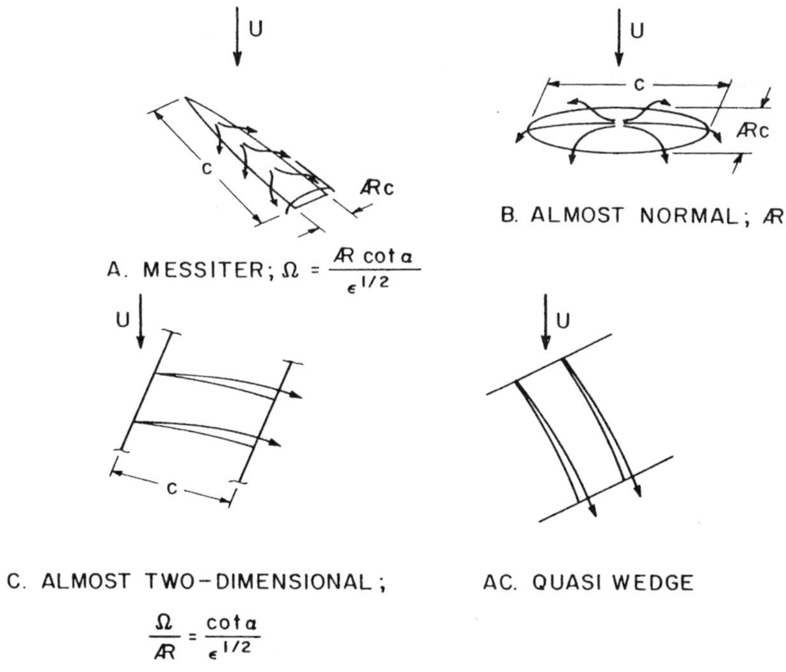

FIG. 5–25. Blunt-faced bodies.

reminiscent of that of Sychev discussed in Section 2.9. The similitude obtained has τ^* and Ω as similarity parameters, and is the similitude obtained by Messiter [1]. Messiter expressed the similitude for the flat wing case, with $\tau^* = 0$. The conical case of Messiter's similitude has already been given in Section 5.6, and forms a special subcase of similitude A. The notation and reduction of dependent variables is different, of course.

In special case B we assume that the parameter Ω is small and that $\mathcal{R}$ is of the order of one. The body in this case is almost normal to the free stream. The similitude obtained has τ^* and $\mathcal{R}$ as similarity parameters. We have gone through one example of a theory in this case, that of axisymmetric bodies in Sections 5.3 and 5.4. The similitude of subsection (g) in Section 5.4 fits similitude B with $\mathcal{R} = 1$. Note that with $\Omega = 0$ we may carry out a reflection transformation, exchanging x^* with z^* and u^* with w^* and replacing $\mathcal{R}$ by $\mathcal{R}' = 1/\mathcal{R}$.

The invariance of the equations shows that in this case there is no essential difference between the principal and lateral directions.

In special case C we assume that 1 is much smaller than Ω and $\mathcal{R}$ and that $\Omega/\mathcal{R}$ is of the order of one. In this case $\mathcal{R}$ is much greater than one and is relabeled $\mathcal{R}'$. The basic similarity parameter in this case is $\Omega/\mathcal{R}'$ or $\mathcal{R}'/\Omega$. The aspect ratio really does not enter the problem, as the parameter $\Omega/\mathcal{R}'$ is

$$\frac{\Omega}{\mathcal{R}'} = \frac{\cot \alpha}{\epsilon^{1/2}} . \tag{5.7.20}$$

Note that this parameter is essentially the inverse of the variable η of Section 4.1. When this parameter is of the order one or smaller we must keep in mind that it is not fully independent of τ^*, that the angle $\pi/2 - \alpha$ of inclination of the normal from the flow direction may be interpreted alternatively as an alteration of body shape. Thus if $\Omega/\mathcal{R}'$ is of the order of one we may consider $\Omega/\mathcal{R}'$ and τ^* together as a single parameter if desired, treating a change of inclination as a change in body shape. The body in this case is almost two-dimensional, and follows the "tangent-wedge" theory discussed in Sections 5.3 and 5.4. There is a difference in notation, as z and w of Section 5.4 are to be replaced by x and u, and the reduction of variables is different. Limitations on body angle for this theory must be satisfied, of course, and this puts a definite lower limit on the allowed values of $\Omega/\mathcal{R}'$. The variable z^* appears as a parameter, and the theory is essentially a strip theory.

A fourth special case appears, denoted AC, corresponding to the assumption that Ω is very large. The resulting theory corresponds to that given in Section 5.5 for quasi wedges, and is basically a two-dimensional one in the variables x^* and y^*. However, as with case C, z^* appears as a parameter in this case, and the theory is again essentially a strip theory. The difference is that in the quasi-wedge limit $\mathcal{R}$ does not have to be large compared with one; it must be large compared with $\epsilon^{1/2} \tan \alpha$, which must itself be small.

Formally, two other limiting special cases appear, labeled AB and BC. These both correspond to bodies which are almost two-dimensional and almost normal with respect to a cross flow. In the first case the limit is one of cylindrical flow about a body of very small $\mathcal{R}$; $\cot \alpha$ may be finite, and the flow is essentially a two-dimensional one in v^* and w^* corresponding to a free stream velocity $U \sin \alpha$. In the second case $\cot \alpha = 0$ and the body is like the almost two-dimensional one of case C; we consider the distinction to be that in this case the flow must be detached.

In these two cases a solution without taking the inner layer into account is either impossible or invalid. We require a separate treatment of the inner layer with a matching of the type discussed in Section 5.4.

Before going into the inner layer analysis we should point out that the second of these cases (case BC) is really a special case of the first (case AB). To see this we first interpret $\cot \alpha$ as an alteration of body shape, as discussed under case C above. Then we carry out a reflection transformation, as discussed under case B above. The result is then the special case AB corresponding to $\Omega = 0$ or $\cot \alpha = 0$. The $1/\text{Æ}'^2$ value of the fourth coefficient of D under BC in Table 5–1 becomes the Æ^2 value of the second coefficient under AC. Thus it is unnecessary to treat both cases, and a treatment of case AB will be sufficient.

(c) Inner layer analysis and similitudes. In the case in which both Ω and Æ^2 are small, the main flow field is one in v^* and w^*, slightly modified by the "principal" flow in the x direction. Following the lead of the analysis of Cole and Brainerd [1] described in Section 5.4, in analogy with (5.4.41) and (5.4.42), we set

$$v^* = \epsilon^{1/2}\tilde{v}, \tag{5.7.21a}$$

$$w^* = \epsilon^{1/2}\tilde{w}. \tag{5.7.21b}$$

The question remains as to how, if appropriate, the component u^* or $u^* + \Omega/\text{Æ}^2$ should be transformed.

If $\text{Æ}^2/\Omega$ is negligibly small, the second coefficient in $\text{Æ}^2/\Omega\Lambda$ is to be neglected in all the equations, and no question of transforming u^* arises. The term in Ω remains to give a time-like term in the operator D. If $\Omega/\text{Æ}^2$ is small enough that the flow in the x direction appears detached, with $u_s^* + \Omega/\text{Æ}^2 = 0$ somewhere on the shock, the value of $u^* + \Omega/\text{Æ}^2$ on and near the surface is governed by the Bernoulli relation. This would suggest the transformation

$$u^* + \frac{\Omega}{\text{Æ}^2} = \epsilon^{1/2}\tilde{u}. \tag{5.7.22}$$

This transformation is not appropriate for the inner layer, however. The key to this question may be found in a study of the quasi-cylindrical regime of the general stagnation point flow treated in Section 4.5. The dividing level between inner and outer layer is different for the main flow field (v^*, w^*) from what it is for the axial flow field (u^*). Throughout the main part of the inner layer for which (5.7.21) are appropriate we are in the outer layer for u^*, so that u^* should remain untransformed. One simplification which may be noted is that u^* in the inner layer is a function $u^*(x_e^*)$ only of the coordinate x^* at entry. Only within an inner-inner layer in which we may set $\tilde{w} = \tilde{w}_b$ is the transformation (5.7.22) appropriate.

We may now carry out the inner layer analysis for case AB. We make the transformation (5.7.21), and introduce the inner operator

$$\tilde{D} = \frac{\Omega}{\epsilon^{1/2}} D = \left(\frac{\Omega}{\epsilon^{1/2}} + \frac{\text{Æ}^2}{\epsilon^{1/2}} u^*\right) \frac{\partial}{\partial x^*} + \tilde{v} \frac{\partial}{\partial y^*} + \tilde{w} \frac{\partial}{\partial z^*}. \tag{5.7.23}$$

The continuity equation (5.7.6) becomes

$$\frac{\text{Æ}^2}{\epsilon^{1/2}}\frac{\partial u^*}{\partial x^*} + \frac{\partial \tilde{v}}{\partial y^*} + \frac{\partial \tilde{w}}{\partial z^*} = 0. \tag{5.7.24}$$

The momentum equations (5.7.8) become

$$\tilde{D}u^* = 0, \tag{5.7.25a}$$

$$\frac{\partial p^*}{\partial y^*} = 0, \tag{5.7.25b}$$

$$\tilde{D}\tilde{w} + \frac{\partial p^*}{\partial z^*} = 0. \tag{5.7.25c}$$

The body boundary condition (5.7.12) takes two forms. If $\text{Æ}^2/\Omega$ is sufficiently small that no inner-inner layer is present, it reads

$$\tilde{v}_b = \tau^* \left[\left(\frac{\Omega}{\epsilon^{1/2}} + \frac{\text{Æ}^2}{\epsilon^{1/2}} u_b^*\right) f_1 + \tilde{w}_b f_2\right], \tag{5.7.26}$$

with u_b^* a constant in general. If an inner-inner layer is present, $u^* + \text{Æ}^2/\Omega$ gives but a negligible contribution to the boundary condition, which then becomes

$$\tilde{v}_b = \tau^* \tilde{w}_b f_2 . \tag{5.7.27}$$

Although this equation corresponds to a different case, it is often helpful to consider it as a special case of (5.7.26) with the coefficient of f_1 zero. Outer conditions for the inner layer involve matching with the outer layer. In the matching region, as in the outer layer, the flow is essentially the same as a steady two-dimensional flow in v^* and w^*. The matching must follow essentially the theory of Cole and Brainerd given in Section 5.4 for two-dimensional detached flows, with x^* as a parameter. The details of the matching remain to be carried out.

The structure of the inner-inner layer is obtained, if desired, from (5.7.22) and the transformed form of (5.7.8a)

$$\tilde{\tilde{D}}\tilde{u} + \frac{\partial p^*}{\partial x^*} = 0. \tag{5.7.28}$$

Here

$$\tilde{\tilde{D}} = \tilde{v}\frac{\partial}{\partial y^*} + \tilde{w}\frac{\partial}{\partial z^*}, \tag{5.7.29}$$

and $\tilde{v}$, $\tilde{w}$, and p^* are known functions in the space.

Examination of (5.7.23) to (5.7.27) shows that they are invariant to a transformation in which ϵ changes provided $\Omega/\epsilon^{1/2}$, $A\!\!R^2/\epsilon^{1/2}$, and τ^* are held fixed. This result provides a similitude for the inner layer which extends to a similitude for the entire flow. The similitude is of the same type discussed at the end of Section 5.4(g) and fails only with respect to the matching. As with the two-dimensional separated flows of the Cole and Brainerd type the similitude applies to shock shape and to all flow variables within either outer or inner layer. The total stand-off distance, which must be obtained through the matching conditions, depends logarithmically on $\epsilon^{1/2}$ and does not follow the similitude. But the quantity $\Delta + \epsilon R_s(x^*) \ln \epsilon^{1/2}$ does follow the similitude. A modification in the specification of the body shape is required, with y_b in (5.7.9) replaced by $y_b + \epsilon R_s(x^*) \ln \epsilon^{1/2}$. Instead of modifying the body boundary conditions (5.7.26), it is preferable to skew the coordinate system in the inner layer by the angle $\epsilon \ln \epsilon^{1/2}\, dR_s/dx$. In this manner the similitude may be kept for $\tilde{v}$. The details of the matching to the outer solution will be required to confirm the similitude.

If $A\!\!R^2/\epsilon^{1/2} \ll 1$ it may be neglected, and the parameters $\Omega/\epsilon^{1/2}$ and τ^* govern the flow. This more special similitude (with $\tau^* = 0$) was suggested by Messiter [1] in the conical case. In this case the modification to (5.7.9) may be omitted, with u^* not following the similitude with this omission. An example fitting the similitude appears at the end of Section 4.5. We identify $\epsilon^{1/2}R_s/R$ appearing in (4.5.33) as $A\!\!R$ to fit the theory of this section. We neglect $\epsilon^{1/2}$ compared with 1, and can rewrite (4.5.33) as

$$\frac{\Delta}{\epsilon R_s} + \ln \epsilon^{1/2} = \ln \frac{2}{\sqrt{3}} \tanh \frac{\sqrt{3}\,\epsilon^{1/2}}{2\Omega}. \tag{5.7.30}$$

This form fits the similitude. In this example $R_s(x^*)$ is constant, and no modification of (5.7.26) or the coordinate system is required.

If $\Omega/\epsilon^{1/2} \ll 1$ it may be neglected, and the parameters $A\!\!R^2/\epsilon^{1/2}$ and τ^* govern the flow. In this case the body boundary condition (5.7.27) applies, and no modification of (5.7.26) is needed. The skew coordinate system in the inner layer may still be helpful, however. An example fitting this more special similitude is that of the quasi-cylindrical stagnation point treated in Section 4.5. We identify κ or $\kappa - 1$ as $A\!\!R^{-2}$ where κ is very large, and α_1 as $\sqrt{3}\epsilon^{1/2}$. Equation (4.5.16) may be put into the form

$$\frac{\Delta_0}{\epsilon R_s} + \ln \epsilon^{1/2} = \ln \frac{\epsilon^{1/2}}{A\!\!R^2} - \sqrt{3}\,\frac{\epsilon^{1/2}}{A\!\!R^2}\, G\left(\sqrt{3}\,\frac{\epsilon^{1/2}}{A\!\!R^2}\right), \tag{5.7.31}$$

valid in what we there termed the quasi-cylindrical range with $\epsilon^{1/2} \ll 1$. This form fits the similitude. The quantity $R_s(x^*)$ is constant in this example also.

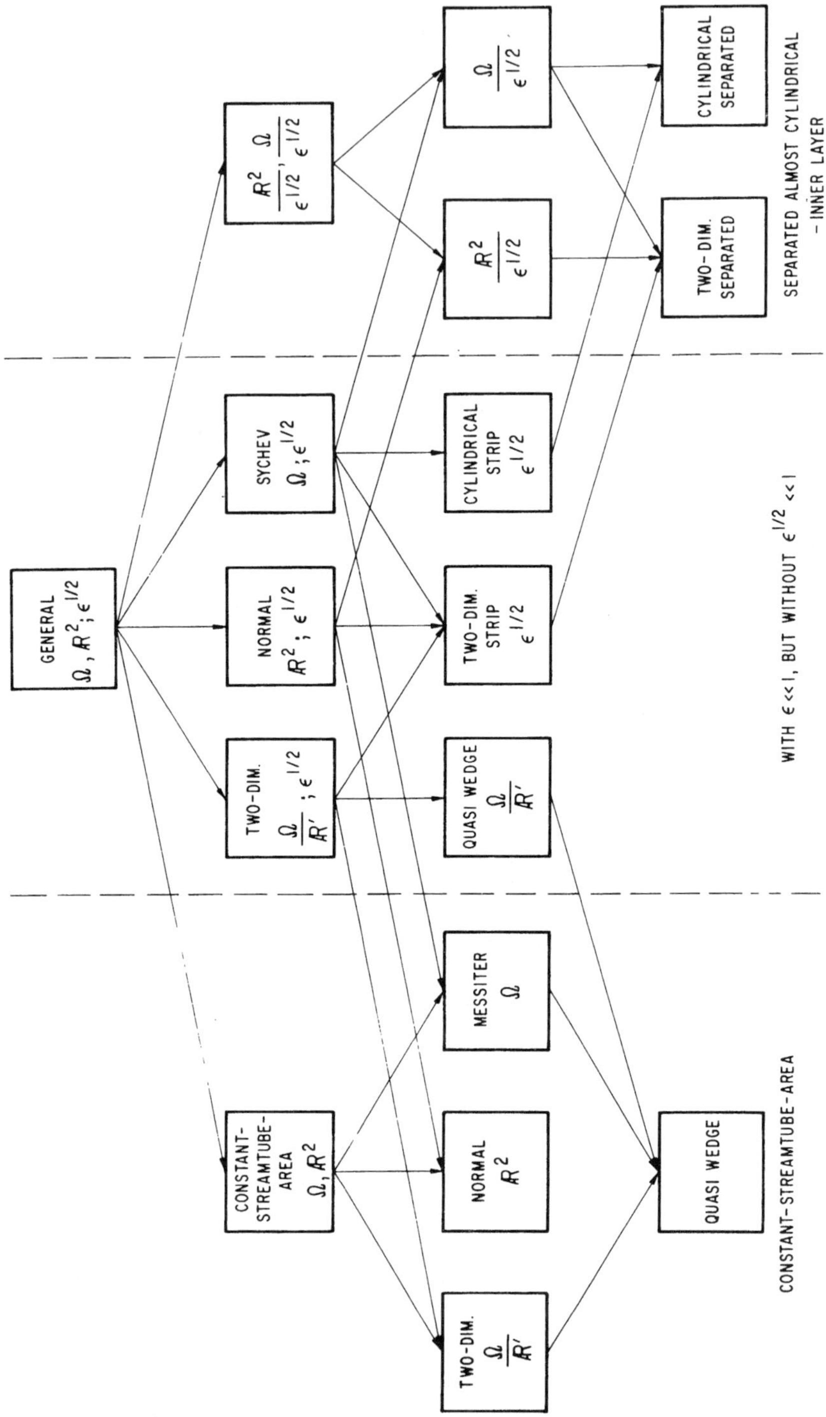

FIG. 5–26. Similitudes for blunt-faced bodies.

(d) Similitudes for blunt-faced bodies. For the similitudes we have considered, both constant-streamtube-area ones and the ones based upon an inner layer analysis, the assumption is required that $\epsilon^{1/2} \ll 1$. Thus we neglect, for example, the inner layer on the blunt-faced axisymmetric body treated in Section 5.4. If we make the less restrictive assumption that $\epsilon \ll 1$ but do not assume that $\epsilon^{1/2} \ll 1$, these similitudes do not hold. However, with $\epsilon^{1/2}$ as an additional parameter, similitudes are obtained through the geometric limiting processes that led to the special similitudes A, B, and C above.

As an example, if we keep Ω and $\epsilon^{1/2}$ fixed and require $Æ^2 \ll 1$ we obtain a similitude in which Ω, $\epsilon^{1/2}$, and τ^* are the basic parameters. This similitude is none other than the similitude of Sychev presented in Section 2.9, applied to blunt-faced bodies with $\epsilon \ll 1$. The further condition $\epsilon^{1/2} \ll 1$ reduces this to Messiter's.

With similitudes of this variety in mind, we can formulate a diagram showing the interrelationships of the similitudes available in thin-shock-layer theory on blunt-faced bodies. This diagram is given in Fig. 5–26. In the center are shown the parameters involved in the similitudes without the restriction that $\epsilon^{1/2} \ll 1$. To the left are the constant-streamtube-area similitudes, on the right those based upon an inner layer analysis. The limiting processes going from the center group to either side are based upon the condition $\epsilon^{1/2} \ll 1$ subject to various geometric conditions. The limiting processes within any one group are basically geometric ones, corresponding to a number of the various parameters becoming very large or very small. The dependence upon the parameter τ^* is considered to be understood, and is not explicitly indicated in Fig. 5–26.

The various arrows leading out of a single box correspond to different limiting processes. To illustrate this let us consider the box labeled "Sychev", with parameters Ω; $\epsilon^{1/2}$. The limiting process $\epsilon^{1/2} \to 0$ with Ω held of the order of one takes us to Messiter's similitude. The same limiting process but with $\Omega/\epsilon^{1/2}$ held of the order of one takes us to the inner layer similitude based upon the parameter $\Omega/\epsilon^{1/2}$. Two limiting processes in which $\epsilon^{1/2}$ is fixed and $\Omega \to 0$ are considered. In one of these we make $\cot \alpha \to 0$, keeping $Æ$ small but fixed. In the other we make $Æ \to 0$, keeping $\cot \alpha$ small but fixed.

Another case involves the two processes which lead from the two-dimensional box to the two boxes with parameter $\Omega/Æ'$. The process must be one which establishes definitely an attached flow without a significant Bernoulli effect. This process may be accomplished by making $\epsilon^{1/2} \to 0$, keeping $\cot \alpha$ fixed. Alternatively, it may be accomplished keeping $\epsilon^{1/2}$ fixed by making sure $\cot \alpha$ is sufficiently large.

8. Integral methods

The theory developed in this chapter and in Chapters IV and VI has in common with classical boundary layer theory the assumption that the phenomena of interest appear in a thin layer near a body surface or other appropriate reference surface. Naturally, analogous methods of attack appear in both theories. Thus, for example, the use of similar solutions and of related series expansions appears in both theories. The present section is devoted to the analogue in the theory of thin shock layers of the classical integral methods of boundary layer theory. Series-expansion schemes are considered in Sections 2.7 and 6.5, while the method of assuming local similarity is discussed briefly later in this section. As in boundary layer theory, we first establish basic integral results based upon the fundamental conservation laws. These results are then used with various approximations about the nature of the flow to obtain approximations to certain unknowns of the flow.

Integral methods have inherent limitations. Inasmuch as the integral results used are global in nature, little information as to the structure of the flow field can be expected. The accuracy of these methods can depend quite critically upon the accuracy of the assumptions employed. In the theory of thin shock layers, however, the lowest-order limit as $\epsilon \rightarrow 0$ of an integral method in this case turns out to be essentially no different from Newtonian theory. We may hope, then, by using assumptions suggested by Newtonian theory to obtain fairly accurate results with ϵ only moderately small.

The work which is of principal interest to us here is that of Chernyi [3; 4; 5]. In his work unsteady motion is usually the basic motion studied, and this is related to steady hypersonic flows through the equivalence principle. A modification of Chernyi's approach has been given by Cheng, Hall, Golian, and Hertzberg [1], who included problems involving boundary layer displacement thickness in their study. Mirels [2] summarizes a number of the integral methods and offers a comparison of their accuracy as well as suggestions for alternative methods. Maslen and Moeckel [1] (Integral Method) used integral methods in their early study of inviscid hypersonic flows.

Accepting the fact that integral methods are limited in the degree to which they can be defended as rational, we may ask to what type of problems they should be applied. The obvious answer is that they should be applied to problems for which we have available no feasible more rational approach. Such problems include the transition problems discussed in Section 2.7, and it is primarily these problems for which approximate solutions have been obtained by Chernyi and by Cheng, Hall, Golian, and Hertzberg.

An alternative approach to problems of this transition type is that of assuming local similarity. The profiles of the various flow variables with respect to a lateral variable are assumed to be described by suitably chosen profiles of the similar type, of the form studied in Section 2.6. The flow equations are satisfied approximately, with derivatives with respect to x arising from the variation in profile shapes dropped. Such a method has been developed and applied by Oshima [1], who studied primarily the transition with finite counterpressure between a blast wave and a weak Mach wave. Lewis [1; 2] extended Oshima's calculations to other cases.

We begin our discussion by writing down the integral conservation equations of mass, momentum, and energy with the restriction to plane or axisymmetric flows. The conservation relations are expressed in terms of integrals across the

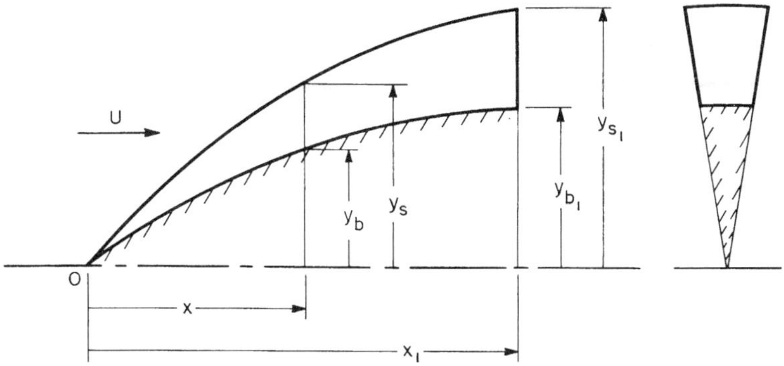

FIG. 5–27. Contour for integral conservation equations.

flow field, where the contour used is shown in Fig. 5–27. The results for the mass, axial momentum, lateral momentum, and energy are, respectively:

$$\rho_\infty U \frac{y_{s_1}^{1+j}}{1+j} = \int_{y_{b_1}}^{y_{s_1}} \rho u y^j \, dy, \tag{5.8.1}$$

$$(p_\infty + \rho_\infty U^2) \frac{y_{s_1}^{1+j}}{1+j} = \int_{y_{b_1}}^{y_{s_1}} (p + \rho u^2) y^j \, dy + \int_0^{y_{b_1}} p_b y_b^j \, dy_b \,, \tag{5.8.2}$$

$$p_\infty \int_0^{x_1} y_s^j \, dx = -\int_{y_{b_1}}^{y_{s_1}} (\rho u v) y^j \, dy + \int_0^{x_1} p_b y_b^j \, dx + j \int_0^{x_1} \int_{y_b}^{y_s} p y^{j-1} \, dy \, dx, \tag{5.8.3}$$

$$\rho_\infty U \left(e_\infty + \frac{p_\infty}{\rho_\infty} + \tfrac{1}{2} U^2\right) \frac{y_{s_1}^{1+j}}{1+j} = \int_{y_{b_1}}^{y_{s_1}} \rho u \left(e + \frac{p}{\rho} + \tfrac{1}{2} u^2 + \tfrac{1}{2} v^2\right) y^j \, dy. \tag{5.8.4}$$

The second term on the right-hand side of (5.8.2) represents the net drag (divided by $2\pi^j$) up to the station considered. It may include a part due to a finite drag addition at the nose. Similarly the second term on the right-hand side of (5.8.3) represents the lateral force or lift acting on the body cross section, or on a small segment of it in the axisymmetric case. It also may include a finite contribution at the origin. For a complete axisymmetric body the net lateral force must be zero, of course.

A more convenient form of the energy equation may be given, one in which the net drag up to a given station is expressed in terms of an integration with respect to y at that station. This form is given by

$$(5.8.5)\qquad U\int_0^{y_{b_1}} p_b\, y_b^j\, dy_b = \int_{y_{b_1}}^{y_{s_1}} \left\{\rho u e + \rho(u-U)\frac{p}{\rho} + \tfrac{1}{2}[(u-U)^2+v^2]\rho u\right\} y^j\, dy - \rho_\infty U e_\infty \frac{y_{s_1}^{1+j}}{1+j},$$

and represents conservation of energy where the energy is that defined relative to an observer fixed in the undisturbed fluid. This relation is obtained with the aid of (5.8.1) and (5.8.2) and corresponds to (2.8.1). In writing (2.8.1) the subscript 1 was not used since there was no possibility of confusion between variable and fixed y coordinates as is the case when the lateral momentum equation is considered. An alternative form of the momentum equation (5.8.2) which is of some use is obtained with the aid of (5.8.1), and is

$$(5.8.6)\qquad \int_0^{y_{b_1}} (p_b - p_\infty)\, y_b^j\, dy_b + \int_{y_{b_1}}^{y_{s_1}} (p - p_\infty)\, y^j\, dy = U\int_{y_{b_1}}^{y_{s_1}} \rho(U-u)\, y^j\, dy - \int_{y_{b_1}}^{y_{s_1}} \rho(U-u)^2\, y^j\, dy.$$

The basic idea behind all integral approaches is that the distribution of various flow variables with respect to y across the shock layer is approximated by functions containing one or more undetermined parameters which depend on the axial distance x. The parameters are determined by satisfying one or more of the integral conservation equations, as well as the boundary conditions or other constraints to be satisfied by the approximating relations. In any integral method the number of parameters to be determined is arbitrary. In addition, not all of the conservation relations or boundary conditions may be satisfied. Conversely, an integral method may be formulated involving additional conservation integrals obtained by taking higher moments of the basic flow equations. Such relations may be obtained, for example, by multiplying the basic flow equations by a power of u and then integrating with respect to y. Integral methods have the advantage of not being restricted to the analysis of a perturbation effect on a basic solution, as was inherent in the approach of

Section 2.7. The disadvantage, however, is that their accuracy can only be evaluated on an empirical basis.

The continuity integral (5.8.1) we have used before, of course. The integral is equivalent to (5.2.7), and in this form is basic to all hypersonic theories except in the Newtonian limit. For example, the integral served as the basis for the displacement thickness concept used in Section 2.7. Application of this integral alone in an integral method in the sense of this section has been made by Moeckel [1] and Serbin [1] in the calculation of stand-off distances for shocks on blunt bodies. The distributions assumed by these authors are uniformly sonic at the critical shoulder of the blunt body.

In applying the integral method to slender bodies, with or without nose blunting, one possibility has been suggested by Chernyi [5] and Mirels [2]. In this possibility we assume the profiles for the flow variables are the same as those corresponding to the similar solutions of Section 2.6. This choice is directly analogous to a standard integral method in boundary layer theory (see, for example, Section 8.5 of the first edition). The parameter k (among others) characterizing the assumed local solution is not assumed to be known at any station. Instead, the value of k comes out as a consequence of the analysis. The difference between this approach and that of Oshima mentioned above is that it is the imposition of the integral conditions (together with certain boundary conditions) which determines the course of the characteristic parameters rather than an approximate fulfilling of the differential equations of motion. In the approach as proposed by Chernyi [5] there are five variables, ρ_s, v_s, p_s, y_s, and k. The three integral relations and the three shock conservation conditions yield six equations. The problem is thus overdetermined, and one of the six conditions must be dropped.

The more fruitful applications of integral methods have involved assumed profiles based upon thin-shock-layer theory. This fact explains the position of this section in Chapter V. The physical idea is that most of the flow is concentrated in a thin layer next to the shock. In the region between this layer and the body or axis the pressure is approximately constant with respect to the lateral variable y or r, and only varies with respect to the axial variable x. The energy integral and the lateral momentum integral give two relations connecting this pressure with the shock shape. The equation of state and the continuity integral give an additional relation of the displacement thickness type. The result is a differential equation for the shock shape. This approach can be made fully rational in the limit $\epsilon \to 0$. In its application, however, approximations are chosen with an aim to making the method usable appreciably away from this limit. Chernyi [4; 5], for example, is able to claim satisfactory results up to $\epsilon = 0.2$ or 0.3.

The thin-shock-layer assumptions do not change the form of (5.8.1). The principal effect is that the primary contribution to the integral comes from a narrow range of y near y_{s_1}. The integral (5.8.2) is not used as such. In the integral (5.8.3) for lateral momentum two changes are made. The pressure in the double integral of the third term on the right-hand side is approximated by $p_b(x)$. The integral of ρuv is approximated by v_{s_1} times the integral of ρu, this approximation being essentially that of the constancy of v across the thin layer near the shock. Following Chernyi, we may also include a term I_N corresponding to a concentrated lateral impulse at the nose, with a change in the designation of the lower limit of the integral from 0 to 0+. The result is

$$\frac{\rho_\infty U v_{s_1} y_{s_1}^{1+j}}{1+j} = I_N + \int_{0+}^{x_1} (p_b - p_\infty) y_s^j \, dx. \tag{5.8.7}$$

The only general change we can make in (5.8.5) is the replacement of v^2 by $v_{s_1}^2$.

To proceed further we simplify the problem in two ways. We assume that the gas is a calorically perfect one, and we assume that the equivalence principle may be applied. With this second simplification $(U - u)$ is a small quantity, and most terms in $(U - u)$ may be dropped. As pointed out by Freeman [6], however, the integral $\int \rho(U - u) y^j \, dy$ appearing in (5.8.6) is the primary term balancing the body drag term of that equation and may not be dropped there. Equation (5.8.5) then becomes, with a nose drag term D_N separated,

$$\frac{1}{2\pi^j} D_N + \int_{0+}^{y_{b_1}} p_b y_b^j \, dy_b = \frac{p_{b_1}}{\gamma - 1} \frac{y_{s_1}^{1+j} - y_{b_1}^{1+j}}{1+j} + \frac{1}{2} \frac{\rho_\infty v_{s_1}^2 y_{s_1}^{1+j}}{1+j} - \frac{p_\infty}{\gamma - 1} \frac{y_{s_1}^{1+j}}{1+j}. \tag{5.8.8}$$

We next express v_s in terms of the shock shape $y_s(x)$. The shock continuity relation is applied, giving $v_s = (1 - \epsilon) U y_s'$, or with the aid of (1.4.17),

$$v_s = \frac{2Uy_s'}{\gamma + 1}\left(1 - \frac{1}{M_\infty^2 y_s'^2}\right). \tag{5.8.9}$$

The pressure is given by the derivative of (5.8.7), by

$$p_b - p_\infty = \frac{2\rho_\infty U^2}{\gamma + 1}\left[y_s'^2 - M_\infty^{-2} + \frac{y_s y_s''}{1+j}(1 + M_\infty^{-2} y_s'^{-2})\right]. \tag{5.8.10}$$

This expression is simplified by Chernyi to the form

$$\frac{p_b - p_\infty}{\rho_\infty U^2} = \frac{2}{\gamma + 1}(y_s'^2 - M_\infty^{-2}) + \frac{y_s y_s''}{1+j} \tag{5.8.11}$$

or to the same expression retaining the $2/(\gamma + 1)$ factor in the y_s'' term. In the limit $M_\infty^{-2} \to 0$, $\gamma \to 1$, both (5.8.10) and (5.8.11) reduce to the slender-body Newtonian result (3.4.38). A factor $1 - (\psi/\psi_s)$ may be included with the y_s'' term to give an approximate expression for p instead of for p_b.

If now (5.8.8) is differentiated with respect to x and the value for p_b from (5.8.11) substituted, the result is a third-order differential equation for y_s. Chernyi [2; 3] has applied his method to the flows on slightly blunted wedges and cones at $M_\infty = \infty$. In the case of the blunted wedge ($j = 0$) the pressure integral in (5.8.8) may be eliminated by (5.8.7) instead of by differentiation, because of the fact that y_b' is constant. The result, with $p_\infty = 0$, $a_\infty = 0$, $j = 0$, and with the factor $2/(\gamma + 1)$ retained in (5.8.11), is

$$\frac{1}{2} y_s \left(\frac{2}{\gamma + 1} y_s'\right)^2 + \frac{1}{\gamma - 1} (y_s - y_b' x) \frac{d}{dx} \left(\frac{2}{\gamma + 1} y_s y_s'\right) = \frac{\frac{1}{2} D_N - y_b' I_N}{\rho_\infty U^2} + \frac{2}{\gamma + 1} y_b' y_s y_s'. \tag{5.8.12}$$

This is a first integral of the general third-order differential equation. In the case of the blunted cone this elimination of the integral is impossible. Figures 5–28 and 5–29 show the shock shape and total frontal drag coefficient for a slender blunted wedge of half-angle θ_w, bluntness thickness d (leading edge drag coefficient C_{D_0}), and length l, in a gas flow with $\gamma = 1.4$ and $M_\infty = \infty$. Shown for comparison are the corresponding results for a wedge with a sharp leading edge. It may be seen that as a result of the positive displacement thickness the total frontal drag coefficient of the blunted wedge is larger than the drag coefficient obtained from the sum of the drag of the blunted leading edge plus the drag of the sharp leading edge wedge.

Figure 5–30 shows the total frontal drag coefficient for a slender blunted cone of half-angle θ_c and nose drag coefficient C_{D_1} in the same flow. This last figure shows the interesting feature of the method that it predicts that the drag of a sharp cone may be decreased by a suitable slight blunting. Our results on optimum Newtonian bodies of revolution (Section 3.4) keep this result from being surprising. These optimum bodies are blunted, and have shapes quite different from conical. The method gives the asymptotic result for the shock angle

$$\tan \sigma = \tan \theta_c \sqrt{\frac{\gamma + 1}{2} + (M_\infty \tan \theta_c)^{-2}} \tag{5.8.13}$$

valid for a sharp cone. This formula is in good agreement with exact numerical results and with the constant-density results of Section 4.2.

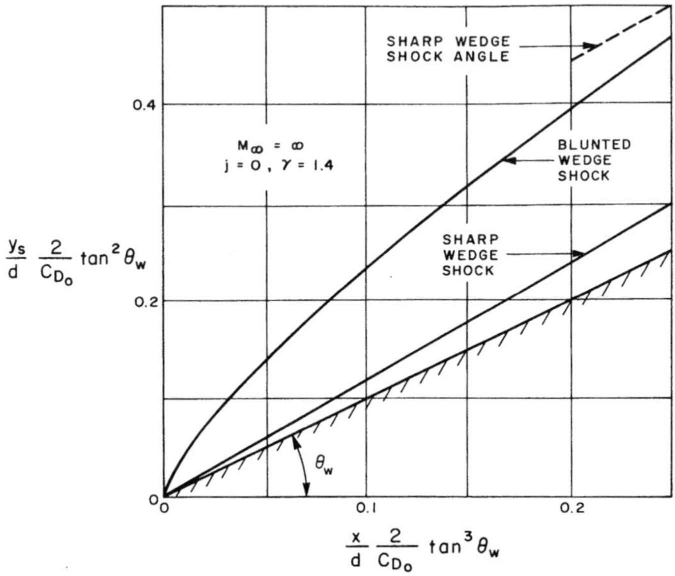

FIG. 5–28. Shock shape on blunted slender wedge (Chernyi [2; 3]).

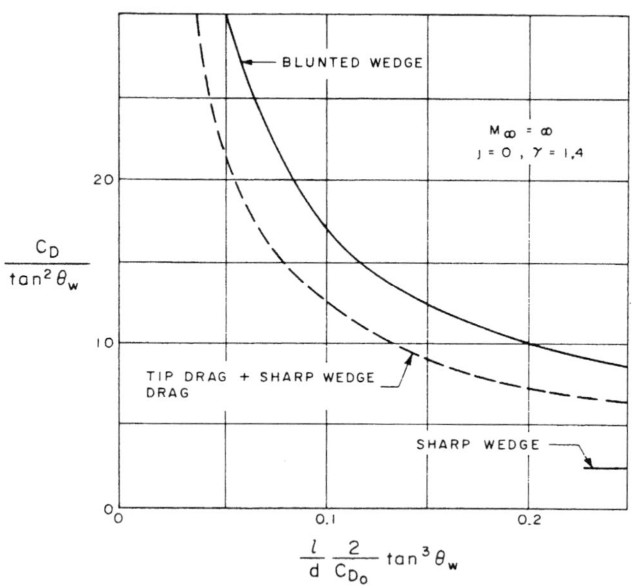

FIG. 5–29. Drag coefficient for blunted slender wedge (Chernyi [2; 3]).

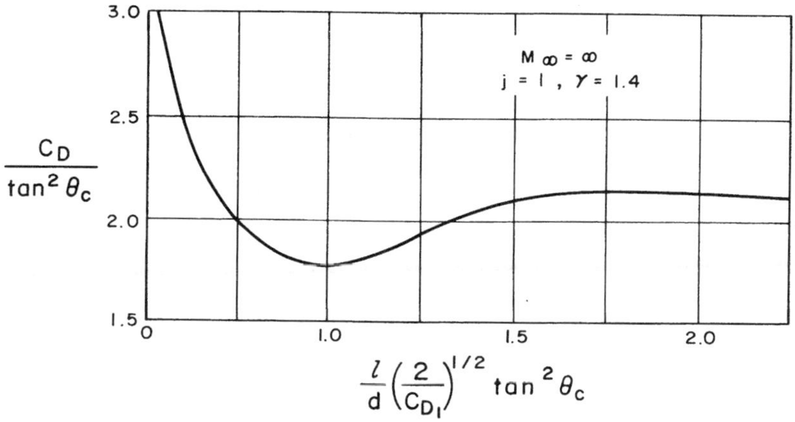

FIG. 5–30. Drag coefficient for blunted slender cone (Chernyi [2; 3]).

As we have indicated, in the limit $\epsilon \to 0$ the basic equation reduces to the analogous form given by Newtonian theory. The kinetic energy term, the one in $v_{s_1}^2$ in (5.8.8), is missing in the Newtonian limit and represents a term of higher order in ϵ. Alternative choices of the basic assumptions in the integral method give differing choices of the terms of higher order in ϵ. With such an alternative choice, assuming the shock pressure p_s throughout the shock layer in calculating the internal energy there, Lunev [1] obtains an equation similar to those of Chernyi but with the kinetic energy term doubled. With this alternative he obtains numerical results for the point explosion flow closer than Chernyi's to those of the exact solutions, and does not predict any possible drag reduction on a cone by blunting. There is no rational basis, however, for considering Lunev's alternative as an improvement.

Of the numerous other applications of the method we mention that of Chernyi [5] to the blast wave or point explosion problem with counterpressure. In this problem ϵ varies from very close to ϵ_{11m} to values close to one. The method clearly must fail when ϵ can no longer be considered reasonably small, but with γ close enough to one should give useful results over a quite large range of ϵ. The interesting feature of this problem which Chernyi points out is that the basic equation can be reduced to an ordinary differential equation of first order in which the dimension parameter j does not appear. Note that the integral of (5.8.8) does not appear in this problem, so that (5.8.8) with (5.8.11) already serves as a first integral to the more general third-order differential equation. Thus in this formulation of the problem the volume dependence of the propagation velocity of the shock and of the pressure p_b behind the shock layer is the same for $j = 0$ as for $j = 1$ (or for $j = 2$ in the unsteady analogue).

The phenomenon of the free layer cannot be studied in any of the formulations of the integral method described above. Mathematically, the kinetic energy term does not behave properly. Physically, the reason for the difficulty is easy to see—Newtonian theory is based upon momentum conservation, and energy conservation plays no role. Energy must be conserved of course; the internal energy has a distribution through the shock layer quite unlike that assumed in the integral methods. In the integral methods the main contribution to the internal energy flow is from the space between the shock layer and body rather than from the shock layer itself.

We may choose among the alternatives a set of assumptions leading to an equation in which no kinetic energy term appears. The simplest way of doing this is to simply take the $\gamma \to 1$, $M_\infty \to \infty$ limit in the equations above. In this limit (5.8.11) reduces to (3.4.38), $p_\infty \to 0$, and (5.8.8) reduces to

$$(y_s^{1+j} - y_b^{1+j}) \left(y_s'^2 + \frac{y_s y_s''}{1+j} \right) = \epsilon \frac{(1+j)D_N}{\pi^j \rho_\infty U^2} . \tag{5.8.14}$$

This equation can treat flows asymptotic upstream to free layer solutions, but only approximately. The right-hand side should not generally be identified with D_N in the manner that is shown in (5.8.14). The rational use of an equation of this form in treating transitions beginning with a free layer solution will be discussed in the last part of the next section.

Cheng, Hall, Golian, and Hertzberg [1] have derived and used an equation similar to (5.8.14) for inviscid problems, together with an extension of it to take the boundary layer displacement thickness into account. Where y_s appears in (5.8.14) these authors use a quantity y_e, the value of y at the inner edge of the thin shock layer. In the first parenthesis this change has about the same effect as the factor $2/(\gamma + 1)$ which has been dropped in arriving at (5.8.14). In making this change in the second parenthesis, that representing the pressure behind the shock layer, these authors have made an erroneous approximation which weakens their analysis. As we have seen repeatedly earlier, the results of Newtonian theory are applicable with the least error when interpreted in terms of the actual shock shape.

Cheng, Hall, Golian, and Hertzberg [1] applied their version of (5.8.14) to the problem of the blunted wedge. Their solutions show an oscillatory behavior of the pressure on the wedge not shown by Chernyi's solutions for the same problem. As they note, the physical reality of this oscillatory behavior is doubtful although not unbelievable. It is likely that it is the kinetic energy term in Chernyi's formulation which prevents this phenomenon in his results.

In all the equations above the body shape enters in terms of y_b^{1+j} or its differential. This quantity is proportional to the cross-sectional area of the body. This fact suggests that it may be only the cross-sectional area of the body which is important. The area rule of Ladyzhenskii [1] for small ϵ exploits this idea. It applies to flows which are initially axisymmetric, on slender three-dimensional bodies which are blunted. The rule is based upon Chernyi's formulation of the integral method with $M_\infty = \infty$. According to this rule two slightly blunted bodies slender in both lateral directions but not necessarily axisymmetric will have the same axial pressure distribution, the same drag, and the same axially symmetric shock waves provided:

(a) The nose drag on the blunted tip is the same for the two bodies.

(b) The two bodies have the same axial distribution of body cross-sectional area.

(c) No part of the body extends laterally beyond the shock wave.

The proof is directly analogous to that given in connection with the entropy wake area rule presented at the end of Section 2.8. In the model of Chernyi the pressure is taken constant across the entire shock layer, and to extend his formula to Ladyzhenskii's case it may be shown that nonabrupt changes in cross-sectional shape do not affect this principle.

Ladyzhenskii [3] has proposed a generalization of his area rule to take into account those cases in which the body may extend beyond the confines of the axially symmetric shock wave created by the initial blunted portion of the body. The restriction on the geometry is only that the body have no sharp corners either on the part extending beyond the nose shock wave or at the intersection of this part with the nose shock. His basic assumption is that the pressure on the extended surfaces is determined by the three-dimensional Newton-Busemann relation, and that on the extended surfaces, termed "Newtonian edges", the shock and body are contiguous. With this assumption it is evident that if the two bodies considered have the same shape beyond the confines of the axially symmetric portion of the shock and satisfy restrictions (a) and (b) (with restriction (b) now referring to the portion of the body contained within the separated nose shock), there is an immediate generalization to the area rule. The generalized rule is simply that the two bodies will again have the same axial pressure distribution and drag, as well as the same nose shock shapes. Again the limitations on this rule need more detailed investigation, since the Newtonian flow assumption implies a limiting case in which the shock layer cross-sectional area is zero. It is possible that some of the complexities of three-dimensional Newtonian flow may interfere with the validity or may require some modification of the statement of the rule. Bogacheva and Ladyzhenskii [1] have applied the equations developed under the assumptions noted above to the calculation of flows past slightly blunted elliptic cones.

Integral methods could be exploited to a far greater extent than they have been. The only approaches which have been well explored have been those based on thin-shock layer concepts. The approach mentioned earlier involving similar solutions has only been a suggestion. Integral methods not based upon thin-shock-layer concepts will necessarily entail numerical computations which are more onerous than those needed in the methods which are based upon these concepts.

9. Newtonian separation

This section is concerned with problems connected with the appearance of a free layer. The free layer with zero pressure behind it exists strictly only in the Newtonian approximation. In actual gas flows with the Newtonian assumptions only approximately fulfilled, the free layer concept is also approximate. A low-density flow appears in the void between the layer and the body, and the pressure on the body does not vanish.

The principal problem of interest here is that of determining the flow near the point at which the Newtonian shock layer separates and becomes free, i.e. the Newtonian separation point. The problem was pointed out by Freeman [1] in his analysis for improving the Newtonian solution on a blunt body. In his lowest-order theory for the improvement over Newtonian the Newtonian separation point appeared as a singular point near which the solution blew up.

The problem was resolved by Freeman [5] for the case of a sphere. Subsequently, Freeman's approach was applied to other cases by Bausch [1]. Here, though we take full advantage of the results and concepts of Freeman, we approach the problem quite differently. The aim of our approach is to set up a method which can be applied to quite general cases, while at the same time keeping the approach as simple as possible. Our main arguments and estimates are analogous to those advanced by Freeman.

The problem is one of the direct type, in which the body shape is given and does not represent the shock shape accurately enough. Boundary layer coordinates of the type shown in Fig. 5–1 are used, with x distance along the body. See Fig. 5–31. Body orientation is used, as is generally more appropriate in direct problems.

There is an alternative to simple body orientation which should be advantageous in the zone downstream of the Newtonian separation point. In this alternative the reference surface is the body upstream of the separation point and is the position of the classical Newtonian free layer downstream of the separation point. This choice has the disadvantage of introducing a discontinuity in the derivative of the curvature where none exists in the body shape. It has the advantage of ignoring irregularities in the body shape downstream of the separation point and of doing less violence to the thin-layer assumptions in this downstream region. We shall comment further on this alternative later.

As we have done in previous sections, we use the Newtonian theory in terms

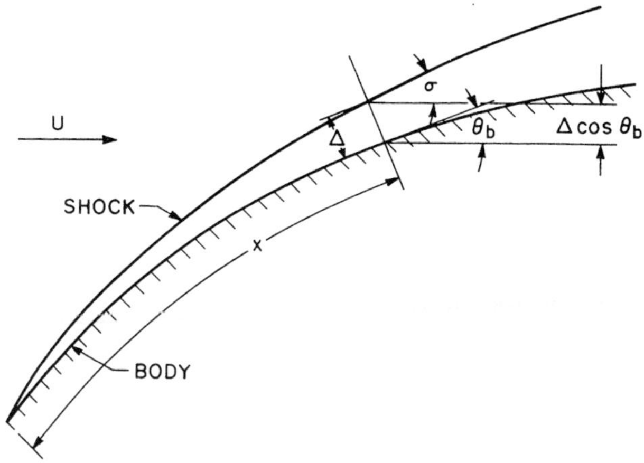

FIG. 5–31. Coordinate system.

of shock shape to obtain an approximation to the pressure. The pressure at the base of the shock layer is given by (3.2.7) as

$$\tfrac{1}{2}C_{p_b} = \sin^2 \sigma - KP, \tag{5.9.1}$$

where

$$K = \frac{d \cos \sigma}{dr} \approx \frac{d \cos \sigma}{\sin \theta_b \, dx}, \tag{5.9.2a}$$

$$P = \frac{1}{r^j} \int_0 \cos \sigma \, r^j \, dr \approx \frac{1}{r^j} \int_0 \cos \sigma \sin \theta_b \, r^j \, dx. \tag{5.9.2b}$$

Here r enters only in the axisymmetric case ($j = 1$), in which it is the classic radial coordinate. The shock angle σ is related to the body angle θ_b by the relation

$$\sigma = \theta_b + \Delta' \tag{5.9.3}$$

within the accuracy of thin-shock-layer theory, with primes denoting differentiation with respect to x. The pressure coefficient which would appear on the body if the shock layer were infinitesimally thin is denoted C_{p_0} and is given by

$$\tfrac{1}{2}C_{p_0}(x) = \sin^2 \theta_b - K_0 P_0, \tag{5.9.4}$$

where

(5.9.5a) $$K_0(x) = \frac{d \cos \theta_b}{\sin \theta_b \, dx},$$

(5.9.5b) $$P_0(x) = \frac{1}{r^j} \int_0 \cos \theta_b \sin \theta_b \, r^j \, dx.$$

The quantity $\frac{1}{2}C_{p_0}(x)$ is negative downstream of the separation point, and is a known function over the entire range.

We now substitute (5.9.3) into the equations preceding and take one term in the Taylor expansions arising. The result is

(5.9.6) $$\begin{aligned} \tfrac{1}{2}C_{p_b} = \tfrac{1}{2}C_{p_0} &+ P_0 \Delta'' + \Delta' \sin 2\theta_b \\ &+ K_0 \left(\frac{1}{r^j} \int_0 \Delta' \sin^2 \theta_b \, r^j \, dx - P_0 \Delta' \frac{\cos \theta_b}{\sin \theta_b} \right) \\ &+ O(K_0 \Delta). \end{aligned}$$

In the cases of interest the order of magnitude of $K_0 \Delta$ is smaller than that of Δ', which is smaller than that of $K_0^{-1} \Delta''$. For pointed bodies in the Newtonian separation region we will be able to show that generally

(5.9.7) $$K_0 \Delta \sim \epsilon^{1/3} \Delta' \sim \epsilon^{2/3} K_0^{-1} \Delta''.$$

Thus in (5.9.6) we will drop all the terms on the right-hand side except the first two. It may be checked afterwards that the neglected terms are indeed of smaller order. The first basic equation for the method is then

(5.9.8) $$\tfrac{1}{2}C_{p_b} = \tfrac{1}{2}C_{p_0} + P_0 \Delta''.$$

The second basic equation which must be obtained is a relation connecting Δ and $\frac{1}{2}C_{p_b}$, of the form

(5.9.9) $$\Delta = D(\tfrac{1}{2}C_{p_b}\, ; x, \epsilon).$$

In the function D, the dependence on x comes from the fairly gradual change in the structure of the shock layer as described by $u(\psi)$ and $S(\psi)$. This dependence on x is weak in the sense that the phenomenon we are investigating extends only over a range of x small enough that D changes very little across the range. We thus may neglect the explicit derivatives of D with respect to x. The dependence upon ϵ is also generally weak, and takes into account only the influence of the Bernoulli effect upon the velocity profile. Elimination of Δ from the last two equations yields

$$P_0\dot{D}(\tfrac{1}{2}C_{p_b})'' + P_0\ddot{D}(\tfrac{1}{2}C'_{p_b})^2 = \tfrac{1}{2}C_{p_b} - \tfrac{1}{2}C_{p_0}, \tag{5.9.10}$$

where the dots indicate differentiation with respect to $\frac{1}{2}C_{p_b}$.

In approximate analyses of (5.9.10) in the separation region we may generally set x equal to its value x_0 at the separation point in evaluating both the functions P_0 and D. The variation of $\frac{1}{2}C_{p_0}$ with x is essential, of course, but $\frac{1}{2}C_{p_0}$ may be approximated by a term linear in x in most cases. Here again, the reason is that D and P_0 change very little across the range of x under consideration. The general behavior of the solution in this region is indicated in Fig. 5–32.

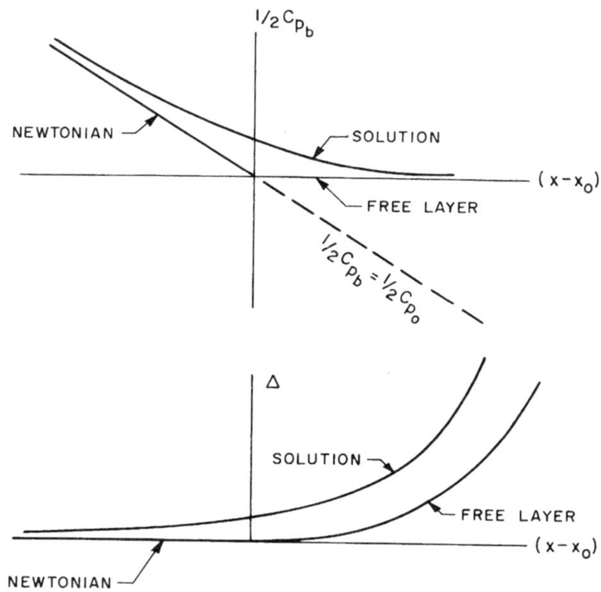

FIG. 5–32. General behavior of solution near Newtonian separation point.

With the alternative choice of a reference surface which follows the free layer, the quantity $\frac{1}{2}C_{p_0}$ is identically zero downstream of the separation point. The distance between the body and the free layer in the Newtonian analysis is denoted $\delta(x)$ and is then a known function of x. In (5.9.9) the quantity $\Delta + \delta$ replaces Δ. The result of these changes is an equation for $x > x_0$ of the same form as that of (5.9.10), with the known function $-\frac{1}{2}C_{p_0}$ replaced by the known function $P_0\,\delta''$. Thus it is clear that the choice of reference surface has but a minor effect on the basic equation to be solved.

Since (5.9.10) is an equation of the second degree its solution involves two free parameters. Assuming that a solution which is well behaved upstream is known, the effect of the free parameters may be studied by a perturbation analysis. The perturbation pressure solutions have a behavior upstream of the separation point which involves physically unacceptable oscillations. The choice of solution and the determination of the free parameters is determined by the condition that these oscillations be absent, and thus by upstream conditions. This conclusion is consistent with the implicitly assumed parabolic nature of the shock layer, with no upstream influence taken into account.

We turn now to the question of the determination of the function D of (5.9.9). The shock layer thickness Δ or D is given by the quadrature (5.2.7), expressed with the velocity denoted uU as

$$D = \frac{1}{r^j \rho_\infty U} \int_0^{\psi_s} \left(\frac{\rho_\infty}{\rho}\right) \frac{d\psi}{u}. \tag{5.9.11}$$

The velocity is a known function of ψ, with

$$u = u(\psi; x, \epsilon). \tag{5.9.12}$$

Here the dependence of u on ϵ and x represents the Bernoulli effect, to be neglected in most cases. The dependence of u on x is very weak, and vanishes in the Newtonian limit $\epsilon \to 0$. The density is given by an equation of state of the form (5.2.4) as a function $\rho(S, p)$, while from (5.2.2) S is a function $S(\psi)$ of ψ alone.

The pressure is given by

$$p = p_b + \frac{UK_0}{r^j} \int_0 u \, d\psi, \tag{5.9.13}$$

so that p takes the functional form

$$p = \frac{K_0}{r^j} \Pi \left(\frac{r^j p_b}{K_0}, \psi; x, \epsilon\right) \tag{5.9.14}$$

or

$$p = p(p_b, \psi, x; \epsilon). \tag{5.9.15}$$

The quadrature of (5.9.11) then leads to the function D of (5.9.9). We assume here that ρ_∞ and U are fixed. In separation problems it is the contribution corresponding to values of p and p_b which are small with respect to $\rho_\infty U^2$ which are of importance, corresponding in the integral of (5.9.13) to small values of ψ.

We now specialize the problem to the case in which the gas is calorically perfect and $M_\infty = \infty$. Here we define ϵ specifically as $(\gamma - 1)/(\gamma + 1)$. The approximation of (5.2.13) applies, and we write

$$\frac{\rho_\infty}{\rho} = \epsilon \frac{p_s}{p}, \tag{5.9.16}$$

with $p_s(\psi)$ the pressure at entry. The integral (5.9.11) becomes

$$D = \frac{\epsilon}{r^j \rho_\infty U} \int_0^{\psi_s} \left(\frac{p_s(\psi)}{p_b + \dfrac{UK_0}{r^j} \displaystyle\int_0 u \, d\psi} \right) \frac{d\psi}{u}. \tag{5.9.17}$$

As a first example we consider the case for which the velocity profile $u(\psi)$ is approximately constant, corresponding to a forebody which is close to a wedge or cone. In this case we write

$$u(\psi) = u(0) = u_b. \tag{5.9.18}$$

We recognize $p_s(\psi)$ as approximately equal to

$$p_s(0) = \rho_\infty U^2(1 - u_b^2), \tag{5.9.19}$$

and $d(\frac{1}{2}C_p)$ in the shock layer as

$$d(\tfrac{1}{2}C_p) = \frac{K_0}{r^j \rho_\infty U} u \, d\psi. \tag{5.9.20}$$

In these terms we can evaluate D as

$$D = \frac{\epsilon(1 - u_b^2)}{K_0 u_b^2} \ln \left(\frac{C_{p_s}}{C_{p_b}} \right). \tag{5.9.21}$$

Here $\frac{1}{2}C_{p_s}$ is half the pressure coefficient at the shock, equal to $\frac{1}{2}C_{p_b} + K_0 P_0$. If $u(\psi)$ is not constant but u_b is of the order of one (corresponding to a pointed body), equation (5.9.21) still gives the correct behavior of D when $\frac{1}{2}C_{p_b}$ is small. For correct quantitative results a small correction term needs to be added to (5.9.21), to be obtained from an accurate evaluation of D from (5.9.17).

To investigate the solution near the Newtonian separation point in this case we treat P_0, K_0, and C_{p_s} as constants. We approximate C_{p_0} by the linear distribution

$$\tfrac{1}{2}C_{p_0} = a(x_0 - x), \tag{5.9.22}$$

where x_0 is the Newtonian separation point. The quantity a is a measure of the Newtonian pressure gradient upstream of the separation point. We next make the transformation

$$\xi = a^{1/3}\,\epsilon^{-1/3}\left(\frac{P_0(1-u_b^2)}{K_0 u_b^2}\right)^{-1/3}(x - x_0), \tag{5.9.23a}$$

$$\pi(\xi) = a^{-2/3}\,\epsilon^{-1/3}\left(\frac{P_0(1-u_b^2)}{K_0 u_b^2}\right)^{-1/3}\tfrac{1}{2}C_{p_b}, \tag{5.9.23b}$$

intended primarily to eliminate ϵ and a from (5.9.10). With (5.9.21) substituted into (5.9.10) and this transformation applied we obtain the equation

$$\frac{d^2}{d\xi^2}\left(\ln\frac{1}{\pi}\right) = \pi + \xi. \tag{5.9.24}$$

The solution of interest is the one for which π behaves as $-\xi$ in the limit $\xi \to -\infty$. This solution corresponds to the unique nonsingular solution upstream. The asymptotic behavior for large positive ξ is given by

$$-\ln\pi \sim \tfrac{1}{6}\xi^3 + A\xi + B, \tag{5.9.25}$$

where A and B are constants to be determined. The two terms bearing these constants give a measure of the displacement of the shock from the free layer position in the region downstream from the separation point. The results of this analysis are in general accord with those of Bausch [1], and substantiate the statement on orders of magnitude of (5.9.7). The exponent $\frac{1}{3}$ which appears on the parameter ϵ is small enough to indicate a severe practical limitation on the analysis. In order that $\epsilon^{1/3}$ be small it is necessary that ϵ be extremely small.

With blunted bodies u_b is of the order of $\epsilon^{1/2}$ instead of the order of one. With axisymmetric bodies this magnitude is small enough to be neglected within the order of accuracy of the analysis. It may be shown that u_b would have to be of the order of $\epsilon^{1/11}$ or larger to be taken into account. The Newtonian result for a sphere may be put in the form

$$\frac{d\psi}{\rho_\infty U} = R_s^2\, u\, du. \tag{5.9.26}$$

This result holds also for u small with other blunted axisymmetric bodies, with R_s the shock radius of curvature at the nose. We replace $p_s(\psi)$ by the stagnation pressure $\rho_\infty U^2$, and express (5.9.17) for the sphere in the form

$$D = \frac{3\epsilon}{K_0}\int_0^{u_s}\frac{du}{u^3 + \dfrac{3r}{K_0 R_s^2}\tfrac{1}{2}C_{p_b}}. \tag{5.9.27}$$

Since we are interested in D only where $\frac{1}{2}C_{p_b}$ is small, we may replace the upper limit of the integral by ∞. With $\frac{1}{2}C_{p_b}$ small, we may also apply (5.9.27) for any blunt axisymmetric body instead of only for a sphere. We may then evaluate D as

$$D = \frac{3\epsilon}{K_0}\left(\frac{3r}{K_0 R_s^2}\,\tfrac{1}{2}C_{p_b}\right)^{-2/3}\int_0^\infty \frac{d\eta}{1+\eta^3} \tag{5.9.28}$$

$$= \frac{2\pi\epsilon}{\sqrt{3}\,K_0}\left(\frac{3r}{K_0 R_s^2}\,\tfrac{1}{2}C_{p_b}\right)^{-2/3}.$$

The transformation

$$\xi = a^{5/11}\,\epsilon^{-3/11}\left(\frac{2^3\pi^3 P_0^3 R_s^4}{3^{7/2} K_0 r^2}\right)^{-1/11}(x - x_0), \tag{5.9.29a}$$

$$\pi = a^{-6/11}\,\epsilon^{-3/11}\left(\frac{2^3\pi^3 P_0^3 R_s^4}{3^{7/2} K_0 r^2}\right)^{-1/11}\tfrac{1}{2}C_{p_b} \tag{5.9.29b}$$

is made, with which (5.9.10) becomes

$$\frac{d^2}{d\xi^2}(\pi^{-2/3}) = \pi + \xi. \tag{5.9.30}$$

This equation may be recognized as a transformed form of the reduced equation (A3) of Freeman [5]. In analogy with (5.9.25), the asymptotic behavior of π for large positive ξ is given by

$$\pi^{-2/3} \sim \tfrac{1}{6}\xi^3 + A\xi + B. \tag{5.9.31}$$

We may note that 3/11 is smaller than 1/3, and earlier remarks on the practical limitation of the analysis hold even more strongly here.

In the case of a blunt two-dimensional body the Bernoulli effect on the velocity profile cannot be neglected, for the simple reason that no velocity profile exists according to Newtonian theory (cf. the discussion following (3.5.6)). We may, however, neglect the Bernoulli effect in the Newtonian separation region, and simply accept the velocity distribution as given somewhat upstream of the separation point. To obtain a picture of the phenomenon in this case, let us investigate a velocity distribution corresponding to the constant-density solution of Section 4.3 for a circular cylinder. This distribution is expressed parametrically as

$$u = \epsilon^{1/2}u_0 \cosh \eta, \tag{5.9.32a}$$

$$\psi = \epsilon^{1/2}u_0\,\rho_\infty U R_s \sinh \eta. \tag{5.9.32b}$$

Here the parameter η corresponds to the $y/\epsilon R_s$ of (4.3.24), and u_0 is defined so that $\epsilon^{1/2}u_0$ is u_b for the velocity profile. In (5.9.17), $j = 0$ and $p_s(\psi)$ is approximated by $\rho_\infty U^2$; the result is

$$D = \frac{1}{K_0 u_0^2} \int_0^{\sinh^{-1}(1/\sqrt{3\epsilon})} \frac{d\eta}{\frac{1}{2}\eta + \frac{1}{4}\sinh 2\eta + \left(\frac{\frac{1}{2}C_{p_b}}{\epsilon K_0 R_s u_0^2}\right)}. \tag{5.9.33}$$

If the parenthesis appearing in this integral is large compared with ϵ^{-1}, D is equal to $\epsilon R_s \sinh^{-1}(1/\sqrt{3\epsilon})$ divided by $\frac{1}{2}C_{p_b}$. This result reduces to a form consistent with (4.3.22) near the stagnation point. Nearer the separation point, where we can consider the parenthesis of (5.9.33) to be of smaller order than ϵ^{-1}, we may replace the upper limit in (5.9.33) by ∞. The result is that D has the functional form

$$D = \frac{1}{K_0 u_0^2} F\left(\frac{\frac{1}{2}C_{p_b}}{\epsilon K_0 R_s u_0^2}\right). \tag{5.9.34}$$

The function F here has the asymptotic behavior

$$F(z) \sim \frac{1}{2z} \ln 8z \tag{5.9.35}$$

for large values of z, and

$$F(z) \sim \ln \frac{A}{z} \tag{5.9.36}$$

for small values of z, where A is an undetermined constant.

A general simplifying transformation analogous to (5.9.23) or (5.9.29) is not available in this case. In the upstream region where (5.9.35) may be used we may set

$$\xi = a^{1/2}\,\epsilon^{-1/4}(\tfrac{1}{2}R_s P_0)^{-1/4}(x - x_0), \tag{5.9.37a}$$

$$\pi = a^{-1/2}\,\epsilon^{-1/4}(\tfrac{1}{2}R_s P_0)^{-1/4}\,\tfrac{1}{2}C_{p_b}, \tag{5.9.37b}$$

$$B = \frac{8}{K_0 R_s u_0^2}\, a^{1/2}\,\epsilon^{-3/4}(\tfrac{1}{2}R_s P_0)^{1/4}. \tag{5.9.37c}$$

The basic equation (5.9.10) then becomes

$$\frac{d^2}{d\xi^2}\left(\frac{\ln B\pi}{\pi}\right) = \pi + \xi. \tag{5.9.38}$$

This equation is valid where $B\pi \gg 1$, this being the condition that (5.9.35) hold. If ϵ is sufficiently small, so that B from (5.9.37c) is sufficiently large, the condition may still be met with π very small. This permits the extension of the validity of (5.9.38) to appreciably large negative values of ξ, and hence through the central part of the Newtonian separation region into the free layer region.

In the far downstream region where (5.9.36) may be used we may set

(5.9.39a)
$$\xi = a^{1/3}\left(\frac{P_0}{K_0 u_0^2}\right)^{-1/3}(x - x_0),$$

(5.9.39b)
$$\pi = a^{-2/3}\left(\frac{P_0}{K_0 u_0^2}\right)^{-1/3}\tfrac{1}{2}C_{p_b},$$

for which the basic equation becomes again (5.9.24). However, in this case the approximations made in obtaining (5.9.8) and (5.9.10) become invalid. The basic equation is thus (5.9.38), with a range of validity which is limited by the condition $B\pi \gg 1$.

By examining the equations above for the two-dimensional blunt-body case, we can reach a general conclusion. Although mathematically the solution does approach the Newtonian solution with zero pressure behind a free layer in the limit $\epsilon \to 0$, this approach is very slow. The characteristic exponent 1/4 is still smaller than the 3/11 of the axisymmetric case. We may expect that although ϵ may be very small the pressure behind the Newtonian separation point will be appreciably larger over an appreciably larger region even than in the cases of a blunt axisymmetric body or of a sharp-nosed body.

We now turn to a different but not unrelated problem. If ϵ is very small and the body is finite, the shock wave leaves the body and has approximately a free layer shape downstream of the Newtonian separation point and, of course, downstream of the body. Downstream of the body the pressure p_b on the axis in the cavity inside the shock layer is small compared to the pressure p_s at the shock just within the shock layer. This pressure, though negligible in its effect on the local shape, is not zero. However, far downstream the blast wave solution should apply if $M_\infty = \infty$, with a shape essentially different from that of a free layer and with p_b approximately equal to $\tfrac{1}{2}p_s$. The problem here is to determine the nature of the transition from the free layer solution to the blast wave solution.

Freeman [6] has investigated this problem in the axisymmetric case. We follow here his approach. The analysis assumes that the shape of the forward part of the shock is known, and that it corresponds to the body shape up to the Newtonian separation point, and to the appropriate free layer shape thereafter. Aft of this forward portion the slender body Newtonian formula

(3.4.38) for the pressure in the cavity is assumed, and this pressure is assumed constant across the cavity. The shape of the shock is related to the pressure in the cavity through (5.2.7) or (5.8.1) and the assumptions that the velocity u and the ratio p/ρ are constant along streamlines. The resulting basic equation is of a form similar to that of (5.8.14).

Although the basic equation obtained is similar to one obtained by an application of integral methods there is an essential difference between the two. The equation of Freeman is not obtained from an energy integral and does not make the equivalence assumption $u = U$ in the cavity; the constant on the right-hand side is not to be identified in terms of the drag of the body. The analysis is related, however, to that for the entropy wake in Section 2.7. The assumptions made are examined later, and it may be shown that the theory is a rational one in that it is correct to the lowest order in the parameter ϵ. Of the modifications to the analysis required for a relaxation of the various assumptions one is essential. This modification is needed because of the failure of the assumption $u =$ *constant* far downstream, and is related to the velocity-defect displacement effect of Section 2.7.

The pressure in the cavity is given by (3.4.38) as

$$p_b = \rho_\infty U^2 \left(y_s'^2 + \frac{y_s y_s''}{1+j}\right). \tag{5.9.40}$$

The velocity u is the Newtonian value

$$u = U \cos \sigma, \tag{5.9.41}$$

where σ is the shock angle at the point of entry. The density ρ_s at entry is ρ_∞/ϵ, while the pressure at entry is the Newtonian value

$$p_s = \rho_\infty U^2 \sin^2 \sigma. \tag{5.9.42}$$

Under the assumption that p/ρ is constant, corresponding to the limit $\gamma \to 1$ with a perfect gas, we have

$$\frac{1}{\rho} = \frac{\epsilon U^2 \sin^2 \sigma}{p_b}. \tag{5.9.43}$$

The continuity integral then yields

$$\frac{y_s^{1+j}}{1+j} = \int_0^{\psi_s} \frac{d\psi}{\rho u} = \frac{\epsilon U}{p_b} \int_0^{\psi_s} \frac{\sin^2 \sigma}{\cos \sigma}\, d\psi. \tag{5.9.44}$$

With the known shape for the forward part of the shock inserted, the integral with ψ_s replaced by ∞ is convergent and the error introduced by this change is of negligible order in the theory. We obtain thereby

$$\frac{y_s^{1+j}}{1+j}\left(y_s'^2 + \frac{y_s y_s''}{1+j}\right) = \epsilon A, \tag{5.9.45}$$

where

$$A = \frac{1}{\rho_\infty U}\int_0^\infty \frac{\sin^2\sigma}{\cos\sigma}\,d\psi. \tag{5.9.46}$$

We recall that this known shape consists of part of the body plus a free layer shape, and does not take into account the actual (as yet unknown) change in the shock shape from the free layer shape.

Equation (5.9.45) is of the same form as (5.8.14), but with a different constant on the right-hand side. Equation (5.8.14) obtained by calculating the drag D_N with the Newtonian assumptions is the same as (5.9.45) with A replaced by A_D, where

$$A_D = \frac{1}{\rho_\infty U}\left[\psi_1 - \cos\sigma_1 \int_0^{\psi_1} \cos\sigma\,d\psi\right], \tag{5.9.47}$$

and point 1 is any point on the free layer. The bracket in (5.9.47) will be recognized as coming from the Newtonian formula (3.2.8) for the drag. It may be readily shown that

$$A > A_D. \tag{5.9.48}$$

The fact that these constants are not equal will be discussed later.

Following Freeman, the parameter ϵ in (5.9.45) is eliminated through the transformation

$$x = \epsilon^{-\frac{2+j}{1+j}} z, \tag{5.9.49a}$$

$$y_s(x) = \epsilon^{-\frac{1}{1+j}} f(z). \tag{5.9.49b}$$

The resulting equation is

$$f(f^{1+j}f')' = (1+j)^2 A, \tag{5.9.50}$$

with primes here indicating differentiation with respect to z. A first integral for (5.9.50) is

$$(f^{1+j}f')^2 = 2(1+j)\,A(f^{1+j} + b^{1+j}), \tag{5.9.51}$$

with b an arbitrary constant. To evaluate this constant we must evaluate the coefficient describing the free layer shape emanating from the body. From (3.3.31) the quantity

$$\frac{P_0}{\rho_\infty U} = \sin\sigma_1 \int_0^{\psi_1} \cos\sigma \, d\psi, \tag{5.9.52}$$

with point 1 again any point on the free layer part of the known forward shock shape, is a constant. This constant appears in the equation for the free layer shape. From (3.3.33), with the y_0 term dropped, we have

$$y_s^{1+j} y_s' = f^{1+j} f' = (1+j) \frac{P_0}{\rho_\infty U} . \tag{5.9.53}$$

In (5.9.51) the free layer shape applies where f is negligibly small compared with b. We obtain then

$$b^{1+j} = \frac{1+j}{2A} \left(\frac{P_0}{\rho_\infty U}\right)^2 . \tag{5.9.54}$$

In the two-dimensional case ($j = 0$) the quadrature of (5.9.51) is

$$z = \frac{2b^{3/2}}{\sqrt{2A}} \left[\frac{1}{3}\left(1 + \frac{f}{b}\right)^{3/2} - \left(1 + \frac{f}{b}\right)^{1/2} + \frac{2}{3}\right], \tag{5.9.55}$$

with the arbitrary constant evaluated to satisfy $z = 0$ at $f = 0$. This case was not treated specifically by Freeman.

In the axisymmetric case ($j = 1$) the quadrature of (5.9.51) is

$$z = \frac{b^2}{2\sqrt{4A}} \left[\frac{f}{b}\sqrt{1 + \frac{f^2}{b^2}} - \ln\left(\frac{f}{b} + \sqrt{1 + \frac{f^2}{b^2}}\right)\right], \tag{5.9.56}$$

with the arbitrary constant again chosen to satisfy $z = 0$ at $f = 0$. This solution and that of (5.9.55) approach those for the free layer shape for f small and those for the blast wave shape for f large.

All but one of the approximations made appear to have the property that they approach correctness in some adequate sense in the limit $\epsilon \to 0$. Most of the errors entailed may readily be shown to be small. There is an error in the slender body Newtonian formula because the impulse in the shock layer (the term f^{1+j} in (5.9.50)) is reduced by the velocity defect. This gives an error of order ϵ in (5.9.50). There is an error because with a perfect gas it is the quantity p/ρ^γ which is constant along a streamline rather than the quantity p/ρ.

This gives a correction analogous to the perturbation for the constant-velocity displacement effect from the entropy layer discussed in Section 2.7. This correction appears also to approach zero in the limit $\gamma \to 1$, although this has not been demonstrated. And of course there are all the errors inherent in any Newtonian analysis.

The one assumption that clearly gives an error which does not vanish in the limit $\epsilon \to 0$ is the assumption that the velocity u is constant along streamlines. The Bernoulli effect which is neglected causes u to approach U on each streamline as the flow goes toward downstream infinity (with $M_\infty = \infty$, of course). The innermost streamlines in the flow have a finite velocity defect determined by the given shape of the forward part of the shock; this defect does not vanish in the limit $\epsilon \to 0$. It is these same innermost streamlines which provide the principal contribution to the integral (5.9.44).

Freeman corrects for this effect by replacing the constant on the right-hand side of (5.9.45) by a variable, one which changes slowly far downstream. The right-hand side of (5.9.45) varies from ϵA for moderate values of z to ϵA_D for very large values of z in this modification. Freeman's modification is represented in a form which is asymptotically correct by replacing (5.9.49b) by

$$y_s = \epsilon^{-\frac{1}{1+j}} f(z) \frac{g(z^\epsilon)}{g(1)}, \tag{5.9.57}$$

where $f(z)$ is the function given by the unmodified analysis. For large values of z^ϵ, g approaches the value appropriate with A replaced by A_D. This approach is consistent in manner with the $2N$ exponent characteristic of the velocity-defect displacement effect discussed in the paragraph following (2.7.47).

Freeman's modification, though informative, is still open to some question in the opinion of the authors. This question lies in the fact that the error in the assumption of the constancy of p/ρ has not been investigated. It is possible that this error remains also of finite magnitude in the $\epsilon \to 0$ limit and needs to be taken into account with the velocity defect in modifying the basic analysis. If this turns out to be the case, we can expect this effect to extend further downstream than the effect of the velocity defect, and to be characterized by the exponent N in some sense rather than by the exponent $2N$ of the velocity defect.

Another possibility suggested by Freeman (private communication) is that another error may be important in this asymptotic region far downstream. This is the error which comes from the use of the assumed forward shock shape with free layer in the displacement effect calculation, without taking into account the development of the shock shape following (5.9.45). It is possible that the correction for this error may entail a perturbation characterized by the exponent N. In any case, the problem deserves further study.

10. Nonequilibrium flows

In a nonequilibrium flow, thin-shock-layer theory is applicable if the densities in the shock layer are of a larger order of magnitude than the free stream density. The conditions for validity and the basic approaches used are essentially no different from those discussed for equilibrium flows in Chapters IV and V. The principal difference is that the density follows a different law, and, for example, cannot be considered constant in a stagnation region flow or a blunt-faced body flow.

The same distinction between direct and inverse problems exists, of course. Inasmuch as the solution of direct problems depends critically on the laws governing the density, direct problems with nonequilibrium flow may be expected to be more difficult than those for blunt-faced bodies treated earlier in this chapter for which the assumption of constant density holds. The same general approach should apply to such problems, however. In this section we limit ourselves to the simpler inverse problems.

The distinction between general flows in a thin shock layer and the more restricted stagnation-region flows or blunt-faced body flows applies here also. In the more general flows the approximation of constant pressure may not be made. In addition to this distinction, the conceptual division of the solution for blunt bodies into an outer solution and an inner solution applies. In the outer solution Newtonian theory is applicable, while in the inner solution a Bernoulli correction is needed.

Freeman [2] investigated a nonequilibrium ideal dissociating gas with thin-shock-layer theory, obtaining a distribution $u(\psi)$ of velocity with stream function from Newtonian theory; the velocity u and hence also the enthalpy are constant along streamlines. Freeman's solution is for the outer layer; he discussed briefly the problem of finding an inner solution. As was indicated in the first edition of this book (p. 395), Freeman's solutions agree with the corresponding constant-density solutions if expressed in terms of the Howarth-Dorodnitsyn variable. Freeman obtained dissociation parameter profiles and other results for the flow of his model gas about a sphere.

The basic physical independent variable is the time measured for a particle from its entry into the shock layer. We denote x and x_1 as the boundary layer coordinates at entry and at the point of interest, in a notation consistent with that used through Chapters IV and V. The variable x also serves in place of the stream function. With the Newtonian result that $u(x)$ is constant with x_1, the time t may be expressed

$$t(x, x_1) = (x_1 - x)/u(x). \tag{5.10.1}$$

The pressure is known from Newtonian theory as a function of x and x_1, and hence of x and t. With the thermodynamic state behind the shock ($t = 0$) and the pressure–time history known, the thermodynamic laws governing the reaction or reactions in question yield the history of the density $\rho(x, t)$ with respect to time. The variable x is a parameter in this computational process.

One simplification does appear here, noted by Freeman. Within the assumptions of Newtonian theory the enthalpy is constant along a streamline, and this condition may be used in place of the condition that the flow is adiabatic in calculating the density–time history. Inasmuch as the temperature cannot consistently be considered constant, this gives little or no real simplification in a direct computation.

A real simplification appears only when a relation between the actual density history for a streamline and the density history behind a normal shock is to be utilized. In a stagnation region, or in the flow field on a blunt-faced body, this useful and important simplification is available. The simplification is that we may assume the pressure (as well as the enthalpy) to be constant for the purpose of computing the density history. The same simplification applies directly to a normal shock with the same normal velocity. The approximate constancy of the pressure in these cases is a consequence of the assumptions that the shock layer is thin and that the velocity component normal to the shock (or to the reference surface for a blunt-faced body) is very small. In these cases the density $\rho(t)$ is a function of the time variable t alone, the same function for all streamlines and for all cases with the same normal velocity. As was indicated in the first edition of this book (p. 394), the time variable t is the basic independent variable to be considered in this case, as it gives a direct correspondence between the density histories in the normal shock and stagnation-region cases. The correspondence of stagnation-region solutions with that for a normal shock was noted by Gibson and Marrone [1]. Within the class of inverse problems we are considering, for the outer solutions, at each point on the body t is a function of the Howarth-Dorodnitsyn variable which does not depend upon the particular density law $\rho(t)$ applicable.

In the more general case, with the pressure not approximately constant, the effect of the changing pressure upon the rates of reaction and upon the equilibrium constitution should be taken into account. In general, the results for a normal shock are not directly useable, although relations between rates of reaction may be available. In one special case, that in which the reaction or reactions involved follow a simple scaling law, most of the simplicity of the constant-pressure case can be retained. The assumption of constant enthalpy is still needed. Following the concepts introduced by Gibson and Marrone [1; 2], we may introduce the new variable

$$\chi = \int_0 p^{N-1} \, dt \tag{5.10.2}$$

to replace the time with Nth order scaling. The usual case considered is that of binary scaling, with $N=2$. The hidden state variables are then functions of χ alone. The density for fluid which has passed through a shock of given strength is then expressible as $\rho(\chi, p)$, with the pressure dependence at fixed χ the same as that for isentropic changes in a frozen state. The dependence with χ at fixed pressure equal to that behind a normal shock is then the same as that for the corresponding normal shock. Gibson and Marrone refer to their method using χ as basic variable as "shock mapping". For a stagnation region shock mapping reduces to the approach already described, using t as basic variable. Their observation that $\rho(t)$ or $\rho(\chi)$ may be obtained directly from a normal shock solution applies in this case.

To illustrate the discussion above for the outer solution we consider a simple relaxation process, following essentially Blythe [1]. The process assumed is a model for vibrational relaxation, similar to that suggested in Section 2.11. The enthalpy is assumed to be given by the expression

$$h = c_{p_f} T + e_{\text{vib}}, \tag{5.10.3}$$

where c_{p_f} is the frozen specific heat at constant pressure and e_{vib} is the energy in the hidden mode, here considered as vibrational. Derivation of (5.10.3) as applied to equilibrium conditions leads to

$$c_p = c_{p_f} + \left(\frac{\partial e_{\text{vib},e}}{\partial T}\right)_p, \tag{5.10.4}$$

where c_p is the total equilibrium specific heat and $e_{\text{vib},e}$ is the equilibrium value of e_{vib} at the temperature T.

The relaxation equation is assumed to have the form

$$\frac{De_{\text{vib}}}{Dt} = \left(\frac{e_{\text{vib},e} - e_{\text{vib}}}{t_{\text{vib}}}\right), \tag{5.10.5}$$

where D/Dt indicates a time derivative along a streamline and $e_{\text{vib},e}$ is taken at the local translational temperature T. The quantity t_{vib} is the relaxation time. The process considered is one at constant enthalpy and pressure. Thus, from (5.10.3), we have

$$c_{p_f} \frac{DT}{Dt} = -\frac{De_{\text{vib}}}{Dt}, \tag{5.10.6}$$

and from (5.10.4),

$$\frac{De_{\text{vib},e}}{Dt} = (c_p - c_{p_f}) \frac{DT}{Dt}. \tag{5.10.7}$$

A suitable combination of (5.10.5), (5.10.6), and (5.10.7) gives

$$\frac{D(e_{\mathrm{vib},e} - e_{\mathrm{vib}})}{Dt} + \frac{c_p}{c_{p_f}}\left(\frac{e_{\mathrm{vib},e} - e_{\mathrm{vib}}}{t_{\mathrm{vib}}}\right) = 0. \tag{5.10.8}$$

With the additional simplifying assumptions that t_{vib} and c_p are constant, we obtain

$$e_{\mathrm{vib},e} - e_{\mathrm{vib}} = (e_{\mathrm{vib},e} - e_{\mathrm{vib}})_s \exp\left(-\frac{c_p}{c_{p_f}}\frac{t}{t_{\mathrm{vib}}}\right). \tag{5.10.9}$$

With the related assumption that e_{vib} varies linearly with T in the region of interest, $T - T_e$ follows the same law as does $e_{\mathrm{vib},e} - e_{\mathrm{vib}}$ in (5.10.9). With the gas assumed to behave as a perfect gas with respect to the translational temperature, the constant-pressure assumption leads to the same law for $\rho^{-1} - \rho_e^{-1}$.

We write $\epsilon = \rho_\infty/\rho$ as a dependent variable in place of ρ^{-1}. We write ϵ_f for the frozen value of ϵ immediately behind the shock, and express

$$\epsilon = \rho_\infty/\rho, \tag{5.10.10a}$$

$$\epsilon_f = \rho_\infty/\rho_s, \tag{5.10.10b}$$

$$\epsilon_e = \rho_\infty/\rho_e. \tag{5.10.10c}$$

The result (5.10.9) in terms of ϵ is then

$$\epsilon - \epsilon_e = (\epsilon_f - \epsilon_e) \exp(-t/\tau), \tag{5.10.11}$$

where

$$\tau = \frac{c_{p_f}}{c_p} t_{\mathrm{vib}}. \tag{5.10.12}$$

Despite the various simplifying assumptions, (5.10.11) may be considered representative of the laws which may govern $\rho(t)$, and may be proposed by itself as a model. Note that the characteristic time τ of the density law is not the same as the characteristic time t_{vib} of the relaxation law (5.10.5).

We apply the density law (5.10.11) to the stagnation region of a blunt body, following the method and notation of Sections 4.3 and 4.4 except that the velocities are not reduced by the factor U. We examine the outer layer solution first. The entering value of the velocity u (cf. (4.3.20) or (4.4.14)) is given by

$$u = \frac{xU}{R_s}. \tag{5.10.13}$$

From (5.10.1) we obtain

$$\frac{t}{\tau} = \Lambda \left(\frac{x_1}{x} - 1\right) \tag{5.10.14}$$

where

$$\Lambda = R_s/U\tau \tag{5.10.15}$$

is Blythe's parameter, analogous to the parameter Λ of Freeman. This parameter is the ratio of a time characteristic of the macroscopic flow to a characteristic time for the density law. Equilibrium flow corresponds to $\Lambda = \infty$, and frozen flow corresponds to $\Lambda = 0$. The new variable s is introduced by

$$s = \Lambda \frac{x_1}{x} = \frac{t}{\tau} + \Lambda. \tag{5.10.16}$$

The continuity condition of the form of (4.3.21) or (4.4.15) may be written, using (5.10.13), as

$$dy = \epsilon R_s \left(\frac{x}{x_1}\right)^j \frac{dx}{x}, \tag{5.10.17}$$

with $j = 0$ in the two-dimensional case and $j = 1$ in the axisymmetric case. The quantity ϵ here is now a variable, for which (5.10.11) may be written as

$$\epsilon = \epsilon_e + (\epsilon_f - \epsilon_e)e^{\Lambda - s}. \tag{5.10.18}$$

The indefinite quadrature of (5.10.17) may now be expressed as

$$y_s - y = R_s\Lambda^j \int_\Lambda \epsilon \frac{ds}{s^{1+j}}, \tag{5.10.19}$$

or, with (5.10.18), as

$$y_s - y = (y_s - y)_e + (\epsilon_f - \epsilon_e)R_s\Lambda^j e^\Lambda \int_\Lambda \frac{e^{-s}\,ds}{s^{1+j}}. \tag{5.10.20}$$

The inner solution, which we have not investigated, corresponds to large values of s. If we may approximate $\epsilon = \epsilon_e$ in the inner layer, i.e. if $e^{\Lambda - s}$ is negligibly small there, the inner layer may be shown to be taken care of by the term $(y_s - y)_e$ in (5.10.20). With this limitation we may write from (5.10.20)

$$\Delta - \Delta_e = (\epsilon_f - \epsilon_e)R_s\Lambda^j e^\Lambda \int_\Lambda^\infty \frac{e^{-s}\,ds}{s^{1+j}} \tag{5.10.21}$$

as a result for the total detachment distance Δ. In the case $j = 0$ this result is

$$\Delta - \Delta_e = (\epsilon_f - \epsilon_e)R_s e^{\Lambda} ei(\Lambda), \tag{5.10.22}$$

where

$$ei(\Lambda) = \int_{\Lambda}^{\infty} \frac{e^{-s}\,ds}{s}. \tag{5.10.23}$$

In the case $j = 1$ we obtain

$$\Delta - \Delta_e = (\epsilon_f - \epsilon_e)R_s[1 - \Lambda e^{\Lambda} ei(\Lambda)]. \tag{5.10.24}$$

If the effect of the inner layer is neglected in this case, we may write $\Delta_f - \Delta_e$ in place of $(\epsilon_f - \epsilon_e)R_s$ in (5.10.24), and thereby obtain Blythe's result.

We turn next to the structure of an inner layer in the stagnation region of a blunt body, with $\epsilon(t)$ a known function of time. The velocity u is no longer constant, but obeys the Bernoulli law

$$u\,du = 2\pi_j \epsilon \frac{U^2 x_1\,dx_1}{R_s^2} \tag{5.10.25}$$

along a streamline. Here $\pi_j = \frac{3}{2}$ for $j = 0$ and $\frac{4}{3}$ for $j = 1$, and (5.10.13) serves as an initial condition at $x_1 = x$. The relation

$$dx_1 = u\,dt \tag{5.10.26}$$

along a streamline replaces (5.10.1), with $t = 0$ at $x_1 = x$. From the last two equations we may write

$$\frac{Du}{Dt} = \frac{2\pi_j U^2}{R_s^2} \epsilon x_1. \tag{5.10.27}$$

Differentiation leads to the equation

$$\frac{D}{Dt}\left(\frac{1}{\epsilon}\frac{Du}{Dt}\right) - \frac{2\pi_j U^2}{R_s^2} u = 0, \tag{5.10.28}$$

for which (5.10.13) and (5.10.27) give the initial conditions at $t = 0$, $x_1 = x$. Solution of this equation gives the inner layer structure. If ϵ is not constant in the inner layer the solution of this equation is essential in the two-dimensional case $j = 0$ for the shock layer structure.

To complete our discussion of flows on blunt bodies, we give a classification of the solution regimes for nonequilibrium stagnation region flows on such bodies. Although the model solutions given above may be used directly for this purpose and certainly provide much insight, we use them only to obtain characteristic times for the flow. The classification will apply to any case for which the density history $\rho(t)$ or $\epsilon(t)$ is characterized by a characteristic time τ. An immediate result from the Newtonian solutions is that the characteristic time of penetration of a fluid particle into the outer layer is R_s/U. An analysis of the inner layer based, for example, on (5.10.28) gives the result that the characteristic time of penetration into the inner layer is $R_s/\epsilon^{1/2}U$.

We distinguish five cases according to the magnitude of the parameter Λ defined by (5.10.15). This classification is given in Table 5–2. In case *A* the entire flow regime is in equilibrium with the exception of a very thin reaction zone immediately behind the shock. In this zone the solution is essentially that for a normal shock. In case B the outer solution comprises the reaction zone and the inner flow is in equilibrium. It is this case which is considered above in our treatment of the outer solution. In both case A and case B the Howarth-Dorodnitsyn variable is unchanged relative to an equilibrium solution.

In case C the reaction is proceeding within both the outer solution and the inner solution, while in case D the reaction is effectively confined to the inner

TABLE 5–2

Classification of nonequilibrium stagnation regimes

Case	Condition	Description
A	$\Lambda \gg 1$	equilibrium outer
B	$\Lambda \gg \epsilon_e^{1/2}, \quad \Lambda \not\gg 1$	equilibrium inner
C	$1 \not\gg \Lambda \not\gg \epsilon_e^{1/2}$	mixed
D	$1 \gg \Lambda, \ \epsilon_f^{1/2} \not\gg \Lambda$	frozen outer
E	$\epsilon_f^{1/2} \gg \Lambda$	frozen inner

solution. There is little practical difference between these two cases. If $\epsilon_e^{1/2}$ is sufficiently small it is possible to eliminate case C by a suitable assignment of the reacting part of the overlap region to either the inner or outer solution, thus transferring the case to case D or to case B, respectively. In case E the entire flow regime is frozen with the exception of a very thin reaction zone next to the body. This classification is illustrated in Fig. 5–33, in which the reaction zone is shown shaded.

Finally, we apply the model of (5.10.11) to the quasi wedge and quasi cone analysis of Section 5.5. We introduce the variable

$$\lambda = \frac{x}{u\tau} = \frac{x}{U\tau \cos \alpha}, \tag{5.10.29}$$

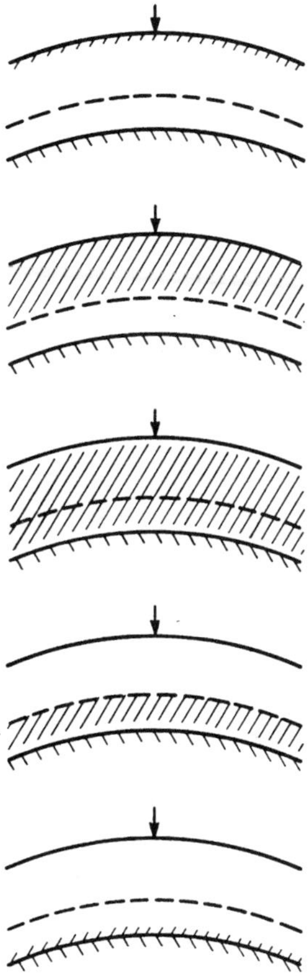

FIG. 5–33. Nonequilibrium stagnation regime classification.

and identify the time as

(5.10.30) $$t = \frac{x_1 - x}{u} = \tau(\lambda_1 - \lambda).$$

The formula (5.10.11) for ϵ becomes

(5.10.31) $$\epsilon = \epsilon_e + (\epsilon_f - \epsilon_e)\, e^{\lambda-\lambda_1}.$$

The form of (5.5.2) holds, and in differential form for x_1 constant is

(5.10.32) $$dm = U\tau \sin \alpha \left(\frac{\lambda}{\lambda_1}\right)^j d\lambda.$$

In place of (5.5.3) we have the differential relation

$$dy = -\epsilon \, dm. \tag{5.10.33}$$

Substitution of (5.10.31) and (5.10.32), with the condition $y = y_b$ at $\lambda = 0$, leads to the quadrature

$$y = y_b - \frac{U\tau \sin \alpha}{\lambda_1^j}\left[\frac{\epsilon_e \lambda^{1+j}}{1+j} + (\epsilon_f - \epsilon_e)\, e^{-\lambda_1}\{(\lambda - 1)^j e^{\lambda} - (-1)^j\}\right]. \tag{5.10.34}$$

This equation replaces (5.5.3). In place of (5.5.4) we have for the shock layer thickness

$$y_b - y_s = U\tau \sin \alpha \left[\frac{\epsilon_e \lambda_1}{1+j} + (\epsilon_f - \epsilon_e)\{(1 - \lambda_1^{-1})^j - (-\lambda_1^{-1})^j e^{-\lambda_1}\}\right]. \tag{5.10.35}$$

To permit comparison we recast (5.10.35) in the form

$$y_b - y_s = \frac{x_1 \tan \alpha}{1+j}\,[\epsilon_e + (\epsilon_f - \epsilon_e) F(\lambda_1)], \tag{5.10.36}$$

where

$$F(0) = 1, \tag{5.10.37}$$

and

$$F(\lambda_1) \sim \frac{1+j}{\lambda_1} \tag{5.10.38}$$

as $\lambda_1 \to \infty$. The asymptotic value of the shock layer thickness is thus that given by (5.5.4) with the equilibrium value of ϵ, but with an additional term obtainable from (5.10.38) interpretable as coming from the structure of the relaxation zone behind a normal shock.

Differentiating (5.10.34) leads for the streamline slope to

$$\begin{aligned} y' &= \frac{1}{U\tau \cos \alpha}\frac{\partial y}{\partial y_1} \\ &= y_b' + \frac{\tan \alpha}{\lambda_1^{1+j}}\left[\frac{j\epsilon_e \lambda^{1+j}}{1+j} + (j + \lambda_1)(\epsilon_f - \epsilon_e)\, e^{-\lambda_1}\{(\lambda - 1)^j e^{\lambda} - (-1)^j\}\right] \end{aligned} \tag{5.10.39}$$

in place of (5.5.5), and for the curvature to

$$y'' = y_b'' - \frac{\tan\alpha}{U\tau\cos\alpha\,\lambda_1^{2+j}}[j\epsilon_e\lambda^{1+j} + \{j(1+j+2\lambda_1) + \lambda_1^2\}(\epsilon_f - \epsilon_e)\,e^{-\lambda_1}\{(\lambda-1)^j e^{\lambda} - (-1)^j\}] \tag{5.10.40}$$

in place of (5.5.6). In place of (5.5.7) by integrating $\rho_\infty U^2 \cos^2\alpha\, y''\, dm$ we obtain the pressure

$$\begin{aligned} p = p_b &+ \rho_\infty U^2 \sin\alpha\cos\alpha \left(\frac{\lambda}{\lambda_1}\right)^{1+j} \frac{x_1}{1+j} y_b'' \\ &- \frac{\rho_\infty U^2 \sin^2\alpha}{\lambda_1^{2+2j}}\left[\frac{j\epsilon_e\,\lambda^{2+2j}}{2(1+j)} + \{j(1+j+2\lambda_1) \right. \\ &\left. + \lambda_1^2\}(\epsilon_f - \epsilon_e)\,e^{-\lambda_1}\{(\lambda^2 - 3\lambda + 3)^j e^{\lambda} - 3^j + \frac{(-\lambda)^{1+j}}{1+j}\right], \end{aligned} \tag{5.10.41}$$

while in place of (5.5.9) we have for the pressure at the shock

(5.10.42)

$$\begin{aligned} p_s = p_\infty &+ \rho_\infty U^2 \sin^2\alpha - 2\rho_\infty U^2 \sin\alpha\cos\alpha\, y_b' \\ &+ \rho_\infty U^2 \sin^2\alpha \left[-\epsilon_f + \frac{2\epsilon_e}{1+j} + 2(\epsilon_f - \epsilon_e)\{(-\lambda_1^{-1})^j e^{-\lambda_1} + j\lambda_1^{-2}(1 - e^{-\lambda_1})\}\right]. \end{aligned}$$

Evaluation of (5.10.41) on the shock and combination with (5.10.42) leads to the analogue of (5.5.10) for the pressure on the body

(5.10.43)

$$\begin{aligned} p_b - p_\infty - \rho_\infty U^2 \sin^2\alpha = &-\rho_\infty U^2 \sin\alpha\cos\alpha\left(2y_b' + \frac{x_1 y_b''}{1+j}\right) \\ &+ \rho_\infty U^2 \sin^2\alpha\left[-\epsilon_f + \frac{4+j}{2(1+j)}\epsilon_e \right. \\ &+ (\epsilon_f - \epsilon_e)\left\{\frac{j(1+j+2\lambda_1) + \lambda_1^2}{\lambda_1^{2+2j}}\left((\lambda_1^2 - 3\lambda_1 + 3)^j \right.\right. \\ &\left. - 3^j e^{-\lambda_1} + \frac{(-\lambda_1)^{1+j}}{1+j}\epsilon^{-\lambda_1}\right) \\ &\left.\left. + 2\left((-\lambda_1^{-1})^j e^{-\lambda_1} + j\lambda_1^{-2}(1 - e^{-\lambda_1})\right)\right\}\right]. \end{aligned}$$

We put this into the form

(5.10.44)

$$p_b - p_\infty - \rho_\infty U^2 \sin^2 \alpha = -\rho_\infty U^2 \sin \alpha \cos \alpha \left(2y_b' + \frac{x_1 y_b''}{1+j}\right) + \rho_\infty U^2 \sin^2 \alpha \frac{2+j}{2(1+j)} [\epsilon_e + (\epsilon_f - \epsilon_e) G(\lambda_1)].$$

where

(5.10.45) $$G(0) = 1,$$

and

(5.10.46) $$G(\lambda_1) \to 0$$

as $\lambda_1 \to \infty$. Equations (5.10.36) and (5.10.44) show that the results of Section 5.5 hold with our model relaxing fluid with correction terms that are functions of the parameter $\lambda_1 = x_1/U\tau \cos \alpha$ alone. In the case $j = 0$ they may be interpreted directly in terms of the solution for a normal shock. With $j = 1$ the expression for the shock shape (5.10.36) reproduces a result of D. Chapman (see Stephenson [1]). Sedney and Gerber [1] have compared the Chapman result with characteristics calculations and find excellent agreement. This is shown in Fig. 7–25. A comparison with numerical calculations of the pressure relations and the $j = 0$ case would also be of interest.

CHAPTER VI

NUMERICAL METHODS FOR BLUNT-BODY FLOWS

1. Nature of the problem

In our work so far we have examined approximate solutions for the inviscid hypersonic flow over blunt bodies based on the assumption of constant density in the shock layer or on the assumption of a very thin shock layer. It seems natural for us to now inquire into the possibility of computing such flows exactly with the aid of high speed computers. We shall generally treat the fluid as if it were in thermodynamic equilibrium and shall look into the changes associated with the flow field geometry. In Section 6.6 we indicate how non-equilibrium effects may be accounted for. In spite of the amount of effort that has gone into this problem in recent years, at present no single numerical method has been agreed on as being the best for calculating the hypersonic flow past general blunt shapes. This is particularly true with asymmetry, such as that associated with angle of attack.

The problem which can be considered to be the important one met in practice is what we term the direct problem (cf. Section 5.1), in which the shape of the body is given and the details of the flow field are unknown. We shall also consider in Section 6.5 the inverse problem, in which the shock shape is given and the shape of the body and the pressure on it are unknown. We take the attitude that the direct problem is more important than the inverse problem, and we shall accordingly emphasize the direct problem. This point of view was also taken in Chapter V. There is some possibility of using methods developed for the inverse problem for an attack on the more difficult direct problem by a successive approximation scheme. However, such an approach would seem to involve procedures of trial and error, in which it is not always necessarily clear how a solution which gives an approximation to the desired body is to be improved.

We first consider the nature of symmetric two-dimensional and axisymmetric flows, deferring the discussion of asymmetry until later. The difficulty in solving the mathematical problem of determining the detached shock wave and the flow field behind the shock for a given body (direct problem) lies principally in the fact that the flow around the body is a nonlinear mixed subsonic-supersonic flow, with a free boundary which is not known in advance. This free boundary is the bow shock wave in front of the body. Also unknown is the location of what we shall term the "limiting characteristic", which generally intersects the sonic line at one point (see Fig. 6–1). The limiting characteristic in an accelerating flow may be defined as the locus of points each of which has only one point of the sonic line in its zone of action. The transonic or pseudoelliptic character of the supersonic flow in the region between the sonic line and the limiting characteristic must be taken into account; a disturbance in this region would affect the shape of the sonic line and hence also the subsonic flow field upstream. This particular supersonic region will be referred to in this chapter as the "transonic zone". In Fig. 6–1 we have shown the structure of this region

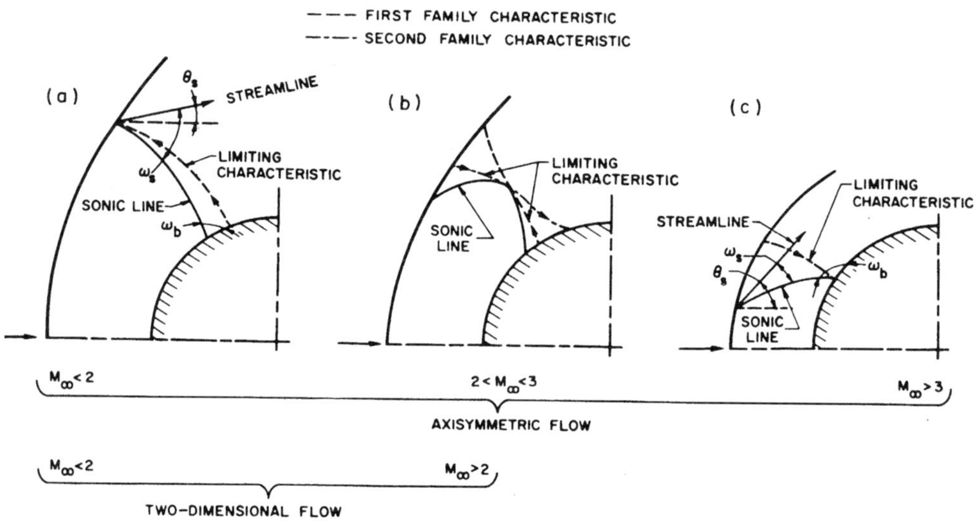

FIG. 6–1. Sketch of sonic line and limiting characteristic shapes for a sphere and a circular cylinder with $\gamma \approx 1.4$. Values of M_∞ are approximate.

schematically for three different Mach number ranges, for the flow of air past a hemispherical nose and a circular cylinder normal to the stream direction. We shall justify the details of the diagrams below. We note that in the moderate M_∞ case shown in Fig. 6–1(b) the limiting characteristic is divided by the point of tangency with the sonic line into two characteristics of different families.

The supersonic domain downstream of the limiting characteristic can always be handled by the method of characteristics (see Section 7.1) once the sonic line or limiting characteristic shape and flow conditions on this shape are known. However, it is not the solution in this purely supersonic downstream region which is of the greatest importance to our problem, but rather that in the initially undetermined subsonic region and the transonic zone.

In order to illustrate in the simplest manner possible how one arrives at the limiting characteristic shapes which are sketched in Fig. 6–1, let us examine some of the geometrical characteristics of the sonic line for supersonic and hypersonic flows past blunt bodies (cf. Probstein [2]). Such considerations can help us to obtain a better appreciation of how a body in hypersonic flow influences the upstream flow in the shock layer, and how this influence is different from that ordinarily associated with detached shock waves at low and moderate supersonic speeds. For most of these considerations we shall use the assumption of a perfect gas of constant γ. We may expect that the conclusions which we reach should hold for most general fluids which do not have some anomalous behavior. It is possible to obtain most of the same primary results for a general fluid with some additional labor and a judicious choice of assumptions on the behavior of the material.

Our method will be to estimate the location of the sonic point and the inclination angle of the sonic line both on the body and on the shock. With this information we shall be able to obtain a picture of the geometry of the sonic line and of the limiting characteristic (cf. also Traugott [1]).

The geometrical quantity which best characterizes the behavior of the sonic line near the body is the angle it makes with the body. We define this angle ω_b to be the angle between the sonic line leaving the body and the direction of the body surface downstream of the sonic point. We shall be primarily interested in whether this angle is acute or obtuse, as this question is important in determining whether the body shape has an influence in the transonic zone.

The angle that the sonic line makes with a streamline (in this case the body surface) is given for a perfect gas by

$$\tan \omega = -\frac{\partial q}{\partial s}\bigg/\frac{\partial q}{\partial n}, \tag{6.1.1}$$

where q is the total velocity and s and n are intrinsic coordinates (see Fig. 7–1). These intrinsic coordinates are discussed in detail in Section 7.1. The s axis here lies in the flow direction at the intersection of the streamline and sonic line. The relation (6.1.1) follows directly from the fact that along the sonic line $dq = 0$ for an isoenergetic flow.

To determine $(\partial q/\partial n)_b$ we utilize the expression for the vorticity in intrinsic coordinates

$$\frac{\partial q}{\partial n} = -\zeta - \frac{q}{R}, \tag{6.1.2}$$

where ζ is the vorticity, and R is the local radius of curvature of the streamline equal to $-ds/d\theta$. The quantity R is positive at the surface of a convex body. In a two-dimensional flow on a blunt body the entropy gradient at the body surface is zero, and Crocco's vorticity law (see Section 1.5) gives us the result that the vorticity there is zero (cf. Section 4.3). From (6.1.2) we see that in two-dimensional flow on a blunt convex body $(\partial q/\partial n)_b < 0$, regardless of the value of the Mach number of the flow. Since $(\partial q/\partial s)_b$ is positive, it follows from (6.1.1) that in two-dimensional flow the sonic line is always inclined at an acute angle to the surface, as shown in Figs. 6–1(a) and 6–1(b). In axisymmetric flow the vorticity at the body surface is finite, and we shall find that whether the angle ω_b is acute or obtuse depends strongly on the density ratio ϵ and on the shape of the body. We shall next obtain an expression for the vorticity at the body surface in an axisymmetric flow.

In the vicinity of the axis of a detached shock we may approximate $\sin \sigma$ by 1, and both $-dr_s/d\sigma$ and $r_s/\cos \sigma$ by the shock radius of curvature R_s. The quantity r here is the cylindrical radius in our axisymmetric problem. From (1.5.9) we obtain the local result

$$\frac{dS}{d\psi} = -\frac{(1-\epsilon)^2 U}{\rho_\infty T_s R_s^2}, \tag{6.1.3}$$

for axisymmetric flow. This result is obtained in terms of conditions immediately behind the shock, but since S is a function of ψ alone the result is valid everywhere for the streamtube nearest the body surface. Using Crocco's vorticity law we may calculate the vorticity at a point on the body

$$\zeta_b = \frac{T_b}{q_b}\left(\frac{\partial S}{\partial n}\right)_b = -\frac{(1-\epsilon)^2}{\epsilon}\left(\frac{\rho_b T_b}{\rho_s T_s}\right)\left(\frac{r_b U}{R_s^2}\right). \tag{6.1.4}$$

We have made the assumption that the gas is a perfect one in (6.1.1), and we now express $(\partial q/\partial n)_b$ at the sonic point from (6.1.2) with this assumption as

$$\left(\frac{\partial q}{\partial n}\right)_{b,\mathrm{son}} = \left(\frac{q_b}{R_b}\right)_{\mathrm{son}}\left[\frac{p_b}{p_s}\left(\frac{R_b r_b}{R_s^2}\right)\left(\frac{(1-\epsilon)^2}{\epsilon}\frac{U}{q_b}\right) - 1\right]_{\mathrm{son}} \tag{6.1.5}$$

It should be kept in mind that the subscript s in (6.1.3) to (6.1.5) refers to conditions immediately behind the shock on the axis of the body. The particular characteristic of the quantity expressed in (6.1.5) which is of interest to us is its sign. From (6.1.1) and the fact that $(\partial q/\partial s)_b$ is positive, this sign determines whether ω_b is greater or less than $\frac{1}{2}\pi$. The quantity p_b/p_s is about $\frac{1}{2}$ at the sonic point if ϵ is reasonably small (say less than $\frac{1}{2}$). The quantity $R_b r_b/R_s^2$ is also of the order of $\frac{1}{2}$ for a sphere if ϵ is small enough (say less than $\frac{1}{5}$), but may be much less than one if ϵ is close to one.

The remaining factor in the first term in the bracket of (6.1.5) is very small at low values of M_∞, for which $1-\epsilon$ is small. Thus, with $M_\infty - 1$ sufficiently small, ω_b is less than $\frac{1}{2}\pi$ and the angle of interest to us is acute. At very high values of M_∞, q_b/U is of the order of magnitude $\epsilon^{1/2}$ at the sonic point, ϵ may be approximated by ϵ_{lim}, and the first term in the bracket of (6.1.5) may be greater than 1 if ϵ_{lim} is sufficiently small.

For our perfect gas of constant γ we may derive the formula

$$\frac{(1-\epsilon)^2}{\epsilon}\frac{U}{q_{b,\text{son}}} = \sqrt{\frac{2}{\gamma+1}}\frac{(M_\infty^2-1)^2}{M_\infty\left(1+\frac{\gamma-1}{2}M_\infty^2\right)^{3/2}}, \tag{6.1.6}$$

and we note that this approaches $4\epsilon_{\text{lim}}^{-3/2}/(\gamma+1)^2$ in the limit as $M_\infty \to \infty$. With helium this limiting value is 4.5, with a diatomic perfect gas it is 10.2, while with a perfect gas with $\gamma = 1.2$ it is 30.2.

The quantity $R_b r_b/R_s^2$ depends critically on the geometry of the body, and also on the value of ϵ. For a sphere it is of the order of $\frac{1}{2}$ with ϵ small, but is probably less than $\frac{1}{4}$ for $\epsilon = \frac{1}{4}$. Thus, on a sphere we may conclude that the angle ω_b is always acute in helium even in the limit $M_\infty \to \infty$, and that it is obtuse in air ($\gamma = 1.4$) if M_∞ is greater than about 2.5 or 3.0. On a body with a flattened face and a rounded corner the quantity $R_b r_b/R_s^2$ will be very small and the angle ω_b may remain acute even in a perfect gas with $\gamma - 1$ small. And it is possible to find a body shape such as a blunted ogive for which R_s is small, R_b at the sonic point is large, and the factor $R_b r_b/R_s^2$ is very large. On such a body the angle would be obtuse even at moderate values of M_∞ in helium.

For bodies for which the radius of curvature does not vary greatly over the front of the body, the location of the body sonic point does not depend greatly on either the Mach number or the gas properties. The critical pressure ratio p_{son}/p_0 varies for a perfect gas only from 0.59 at $\gamma = 1.1$ to 0.49 at $\gamma = 5/3$. Here the subscript 0 refers to stagnation conditions. The pressure on the body divided by the stagnation pressure varies roughly as $\sin^2\theta_b$ ("modified Newtonian"), and is somewhat less than this if γ is small. Thus at hypersonic speeds the body sonic point will occur somewhat farther downstream for a gas with a higher value of γ because of the lower pressure required. Nevertheless, the sonic point movement will not be very large (compare for example the experiments of Oliver [1] in air and those of Vas, Bogdonoff, and Hammitt [1] in helium). This fact regarding the sonic point location is significant in determining the structure of the transonic zone.

The distance from the axis to the sonic point on the shock is roughly proportional to the radius of curvature of the shock at the axis R_s times $\cos\sigma_{\text{son}}$. As the Mach number increases, both of these quantities decrease and the shock sonic point moves toward the axis. The limiting value of $\cos\sigma_{\text{son}}$ as $M_\infty \to \infty$ is $\sqrt{(\gamma-1)/2\gamma}$, and this limit varies from $63\frac{1}{2}^\circ$ for $\gamma = 5/3$ to 90° for $\gamma = 1$.

One of the final quantities to be determined is the angle the sonic line makes with the flow direction at the shock wave. To do this we again use (6.1.1) and evaluate the velocity derivatives from the oblique shock relations. The result for plane flow was obtained by Hasimoto [1], Drebinger [1], Belotserkovskii [1], and others. This result is independent of the shock curvature and may be expressed in terms of the shock angle and flow deflection angle as

$$\tan \omega_s = \frac{\tan^3(\sigma - \theta_s)[3(\gamma + 1)\tan^2(\sigma - \theta_s) + 5 - \gamma]}{[1 - \tan^2(\sigma - \theta_s)][(\gamma + 1)\tan^2(\sigma - \theta_s) + 2]}, \tag{6.1.7}$$

where ω_s is negative when measured clockwise from the streamline direction (see Fig. 6–1). The angle ω_s is the angle between the streamline direction and the direction of the sonic line leaving the shock, and is thus the negative of the ω in (6.1.1) evaluated at the shock. In Fig. 6–2 we have plotted both the sonic

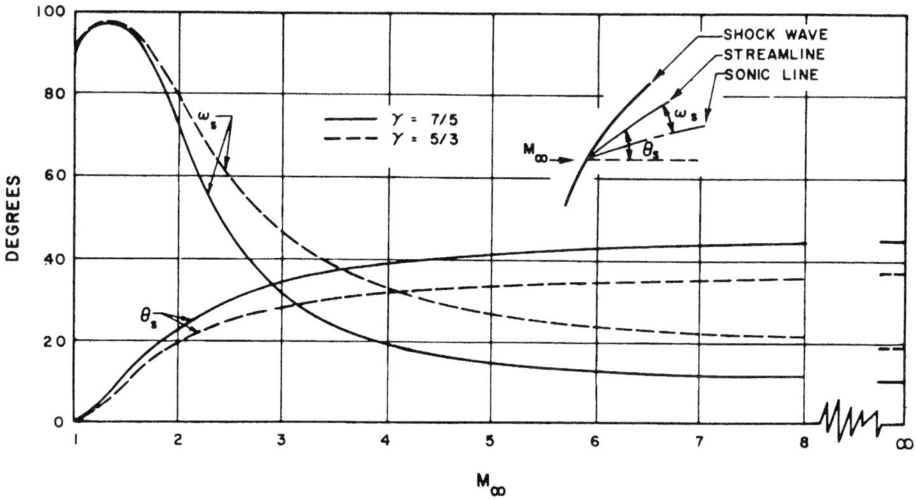

FIG. 6–2. Sonic line and streamline inclination behind a shock wave in two-dimensional flow.

line angle at the shock and the flow deflection angle behind the shock as a function of Mach number for $\gamma = 7/5$ and $5/3$. We note for $\gamma = 7/5$ above a Mach number of about 3 the sonic line always has a positive inclination angle with respect to the free stream direction. This inclination increases with increasing Mach number until at about $M_\infty = 8$ it has almost reached its asymptotic value for $M_\infty = \infty$. This rapid approach to an asymptotic value is characteristic of the behavior of a perfect gas of constant γ. It is not difficult to show, for the type of body we are considering, that this same general type of behavior also holds for axisymmetric flow, although in this case the angle does depend on the ratio of the radius of curvature of the shock to the distance from the axis of symmetry (see e.g. Drebinger [1], Gerber and Bartos [1], or Belotserkovskii [4]).

On the basis of the considerations above and of general experimental evidence we now have a picture as to the location of the sonic point and the inclination of the sonic line on the body surface and on the shock. We make the plausible assumption that the sonic line is a reasonably smooth curve between its intersections with the shock and the body surface. For most cases on bodies of approximately circular cross section, the inclination angles of the sonic line indicate that the sonic line is concave with respect to the upstream subsonic region. We must emphasize that for bodies of widely varying shape and with general fluids of different thermodynamic behavior it is impossible to make such a general statement.

In a diatomic gas ($\gamma = 1.4$) on an almost-spherical body in a symmetric flow we may conclude that the geometry of the sonic line and transonic zone will be as shown in Fig. 6–1(c) at high Mach numbers ($M_\infty > 3$). At low Mach numbers (say, $M_\infty < 2$) this geometry will be as shown in Fig. 6–1(a) for either almost-spherical or almost-cylindrical bodies. At intermediate Mach numbers on an almost-spherical body and at intermediate and high Mach numbers on an almost-cylindrical body the geometry is intermediate in character, as shown in Fig. 6–1(b). The flow pattern shown in Fig. 6–1(c) is a characteristic one, and we may refer to the sonic line in this case as a "hypersonic axisymmetric sonic line".

In the low Mach number case on a body of almost-circular cross section the limiting characteristic intersects the sonic line at the shock. On the other hand, in the high Mach number case on an almost-spherical body (γ about 1.4 or less) the limiting characteristic intersects the sonic line at the body. In this case no disturbance propagated by the first family (left-running) Mach waves from the body can make its influence felt in the subsonic region of the flow field, although first family Mach waves from the sonic line can carry disturbances from the sonic line to the shock wave. Between the sonic line and the limiting characteristic (of the second family, or right-running) is the transonic zone, from a point in which a disturbance will affect the sonic line and hence the entire subsonic flow field.

Except with this characteristic hypersonic axisymmetric sonic line there will be a portion of the body downstream of the sonic point from which a disturbance will affect the subsonic region. At low speeds this portion is appreciable in extent, as indicated in Fig. 6–1(a). In two-dimensional flow this portion of the body always exists, but at very high Mach numbers its extent becomes very small.

In the discussion above we have assumed that the value of γ is of the order of 1.4 or less. The way in which the value of γ influences the flow field lies primarily in its influence upon ϵ, although a number of other influences are important. The primary effect of an increase in γ is thus equivalent to a decrease in M_∞. Thus, in this special sense a flow at $M_\infty = \infty$ and $\gamma = 5/3$ is roughly equivalent in a gas of $\gamma = 7/5$ to a flow with the same value of ϵ, or to a flow with M_∞ about 3.

Our discussion has so far centered on the symmetric blunt-body problem. When an asymmetry is present (as, for example, with an angle of attack) an additional degree of freedom is introduced in the problem. This manifests itself by the shape and location of the stagnation streamline (sometimes termed the dividing streamline) being unknown in advance.

To illustrate the behavior of the stagnation streamline we refer to the analyses of Sections 4.3 and 4.4. Although they are local solutions and assume constant density, we may expect them to be representative of behavior in a stagnation region with compressibility taken into account. We showed there that the stagnation streamline does not pass through the normal part of the shock in general, and that therefore the body is not wetted by the streamline of maximum entropy. Some investigators had thought that the stagnation streamline was necessarily the streamline of maximum entropy. However, in hypersonic flow with ϵ small the point of entry of the stagnation streamline does turn out to be close to the normal point on the shock. The local solutions of Chapter IV show that the lateral displacement from the normal to the shock of the point of entry of the stagnation streamline to be proportional to $\epsilon^{3/2}R_s$ times a suitable measure of the asymmetry. In the case of a blunt body of revolution, for example, this displacement is $\epsilon^{1/2}$ times the shock layer thickness times the measure of the asymmetry. The lateral streamline displacement is small when $\epsilon^{1/2}$ is sufficiently small. Its neglect corresponds in our thin-shock-layer terminology to neglecting the inner layer where Bernoulli effects are important but centrifugal effects are small.

We may illustrate our point by drawing on numerical results obtained by Swigart [3] using an inverse method (to be discussed in Section 6.5). In Table 6–1 we have tabulated as a function of ϵ_{lim} a parameter $\bar{\nu}R_s$ ($\frac{1}{2} - \delta$ in Swigart's

TABLE 6–1

Distance between normal and stagnation streamlines for parabolic shock at 10° *incidence* ($M_\infty = \infty$, *Swigart* [3])

γ	ϵ_{lim}	$\bar{\nu}R_s \cdot 10^2$	$\bar{\nu}R_s/\epsilon_{\text{lim}}^{3/2}$
1.4	0.167	2.32	0.340
1.2	0.091	0.92	0.335
1.1	0.048	0.34	0.324
1.05	0.024	0.12	0.324

notation) which is proportional to the distance between the point of entry of the stagnation streamline and the normal to the shock. The values of $\bar{\nu}R_s$ have been taken from Swigart's calculations for a parabolic shock having a 10° asymmetry in a flow at $M_\infty = \infty$ of a calorically perfect gas. With our characterization of a hypersonic flow through ϵ it is clear how the streamline separation decreases with ϵ. Furthermore, it is clear how the density ratio factor of the local analysis closely correlates the calculated results. The analogous axisymmetric results of Swigart are similarly correlated, although in this case there is some question regarding the accuracy of his result for $\gamma = 1.05$. In this case, $\bar{\nu} R_s$ is given as a small negative value, while in all other cases it is positive.

We note again the two-dimensional result of Muggia [1] that the stagnation point lies on the side of the normal point on the shock for which the body or shock curvature is greater and that the streamline of maximum entropy on the shock turns in the direction of decreasing body or shock curvature. We may also expect this result to hold for the axisymmetric body as well. This behavior is illustrated in Fig. 6–3, where the relative position of the maximum entropy

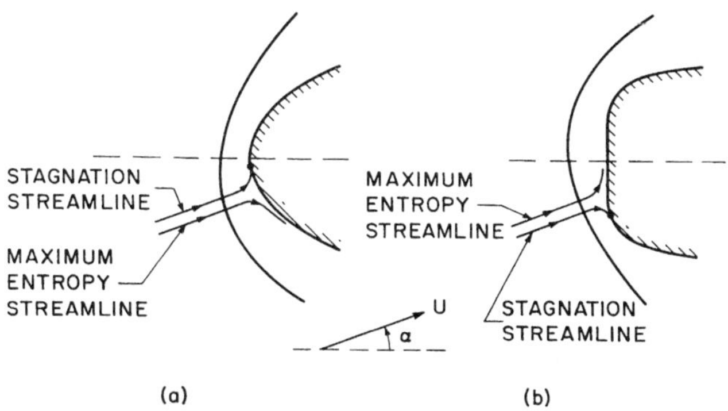

FIG. 6–3. Sketch of asymmetric flow over two kinds of blunt bodies.

streamline and stagnation streamline is shown for a body which is almost parabolic and for a flat-nosed body with rounded corners. The results of Swigart [3] for almost-parabolic bodies are in agreement with the streamline behavior noted.

We would also note here that Swigart's two-dimensional calculation for $\gamma = 1.4$ (cf. Table 6–1) indicates that the normal point on the body is shifted laterally from the normal point on the shock about six times the lateral shift in the stagnation streamline through the shock layer. The result of Muggia [1] indicates a factor of eighteen for the shift. However, we should not expect any better agreement than indicated, since for Swigart's calculation $\sqrt{3\epsilon}$ is by no means small.

With all the ideas we have discussed in mind, we may now inquire what simplifications, if any, are possible in the techniques for obtaining exact numerical solutions as a result of the limiting nature of the hypersonic flow. From all of our discussion it would still appear that the one basic characteristic which can be utilized either to simplify the method of solution, or at least to permit somewhat different techniques, is the fact that the shock lies very close to the body surface. In the present chapter we therefore envisage employing this fact to allow the use of mathematical techniques which might prove to be impractical or inapplicable when applied to flows with large shock layer thicknesses. The limitations involved will be appreciably weaker than were required in Chapter V.

One complication introduced by considering the flow of air at hypersonic speed over a blunt body is the variation in gas properties from those of a perfect gas. We will only briefly indicate in this chapter how such variable fluid properties can be accounted for in both equilibrium and nonequilibrium flows. The introduction of the variable fluid properties only adds somewhat to the computational effort involved. In principle, however, any of the methods given in this chapter may be applied to flows in a fluid with general thermodynamic properties. We shall enlarge upon our discussion of flows with general fluid properties in the next chapter when we discuss purely supersonic flows. There the general methods of solution are both simpler and clearer.

Of the various methods of attack available to us for the direct problem of a given body, we shall consider first a streamtube-continuity technique in which the streamline patterns are assumed and corrections to this pattern are carried out by satisfying mass flow and vorticity criteria. Because of the lengthiness of the computations involved in this method we examine it only in connection with the symmetric flow problem. We shall examine next a numerical scheme which in two-dimensional or axisymmetric flow is based upon dividing the shock layer into curvilinear strips generally parallel or normal to the surface. By putting the hydrodynamic equations in divergence form, integral conservation relations are formulated across the strips. The integrand functions of the integrals are evaluated on the basis of some assumed dependence on the independent variable by employing interpolation expressions of a general form. In this manner the flow equations are reduced to a finite system of ordinary differential equations. Alternatively, the shock layer can be subdivided into rectangular regions defined by the parallel and normal strips. The integration is then carried out in both directions with the integrand functions approximated in both independent variables. The approximating system of equations is thus reduced to a system of nonlinear algebraic or transcendental equations. The integral relation approaches can be generalized to three-dimensional problems. The inherent possibilities will be discussed when the method is considered.

For the direct problem we shall discuss a relaxation technique for determining the solution in the purely subsonic or elliptic region. This technique can be combined with a streamtube-continuity approach for calculating the mixed flow domain in the neighborhood of the sonic line. Finally, for the direct problem we present an unsteady approach method. In this method the steady flow is obtained as the limiting state reached asymptotically in time by an unsteady flow with constant free stream and body boundary conditions and suitably chosen initial conditions. The unsteady flow equations are solved by finite difference techniques. In one approach it is the unsteady integral conservation relations which are solved, while in another approach the finite difference approximation is applied directly to the appropriate partial differential equations.

For the inverse problem of determining the body shape associated with a given shock shape we shall describe essentially two methods, of which one operates in the complex plane. Both methods start at the known shock wave and have the body as an undetermined boundary. To be determined also is the sonic line in two-dimensional or axisymmetric problems or the sonic surface in three-dimensional problems.

The methods to be described by no means exhaust the possible numerical approaches, nor for any particular problem are they necessarily mutually exclusive. Thus in Section 6.4 we describe a combination of the relaxation and streamtube-continuity methods, while in three-dimensional problems a combination of different numerical methods may often prove advantageous.

Although all of the schemes to be discussed are satisfactory under certain flow and geometry conditions no one scheme can in general be considered to be clearly better than the others. We shall therefore consider it one of the primary purposes here to indicate the limitations of the methods and in some cases to offer suggestions for improvements.

To complete the discussion of the nature of the blunt-body problem we consider the Newtonian pressure laws (see Section 3.1) commonly used for comparison purposes, particularly the uncorrected modified Newtonian law. As we pointed out in Section 3.1, the adjective "modified" indicates division of the pressure by the correct stagnation pressure p_0 instead of by $\rho_\infty U^2$. The modification obtained by multiplying the pressure coefficient by $p_0/\rho_\infty U^2$ to ensure agreement at the stagnation point is necessary in order that the formula give the correct result for the maximum pressure on blunt bodies. As we have pointed out in Section 3.1, only the Newton-Busemann pressure law has a rational basis and is a correct result in a limiting process. As we have also pointed out in Chapter IV, if the shock and the body surface are not strictly parallel any result of Newtonian theory is best interpreted in terms of shock angle and shock curvature. This is clearly illustrated in Figs. 6–28 and 6–29, where comparison of exact inverse calculations for the stand-off distance show excellent agreement with the constant-density results expressed in terms of shock radius of curvature. We would emphasize again that the Newtonian pressure law without the centrifugal correction applied in terms of the local angle of incidence of the body is a purely empirical formula.

Many comparisons of experimental results and exact calculations show excellent agreement with the modified sine-squared pressure law expressed in terms of body angle; and they show relatively poorer agreement with the Newton-

Busemann law, expressed also in terms of body angle. The only rational comparison of the Newton-Busemann law with experiment must be in terms of shock angle and for flows with reasonably thin shock layers, although the law must approach correctness in terms of the body angle with ϵ extremely small. Such a rational comparison based on shock angle is made in Fig. 6–26. The improvement of agreement with this law in terms of body angle with decreasing ϵ is indicated in Fig. 6–10. The agreement with the uncorrected law in terms of body angle clearly becomes worse as ϵ becomes smaller (cf. Fig. 6–10). The agreement can be very poor for unusually shaped bodies, as for example bodies with concave shapes (see Chernyi [3, p. 125]). The excellency of the agreement with the uncorrected law depends upon ϵ being moderately large and upon the bodies involved being roughly spherical or cylindrical at the nose.

We should look on the observed agreement with modified Newtonian as fortuitous; it is evident that the centrifugal pressure difference across the shock layer is approximately offset in symmetric flows past smooth convex bodies by the effect of the difference between shock angle and body angle. With concave bodies this situation will not obtain. Chernyi [3, p. 125] discusses this point in detail with illustrative examples. The authors do not seek to imply that the modified Newtonian pressure law is not a useful one in practice. But it is essential that the user recognize its empirical nature and its limitations.

2. Streamtube-continuity methods

For hypersonic flow past a blunt body the shock wave lies close enough to the surface that the streamlines are roughly parallel to the body except in a small region near the stagnation point. This fact immediately suggests the possibility of utilizing an assumed streamline pattern and shock shape for calculating the flow over the nose of a blunt body in a high speed stream. Presumably a method could then be worked out in which these initial assumptions are corrected by an iterative procedure employing mass flow and vorticity considerations. In this manner the flow field variables could be determined, while at the same time many of the difficulties associated with the transonic character of the flow in the neighborhood of the sonic line could be minimized. The general philosophy of such an iterative approach has been discussed in Section 5.2. However, the applications discussed there have been based directly on the assumption of a thin shock layer.

Two such iteration schemes have been reported in the literature for symmetric two-dimensional and axisymmetric flows, one by Maslen and Moeckel [1] (Streamtube Method), the other by Uchida and Yasuhara [1]. A related scheme by Gravalos [1] and Gravalos, Edelfelt, and Emmons [1] has also been reported and will be discussed in connection with the relaxation technique in Section 6.4. The approach of Maslen and Moeckel, although simple to apply since it only requires iteration of the shock shape, is nevertheless a very rough approximation and requires as a starting point of the calculation a knowledge of the surface pressure distribution. As a result, the method lacks an essential necessary feature. This pressure distribution is generally unknown in advance, and is usually what we would like to find out in most problems of interest. On the other hand, the method of Uchida and Yasuhara, although very laborious since it requires iterating both the shock shape and streamline pattern, is nevertheless exact in principle. For the symmetric problem their method can be modified so that for an assumed streamline pattern it is not necessary to iterate the shock shape. With asymmetry this modification is not available, and because of the labor involved in the computation the method may not be practicable for such problems. Provided no difficulty in convergence is experienced, the method provides a way of completely computing the flows under consideration.

In the procedure of Maslen and Moeckel (Streamtube Method), both the pressure distribution on the entire body and the shock shape and stand-off or detachment distance in the stagnation region must be presumed known. Starting with the streamtube nearest the body and an approximate shock shape, a picture of the streamline pattern is built up on the basis of the mass flow calculated from the known entropy distribution $S(\psi)$ and the presumably known pressure. The pressure at a distance from the body is estimated from the pressure on the body through an equation for the pressure gradient normal to the streamlines. The shock location is determined by a continuity condition of equal mass flows, and if the shock location disagrees with the original approximate shock shape a new approximate shock shape is chosen and the procedure is repeated.

That this method is an unsatisfactory one is immediately evident from the fact that the quantities which would generally be considered the principal answers to the problem have had to be assumed as being accurately known before the problem is started. In the details of the method, the stream function describing the mass flow is correctly matched immediately behind the shock, but there is no way of ensuring that the pressure from the analysis matches the pressure from the oblique shock relations immediately behind the shock. It is clearly possible to repair this aspect of the method by providing for an adjustment of the pressure distribution on the body. We would then be led to a procedure following the general lines described in Section 5.2, or to procedures like that of Uchida and Yasuhara or of Gravalos.

In the scheme given by Uchida and Yasuhara, although the computations are laborious, no fixed assumptions are required regarding values of the flow variables either at the surface or in the flow field. The method starts by assuming an approximation for both the shock shape and distribution of streamlines, and a double iteration technique is carried out whereby both the streamlines and the shock are readjusted until a consistent solution is obtained. The philosophy of this procedure is very like that given in Section 5.2, although the details are different.

In the first step of the procedure, an assumed approximate pattern for the streamlines around the body is chosen. Such a streamline pattern can be obtained from one of the more approximate theories which we have discussed previously. In addition to assuming the streamlines, an approximation for the location and shape of the shock wave is also assumed, and this too may also be obtained from an approximate solution.

In the assumed patterns, the streamline deflection θ in general does not coincide with the flow deflection angle immediately behind the shock θ_s obtained from the oblique shock relations for the assumed shock angle. Uchida and Yasuhara use the ratio of the maximum value of these two flow deflections as a parameter to define the incompleteness of the solution at the shock boundary for a fixed detachment distance. The first step in their calculation is to find a shock shape which is consistent with the streamline pattern in satisfying $\theta = \theta_s$ approximately. They set the value of the detachment distance on the axis. With this quantity set it is impossible in general to find a shock shape of the type desired satisfying $\theta = \theta_s$. Instead, they satisfy the relation $\theta = \theta_s Q$, where Q is a constant which satisfies

$$Q = \frac{\theta_{\max}}{\theta_{s,\max}}, \tag{6.2.1}$$

and serves as a measure of the failure of the condition $\theta = \theta_s$. With a given value of the detachment distance they find an appropriate value of Q and the corresponding shock shape by a method of successive approximation. See Fig. 6–4.

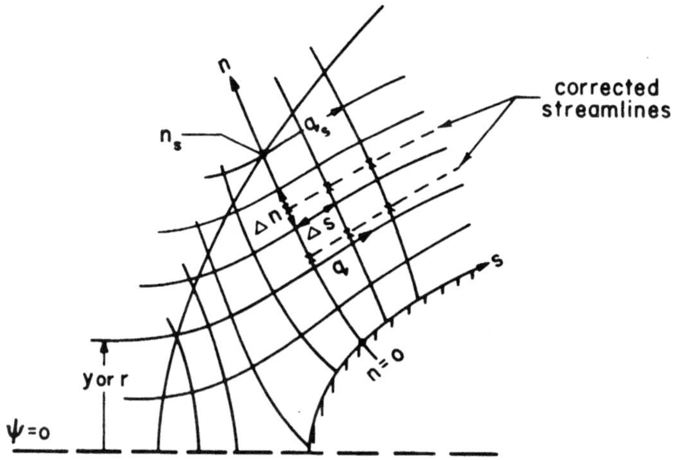

FIG. 6–4. Coordinate system for streamtube-continuity analysis of Uchida and Yasuhara [1].

attempt to increase the accuracy for a given number of strips N. The increased accuracy is obtained through a "smoothing" of the unknown functions corresponding to the basic smoothing achieved through the integration procedure. We shall not discuss this generalization further but instead refer the reader to Dorodnitsyn's articles or the review article of Belotserkovskii and Chushkin [2].

Suppose now that we approximate the integrands in (6.3.2) by interpolation polynomials, for example P_1 by

$$P_1 = \sum_{m=0}^{N} a_m(x) y^m, \tag{6.3.3}$$

where the $a_m(x)$ depend linearly on the values of the function P_1 on the strip boundaries. By substituting relations of the form of (6.3.3) into the integral relations and integrating, we obtain a system of nN simultaneous ordinary differential equations, with the values of the functions u_n at the interpolation points as the dependent variables. Applying the appropriate boundary conditions and numerically integrating this system from $x = 0$ to $x = a$ will then give us the desired solution. The basic question of how large N must be can be answered readily only by carrying out the solution for increasing values of N until a satisfactory convergence of the result is achieved. We would note here that we could approximate the integrands by other interpolation functions appropriate to the problem and that we need not utilize the polynomial form (6.3.3).

An alternative and significantly different approach involves integration with respect to x instead of with respect to y. We transform the coordinate system to a coordinate system x, $\eta = y/\Delta(x)$, in which the boundary $\Delta(x)$ appears parallel to the x axis. The system is then integrated along a line $\eta =$ *constant* and the integrands approximated by polynomials or by other functions in the variable x. The choice of which of these approaches is better in a particular problem should be determined through some knowledge of how the functions behave in both directions and whether interpolation functions of a desired form may be applied. The interpolation approximation should be applied in the direction in which the variation of the dependent variables is either smoother or more predictable in form. The differential equations then appear in the direction for which the variables are more rapidly varying or unpredictable. Of course, the behavior of the dependent functions generally will be determined more accurately in the direction in which the system of ordinary differential equations is integrated than in the direction of interpolation.

Still another variant of the method is to carry out the integration in both directions, after having divided the region into elementary rectangles (Fig. 6–5). The variables are approximated by interpolation functions in both variables. The system of equations then reduces to a system of nonlinear algebraic or transcendental equations. It is to be expected that for a fixed number of strips in either direction the accuracy of this last approach is bound to be lower than with the differential equation approach. Belotserkovskii and Chushkin [2] report that this scheme was applied by Chushkin in 1952 to the calculation of supersonic flow about a wedge with a detached shock wave. We shall not go further into the method. In cases with flow fields having smooth variations in the flow variables in both directions the method might prove attractive. An example might be a sphere or cylinder in a calorically perfect gas flow.

The method of integral relations can also be applied to problems involving three independent variables. One approach is to integrate the system with respect to one of the variables and then represent the integrand functions in terms of this variable by an appropriate interpolation function, the coefficients of which depend on the remaining two independent variables. In this manner the system is reduced to a problem in two independent variables and the previous approach can be applied. This is of course not the only method that can be used to reduce the problem to two dimensions, and schemes combining the method of integral relations with other methods are possible. One scheme for applying the method directly in a three-dimensional problem involves approximation through interpolation functions and integration in two of the three variables. The result is a system of equations in the third variable.

A principal advantage of this method lies in the fact that it is an integrand which is being approximated; the loss in accuracy from the approximation is thereby reduced. Furthermore the divergence form of the method is convenient, since the conservation equations are automatically satisfied and calculations may therefore be carried out with discontinuities present. A disadvantage of the scheme lies in the fact that the results of the method depend upon the choice of coordinate system. In addition, there is a certain degree of arbitrariness not only in the interpolation functions chosen but also in the definitions of the quantities being approximated by them. The choice of coordinates, variables, and functions depends on the judgment of the individual applying the scheme. Of course, the question of convergence also remains.

Dorodnitsyn's method was originally applied to the blunt-body problem by Belotserkovskii [1–3], who calculated the supersonic flow of a perfect gas past a circular cylinder. Following this method of approach, many calculations of more general body shapes in symmetric flows have appeared in the literature. Belotserkovskii [5] and Belotserkovskii and Chushkin [2] provide rather comprehensive bibliographies which will not be repeated here. These survey articles also discuss the extension of the method to nonequilibrium flows. Belotserkovskii and Chushkin [2] also discuss the extension to problems with asymmetry. In this section we shall not discuss calculations with nonequilibrium; this will be considered in Section 6.6.

For the direct problem the most convenient basic coordinate system is a body-oriented orthogonal curvilinear system. With asymmetry the choice is somewhat more arbitrary, though a body-oriented orthogonal system with respect to the geodesics through the stagnation point would be an appropriate choice. As we pointed out earlier, the integration may be carried out in the direction of either of the independent variables.

We shall illustrate the method first for a sphere or cylinder, using polar coordinates. Our reason for presenting the case of the cylinder or sphere lies in the simplicity of form afforded by polar coordinates. Our presentation follows the original approach of Belotserkovskii [1–4]. In carrying out this method we will restrict our considerations to the case where the strips are taken parallel to the body surface. We shall later mention the alternative approach in connection with nonequilibrium flows, and a modification of it in connection with the asymmetry problem.

As in our constant-density solutions for the cylinder and sphere, we use polar coordinates R and ϑ, where R is the cylindrical or spherical radius measured from the center of the body and ϑ is the polar angle measured from the upstream axis (see Fig. 4–6). The equations of continuity and momentum in these coordinates are expressible as

$$\frac{\partial \rho v R y^j}{\partial R} + \frac{\partial \rho u y^j}{\partial \vartheta} = 0, \tag{6.3.4}$$

$$v\frac{\partial v}{\partial R} + \frac{u}{R}\frac{\partial v}{\partial \vartheta} - \frac{u^2}{R} = -\frac{1}{\rho}\frac{\partial p}{\partial R}, \tag{6.3.5}$$

$$v\frac{\partial u}{\partial R} + \frac{u}{R}\frac{\partial u}{\partial \vartheta} + \frac{uv}{R} = -\frac{1}{\rho R}\frac{\partial p}{\partial \vartheta}. \tag{6.3.6}$$

Here $y = R \sin \vartheta$ is the radius from the axis, with $j = 0$ for the cylinder and $j = 1$ for the sphere, while u and v are the velocity components along the ϑ and R axes, respectively.

In order to recast the flow equations into Dorodnitsyn's divergence form, we first rewrite (6.3.5) so that a term $\partial(p + \rho v^2)/\partial R$ appears, then eliminate $\partial \rho v/\partial R$ by means of the continuity equation. The result is

$$\frac{\partial R y^j(p + \rho v^2)}{\partial R} + \frac{\partial \rho u v y^j}{\partial \vartheta} - [(1 + j)p + \rho u^2]y^j = 0. \tag{6.3.7}$$

By using the Bernoulli relation, only one of the momentum equations is required so that (6.3.7) is sufficient. At this point we make the restriction that the fluid is a perfect gas of constant γ. As with our restriction to polar coordinates this restriction is not a fundamental one and is made only to facilitate our presentation. Following Belotserkovskii we now rewrite (6.3.7) in nondimensional form by referring the pressure and density to their free stream stagnation values ($\bar{p} = p/p_{0,\infty}$; $\bar{\rho} = \rho/\rho_{0,\infty}$) and all velocities to the maximum adiabatic velocity denoted by $q_{\max}$ (e.g. $\bar{v} = v/q_{\max}$). Note that $q_{\max}^2 = 2a_0^2/(\gamma - 1)$. The dimensionless momentum equation becomes

$$\frac{\partial \bar{R}\bar{y}^j P_v}{\partial \bar{R}} + \frac{\partial \bar{y}^j P_{uv}}{\partial \vartheta} - \bar{y}^j P_u = 0, \tag{6.3.8}$$

where

$$P_v = \frac{\gamma - 1}{2\gamma}\bar{p} + \bar{\rho}\bar{v}^2, \tag{6.3.9a}$$

$$P_u = (1 + j)\frac{\gamma - 1}{2\gamma}\bar{p} + \bar{\rho}\bar{u}^2, \tag{6.3.9b}$$

$$P_{uv} = \bar{\rho}\bar{u}\bar{v}. \tag{6.3.9c}$$

Here the barred quantities denote the appropriate dimensionless variables, and $\bar{R} = R/R_b$ and $\bar{y} = y/R_b$.

Using the condition of the constancy of entropy along streamlines, the continuity equation can be written as

$$\frac{\partial \bar{R}\bar{y}^j Q_v}{\partial \bar{R}} + \frac{\partial \bar{y}^j Q_u}{\partial \vartheta} = 0, \tag{6.3.10}$$

where

$$Q_v = (1 - \bar{q}^2)^{\frac{1}{\gamma-1}}\bar{v}, \tag{6.3.11a}$$

$$Q_u = (1 - \bar{q}^2)^{\frac{1}{\gamma-1}}\bar{u}, \tag{6.3.11b}$$

and $\bar{q}^2 = \bar{u}^2 + \bar{v}^2$.

Finally, we have the entropy equation for a perfect gas

$$\bar{p} = E(\bar{\psi})\bar{\rho}^{\gamma} \tag{6.3.12}$$

whence we obtain

$$\bar{\rho} = [(1 - \bar{q}^2)/E]^{\frac{1}{\gamma-1}}. \tag{6.3.13}$$

The quantity E is a function only of the stream function $\bar{\psi}$ and may be found from its value just behind the shock wave. The stream function $\bar{\psi}$ is defined by

$$d\bar{\psi} = \bar{\rho}\bar{y}^j(\bar{u}\, d\bar{R} - \bar{v}\bar{R}\, d\vartheta). \tag{6.3.14}$$

We note that with this definition the velocity components are given by

$$\bar{\rho}\bar{u}\bar{y}^j = \frac{\partial \bar{\psi}}{\partial \bar{R}}, \tag{6.3.15}$$

$$\bar{\rho}\bar{v}\bar{y}^j\bar{R} = -\frac{\partial \bar{\psi}}{\partial \vartheta}. \tag{6.3.16}$$

The equations (6.3.8) to (6.3.14) define our problem for the four unknowns $\bar{v}$, $\bar{u}$, E, and $\bar{\psi}$, and are in a form suitable for the application of the method of integral relations.

In order to complete the definition of our problem we must now specify the boundary conditions. On the body where $\bar{R} = 1$ this is quite straightforward, since

(6.3.17a) $$\bar{v}_b(\vartheta) = 0,$$

(6.3.17b) $$\bar{\psi}_b(\vartheta) = 0,$$

(6.3.17c) $$E(\bar{\psi}_b) = E_b(\vartheta) = \textit{constant}.$$

On the shock wave, on the other hand, the boundary conditions must be specified in terms of two unknown parameters describing both the location and shape of the shock, namely the shock inclination angle σ and the shock layer thickness. The shock is taken to be located at $\bar{R}_s(\vartheta) = 1 + \bar{\Delta}(\vartheta)$, where $\bar{\Delta}$ is the unknown shock layer thickness made dimensionless with respect to the body radius and measured from the body to the shock along a ray $\vartheta = \textit{constant}$. From the geometry of the shock wave we obtain the relation

(6.3.18) $$\frac{d\bar{\Delta}}{d\vartheta} = -(1 + \bar{\Delta}) \cot (\vartheta + \sigma).$$

The stream function at the shock is given in terms of $\bar{\Delta}$ by

(6.3.19) $$\bar{\psi}_s = \frac{\bar{\rho}_\infty \bar{U} \bar{y}_s^{1+j}}{1 + j} = \frac{\bar{\rho}_\infty \bar{U}[(1 + \bar{\Delta}) \sin \vartheta]^{1+j}}{1 + j}.$$

The remaining boundary conditions can be found from the oblique shock relations and are

(6.3.20) $$\bar{p}_s = \frac{4\gamma}{\gamma^2 - 1} (1 - \bar{U}^2)^{\frac{\gamma}{\gamma-1}} \left[\frac{\bar{U}^2 \sin^2 \sigma}{1 - \bar{U}^2} - \frac{(\gamma - 1)^2}{4\gamma}\right],$$

(6.3.21) $$E_s = \bar{p}_s(1 - \bar{U}^2)^{-\frac{\gamma}{\gamma-1}} \left(\frac{\gamma - 1}{\gamma + 1}\right)^\gamma \left(\frac{1 - \bar{U}^2 \cos^2 \sigma}{\bar{U}^2 \sin^2 \sigma}\right)^\gamma,$$

(6.3.22a) $$\bar{u}_s = W_1 \sin \vartheta + W_2 \cos \vartheta,$$

(6.3.22b) $$\bar{v}_s = W_2 \sin \vartheta - W_1 \cos \vartheta,$$

where

(6.3.23) $$(\gamma + 1)\bar{U}W_1 = \gamma - 1 + 2\bar{U}^2 \cos^2 \sigma,$$

and

(6.3.24) $$(\gamma + 1)\bar{U}W_2 = \cot \sigma \, [\bar{U}^2(\gamma + 1 - 2 \cos^2 \sigma) - (\gamma - 1)].$$

It is to be recalled that $\bar{U} = U/q_{\max}$ is related to the free stream Mach number by the equation

$$\bar{U}^2 = \frac{\frac{\gamma - 1}{2} M_\infty^2}{1 + \frac{\gamma - 1}{2} M_\infty^2}. \tag{6.3.25}$$

With the problem thus formulated, the method of solution is quite straightforward. In order, however, to carry out Dorodnitsyn's scheme with an unknown shock boundary we introduce the dimensionless variable

$$\eta = \frac{\bar{R} - 1}{\bar{\Delta}}; \qquad 0 \leqslant \eta \leqslant 1. \tag{6.3.26}$$

In terms of this variable the shock layer can be broken up into N strips by drawing equidistant lines $\eta_k = constant$ between the wave and the body. Of course, at this stage of the calculation the actual shock layer thickness is still unknown. As before, let us denote all quantities on the body where $\eta = 0$ by the subscript b. For the remaining strip boundaries we will adopt Belotserkovskii's notation and denote all quantities on the kth line where $\eta = \eta_k = [N - (k - 1)]/N$ by the subscript k, and on the wave where $\eta_s = \eta_1 = 1$ by the subscript 1.

According to the method described at the outset, we must now integrate our partial differential equations (6.3.8) and (6.3.10) along an arbitrary ray $\vartheta = constant$ from the body surface to the boundary of each of the strips. Carrying out these integrations and applying the appropriate boundary conditions at the body surface ($\bar{R} = 1$), we obtain the following $2N$ independent relations:

$$(P_v)_k \bar{R}_k (\bar{R}_k \sin\vartheta)^j - (P_v)_b (\sin\vartheta)^j + \frac{d}{d\vartheta}(\sin\vartheta)^j \int_1^{1+\eta_k\bar{\Delta}} P_{uv}\bar{R}^j \, d\bar{R}$$
$$- (P_{uv})_k \, \eta_k (\bar{R}_k \sin\vartheta)^j \frac{d\bar{\Delta}}{d\vartheta} - (\sin\vartheta)^j \int_1^{1+\eta_k\bar{\Delta}} P_u \bar{R}^j \, d\bar{R} = 0, \tag{6.3.27}$$

$$(Q_v)_k \bar{R}_k (\bar{R}_k \sin\vartheta)^j + \frac{d}{d\vartheta}(\sin\vartheta)^j \int_1^{1+\eta_k\bar{\Delta}} Q_u \bar{R}^j \, d\bar{R}$$
$$- (Q_u)_k \, \eta_k (\bar{R}_k \sin\vartheta)^j \frac{d\bar{\Delta}}{d\vartheta} = 0, \tag{6.3.28}$$

with $k = 1, 2, ..., N$. From the above equations it is apparent that the undetermined boundary enters through the additional unknown function $\bar{\Delta}(\vartheta)$.

Our next step is to reduce these integrodifferential equations to ordinary differential equations. For this purpose we assume P_u, P_{uv}, and Q_u can be represented by polynomials in η. For example, we assume P_u is of the form

$$P_u(\bar{R}, \vartheta) = a_b(\vartheta) + \sum_{m=1}^{N} a_m(\vartheta)\eta^m, \tag{6.3.29}$$

where $a_m(\vartheta)$ will depend linearly on the values of the function P_u on the strip boundaries. For example, for $N = 2$ we would have with the subscript *u* dropped $a_b = P_b$, $a_1 = 4P_2 - P_1 - 3P_b$, and $a_2 = 2(-2P_2 + P_1 + P_b)$. We remind the reader that the subscript 1 denotes conditions at the shock wave, subscript *b* conditions on the body, and in this case the subscript 2 represents conditions along a line midway between the body and the shock. This assumption involves $N + 1$ independent polynomials. In a more general approach we would use $N + 1$ suitably chosen interpolation functions.

If we substitute the appropriate relations of the form of (6.3.29) into our integral relations we obtain $2N$ ordinary differential equations. By writing (6.3.12) and (6.3.14) along each of the $N - 1$ lines $\eta = \eta_q$, where $q = 2, 3, ..., N$, we obtain $N - 1$ additional ordinary differential equations and $N - 1$ ordinary equations. The final relation to complete the system is the differential equation (6.3.18). This then gives us a total of $3N$ ordinary differential equations and $N - 1$ ordinary equations to determine the $4N - 1$ unknowns $\bar{u}_b$, $\bar{\Delta}$, σ, $\bar{u}_q$, $\bar{v}_q$, E_q, $\bar{\psi}_q$. The $N - 1$ ordinary equations are for the quantities $E_k(\vartheta) = E(\psi_k)$, relating them to shock inclination at the points of entry. These are best considered as difference equations. The boundary conditions on the differential equations in asymmetric problems are different in nature from those in symmetric problems. They are discussed separately below.

In order to find the desired solution in the symmetric case, the approximating system of equations must now be integrated with respect to ϑ from the stagnation streamline ($\vartheta = 0$). An examination of the initial conditions at $\vartheta = 0$, namely $\bar{u}(\bar{R}, 0) = 0$, $\bar{\psi}(\bar{R}, 0) = 0$, $\sigma(0) = \frac{1}{2}\pi$, makes it clear that these conditions are by themselves insufficient to determine a unique solution to the problem. Physically, of course, it is obvious that if they were sufficient there would be no influence of the downstream flow on the upstream region. In point of fact, however, the upstream influence is manifested in the differential equations by a singular behavior somewhat analogous to that associated with the critical point or singular point in the thin shock layer theory of Chapter V. We shall term a singular point of this type in this theory a "sonic singular point". The reason for this terminology will appear later. To find that solution which properly traverses the sonic singular points it is necessary to add other requirements based on the nature of the solution near these singularities.

As a result of the imposition of the additional conditions at the sonic singular points the problem becomes a two-point boundary value problem. We start with an assumed solution at the stagnation point, and require that it satisfy the imposed conditions in the critical region. If the conditions governing the singular behavior of the equations are not met, the stagnation point solution must be revised and the integration process repeated until all the requirements in the critical region are satisfied and a single unique solution is determined.

To illustrate the nature of the singularities and the computational technique, let us consider for either a sphere or cylinder the simple case in which we take $N = 1$ and thereby consider the shock layer as a single strip. Then the functions in (6.3.27) and (6.3.28) are linear functions of η determined by their values on the body and on the shock. For this example there are three simultaneous first-order differential equations to solve. These are (6.3.18), and two differential equations of the form

$$\frac{d\sigma}{d\vartheta} + A\frac{d\bar{\Delta}}{d\vartheta} + B = 0, \tag{6.3.30}$$

$$\left(\frac{\gamma - 1}{\gamma + 1} - \bar{u}_b^2\right)\frac{d\bar{u}_b}{d\vartheta} + C\frac{d\sigma}{d\vartheta} + D\frac{d\bar{\Delta}}{d\vartheta} + F = 0, \tag{6.3.31}$$

where A, B, C, D, and F are known functions of $\bar{\Delta}$, σ, $\bar{u}_b$, and ϑ. We note that the first of these equations comes from (6.3.27), and the second from the continuity relation (6.3.28).

The initial conditions at $\vartheta = 0$ for the integration of the differential equations are that $\sigma = \frac{1}{2}\pi$, $\bar{u}_b = 0$, and $\bar{\Delta} = \bar{\Delta}_0$, where $\bar{\Delta}_0$ is an assumed value of the stand-off distance at the axis. The equations (6.3.30) and (6.3.31) are singular at the axis in the axisymmetric case, and their integration requires a preliminary step. In order to start the integration process in the axisymmetric case we require the initial values of the quantities $d\sigma/d\vartheta$ (proportional to shock curvature) and $d\bar{u}_b/d\vartheta$ (proportional to velocity gradient on the body). These must be determined at $\vartheta = 0$ by a procedure using a Taylor series expansion, or by some equivalent procedure. Once these derivatives have been determined from (6.3.30) and (6.3.31) for $\vartheta = 0$ and hence for $\sigma = \frac{1}{2}\pi$, we are prepared to initiate the integration in terms of the unknown but initially estimated stand-off distance.

From (6.3.31) it can be seen that $\bar{u}_b = \sqrt{(\gamma - 1)/(\gamma + 1)}$ is a singular point of the system and that, since $q_{max} = \sqrt{(\gamma + 1)/(\gamma - 1)}\, a_{son}$, this singularity corresponds to the sonic point on the body. By combining the three differential equations we can rewrite (6.3.31) as

$$\frac{d\bar{u}_b}{d\vartheta} = \frac{J_b^{(1)}}{\left(\dfrac{\gamma - 1}{\gamma + 1} - \bar{u}_b^2\right)}, \tag{6.3.32}$$

where $J_b^{(1)}$ is a function of $\bar{\Delta}_0$, $\bar{u}_b$, σ, and ϑ. At this juncture we must now make a distinction between the types of bodies being considered. If the body shape is analytic and smooth so that there are no sharp corners at which the velocity must become sonic, then $\bar{u}_b$ is a continuous function of ϑ and we require that $J_b^{(1)} = 0$ at $\bar{u}_b^2 = (\gamma - 1)/(\gamma + 1)$. In this case the solution is obtained by initially estimating the stand-off distance at the axis using one of the more approximate solutions, and then integrating the differential equations to $\bar{u}_b^2 = (\gamma - 1)/(\gamma + 1)$. If the condition $J_b^{(1)} = 0$ is not satisfied at that point, we must re-estimate $\bar{\Delta}_0$ and integrate again, repeating the process as many times as are necessary until the condition $J_b^{(1)} = 0$ at $\bar{u}_b^2 = (\gamma - 1)/(\gamma + 1)$ is satisfied.

Kao [1] has suggested that the use of $\bar{y}^j Q_u$ in place of $\bar{u}$ should simplify the numerical procedure. The differential equation for $\bar{y}^j Q_u$ does not have the singular denominator of (6.3.32), (6.3.33), or (6.3.34). It is possible that some computational advantage may result from this substitution. However, the quantity $\bar{u}$ is needed to evaluate the coefficients in the differential equations, for example the coefficients A to F in (6.3.30) and (6.3.31). This evaluation requires solving for $\bar{u}$ from the additional algebraic equation (6.3.11b) in the computational procedure.

Kao claims that the singularity does not really exist, and in fact is introduced by the poor choice of $\bar{u}$ as a dependent variable. This claim must be rejected outright. The singular behavior persists in Kao's formulation, but has been transferred from the differential equation to the process of calculating $\bar{u}$ from Q_u. With $\bar{v}$ fixed, Q_u is a maximum at the sonic singular point. The value of $\bar{u}$ must change there continuously from subsingular to supersingular values. If Q_u reaches a maximum in the computation which is lower than permitted by (6.3.11b), $\bar{u}$ remains subsingular. If Q_u is still increasing in the computation when its maximum value permitted by (6.3.11b) is reached, it becomes impossible to solve for $\bar{u}$ thereafter and continue the computation. This behavior is essentially no different from that found using $\bar{u}$ as a dependent variable.

Had one intermediate line been introduced in the problem of the sphere or cylinder, we would then have $N = 2$ and we should have obtained six simultaneous ordinary differential equations and one ordinary equation. In this case three of the differential equations are of the same form as for the first approximation. The other three differential equations involve $d\bar{u}_2/d\vartheta$, $d\bar{v}_2/d\vartheta$, and $d\bar{\psi}_2/d\vartheta$, where the subscript 2 represents conditions along the line midway between the body and the shock. The two equations which come from the continuity relation (6.3.28) are singular and have the form

(6.3.33)
$$\frac{d\bar{u}_b}{d\vartheta} = \frac{J_b^{(2)}}{\left(\dfrac{\gamma - 1}{\gamma + 1} - \bar{u}_b^2\right)},$$

(6.3.34)
$$\frac{d\bar{u}_2}{d\vartheta} = \frac{J_2^{(2)}}{\dfrac{\gamma - 1 + 2\bar{v}_2^2}{\gamma + 1} - \bar{q}_2^2},$$

where $J_b^{(2)}$ and $J_2^{(2)}$ are known functions of ϑ and the dependent variables. We may note that $\bar{a}^2 = \frac{1}{2}(\gamma - 1)(1 - \bar{q}^2)$, and that the denominator on the right hand side of (6.3.34) may be expressed as $2/(\gamma + 1)$ times the quantity $\bar{a}_2^2 - \bar{u}_2^2$.

As with the case $N = 1$ the integration is again carried out from $\vartheta = 0$, where we apply the initial conditions $\bar{u}_b = \bar{u}_2 = 0$, $\bar{\psi}_2 = 0$, $\sigma = \frac{1}{2}\pi$. In this case, however, besides having to specify the parameter $\bar{\Delta}_0$ as an initial unknown, we must also specify $\bar{v}_2(0)$. Thus we see that in the second approximation there are two unknown parameters to be determined by conditions in the critical region. Evidently, if we again require a continuous solution, then in addition to our requirement on $\bar{u}_b$, we also require that $J_2^{(2)} = 0$ when $\bar{u}_2^2 = \bar{a}_2^2$. In fact for each increase in N by one, one parameter, $\bar{v}_k(0)$, and one condition, $J_k = 0$ for $\bar{u}_k^2 = \bar{a}_k^2$, are added. Hence the sonic singular points of the system will be located where the line

$$\bar{u}^2 = \bar{a}^2 \tag{6.3.35a}$$

intersects the strip boundaries. An examination of the topology of the differential equations from a number of numerical calculations indicates that these singularities are of a saddle type. Furthermore, since $q_{\max}^2 = (\gamma + 1)q_{\text{son}}^2/(\gamma - 1)$, we note that the sonic singular points occur along the line

$$\bar{q}^2 = \left(1 + \frac{2}{\gamma - 1}\bar{v}^2\right) q_{\text{son}}^2 . \tag{6.3.35b}$$

The additional $N - 1$ dependent variables E_k consist in this case of the single variable E_2. The corresponding algebraic relation relates its value to the shock inclination angle at the point of entry of the streamline passing through the midpoint of the shock layer. This additional variable and relation do not effectively raise the order of the system, because we can consider E_2 as a known function locally. Its presence does make the system a differential-difference system, analogous to that treated in Section 5.6. A consequence is that the system *must* be integrated from the axis outward if $N > 1$. In the case $N = 1$ the system could in principle be integrated inward from the sonic singular point to the axis, though this would usually be an unclever choice of procedure.

Belotserkovskii's scheme is essentially a finite difference scheme, and it is well established that finite difference schemes in general have certain inherent limitations in solving partial differential equations. Thus, in the method of integral relations we should expect minor divergences in such problems as the precise determination of zones of action and of lines dividing elliptic and hyperbolic regions. With this minor limitation in mind we interpret the line of sonic singular points (6.3.35) directly as the sonic line. This interpretation is exact at the body and at any point for which $\bar{v}_k$ is zero, and is almost exact at other points because the quantity $2\bar{v}_k^2/(\gamma - 1)$ is generally small compared with one.

Because the functions J_k depend upon all the dependent variables in general, we do have an analogue of the transonic zone in this analysis. The value of a function u_k at a point past the sonic singular point for the strip $k = k_1$ may still have an effect on the solution at the same value of ϑ for u_k at $k = k_2$ which is less than its singular value. A change in u_k on this second strip has an upstream influence and would change the entire subsonic region. Let us define ϑ_{lim} as the greatest value of ϑ for which a sonic singular point occurs. Only for $\vartheta > \vartheta_{\text{lim}}$ is there no more upstream influence. The line $\vartheta = \vartheta_{\text{lim}}$ is thus the analogue of the limiting characteristic in the method of integral relations. The roundabout manner by which a point in the transonic zone affects the subsonic region is completely analogous in the method of integral relations and in physical reality.

Belotserkovskii [3] has carried out digital machine computations using the method outlined, to determine the flow field around a circular cylinder. In Fig. 6–6 we show Belotserkovskii's results for the sonic line and shock shape

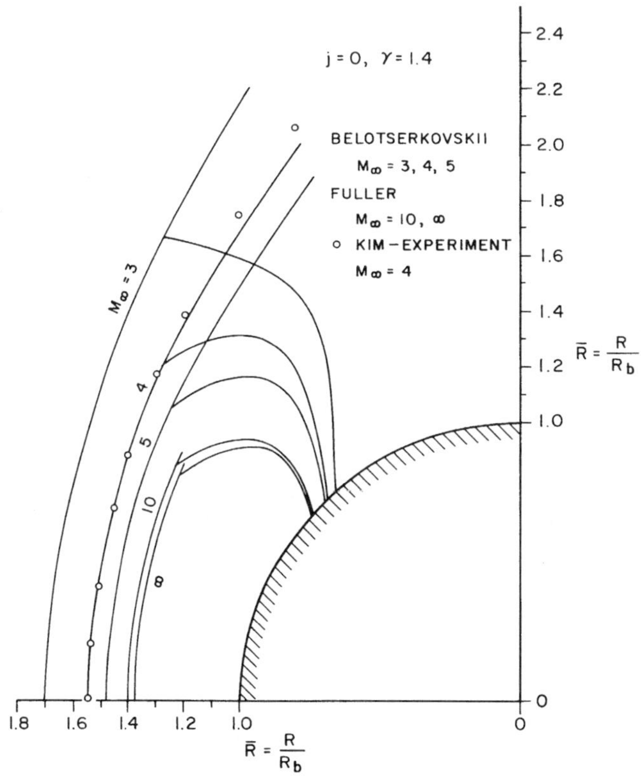

FIG. 6–6. Shock wave and sonic line shapes for a circular cylinder (Belotserkovskii [3]; Fuller [1]).

for free stream Mach numbers of 3, 4, and 5 with $\gamma = 1.4$. For $M_\infty = 3$ and 5 the calculations were carried out with $N = 3$ in the method, while for $M_\infty = 4$ the solution was computed only for $N = 2$. A comparison of an experimentally determined shock shape obtained by Kim [1] from shock tube experiments at $M_\infty = 4$ is also shown. The agreement is seen to be quite good. Also shown in Fig. 6–6 are results for $M_\infty = 10$ and ∞ of an inverse calculation by Fuller [1] (discussed in Section 6.5). Although not shown, calculations given by Belotserkovskii and Chushkin [2] for $M_\infty = 10$ agree closely with those of Fuller.

In all cases, the angle the sonic line makes with the streamline behind the shock is in good agreement with the exact results shown in Fig. 6–2. In addition we note that, consistent with our previous approximate calculations, the sonic point on such a cylindrical body is practically independent of Mach number. We see also that the sonic line at the body always makes an acute angle with the surface, consistent with our observation in Section 6.1 for two-dimensional bodies. Furthermore, because the Mach number is also moderate, the sonic line is peaked well within the shock layer. In this case, then, the first family

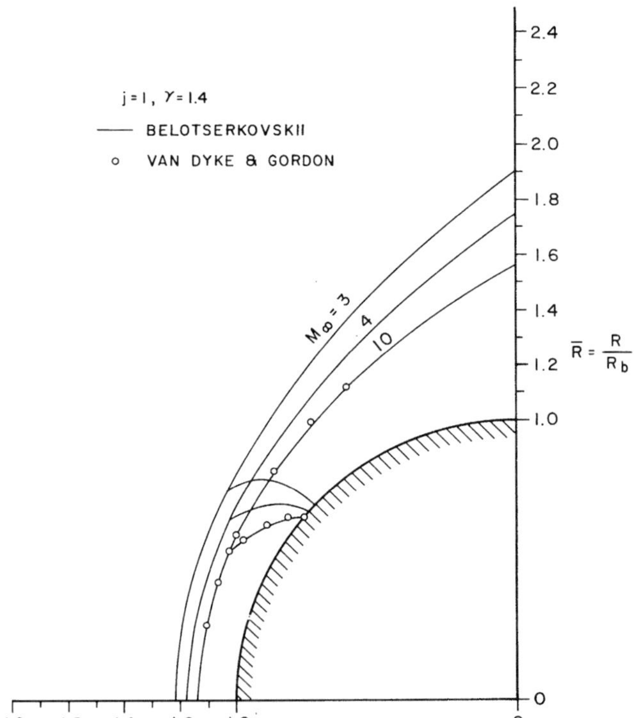

FIG. 6–7. Shock wave and sonic line shapes for a sphere (Belotserkovskii [4]; Van Dyke and Gordon [1]).

(left-running) characteristics from the body can make the influence of the body in the supersonic region felt on the subsonic part of the flow field (cf. Fig. 6–1(b)). From the results shown it is certainly clear that the actual region of body influence in the supersonic region is decreasing with increasing Mach number.

The shock and sonic line shapes as calculated by Belotserkovskii [4] for a sphere with $N = 2$ and $\gamma = 1.4$ are shown in Fig. 6–7. The circles are the results for $M_\infty = 10$ of an inverse calculation by Van Dyke and Gordon [1] (discussed in Section 6.5). Although not shown, the curves for $M_\infty = \infty$ are very close to those for $M_\infty = 10$. In Fig. 6–8 we show similar calculations for a sphere at $M_\infty = 6$ but for different values of γ. These curves were taken from Belotserkovskii and Chushkin [2].

Provided the Mach number is not too low or γ too large we observe in Figs. 6–7 and 6–8 the characteristic hypersonic axisymmetric sonic line associated with smooth bodies in hypersonic flows which we discussed in Section 6.1. As expected from our previous considerations, a comparison of the sonic line for $\gamma = 1.67$ and $\gamma = 1.4$ or less shows that the larger γ is, the more closely the sonic line approaches the typical behavior at moderate Mach numbers. Thus from Fig. 6–8 it is clear that the $\gamma = 1.67$ sonic line for $M_\infty = 6$ has the typical behavior of the moderate Mach number pattern previously shown in Fig. 6–1(b), and is in fact very close in shape to the $M_\infty = 3$, $\gamma = 1.4$ calculation of Fig. 6–7. From both Figs. 6–7 and 6–8 we also note the fact that up to the sonic point

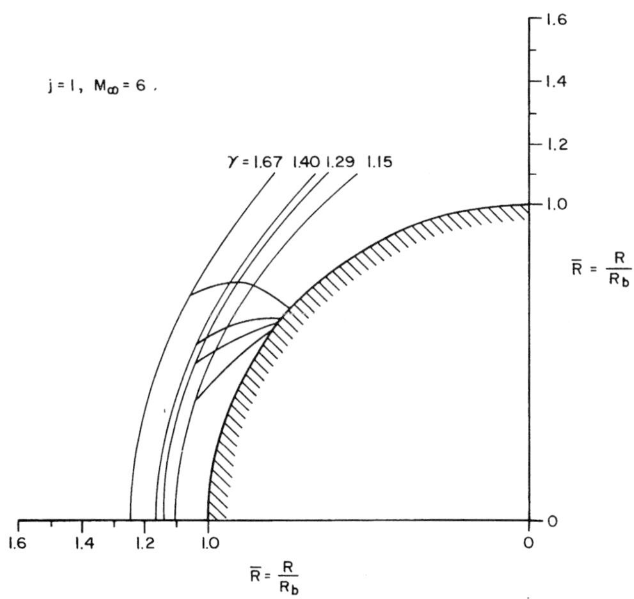

FIG. 6–8. Shock wave and sonic line shapes for a sphere at $M_\infty = 6$ and different γ (Belotserkovskii and Chushkin [2]).

the body is roughly parallel to the shock, as would be predicted by the constant-density analysis of Section 4.4. Finally, the fact that the results for $M_\infty = 10$ and $M_\infty = \infty$ differ but slightly points out the relative insensitiveness at high Mach numbers of the shock shape to the Mach number for a given body at constant γ.

In Fig. 6–9 we show the surface pressure distribution on a cylinder with $\gamma = 1.4$ as calculated by Belotserkovskii [3] for the approximation $N = 3$ at $M_\infty = 3$. Although not shown in the figure, the increase in accuracy with increasing N of Belotserkovskii's calculation is very rapid, with the approximation $N = 2$ giving the final result to what may be considered a desired practical accuracy. Belotserkovskii has compared his calculation with experiment; within experimental accuracy the agreement is exact. In order to complete the picture for the cylinder we also show in Fig. 6–9 the results of an inverse calculation

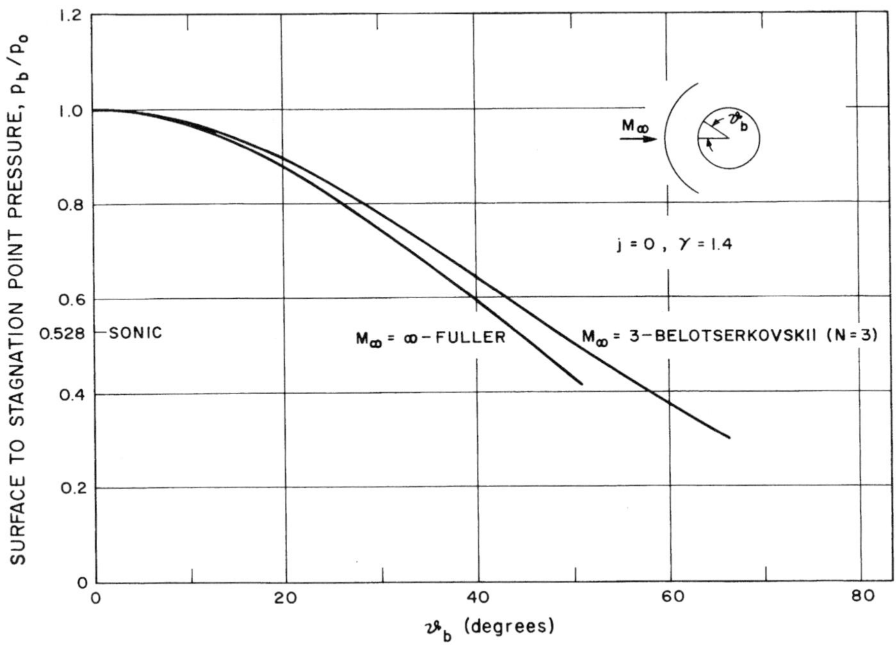

FIG. 6–9. Pressure distribution on a circular cylinder (Belotserkovskii [3]; Fuller [1]).

of Fuller [1] for $\gamma = 1.4$ and $M_\infty = \infty$. By comparing the two pressure distributions we may observe the relative insensitivity to Mach number above about $M_\infty = 3$. This serves as a good example of how the Mach Number Independence Principle (Section 1.6) is operative at relatively low Mach numbers when the body is blunt.

In Fig. 6–10 we show the pressure distributions on a sphere for $\gamma = 1.4$ and

$M_\infty = 3$ and 10 as calculated by Belotserkovskii [4]. For comparison purposes we have also shown a corresponding inverse calculation of Van Dyke and Gordon [1] for $M_\infty = 10$ and $\gamma = 1$. On the same figure we have also plotted the modified Newton-Busemann pressure distribution calculated in terms of the body angle and the modified Newtonian (without the centrifugal correction) distribution,

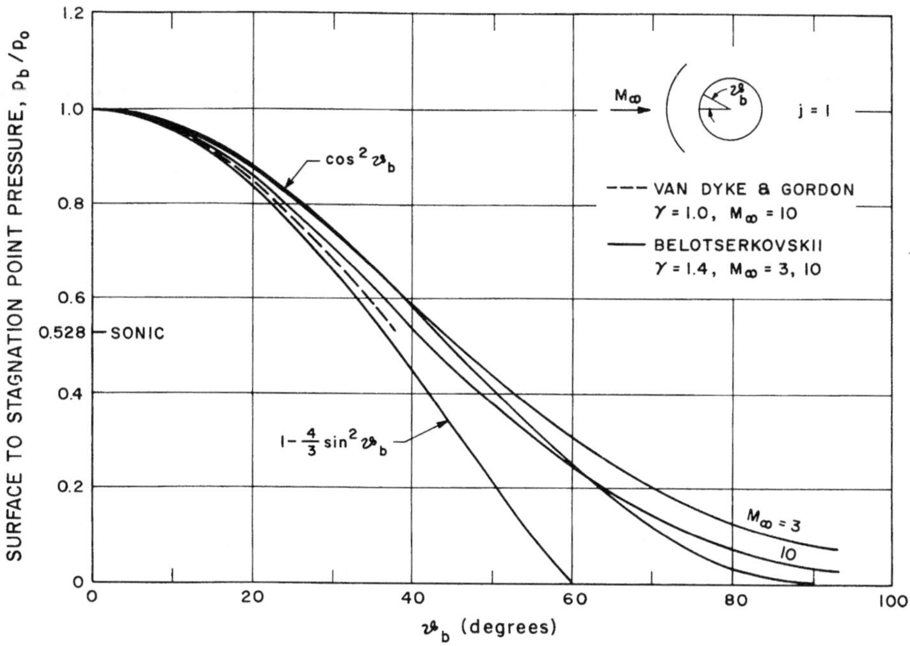

FIG. 6–10. Pressure distribution on a sphere (Belotserkovskii [4] and Van Dyke and Gordon [1]).

also in terms of body angle. From (3.3.4) with $n = 3$ the Newton-Busemann relation is given by

$$\frac{p_b}{p_0} = 1 - \tfrac{4}{3}\sin^2\vartheta_b\,. \tag{6.3.36}$$

At $M_\infty = 10$, the agreement with the modified Newtonian approximation is no better than with the modified Newton-Busemann relation up to about 30°. However, we again remind the reader that any agreement with modified Newtonian must be considered fortuitous foı the case considered. Nevertheless, it is clear from Fig. 6–10 that a decrease in γ results in an increased agreement with Newton-Busemann. The difference between the two distributions seen in Fig. 6–10 is primarily a consequence of the variable-ϵ effect associated with finite Mach number.

As already noted, the application of the present method to symmetric two-dimensional or axisymmetric flows for body shapes other than a cylinder or sphere is quite straightforward and may be accomplished by introducing an orthogonal coordinate system oriented with respect to the body. As a simple example let us consider either a flat-faced cylinder or disc, or a flat plate, oriented normal to the stream (Fig. 6–11). We choose a rectangular coordinate system

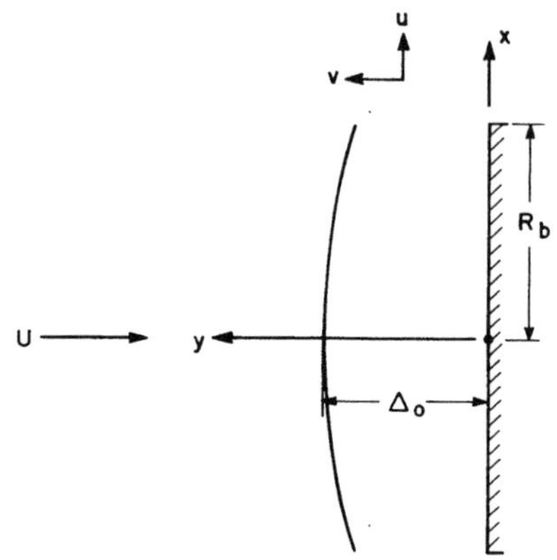

FIG. 6–11. Flat-faced plate or cylinder normal to the stream.

with the x coordinate measured from the symmetry axis and the y coordinate measured parallel to the axis, with origin at the stagnation point. In this case the dimensionless momentum equation replacing (6.3.8) is

$$\frac{\partial x^j P_{uv}}{\partial x} + \frac{\partial x^j P_v}{\partial y} = 0, \tag{6.3.37}$$

while the continuity equation replacing (6.3.10) is

$$\frac{\partial x^j Q_u}{\partial x} + \frac{\partial x^j Q_v}{\partial y} = 0. \tag{6.3.38}$$

The P's and Q's are as previously defined but with u and v now referring to the velocity components in the x and y directions, respectively. The system of equations is closed by (6.3.12) and (6.3.13), expressing the entropy and state equations.

While the procedure for the reduction of the above system of differential equations to ordinary differential form is the same as before, the integration procedure is not the same. We showed with the cylinder or sphere that with one strip the number of independent free parameters to be determined by the differential equations was one. In the case of a flat-faced body the body shape without regard for the corners is invariant to a scale transformation. A consequence of this property is that there are zero essential free parameters. We may select the stand-off distance arbitrarily, and then carry out an initial value procedure in which the integration is started from the axis. The solution is completed when the known inviscid sonic solution at the corner is reached. This solution is characterized by an infinite value of the velocity gradient at the corner. The problem in this case is therefore simplified from a two-point boundary value problem to an initial value problem. This same result holds true for a wedge or cone with a detached shock, and applies to these geometries in the axisymmetric theory of Section 5.3.

For the geometries mentioned there will be one free parameter with two strips instead of two free parameters as before. Similarly for N strips there will be $N - 1$ free parameters instead of N. The fact that the one-strip flat-faced cylinder problem could be handled by an initial value procedure is obvious from the analogous solution for the flat-faced disc of (5.3.29). For the integral equation method applied to this problem, the fact was noted by Xerikos and Anderson [1]. These results are consonant with the discussion of effective orders given in Section 5.4 in connection with the two-dimensional blunt-faced body solution. The geometries discussed along with the number of free parameters for a given number of strips are shown in Fig. 6–12.

One feature of solutions with sonic corners with $N > 1$ must be noted. The sonic singular points off the body generally occur farther from the axis than the sonic point on the body. The solution in the coordinate system described must terminate at the corner, where physically a local Prandtl-Meyer flow

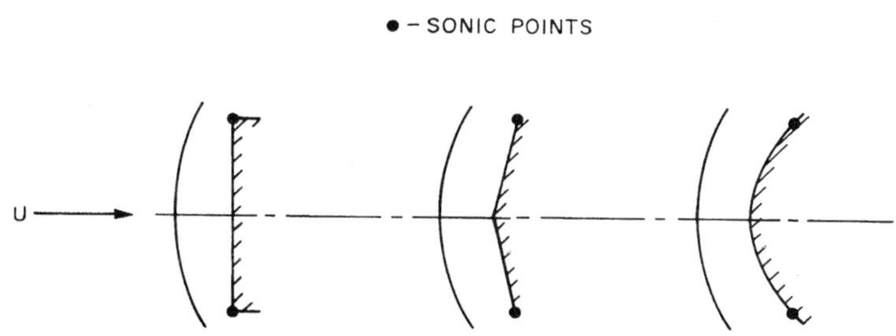

	FREE PARAMETERS		
N = 1	0	0	1
N = 2	1	1	2
N	N−1	N−1	N

FIG. 6–12. Number of free parameters or effective order in method of integral relations for symmetric flow.

appears. This difficulty must be circumvented by shifting to a different coordinate system. One possibility, for example, is a polar coordinate system with the corner as origin (see Belotserkovskii and Chushkin [1]).

The shock boundary conditions previously expressed in terms of shock angle (e.g. (6.3.20) and (6.3.21)) are the same for the plate or cylinder problem. Here

$$\frac{d\Delta}{dx} = \cot \sigma. \tag{6.3.39}$$

The remaining boundary conditions are straightforward. The differential equations derived from (6.3.37) and (6.3.38) with polynomial interpolation functions have been integrated by several authors for one strip, and in one axisymmetric case for two strips. Figure 6–13 shows the shock wave shapes obtained by Bazzhin [1] and Xerikos and Anderson [1] in the two-dimensional case for one strip. The slight discrepancy between these authors probably

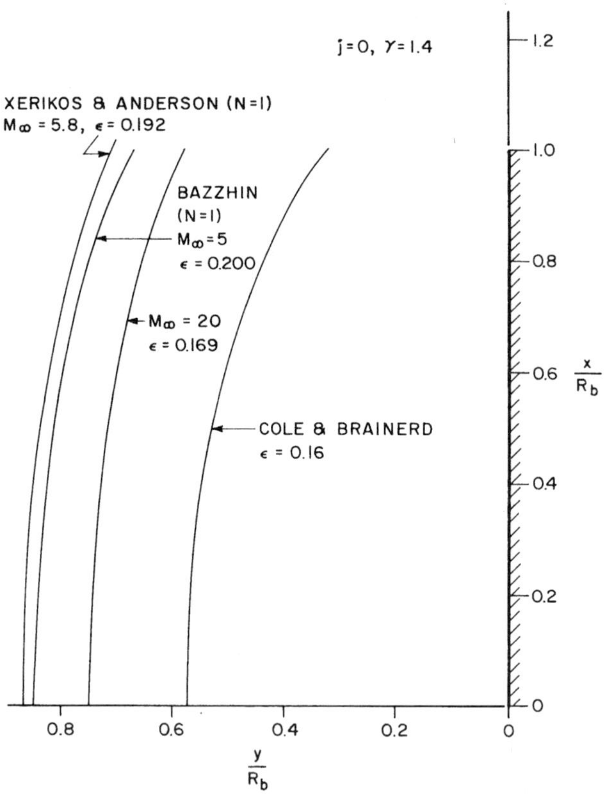

FIG. 6–13. Shock wave shapes for a flat plate normal to the free stream (Bazzhin [1]; Xerikos and Anderson [1]; Cole and Brainerd [1]).

results from a procedural difference in the calculation, since the shock location in the two-dimensional case is quite sensitive at moderate Mach numbers to small changes. Also shown for comparison purposes is the constant-density result of Cole and Brainerd [1] for $\epsilon = 0.16$, taken from their paper, with minor corrections incorporated into the treatment of Section 5.4. This result may also be obtained from the universal curve of Fig. 5–12b. Whether the Bazzhin or Cole and Brainerd solution more closely represents the true solution is not known. For this problem it must be recognized that the stand-off distance is very sensitive to the assumptions made in the theory, because of the very low velocities in the neighborhood of the body near the axis. Neither two-strip calculations nor one-strip calculations with interpolation functions other than straight lines are available to see if the differences from the calculations presented are large or small. For the actual shape of the shock wave itself the Cole and Brainerd assumptions may provide for a more realistic result.

In Fig. 6–14 we show the corresponding pressure distributions. The Cole and Brainerd result was obtained from the universal curve of Fig. 5–12a. The

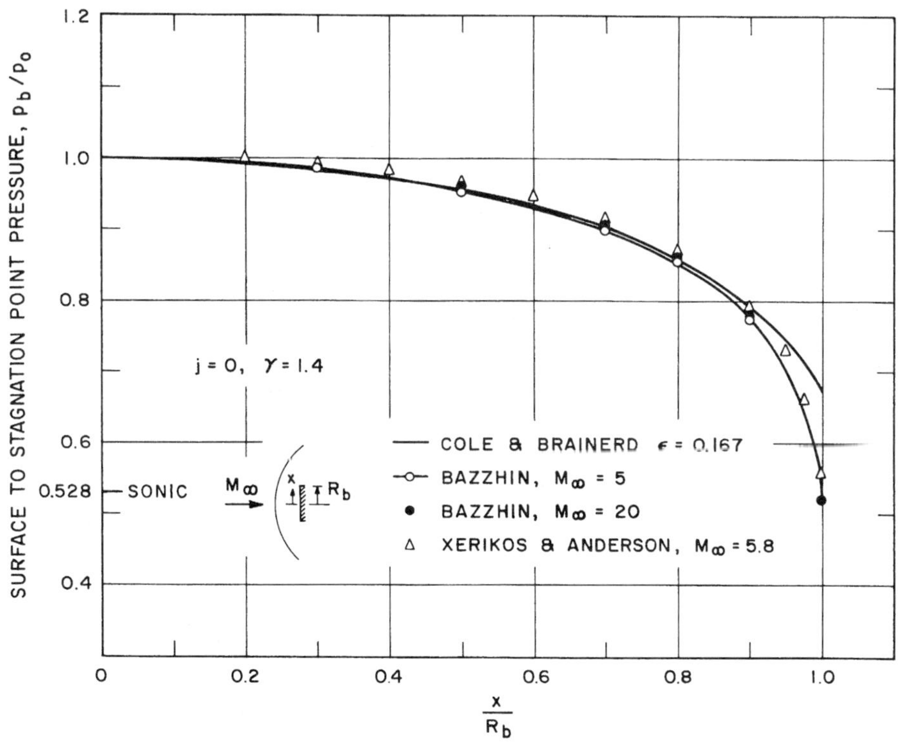

FIG. 6–14. Pressure distribution on a flat plate normal to the free stream (Bazzhin [1]; Xerikos and Anderson [1]; Cole and Brainerd [1]).

dynamic pressure, with which that curve is made dimensionless, was converted to stagnation pressure by multiplying it by $(1 - \epsilon/2)$. It is clear that except very close to the corner all of the pressure curves essentially agree. The deficiency of the constant-density theory in not going to the critical pressure ratio of 0.528 at the corner is evident. The relative insensitivity to Mach number is apparent. This may be expected here because of the relatively small change in ϵ to which the relative pressure drop at a given station is proportional (cf. (5.4.4a)).

Figure 6–15 shows the shock wave shapes obtained for the axisymmetric flat-faced cylinder problem. The $M_\infty = 4$, $\gamma = 1.4$ calculation reported by Belotserkovskii and Chushkin [2] (see also Belotserkovskii [5]) was for two strips, while the $M_\infty = 5.8$, $\gamma = 1.4$ calculations reported by Gold and Holt [1] and Xerikos and Anderson [1] are for one strip. These later calculations are in complete agreement so that only one curve need be drawn. Also shown for comparison is the experimental result of Oliver [1]. The agreement is seen to be

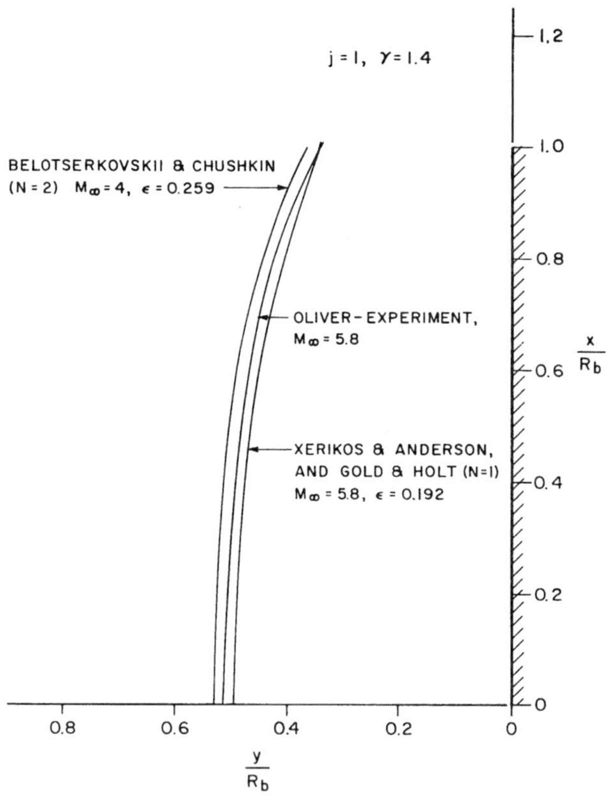

FIG. 6–15. Shock wave shapes for a flat-faced cylinder (Belotserkovskii and Chushkin [2]; Gold and Holt [1]; Xerikos and Anderson [1]) and comparison with experiment (Oliver [1]).

quite good. It is clear that in the axisymmetric case the effect of a second strip is principally to change the shock shape in the region of the sonic corner because of the low density there. However, the profiles assumed in the integral relations method are for this case more realistic than for the two-dimensional case and so the method may be expected to give a better approximation for low N.

Figure 6–16 presents the corresponding pressure distributions. Belotserkovskii and Chushkin [1], on the basis of calculations with $\gamma = 1.4$ and $N = 2$,

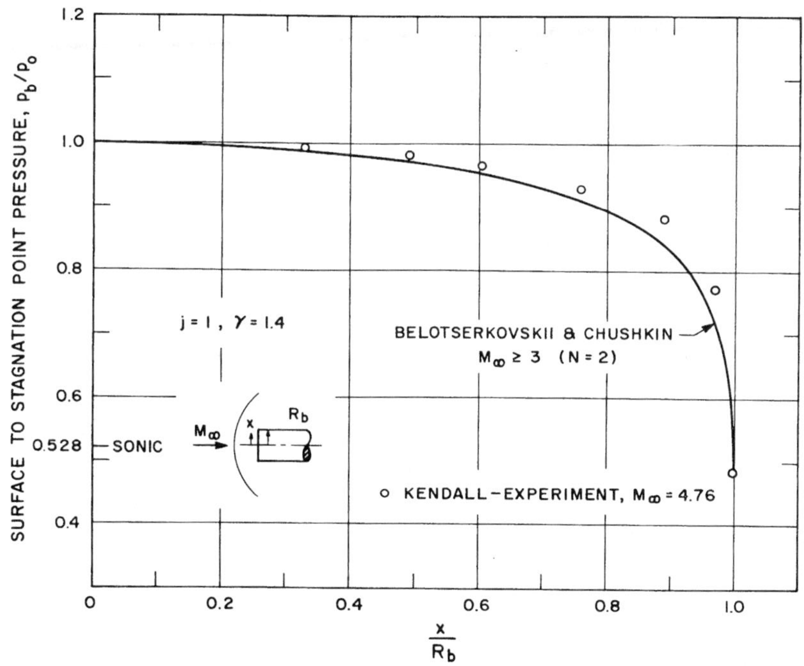

FIG. 6–16. Pressure distribution on a flat-faced cylinder (Belotserkovskii and Chushkin [1; 2]) and comparison with experiment (Kendall [1]).

report the same pressure curve for $M_\infty = 3$, 4, 6, 10 and ∞. The calculations of Gold and Holt [1] and Xerikos and Anderson [1] for $N = 1$ are sufficiently close to the curve shown that they are not presented. The relative insensitivity of the pressure distribution to changes in free stream Mach number with $\gamma = 1.4$ is in agreement with the reported experimental results of Kendall [1]. He shows little difference in pressure on the flat face of a cylinder for the Mach number range of 1.82 to 4.76 in air. In Fig. 6–16 we have shown his pressure measurements for $M_\infty = 4.76$. The agreement is reasonably good although there is a definite difference between the theory and experiment. Unfortunately a complete axisymmetric constant-density theory with Bernoulli effect analogous to Cole and Brainerd's theory for two-dimensional flow is not yet available. In addition, the compressibility correction may also be important for this problem. It is certainly important near the shoulder.

We turn now to the application of the method of integral relations to two-dimensional problems with asymmetry. Two new problems appear here which do not appear in symmetric problems. In the symmetric problem the boundary conditions for the $3N$ differential equations consist of $2N$ conditions dictated by symmetry at the axis, plus N conditions on the proper passage of the solutions through N sonic singular points. In asymmetric pioblems the solutions must pass through N sonic singular points on each side of the body, and there are thus $2N$ boundary conditions of this type. The $2N$ symmetry conditions vanish, and we require N additional conditions to determine the solution in general. The setting of these additional conditions is the first of the two new problems. The second involves the $N - 1$ equations for E_k.

In the case $N = 1$ the additional condition may be found, but gives little insight into the case $N > 1$. At one point on the shock $\bar{u}_1 = 0$, and at that point both P_{uv} and Q_u must be zero. With a body-oriented coordinate system $P_{uv} = 0$ on the body, because $\bar{v}_b = 0$ there. The quantity P_{uv} must be zero over the entire range of y, because of the linear interpolation approximation. Unless the quantity Q_u is also zero over the range of y, the ratio

$$\frac{P_{uv}}{Q_u} = \bar{v}E(\psi)^{-\frac{1}{\gamma-1}} \tag{6.3.40}$$

must be zero. But $\bar{v}$ from (6.3.40) should be consistent at the shock with the shock conditions, with

$$\left(\frac{P_{uv}}{Q_u}\right) = \bar{v}_1 E(\psi_1)^{-\frac{1}{\gamma-1}}. \tag{6.3.41}$$

Since $\bar{v}_1$ is clearly nonzero, $Q_u = 0$ and $\bar{u}_b = 0$. The stagnation streamline is here a straight line along a coordinate line, and enters the body normal to the body. The same result would be obtained with $N = 1$ with any other interpolation functions, provided the same ones are used for both P_{uv} and Q_u. This argument is a modification of one given by R. Vaglio-Laurin (private communication) and Brong and Leigh [1].

The situation for $N > 1$ has not yet been resolved. The differential equations themselves appear to yield a family of solutions with an N-fold degree of arbitrariness. That the difficulty is not connected with the determination of the E_k functions may be seen by the fact that it still appears in the constant-density case ($\gamma_s^{-1} = 0$, $\epsilon =$ *constant*) or in the isothermal case ($\gamma = 1$, $E^{1/(\gamma-1)} =$ *constant*), for which the extra $N - 1$ variables and conditions drop out. The extra conditions must be found in conditions on the mutual consistency of the assumed approximated forms for P_{uv} and Q_u in the stagnation region. But just how this should be done is not clear except in the case $N = 1$.

The second problem concerns the fact that the extra $N - 1$ equations for the E_k are difference equations, and that the elimination of the E_k leaves the other $3N$ equations as differential-difference equations. In the symmetric case the effect is not fundamental. In the asymmetric case the effect is important in the stagnation region. This region we define as the region in which $\bar{u}_b$, $\bar{u}_k$, and $\bar{u}_1$ do not all have the same sign.

The local flow in an asymmetric stagnation region has been discussed in Sections 4.3, 5.4, and 6.1. We can take these solutions as giving us at least a correct qualitative picture for our case with $N > 1$. On the basis of this picture of the flow field (for example, Fig. 4–9), there is an inner part of the stagnation region which is characterized by a feature which does not appear in the symmetric case. In evaluating the extra $N - 1$ variables E_k at a station in this inner region, information is required from the shock shape on both sides. This property defines the inner region. The differential-difference system in this region must be solved for the entire inner region at once rather than step-by-step in a specified direction. An analogous difference exists between the solutions of Fredholm integral equations and Volterra integral equations. Our situation corresponds to an infinite number of free parameters. This is also a property shared by the Fredholm equation.

We can set up an interpolation approximation for the shock shape in the inner region, in terms of M parameters. The effective order or number of free parameters is thereby reduced from an infinite number to $3N + M$. These parameters must describe not only the shape in the region but also the width of the region. Once the solution in this region is obtained, it may be propagated in both directions as a system of differential-difference equations. The solution must, of course, be subject to the N conditions required to resolve the first problem. The large effective order here does not make the method appear encouraging for $N > 1$. An alternative approach is one of successive approximation, with the shock shape assumed for the purpose of providing a first estimate for the E_k. The resulting solution provides an improved approximation of shock shape for the next approximate solution.

A possibility still exists of simplifying the procedure, one that corresponds to the immediate simplification when $N = 1$. Advantage would have to be taken of the fact that the stagnation region is normally of rather small extent. A local approximate solution for the stagnation region must be found which fits the assumed forms of the basic variables, at least at either end. This solution must yield adequate information to permit the integration of the $3N$ differential-difference equations to be propagated in both directions. Whatever the number of initially free parameters chosen for the local solution, the solution must provide relations between them to reduce their number to $2N$. This "reduced" problem appears as an extension of the symmetric problem, one in which the $2N$ symmetry conditions are effectively replaced by N conditions. The reduction of the problem to this form would resolve both of the basic asymmetric difficulties discussed above.

In the case of a flat-faced body or flat plate with its normal inclined relative to the free stream, the number of free parameters is reduced by two. We may pick Δ_0 arbitrarily, and may also pick the point at which the shock is normal or other suitable reference point arbitrarily. The corners of the body are located at the sonic singular points on the body, as in the symmetric case. A scale transformation and translation will transfer the body found to the size and position of a standard body, if desired. This reduction is possible because the body without regard to its corners is invariant both under a scale transformation and a translation in its plane.

In the case of a V-shaped body or double wedge we may pick either Δ_0 or the distance of the starting point from the vertex arbitrarily. The ratio between these two remains as a basic parameter. The corners again appear at the body sonic points, but are subject to the condition that the ratio of their distances from the vertex must have the desired value. Only the scale transformation is available, and the number of free parameters is reduced by one.

The effective order or number of free parameters for reduced asymmetric problems with the method of integral relations is summarized in Fig. 6–17.

We look now at the asymmetric flat plate problem in more detail. See Fig. 6–18. The stagnation streamline is straight and normal to the plate, and thereby is one which is deflected in passing through the shock. It thus does not correspond to maximum entropy but to the entropy corresponding to the given de-

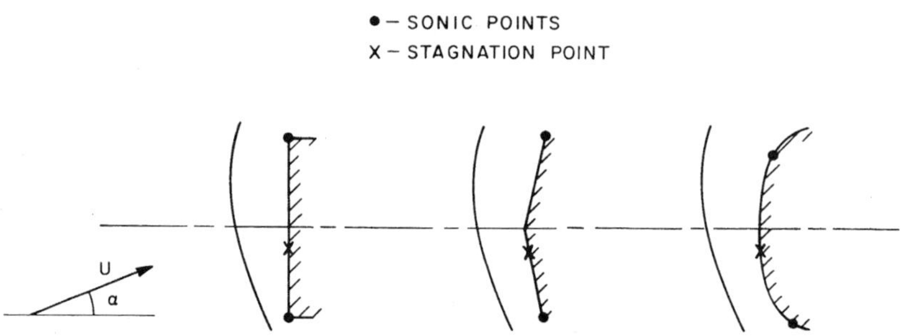

	FREE PARAMETERS IN REDUCED PROBLEM		
N = 1	0	1	2
N = 2	2	3	4
N	2N−2	2N−1	2N

FIG. 6–17. Number of free parameters or effective order with asymmetry in the reduced problem.

flection angle. Note that this entropy is governed by the choice of coordinate system. The stagnation streamline entropy gradient and vorticity are not zero, so that the vorticity is not consistent with the angle the stagnation streamline makes with the body. The oversimplified behavior of the stagnation streamline is analogous to that of the sonic singularity line with $N = 1$.

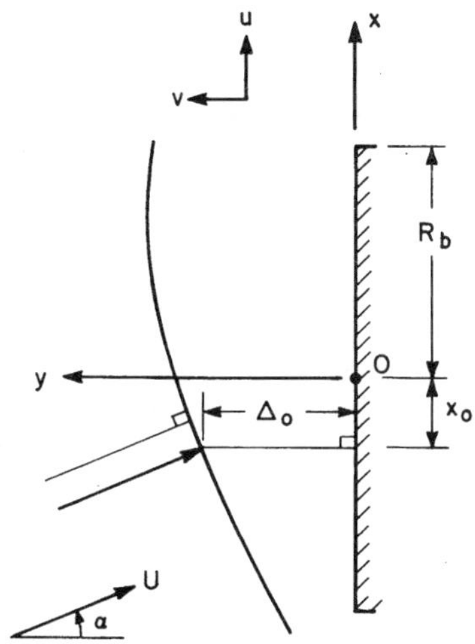

FIG. 6–18. Asymmetric flow on a flat plate with one strip.

With $N = 1$, asymmetric two-dimensional problems have been considered by Vaglio-Laurin [1] and Bazzhin [1]. Vaglio-Laurin studied blunt bodies, and imposed the condition that the stagnation streamline pass through the normal point on the shock. Bazzhin studied the flat plate, and applied a method of successive approximations to try to determine the correct entropy on the stagnation streamline. Neither author in the references cited used the condition indicated above as being correct with $N = 1$.

In the two-dimensional case the complete solution near the stagnation point is always analytic. In using polynomials we are approximating analytic functions by functions which are analytic. In the case of a general three-dimensional stagnation point with asymmetry the complete solution is nonanalytic at the stagnation point, and the stagnation streamline comes in tangent to the body (see Section 4 of Hayes [7]). An approximation to such a solution by a finite number of analytic interpolation functions is limited in how closely it can approximate the actual solution near the body. We must either accept this limitation or use appropriately chosen nonanalytic interpolation functions. If the solution is analytic at the stagnation point, the analysis of Waldman [2] or Hayes [7] shows the stagnation streamline must come in normal to the body. In an integral-relation method with $N = 1$ for a three-dimensional body, the same argument given above for a two-dimensional body would apply and give a stagnation streamline which is a straight line. This line would be normal to the body with a body-oriented orthogonal coordinate system.

The method suggested by Minailos [1] involves the approximation of the basic variables (from the divergence form) in two directions. In the direction normal to the body this approximation is in terms of polynomials, as in the standard method of Belotserkovskii. He uses a polar body-oriented boundary-layer coordinate system on a body of revolution, and approximates the azimuthal behavior of the variables through Fourier series. He uses one strip, and one Fourier term besides the constant term. With one Fourier term it is sufficient to write the equations in the plane of symmetry and integrate across the shock layer to yield a system of ordinary differential equations in the radial coordinate. His conditions on the stagnation streamline are patterned after Bazzhin's, and do not fit the condition of being normal to the body. He tested his method on the case of a sphere, and concludes his approximations were satisfactory for that example. He also carried out calculations for a circular disc.

Waldman [2] has presented a method similar to that of Minailos. He uses a boundary layer coordinate system based on a geodesic polar coordinate system on the body. The origin is placed at the estimated position of the stagnation point. The condition that the stagnation streamline be normal to the body is imposed, and azimuthal variations are assumed to be sinusoidal. He also tested his method on a sphere, with satisfactory agreement.

The complexity of applying the method of integral relations to yawed blunt-nosed bodies should not be underestimated, particularly when more than a single strip is necessary. The presence of these difficulties suggests that perhaps the method of approximating and integrating the variables along the body might be more practicable. This approach would lead to a set of ordinary differential equations to be solved in the coordinate normal to the body, between the shock and the body. The natural coordinate normal to the body, as we have mentioned earlier, is distance divided by distance between shock and body. Thus the body appears at $\eta = 0$ and the shock at $\eta = 1$ (or vice versa).

The question of how the outgoing strips are to be oriented remains. In Scheme II of Belotserkovskii [5] the strips are adjusted to fit the first outgoing characteristic from the body. See also Belotserkovskii and Chushkin [2]. With this approach the geometry depends upon the solution. It would appear preferable from a practical point of view to have a fixed orientation of the strips. Telenin has proposed a scheme which he has applied together with various collaborators (e.g., Gilinskii, Telenin, and Tinyakov [1] and Telenin and Tinyakov [1]) in which variables are approximated along the body and differential equations are established normal to the body. The orientation of their strips is fixed with respect to a spherical polar coordinate system whose origin is chosen conveniently for the given problem. For the symmetric case the polar angle ϑ and a radial variable η, made dimensionless with respect to the unknown shock layer thickness (cf. (6.3.26)), are used as independent variables. The shock layer is divided by rays from the fixed origin located within the body, say near the center of curvature, into segments of equal polar angle. The flow variables and shock shape are represented by polynomials in ϑ in which the coefficients depend on η. The values of the coefficients are linear functions of the values of the flow variables and shock shape on the rays. Substituting these polynomials into the equations of motion written in spherical coordinates yields a system of ordinary differential equations for the flow variables. The integration is carried out from an assumed shock to the body with the shock shape adjusted until the boundary conditions at the body are satisfied. No difficulty was reported in achieving convergence. In the generalization of the method to asymmetry (Telenin and Tinyakov [1]) variations with respect to azimuth angle are eliminated by interpolating between meridional planes. The variation in the flow variables with respect to the azimuth coordinate is represented by trigonometric polynomials. The use of the flow equations not in divergence form, the limitation to a spherical coordinate system and the representation of the flow variables by the particular polynomials used in the calculations might impose limitations on the scheme. However, calculations reported for various shaped blunt bodies show excellent agreement with other symmetric calculations and with experiment in the case of asymmetry.

We outline here a related alternative approach to the problem which has been suggested by Waldman and Probstein (unpublished). This method was suggested by available information on the method of Telenin prior to its publication.

Use is made of a body-oriented coordinate system with x measured along the body surface from the axis, y normal to the body, and φ an azimuth angular coordinate. The momentum and continuity equations are written in divergence form in these coordinates and have the general form

$$\frac{\partial A}{\partial x} + \frac{\partial B}{\partial \eta} + \frac{\partial C}{\partial \varphi} = D. \tag{6.3.42}$$

Here, the normalized distance $\eta = y/\Delta$ is used, with Δ the detachment distance. In accordance with the second approach of the integral relations method the equations are integrated with respect to x from the axis to the edges $x = x_k$ of a number of strips which divide the shock layer laterally (Fig. 6–19). The result is

$$A(x_k) - A(0) + \frac{\partial}{\partial \eta}\int_0^{x_k} B\,dx + \frac{\partial}{\partial \varphi}\int_0^{x_k} C\,dx = \int_0^{x_k} D\,dx. \tag{6.3.43}$$

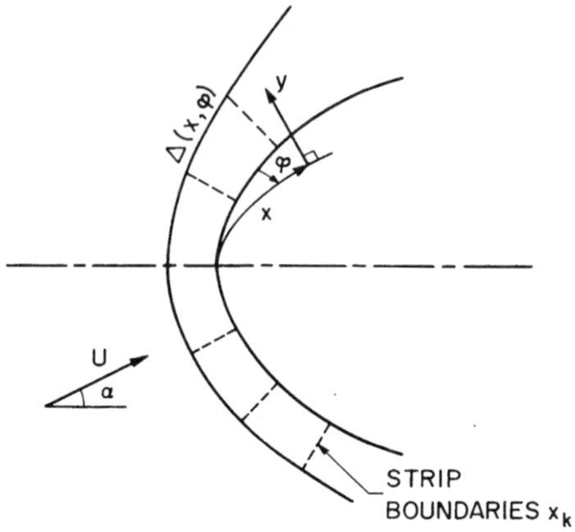

FIG. 6–19. Coordinate system in polynomial approximation method.

The functions B, C, and D are assumed to be expandable in terms of a set of polynomials $P_m(x)$,

$$B = \sum_m b_m(\eta, \varphi)P_m(x), \tag{6.3.44a}$$

$$C = \sum_m c_m(\eta, \varphi)P_m(x), \tag{6.3.44b}$$

$$D = \sum_m d_m(\eta, \varphi)P_m(x). \tag{6.3.44c}$$

Because of this assumption we term this method one of polynomial approximation.

We now substitute (6.3.44) in (6.3.43) and obtain

$$\frac{\partial B_k}{\partial \eta} = -\frac{\partial C_k}{\partial \varphi} + D_k + A(0) - A(x_k), \tag{6.3.45}$$

where B_k, C_k, and D_k have the general form

$$B_k = \sum_m b_m \int_0^{x_k} P_m \, dx. \tag{6.3.46}$$

Following the suggestion of Minailos [1] the azimuthal dependence of the functions $A(x_k)$, B_k, C_k, and D_k is represented by Fourier series in the form

$$B_k = \sum_n B_{nk}^{(1)}(\eta) \cos n\varphi + \sum_n B_{nk}^{(2)}(\eta) \sin n\varphi + B_k^{(0)}(\eta). \tag{6.3.47}$$

Substituting (6.3.47) and the analogous expressions for $A(x_k)$, C_k, and D_k into (6.3.45), the basic ordinary differential equations relating the coefficients in the Fourier series as functions of the normal coordinate are:

$$\frac{dB_{nk}^{(1)}}{d\eta} = -nC_{nk}^{(2)} + D_{nk}^{(1)} - A_{nk}^{(1)}, \tag{6.3.48}$$

$$\frac{dB_{nk}^{(2)}}{d\eta} = nC_{nk}^{(1)} + D_{nk}^{(2)} - A_{nk}^{(2)}, \tag{6.3.49}$$

$$\frac{dB_{k}^{(0)}}{d\eta} = D_{k}^{(0)} + A(0) - A_{k}^{(0)}. \tag{6.3.50}$$

The equation of the shock is also expanded in a Fourier series and written

$$\Delta = \sum_m P_m(x) \left[\sum_n \Delta_{mn}^{(1)} \cos n\varphi + \sum_n \Delta_{mn}^{(2)} \sin n\varphi + \Delta_m^{(0)}\right]. \tag{6.3.51}$$

By inversion, the values of the functions $B_{nk}^{(l)}(1)$ on the shock surface can be related to the values of the original variables $B(x_k, \Delta, \varphi)$ through equations (6.3.44), (6.3.46), and (6.3.47) evaluated at the shock. The shock relations are used to give the original variables in terms of the shock orientation $\Delta(x_k, B)$. The shock orientation can in turn be related to $\Delta_{mn}^{(l)}$ by means of (6.3.51) above and its derivatives. This gives the shock boundary conditions on (6.3.48) to (6.3.50) as functions of $\Delta_{mn}^{(l)}$. The integration of the basic ordinary differential equations is based on an initial guess of the parameters $\Delta_{mn}^{(l)}$. The equations are then integrated from the shock to the body. If the boundary conditions are not satisfied then the shock parameters must be adjusted and the integration repeated. A scheme for doing this adjustment must be found such that the solution will approach one in which the boundary conditions at the body are satisfied.

The set of approximating polynomials $P_m(x)$ has been left unspecified with the choice to be guided by the general nature of the solution. If the maximum order of the polynomials equals the number of strips there is no essential choice. The advantage of methods of this type is clear, since the coordinate system may be fixed in advance and the need of integrating through any sonic singularities is eliminated. This latter point has been made previously by Belotserkovskii [5] and Belotserkovskii and Chushkin [2]. Although it cannot be expected to give satisfactory results for solutions which do not vary smoothly along the body it should be able to provide as accurate results as the original Belotserkovskii approach in most cases, at a great deal less expense in labor. It should be much easier to carry out the scheme with N not very small than in the other approach. In this method details of the stagnation streamline and transonic region are expected to come out as consequences of the solution.

We have not attempted here to give a complete discussion of the method of integral relations other than for the blunt-body problem. For other applications we refer the reader to the survey article of Belotserkovskii and Chushkin [2].

4. Relaxation techniques and the unsteady approach method

The numerical methods for the direct problem we discuss next comprise what are termed relaxation methods and the closely related unsteady approach method. The classical method of relaxation is the method associated with the name of Southwell. We sketch this method for the case of a single partial differential equation with one dependent variable and two independent variables. The plane of the independent variables is covered by a network of squares, and at each net point an assumed estimated value of the dependent variable is assigned. The basic differential equation is written in difference form as a set of difference expressions defined at each point which must be zero at each point. The assumed values of the dependent variable are then substituted into these equations. In general these difference expressions will not be zero. Their value at each net point is known as the residual, and is a measure of the failure of the assumed solution to satisfy the equation. The problem is to reduce the residuals to zero by suitably adjusting the assumed values of the dependent variable. Although this particular relaxation technique is applicable to the problem at hand, it does have certain disadvantages. It is laborious, is generally ill suited for machine computation, and generally requires a good deal of judgment on the part of the operator. It does, of course, involve more than one dependent variable and equation.

An alternative approach to the direct blunt-body problem using almost the same formulation involves examining the corresponding time dependent problem. Here the limiting steady state solution is sought, which we assume is unique and is approached asymptotically in time. Appropriate initial conditions are specified. For example, we may assume that the body is impulsively accelerated from rest to a constant velocity relative to the undisturbed fluid. We term this technique the unsteady approach method. It may be looked upon as a specialization of the more familiar relaxation method where the residual is identified as a time derivative term. The change in the estimate of the dependent variable or variables in a relaxation is proportional to the residual at the same point times an assumed time interval. This interval should be small enough to ensure stability of the process and large enough not to waste computation effort.

We begin by discussing the application to the direct problem of the Southwell method. In this case (as in the unsteady approach method) we shall only consider the symmetric problem. The extension "in principle" to the asymmetric case is clear, though the computational effort would naturally be greater. Several authors have employed relaxation methods for determining the shock wave and inviscid field over blunt-nosed bodies at relatively low supersonic Mach numbers. The first of such investigations was carried out by Maccoll and Codd [1], who computed the supersonic flow past flat-nosed two-dimensional and axisymmetric bodies. Their initial estimations of the location and shape of the shock wave were obtained from experimental photographs. The accuracy of this study was strongly limited by their assumptions that the flow behind the shock was isentropic, and that the streamlines crossed the sonic line at right angles. Also, with these assumptions it is impossible to take into account the influence of the supersonic region in altering the sonic line shape and thereby in affecting the subsonic flow field.

Somewhat later, Drebinger [1], in his doctoral thesis, employed a relaxation technique to determine the flow over a blunt wedge behind a detached shock wave at a Mach number of 1.44. Although Drebinger was correct in taking the rotationality of the flow field into account, he also did not consider the upstream influence of the transonic zone between the sonic line and the limiting characteristic.

Mitchell [1] and Mitchell and McCall [1] again calculated the supersonic flow about flat-nosed two-dimensional and axisymmetric bodies. They automatically took into account the effect of the transonic zone on the upstream flow by carrying the relaxation calculation through to the purely supersonic region. In their calculations, no iteration on the shock shape was carried out, but instead the shock shape measured from an experimental photograph was taken as the shock shape in their solution. That this shock shape may not have been the correct one for their problem can be seen by examining their published results: the slope of the sonic line within and on the boundaries of the flow field and the shape of the streamlines in the sonic region appear to be inconsistent with exact calculations of local flow properties in the sonic region such as those discussed in Section 6.1.

We propose now to indicate a possible way of carrying out a relaxation solution for the symmetric blunt-body problem. This method is the one suggested in the first edition of this book. Independently of each other and of the authors a similar approach was carried out in detail by Hamaker [1] and by Gravalos [1] (see also Gravalos, Edelfelt, and Emmons [1]). Our method is closest to that developed by Gravalos, Edelfelt, and Emmons. Although this approach will apply for a general fluid, we shall again use the restriction to a perfect gas for simplicity. To arrive at the basic differential equation for the interior field let us again introduce the usual stream function through the relations

$$\rho u y^j = \frac{\partial \psi}{\partial y}, \tag{6.4.1a}$$

$$\rho v y^j = -\frac{\partial \psi}{\partial x}, \tag{6.4.1b}$$

where x and y are rectangular cartesian or cylindrical coordinates with origin at the stagnation point. From the definition of the vorticity $\zeta = \partial v/\partial x - \partial u/\partial y$, and the Crocco vorticity law (6.2.5) we may write the differential equation for the stream function in the form

$$\frac{\partial^2\psi}{\partial x^2} + \frac{\partial^2\psi}{\partial y^2} - \frac{\partial\psi}{\partial x}\frac{\partial \ln \rho}{\partial x} - \frac{\partial\psi}{\partial y}\frac{\partial \ln \rho}{\partial y} - \frac{j}{y}\frac{\partial\psi}{\partial y} + \frac{y^{2j}p\rho}{\mathscr{R}}\frac{dS}{d\psi} = 0, \tag{6.4.2}$$

where $\mathscr{R}$ is the gas constant in the perfect gas law. In addition, the pressure, density, and entropy are connected by the relation

$$\frac{p}{p_0} = \left(\frac{\rho}{\rho_0}\right)^{\gamma} \exp\left(\frac{S - S_0}{c_v}\right), \tag{6.4.3}$$

with the subscript 0 indicating stagnation conditions. The final relation required is supplied by the energy equation

$$\frac{q^2}{2} + \frac{\gamma}{\gamma - 1}\frac{p}{\rho} = \frac{\gamma}{\gamma - 1}\frac{p_0}{\rho_0} \tag{6.4.4}$$

By using (6.4.1) and (6.4.3) we may rewrite (6.4.4) as

$$\frac{\gamma - 1}{2a_0^2\rho^2 y^{2j}}(\psi_x^2 + \psi_y^2) + \left(\frac{\rho}{\rho_0}\right)^{\gamma-1} \exp\left(\frac{S - S_0}{c_v}\right) = 1, \tag{6.4.5}$$

with a_0 the stagnation speed of sound.

If we consider (6.4.2) as the basic differential equation for the stream function ψ, and (6.4.3) and (6.4.5) as auxiliary equations, we see that in the subsonic region behind the shock we have an elliptic partial differential equation for the stream function. As we have already observed, this problem is by no means a standard one since the location of the boundaries of the subsonic region, namely the shock and sonic line, are unknown.

The boundary conditions for this problem are as follows: at the shock wave the stream function and tangential velocity component are continuous across the shock. At the body and on the stagnation streamline $\psi = 0$, with the tangential velocity component being specified through the isentropic Bernoulli equation. Finally, the sonic line is determined by

$$q^2 = a^2 = a_0^2\left(\frac{\rho_{\text{son}}}{\rho_0}\right)^{\gamma-1} \exp\left(\frac{S_{\text{son}} - S_0}{c_v}\right). \tag{6.4.6}$$

It might at first appear that we have overspecified the boundary conditions for the elliptic problem. However, this is not the case, since the fact that the boundary location is unknown requires the added information.

The general approach to the relaxation solution of (6.4.2) is to make an initial assumption as to the shock shape and location and the streamline distribution, based, say, on one of the more approximate theories discussed previously. Then, as we have noted, (cf. Green and Southwell [1]) both the subsonic region and the supersonic region in the neighborhood of the sonic line are covered with a network of squares and the differential equations for ψ are written in finite difference form. An assumed value of ψ based on the approximate solution is assigned to each net point (making sure, of course, that the values of ψ at both the shock boundary and the surface satisfy the appropriate boundary conditions). These values of ψ are substituted in the difference equations, and the residuals calculated. The first step consists of obtaining a pattern relating a change in ψ at each individual point to the changes in the neighboring residuals. Derivatives of ψ can be computed from the assumed distribution of ψ. Since S is a known function of ψ for a given shock shape, $\ln \rho$ can then be determined from (6.4.5). Thus ψ, its derivatives, $\ln \rho$, p, and $dS/d\psi$ can be computed, and one can in turn determine the residual of (6.4.2).

Because the bow shock is curved, an interpolation technique such as discussed in Green and Southwell [1] or in Mitchell [1] is required at this boundary. A much more serious problem than this, however, is the breakdown of the relaxation technique in the neighborhood of the sonic line because of the fact that small changes in ψ cause relatively large variations in $\ln \rho$. Under such circumstances the relaxation pattern is no longer workable. Therefore, the main problems which we must consider are that of devising a method for carrying the relaxation calculation through the sonic region and somewhat beyond the limiting characteristic, and that of the establishment of certain consistency relations which are to serve as boundary conditions on the sonic line. These consistency relations can indicate how an incorrect shock shape should be adjusted so that the relaxation process can be repeated to obtain an improved solution.

One method of handling the flow pattern in the transonic zone is to approach the sonic line asymptotically from both the subsonic and supersonic regions by means of the relaxation technique, employing smaller and smaller squares as the sonic line is approached. Now the throat condition fixes ρ exactly at the sonic line. The modification of the distribution of ψ in the sonic region must be made subject to the mass flow requirement at $M = 1$, rather than with only the elimination of the residuals in mind. Therefore, a trial and error technique employing Bernoulli's equation and using first and second differences in ρ can be used (because of the monotonic character of ρ) to extrapolate the location of the sonic line as the sonic line is approached. In this manner one may complete a tentative interior solution. In general, however, this solution will not be compatible with the conditions on the derivatives along the sonic line and along the shock, and the shock wave must be adjusted accordingly and another try made.

At this point any number of schemes are available for determining how the shock is to be altered. Perhaps the simplest of these is a method similar to the one described previously in conjunction with the streamtube-continuity technique. In this approach as many normal trajectories to the streamlines are drawn as are necessary to locate the shock. From the solution the new shock points y_s are then computed by means of (6.2.2) and (6.2.3), which may be combined for a point on the shock in the form

$$\frac{y_s^{1+j}}{1+j} = \int_{\psi=0}^{\psi=\psi_s} \frac{\rho q}{\rho_\infty U} y^j \, dn. \tag{6.4.7}$$

This method is related to that of Gravalos [1] and Gravalos, Edelfelt, and Emmons [1]. However, they first carried out a streamtube-continuity calculation in the transonic region. The calculation was based on an initially assumed surface pressure distribution and on the normal momentum equation expressed in terms of the streamline radius of curvature. The method near the sonic line is thus of the type discussed in Section 6.2. After obtaining a successfully iterated solution they applied the results along an appropriate line to serve as the boundary condition on one of the boundaries of the subsonic region. This subsonic region is the one treated by relaxation. Of course one could also draw in the sonic line in the tentative solution, and a corresponding integral relation along this line could be obtained, in which the flow deflection angle, the angle the sonic line makes with the streamline, and $(\rho q)_{\text{son}}$ are the parameters (Drebinger [1] and Hamaker [1] follow analogous lines of approach). In any event, by this scheme or by one similar to it a new shock shape is determined in which the values of ψ at the shock wave are altered, thereby causing a change in the residuals near the shock. A new interior solution must now be obtained by a repetition of the relaxation process. The shock shape must again be adjusted, and the cycle repeated until a consistent convergent solution is obtained.

Once such a solution is found it can be checked by measuring the angles between the streamlines and constant velocity lines and then comparing these values with the corresponding angles which can be computed from the now known derivatives of q and θ. Of course, in two-dimensional flow the angles the streamlines make with the constant velocity lines such as the sonic line are known at the shock boundary independent of the shock curvature (cf. Fig. 6–2).

The relaxation method has been successfully applied to the flow of a calorically perfect gas over a cylinder at $M_\infty = \infty$ by Hamaker [1]. These calculations were carried out on a desk computer; Hamaker noted that the computations were tedious in the subsonic region and difficult in the region of the sonic line. Gravalos [1] and Gravalos, Edelfelt, and Emmons [1] performed machine calculations about a blunt body of revolution for air in thermodynamic equilibrium. There is little question that the use of a streamtube-continuity approach in the transonic region greatly facilitated their computation.

We turn now to the unsteady approach method described at the beginning of the section. As with any hydrodynamic problem either a Lagrangian or an Eulerian formulation may be used. In the Lagrangian formulation the coordinate system is one which moves with the fluid and is distorted by the motion. Each coordinate point identifies a single fluid particle. In the Eulerian approach the coordinate system is fixed in space, usually with the body also fixed.

Several unsteady approach methods have been reported in the literature. One by Evans and Harlow [1–3] (see also, Harlow [1]) makes use of what is essentially an Eulerian frame in connection with a Lagrangian calculation procedure. These authors have termed their method the "particle-in-cell" method. We have called the procedure essentially Eulerian since the computational net is fixed so that after an arbitrary time it is not possible to distinguish individual particles in the Lagrangian sense. The actual computation is, however, carried out using the equations in Lagrangian form over a time δt which is small. The results are then interpolated to the original Eulerian coordinates and the calculation is repeated. In two short notes they have reported the results of such calculations for a cylinder transverse to the stream and for the flow past a flat-faced cylinder with axis parallel to the stream (Evans and Harlow [2; 3]). On the basis of these results it would appear that the method has limitations. In particular, the reported results have an accuracy which is less than that normally desired in detailed flow calculations. Furthermore, the calculation time appears to be rather long, and any attempt to improve the accuracy by increased mesh fineness is reported to increase the computing time as the inverse cube of the mesh scale.

A second unsteady approach method has been applied by Godunov, Zabrodin, and Prokopov [1] to the flow of a perfect gas with $\gamma = 1.4$ and $M_\infty = 4$ about a sphere. The method generalizes to two-dimensional and axisymmetric unsteady problems an approach originally put forward by Godunov [1] for one-dimensional unsteady motions. The method uses the hydrodynamic conservation laws for plane or axisymmetric flow in divergence form. These equations are integrated over a closed surface in a three-dimensional space consisting of two geometric dimensions and time. The fact that the integral conservation laws are used directly ensures that shock waves or other gasdynamic discontinuities present in the flow do not interfere with the attainment of the proper solution.

The equations governing these conservation integrals in cylindrical coordinates take the form

$$\int y^j[\rho\, dx\, dy + \rho u\, dy\, dt + \rho v\, dx\, dt] = 0, \tag{6.4.8}$$

$$\int y^j[\rho u\, dx\, dy + (p + \rho u^2)\, dy\, dt + \rho uv\, dx\, dt] = 0, \tag{6.4.9}$$

$$\int y^j[\rho v\, dx\, dy + \rho uv\, dy\, dt + (p + \rho v^2)\, dx\, dt] = 0, \tag{6.4.10}$$

$$\int y^j \left[\rho \left(e + \frac{q^2}{2}\right) dx\, dy + \rho u \left(e + \frac{p}{\rho} + \frac{q^2}{2}\right) dy\, dt + \rho v \left(e + \frac{p}{\rho} + \frac{q^2}{2}\right) dx\, dt\right] = 0, \tag{6.4.11}$$

with $q^2 = u^2 + v^2$. The system is completed by the equation of state. The mesh network appropriate to the geometry being considered is chosen and integrals in (6.4.8) to (6.4.11) are written as appropriate sums over the values of the variables at the mesh points. The coordinate system is not assumed to be steady. These integrals are then applied to individual cells in the coordinate net to obtain difference equations appropriate for the moving coordinate system. In writing the difference equations the quantities ρ, ρu, ρv, etc. in the integrals are assumed constant on each cell boundary throughout each time step. The resulting system of difference equations obtained from the integral conservation equations then approximates the corresponding exact system of differential equations. Godunov, Zabrodin, and Prokopov [1] chose a network of the form shown in Fig. 6–20 for calculating the unsteady approach past a sphere. The limiting ray BC is selected to encompass the region in which the supersonic flow affects the subsonic domain in the steady state solution. The initial outer boundary AB may be chosen arbitrarily, but was chosen in the example to be a sphere of radius 50% larger than that of the body. It is displaced in the calculations with advancing time as a consequence of the calculation method.

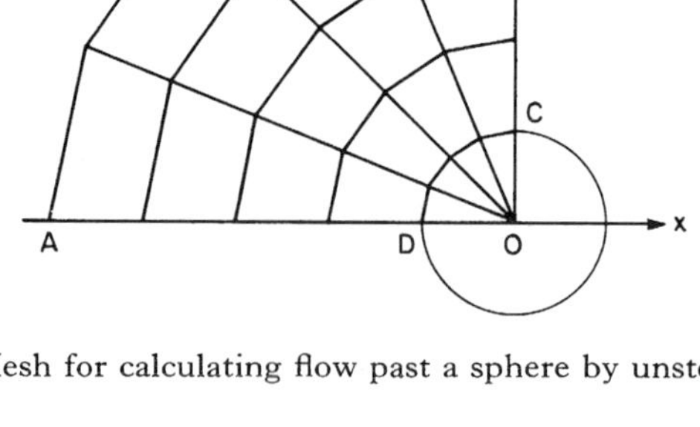

FIG. 6–20. Mesh for calculating flow past a sphere by unsteady approach.

When the steady state is reached the outer boundary corresponds to the position of the detached shock.

As shown in Fig. 6–20, in the finite difference scheme both the outer boundary *AB* and body surface *DC* are approximated by straight line segments. Symmetry conditions for the finite difference equations are applied along the axis *AD*, while the boundary conditions along the surface *DC* are determined by the vanishing of the normal velocity component.

Initially, at time $t = 0$, the fluid about the sphere is assumed to be at the uniform steady state free stream conditions. What happens physically under such conditions is that a shock is formed at the surface of the body and moves outward. With increasing time the shock asymptotically approaches its steady state position. The procedure adopted for the calculation of this process is to assume that at the outer boundary *AB* the Hugoniot shock jump conditions always apply. Initially, then, the boundary *AB* propagates into the undisturbed fluid in front with sonic speed, along a characteristic surface in the space-time space. The actual motion of the boundary *AB* depends on the motion of the undisturbed fluid as well as on the propagation velocity of the surface relative to the undisturbed fluid. Eventually, the boundary propagates into the undisturbed fluid as the shock wave of the final solution, and thereby becomes stationary. The treatment of the shock which leaves the body and lies within the region *ABCD* is carried out quite differently. The authors report that they use a "smearing" technique. We assume that such a technique is effectively similar to that of von Neumann and Richtmeyer [1], in which artificial dissipative terms (similar to terms from an artificial bulk viscosity) are introduced into the equations of motion. With the presence of such dissipative or artificial viscosity terms a continuous solution can then be obtained which represents a shock wave. The shock wave then appears as a layer of finite thickness.

In the calculational procedure an increment of time δt is selected. For each time interval the new position of the outer boundary is calculated by means of the shock conditions at the outer boundary. The internal net is then relocated. Then for the time considered the values of the flow variables at the net points are calculated using the basic finite difference equations already obtained from the conservation integrals. In the first of these steps the "smeared" shock wave appears within the field. As the procedure is repeated the outer boundary and the smeared shock merge into a discontinuous shock matched to the free stream conditions. The procedure is repeated until the shock becomes stationary and until the flow variables in the field are also stationary.

On Fig. 6–21 we have shown the computed density as a function of increasing dimensionless time for a net point about 38° from the axis and located approximately half way through the shock layer. The approach to steady state as well as the oscillatory character of the interior flow is clear. The oscillations may possibly be associated with reflection processes within the flow field. In

Fig. 6–22 is shown the pressure distribution on the surface of the sphere as computed by the unsteady approach method compared with the results of Belotserkovskii [4]. Although not shown, a very much coarser mesh does not appear to give significantly poorer results.

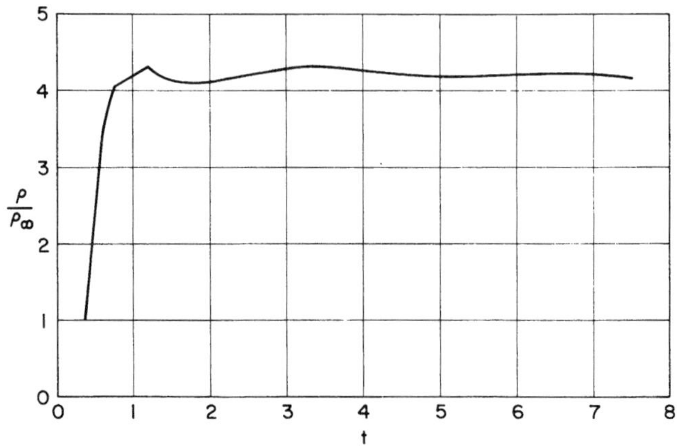

FIG. 6–21. Density at an interior point as a function of time (Godunov, Zabrodin, and Prokopov [1]).

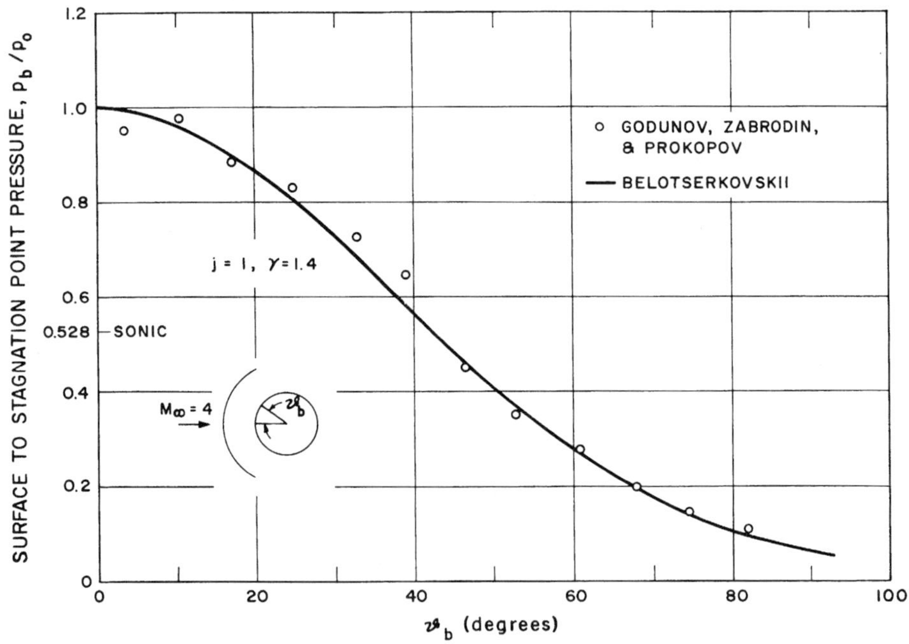

FIG. 6–22. Pressure distribution on a sphere (Godunov, Zabrodin, and Propokov [1] and Belotserkovskii [4]).

The error in the calculation appears to increase as the stagnation region is approached. This may be associated with the fact that the fluid there has low velocities and possesses to some extent the initial incorrect thermodynamic properties from the early part of the solution. This fluid is stagnant in a sense, and unable to escape fast enough. If this is the case the calculation scheme could be modified by introducing a sink in the stagnation region in order to help drain off the low velocity fluid with incorrect properties. The approach to steady state might be appreciably speeded up.

An unsteady approach method with features similar to the one proposed by Godunov [1] is due to Lax [1], with subsequent modification by Lax and Wendroff [1]. The basic idea of the method is that the hydrodynamic equations are written in divergence form and serve as the differential analogues of the conservation relations (6.4.8) to (6.4.11). The particular differencing scheme which is applied to the differential equations is not of fundamental importance (see, e.g., Burstein [1]).

The conservation equations in two independent space variables may be expressed in the form

$$\frac{\partial L}{\partial t} = \frac{\partial P}{\partial x} + \frac{\partial Q}{\partial y}, \tag{6.4.12}$$

where P and Q are functions of L. In the Lax-Wendroff differencing scheme the function $L(x, y, t + \Delta t)$ is expanded in a Taylor series and includes the second-order term $\frac{1}{2}\, \partial^2 L/\partial t^2\, \Delta t^2$. It is this term transformed into space derivatives which is added to the right-hand side of the straightforward difference analogue of (6.4.12)

$$L(x, y, t + \Delta t) = L(x, y, t) + \left[\frac{\partial P}{\partial \bar{x}} + \frac{\partial Q}{\partial \bar{y}}\right] \Delta t, \tag{6.4.13}$$

where the bar over the space variable denotes the appropriate centered difference quotient. The second central difference has the form and effect of an artificial viscosity term similar to that discussed by von Neumann and Richtmeyer [1], with a viscosity coefficient proportional to a square of the step length divided by the forward time difference. This effective artificial viscosity, introduced as a step width, permits any discontinuities to be represented calculationally by finite width regions of steep gradients. A modification of the approach described has been applied by Bohachevsky, Rubin, and Mates [1] to the calculation of blunt-body flows with detached shock waves, including the effects of asymmetry. Other unsteady approach calculations have been carried out. Some of these are reported, for example, in Babenko, Voskresenskii, Lyubimov, and Rusanov [1].

5. The inverse problem

In the preceding sections of this chapter we have concerned ourselves with methods for finding exact numerical solutions for the direct problem, in which the shape of a blunt body in a uniform supersonic or hypersonic stream is given, and the shape and location of the detached bow shock and the flow in the shock layer are unknown. We turn now to what is termed the inverse problem (cf. Section 5.1), in which the shape of the shock is given and the shape of the body and the details of the flow in the shock layer are unknown. We shall restrict the principal part of our discussion to symmetric flows, though the asymmetric problem will be considered. In this inverse problem, fundamental questions arise with respect to the uniqueness and existence of a solution and with respect to stability and convergence of calculation procedures.

The determination of the subsonic part of the flow field in this problem is governed by an elliptic partial differential equation, and is carried out as a Cauchy or initial value problem with Cauchy data (value of a function and of its normal derivative) specified along the shock wave; the boundaries of the subsonic region, namely the sonic line and the body, are unspecified. From the mathematical point of view a Cauchy problem for an elliptic differential equation is an improperly posed problem, and stringent restrictions must be placed on the boundary values in order that a solution exist. If a solution does exist, then the question of the uniqueness of the numerical solution to a given problem and the question of the stability of a calculation procedure for finding the solution are closely connected. The most important single feature of the inverse problem is the insensitiveness of the shock shape to local changes in body shape, and this feature leads to essential difficulties in the inverse problem. A minute local change in the shock shape will in general cause a large change in the body shape and may even preclude the existence of a solution.

Among the first to suggest the idea of approaching blunt-body flows as a Cauchy problem were Lin and Rubinov [1]. Their idea was simply that if the form of the detached shock wave were known precisely, the values of the flow variables behind the shock could be found from the oblique shock relations, while the first and higher derivatives of the flow quantities could be obtained from the equations of motion. The flow field is known to be analytic in a region immediately downstream of the shock if the shock shape is analytic. Lin and Rubinov pointed out that if the flow could be assumed to be analytic all the way to the body, then one could represent the flow between the shock wave and the body by an appropriate power series expansion. Lin and Shen [1], following these original ideas, carried out such an analysis for axisymmetric flow by means of a double power series expansion from an assumed parabolic shock. Their treatment was an essentially local one which was restricted to the stagnation region of the body.

Independently and at about the same time, Cabannes [1] performed a similar analysis for axisymmetric flow in which he assumed the shock wave to be described in rectangular cartesian coordinates by a power series of the form $x_s = \Sigma\, a_n y_s^n$. With this form of the shock wave he determined, for $\gamma = 1.4$ and arbitrary free stream Mach number, the coefficients in the double power series expansion for the flow variables to the x^4, x^2y^2, and y^4 terms, and for the stream function to the x^6, x^4y^2, ... terms. Cabannes [2] later extended his calculations to terms of order two higher in the distance in the special case $M_\infty = 2$. An extensive use of such double power series expansions is to be found in the numerical work of Richtmeyer [1] and G. Lewis [1]. By programming for a computing machine the algorithms for performing formal operations on power series, they have carried out double power series expansions to the nineteenth degree. In this procedure the expansion can contain as many as 210 terms. The particular case reported by Richtmeyer is for a shock in the shape of a hyperboloid of revolution, with $M_\infty = 12$ and $\gamma = 1.4$.

In both the work of Lin and Shen [1] and Cabannes [2] numerical calculations were made only for low supersonic Mach numbers. From unpublished calculations of Probstein which extended Lin and Shen's work to higher order terms, and from the calculations of Cabannes, the series expansions from the shock diverge in the region including the stagnation point. The existence of this divergence was pointed out by Van Dyke [6] through a recasting of Cabannes' series into parabolic coordinates. Richtmeyer [1] also points out that the series he obtains diverges outside of a domain whose boundary is about eight-tenths of the way from the shock to the body. Van Dyke [5; 6] has observed that the source of the divergence lies in the fact that the power series describes not only the flow downstream, but also its analytic continuation upstream. He points out that such a fictitious flow contains a limiting line at which the flow variables are nonanalytic. Therefore if the shock wave is closer to the limiting line than to the body, the power series will not include the body in its radius of convergence.

Actually for high Mach numbers and low ϵ where the shock layer is thin we might expect such methods to yield better results than for the lower Mach number cases, but even there the series-expansion techniques which have been developed are probably still not too satisfactory. For sufficiently small ϵ the likelihood exists that a power series expansion will not diverge. This is particularly likely to be so in the axisymmetric case. The solutions of Sections 4.3 and 4.4 show the differences between the axisymmetric and two-dimensional cases. It should be pointed out, however, that the divergence of the power series at the stagnation point does not exclude the possibility of the use of this method. As Lin and Shen observed, the justification of the method is really to be found in the idea of polynomial approximation. Therefore it seems reasonable that the method, if properly modified, might work just as the Kármán-Pohlhausen method works in boundary layer theory, where a series expansion of the Blasius equation diverges for large values of the independent variable. One reasonable modification might be to introduce in the scheme one (or more) integral relations, as for example an overall continuity condition, which would then serve to bring in the downstream influence on the upstream flow as well. Compare the Kármán-Pohlhausen approach of Maslen and Moeckel [1] (Integral Method). Van Tuyl [1] and Van Dyke [7] re-expressed Cabannes' series in terms of rational fractions and showed that a result close to the true computed solution can be obtained. Such a procedure, though possibly quite useful, must nevertheless be considered empirical to a large extent, particularly when the source of the divergence is not considered.

Another series expansion technique is to expand in only one coordinate, not from the shock toward the body but from the symmetry axis of the shock outward. Such procedures have been used for problems in which viscosity is important, including boundary layer theory, and form the basis of the viscous layer theory described in Section 10.2 of the first edition of this book. In the inviscid case they had been used by Swigart [3] and by Bazzhin and Gladkov [1]. Swigart considers parabolic shocks and uses parabolic coordinates oriented with respect to the shock. Bazzhin and Gladkov use cylindrical coordinates with the origin at the shock on its symmetry axis. The idea behind the two procedures is fundamentally the same, and involves reducing the partial differential equations of the problem to a series of ordinary differential equations describing the coefficients in the expansions.

The method is easily described with reference to the inviscid equations (5.1.6), where the reference surface for the coordinate system is taken to be the prescribed shock wave. The procedure is to expand the velocity component u tangent to the shock in odd powers of the distance x measured along the shock surface. The coefficients in the expansion are functions of the distance y normal to the shock. The other dependent variables are similarly expanded in even powers of x. This leads to a succession of ordinary differential equations in y.

There is a basic difficulty in this approach, which lies in the process of truncation. To illustrate this we note that in solving (5.1.6a) and (5.1.6b), with ρ assumed constant for simplicity, we need the second coefficient of the pressure p in order to be able to calculate the first coefficients in u and v. Calculation of this coefficient of p from (5.1.6c) involves the second coefficient of v, which is in turn related to the x^3 component of u by (5.1.6a). There is no way of stopping this argument with a finite number of terms, of truncating the series without inconsistency. This is simply a manifestation of the influence of the downstream transonic flow in the elliptic region, as discussed in Section 6.1. The only solution to the problem is therefore to arbitrarily terminate the series without attempting to obtain full self-consistency. The consistency is then obtained only for the lowest-order terms in x which are retained.

If ϵ is small a double power series in x and ϵ may be considered, in a theory for thin shock layers. In this case the truncation difficulty is resolved, provided phenomena such as the inner layers discussed in Chapter V are properly taken care of. Thus with ϵ small the termination of the series need not be done in an arbitrary manner. The truncated power series corresponds to a stage in a successive approximation scheme of the type discussed in Section 5.2.

The approach using a power series in x has been used by Swigart in connection with the asymmetric problem, and we shall discuss it further later. A procedure of this type can be quite satisfactory for bodies which do not depart very far from conic sections. In spite of this fact the procedure cannot be expected to be too satisfactory in the neighborhood of the corners of a blunt-nosed body with sharp corners. Van Dyke [7] has suggested improving the radius of convergence of such series, essentially by expanding in powers of $x^2/(1 + x^2)$, rather than of x^2 itself.

We would note in concluding our discussion of series expansions that Richtmeyer [1] and G. Lewis [1] have actually used the method to accurately determine a body shape. The procedure involves obtaining new Cauchy data in the shock layer along a line within the domain of convergence of the original expansion about the shock. The classical process of analytic continuation of power series is actually carried out numerically. With this new data a new expansion is carried out with the coefficients again determined by the equations of motion. Hopefully the body, whose location is determined by the surface boundary conditions, is contained within the domain of convergence of this second expansion. An extremely high degree of accuracy was employed in these calculations. A technique termed "significance arithmetic" was used to suppress numerical instabilities, and this must act as a smoothing process of the type discussed below. Such a procedure is essential in a numerical analytic continuation of a power series. Any truncation of a power series is a polynomial, which has an infinite radius of convergence. A recasting of a polynomial with new origin does not get around a convergence difficulty in a power series approximated by the original polynomial. Richtmeyer himself points out that the method is not to be recommended for practical calculations.

Returning to the general inverse problem which we have set ourselves, we know that as far as elliptic differential equations are concerned, the initial value problem is improperly posed and leads to an unstable solution when treated by finite differences. If ϵ is small the shock layer is thin compared with the radius of curvature of the shock. In a "marching-ahead" procedure which starts with Cauchy data on the shock, only the relatively small thickness of the shock layer must be traversed. We may expect that there are fundamental singularities in the analytic continuation of the flow field past the body, and that these singularities are somewhere near the natural foci of the shock curve (e.g. near the center of a spherical shock). With a thin shock layer such singularities would be imbedded deeply within the body, far from the region in which the solution is to be obtained. The possibility presents itself of using a fundamentally unstable procedure to obtain a realistic solution. The idea here is that the essential instability is inescapable in the vicinity of one of these fundamental singularities, but that the instability may be circumvented in a narrow shock layer.

In order to better understand the nature of the instability which arises in an improperly set elliptic problem, let us briefly review the classic example of Hadamard [1, pp. 32–34] of solutions of Laplace's equation satisfying Cauchy conditions. Following Hadamard, let us consider Laplace's equation in cartesian coordinates

$$\phi_{xx} + \phi_{yy} = 0, \tag{6.5.1}$$

subject to initial conditions on the y axis

$$\phi(0, y) = 0; \qquad \phi_x(0, y) = A_n \sin ny, \tag{6.5.2}$$

where A_n is a function of n equal to n^{-q}, where q is a positive integer. The quantity n is a parameter which may be very large.

We now note that the initial value of $\phi_x(0, y)$ can be made to differ as little as we like from zero by making n sufficiently large. For any value of $A_n > 0$, the solution of Laplace's equation satisfying (6.5.2) is

$$\phi(x, y) = n^{-1}A_n \sin ny \sinh nx. \tag{6.5.3}$$

Now on the other hand, suppose we were to have set $A_n = 0$ (n arbitrarily large) in the initial conditions so that $\phi_x(0, y) = 0$. In this case we should have obtained the solution $\phi(x, y) = 0$. Clearly this result does not agree with (6.5.3) in the limit $n \to \infty$ for any x not equal to zero, since $|\phi(x, y)| \to \infty$ for $n \to \infty$; this is because $\sinh nx$ behaves like $\frac{1}{2}e^{nx}$ for large n. One interpretation of this from a practical point of view is that slight inaccuracies in the initial conditions can lead to large deviations in the solution. In fact, as Hadamard observed, the factor $\sin ny$ produces a fluting of the solution surface in this problem. This fluting, no matter how small in the immediate neighborhood of the y axis, will "blow up" at any given distance away no matter how small this distance is, provided n is sufficiently great. We should note that it is the higher frequencies here which cause the more rapid instability.

In order to take care of the inherent instability associated with high frequencies in the hypersonic blunt-body problem, Zlotnick and Newman [1] consider the possibility of eliminating this magnification of errors. Such high frequencies are considered to be physically inadmissible in the blunt-body problem. We may expect this possibility to be realizable for smooth bodies at hypersonic speeds for which sudden large changes in shock or body curvature do not occur. These authors therefore propose to put an additional constraint on the solution wherein at every step of their marching-ahead process they filter out the high frequencies before proceeding by finite differences to the next step.

In order to estimate the error involved in such a filtering process, Zlotnick and Newman examined Laplace's equation subject to initial conditions of the type (6.5.2) on the y axis, under the assumption that no frequencies higher than a given value, say λ, occur in the solution. They find that the absolute value of the error in calculating ϕ_{x_i}, where $x_1 = x$ (for $i = 1$) and $x_2 = y$ (for $i = 2$), will be less than or equal to the quantity

$$\text{(6.5.4)}\quad \max|\text{error}\,\phi_{x_i}(x)| = e^{\lambda x}\left[\sum_{j=1}^{2}\max|\text{error}\,\phi_{x_j}(0)| + \lambda^{-1}\,\delta\,\max|\phi_{x_i xx}|\right].$$

Here ϕ refers to the true solution of the problem, δ is the step length in the finite difference process, and max|error $\phi_{x_j}(0)$| is the maximum absolute value of the initial error in ϕ_{x_j}. From this analysis we should have a good approximation as long as the step length, initial errors, and maximum allowable frequency are small. Although this estimation was carried out using Laplace's equation with a particular kind of smoothing process, we may reasonably assume that its validity would extend to the hypersonic blunt-body problem with other types of smoothing. We emphasize, however, that such an imposed constraint may by itself restrict the determinable bodies to smooth shapes with no sharp discontinuities or sudden changes in the curvature. Furthermore, such a smoothing process also raises the question of whether a small change in the shock shape would ever show up in an alteration of the body shape. The smoothing process gives us a spurious uniqueness to the problem, because we know that indistinguishably different shock shapes may correspond to radically different body shapes.

Many approaches are possible for solving the inverse problem numerically. For the symmetric problem we shall present the basic method of Zlotnick and Newman [1], Van Dyke [6] (see also Van Dyke and Gordon [1]), and Mangler and Evans [1] (see also Mangler [1]), which is carried out in the real physical plane, and that of Garabedian [1] and Garabedian and Lieberstein [1], which is carried out in a complex three-dimensional space on planes that intersect the real physical plane. No essential difference exists between the Zlotnick-Newman procedure, that of Van Dyke, and that of Mangler and Evans, all of which were developed independently. Zlotnick and Newman point out specifically that the basic instability in the subsonic region is suppressed by filtering out higher harmonics. A smoothing process is implicit in the procedure of Van Dyke. Both Lieberstein and Zlotnick have pointed out to the authors that in a finite difference method the use of differentiation schemes involving a large number of points is equivalent to applying a smoothing process. These procedures use shock-oriented coordinates; Van Dyke, who concerns himself specifically with shock waves described by conic sections, introduces a natural coordinate system appropriate to conic sections.

Many papers using the above methods and modifications of them (assuming equilibrium) have appeared subsequently. Examples are Vaglio-Laurin and Ferri [1], Fuller [1], Inouye and Lomax [1], Batchelder [1], and Flemming [1]. These references are intended to be illustrative rather than complete. Developments of the marching-ahead approach for problems involving asymmetry have also been put forward by Vaglio-Laurin and Ferri [1] and Swigart [3; 4]. Both of these methods are carried out in the real physical plane and are considered below.

The method of Garabedian and Lieberstein differs from the other approach in that in it the initial data for a known analytic shock curve are first analytically continued into a fictitious three-dimensional space composed of the real value of one of the independent variables and a complex value of the other. In this manner the basic equation is transformed from elliptic to hyperbolic form, and thus the essential instability of the marching-ahead procedure is avoided through the stable numerical method of characteristics. The essential instability of the procedure appears in the analytic continuation step and is avoided by the choice of shock shapes expressed in terms of simple closed-form analytic functions. Although the method is mathematically rigorous it does possess the drawbacks of involving more computational effort than the other method mentioned and of being restricted in practice in the choice of shock shapes to those for which the required analytic continuation is feasible. However, the method does permit a control of the fundamental singularities mentioned above.

In presenting the inverse method in the real physical plane for the symmetric problem we shall follow the approach of Zlotnick and Newman in which the appropriate hydrodynamic equations are written in shock-oriented orthogonal curvilinear coordinates. These equations (5.1.1) and (5.1.6) have been discussed previously in Section 5.1 and are rewritten here for convenience:

(6.5.5a) $$uu_x + \mathscr{H} vu_y - Kuv = -\frac{p_x}{\rho},$$

(6.5.5b) $$uv_x + \mathscr{H} vv_y + Ku^2 = -\mathscr{H}\frac{p_y}{\rho},$$

(6.5.5c) $$(\rho ur^j)_x + (\mathscr{H}\rho vr^j)_y = 0,$$

where

(6.5.6) $$\mathscr{H} = 1 - Ky$$

and $K(x)$ is the curvature of the shock. In place of the entropy equation (5.1.7) we use the energy equation for a perfect gas

$$\frac{\gamma}{\gamma - 1}\frac{p}{\rho} + \frac{u^2 + v^2}{2} = \frac{\gamma}{\gamma - 1}\frac{p_0}{\rho_0}. \tag{6.5.7}$$

Again, the extension to arbitrary equilibrium equations of state presents no essential difficulty. Here x is the coordinate along the shock wave, y is the coordinate at right angles to the shock wave and directed inward toward the center of curvature, and u and v are the respective velocity components.

By differentiating the energy equation with respect to y and solving simultaneously for the y derivatives of u, v, p, and ρ, we obtain

$$u_y = -\frac{1}{\mathscr{H}v}\left(\frac{p_x}{\rho} + uu_x - Kuv\right), \tag{6.5.8}$$

$$v_y = \frac{\dfrac{\gamma - 1}{\gamma}\mathscr{H}uu_y - uv_x - Ku^2 + \dfrac{p(\rho ur^j)_x}{\rho^2 vr^j} + \dfrac{p(\mathscr{H}r^j)_y}{\rho r^j}}{\mathscr{H}\left(\dfrac{v}{\gamma} - \dfrac{p}{\rho v}\right)} \tag{6.5.9}$$

$$\rho_y = -\frac{(\rho ur^j)_x + \mathscr{H}\rho v_y r^j + \rho v(\mathscr{H}r^j)_y}{\mathscr{H}vr^j} \tag{6.5.10}$$

$$p_y = -\frac{\rho}{\mathscr{H}}(uv_x + \mathscr{H}vv_y + Ku^2). \tag{6.5.11}$$

These relations give us four equations for the determination of the four unknowns u, v, ρ, and p.

If we now start with a known shock shape the unknown functions u, v, ρ, and p immediately behind the shock are determined by the oblique shock relations. With the functions known on this line $y = 0$ their x derivatives are determined by differentiation. We may solve (6.5.8) to (6.5.11) successively for the values of the (normal) derivatives. Note that in (6.5.8) to (6.5.11) only y derivatives determined in an earlier equation of the sequence are needed. The same procedure may be applied on any line $y =$ *constant* if the values of the functions are known. With the values of the unknown functions and their y derivatives known as functions of x at any value of y, say at y_n, the new values of the unknowns are determined in a finite difference scheme at $y_{n+1} = y_n + \delta y$ by taking $f(y_{n+1}) = f(y_n) + f_y(y_n)\,\delta y$, where f is any one of the unknowns. Following Zlotnick and Newman's idea, the new value of the functions can be "appropriately smoothed". The process is then repeated until the unknown body is reached as determined by boundary conditions on the surface.

Zlotnick and Newman have performed automatic digital computations using the preceding scheme to determine the body shapes and shock layer profiles corresponding to a spherical shock for different values of the free stream Mach number and perfect gas specific heat ratio. They carried out the smoothing process by replacing the data at every forward step by new data obtained by fitting polynomials in x to the old data. For their particular cases they found that a smooth fit of the data was necessary over an arc length no less than 0.16 of the shock radius of curvature in order to obtain a reasonably stable solution. The first step in determining the body shape was to note where the velocity on the stagnation streamline vanished, and thus to locate the stagnation point on the body. The body was then traced from the stagnation point along a line which was tangent to the streamline slope at each point, and for which the entropy had the same value as that of the stagnation streamline. In order to test the reasonableness of their calculated results, several streamlines were drawn in addition to the body streamline, and it was found that the two imposed requirements for a streamline were always fulfilled simultaneously: lines which passed through points of constant entropy were tangent to the streamline slopes.

Van Dyke, rather than using u, v, ρ, and p as the independent variables, chose instead the density, a modified stream function, and its derivatives. He employed an 11-point differentiation scheme, which had the effect of a smoothing process in the same manner as in Zlotnick and Newman's procedure. It appears that Van Dyke's numerical calculations were somewhat more accurate than those of Zlotnick and Newman. As we have mentioned, Van Dyke's procedure is essentially equivalent to that we have presented, and we shall not discuss it separately. A very extensive series of calculations for axisymmetric shock waves was carried out by Van Dyke and Gordon [1] by this procedure for different values of γ. The emphasis in this work was on bodies which closely approximate spheres and paraboloids.

In Mangler and Evans' procedure, the instability inherent in the marching-ahead procedure is overcome by using higher order differences in the marching direction, with values of the functions at five or three preceding points used at each step. Results of digital computations for several axisymmetric shapes with different values of γ and M_∞ may be found in Mangler [1].

In Fig. 6–23 we have shown two computations of Zlotnick and Newman for the body and sonic line shape in axisymmetric flow, with a spherical shock in a flow at a free stream Mach number of 10 and with constant specific heat ratios $\gamma = \gamma_s = 1.2$ and 1.4. Although not shown, the results of the calculations for $M_\infty = 20$ and $M_\infty = \infty$ are very close to those for $M_\infty = 10$. We should point out that the derivative $r_y = \sin \vartheta$ in (6.5.9) and (6.5.10) was taken to be zero in the original calculations in order to simplify the calculations. This approximation appears in the results for $\gamma = 1.4$. It does not appear in the results for $\gamma = 1.2$ shown in Figs. 6–23, 6–25, and 6–26, which are based on unpublished calculations of the same authors carried out without the approximation. The results for $\gamma = 1.2$ are very close to those calculated with $r_y = 0$.

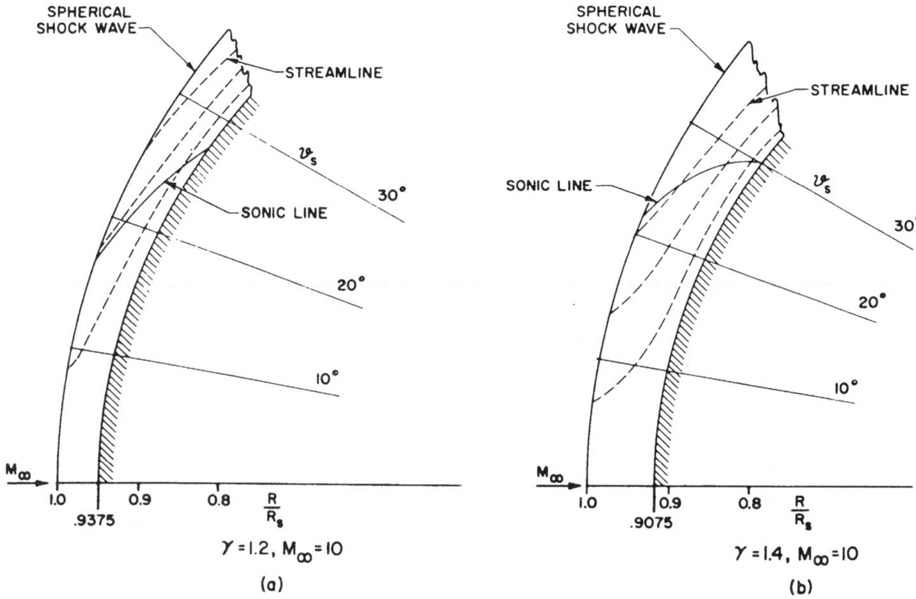

FIG. 6–23. Flow field and body for a spherical shock wave. (a) $\gamma = 1.2$, $M_\infty = 10$ (Zlotnick and Newman, unpublished). (b) $\gamma = 1.4$, $M_\infty = 10$ (Zlotnick and Newman [1]).

More accurate calculations of Flemming [1] indicate better than ten percent accuracy for the early Zlotnick and Newman calculations reported here.

In Fig. 6–24 we show the results of an inverse calculation reported by Inouye and Lomax [1] using the method of Fuller [1], in which the object was to obtain a hemispherical body from the assumed shock at $M_\infty = 4.76$ with $\gamma = 1.4$. The extent to which this was achieved is readily seen in the figure. Also shown is the calculated sonic line. This is compared with measurements of Kendall [1], who determined the sonic line by locating the points at which the disturbance generated by a small probe vanished. The fact that all of the measured points lie within the subsonic region indicates the possibility of limitations of the experimental technique. The remarks made in Section 6.3 on the variation of the sonic line shape and flow field with γ and M_∞ are also seen to be verified by the results of Figs. 6–23 and 6–24.

We shall use for comparison purposes with the numerical solutions the constant-density or Newtonian solutions because of the simple basis for comparison which they offer. The extent to which the numerical solutions of Zlotnick and Newman approximate a constant-density flow is shown in Fig. 6–25. Here we have plotted the profiles of pressure, density, and velocity components parallel and normal to the shock at two shock locations ($\vartheta_s = 5°$ and $32°$) for

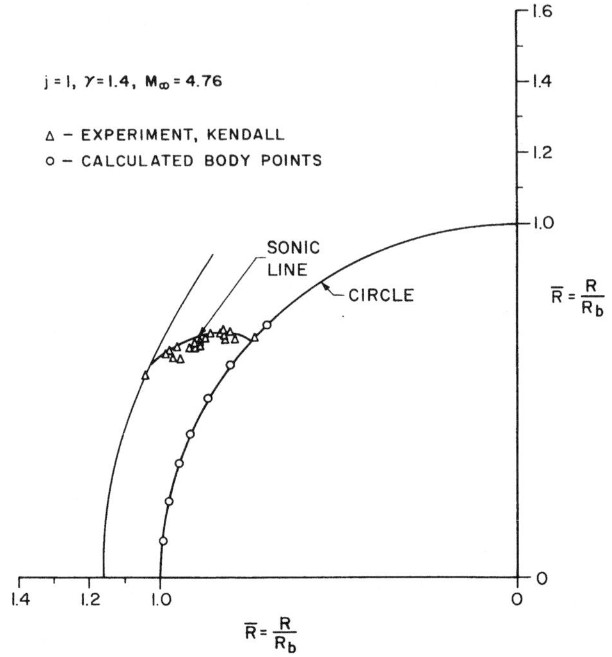

FIG. 6–24. Shock wave producing a spherical body with $\gamma = 1.4$, $M_\infty = 4.76$ (Inouye and Lomax [1]) and comparison of calculated sonic line with experiment (Kendall [1]).

the case of $\gamma = 1.2$, $M_\infty = 10$. From these profiles we observe that near the stagnation point the assumption of constant density across the shock layer is indeed excellent. On the other hand, from the results at $\vartheta_s = 32°$ (somewhat beyond the sonic point on the body) it is equally clear that the constant-density assumption is beginning to break down. For example, according to the constant-density sphere solution the tangential velocity profile is approximately linear in y, and in fact from (4.4.18)

$$\frac{u}{u_s} = \frac{\kappa(\Delta - y) + \epsilon R_s}{\kappa\Delta + \epsilon R_s}, \tag{6.5.12}$$

with $\kappa = (1 - 8\epsilon/3)/\sqrt{8\epsilon/3}$. We remind the reader again that here y is measured at right angles to the shock toward the body and equals zero on the shock. In Fig. 6–25(a) we have plotted for $\vartheta_s = 5°$ the constant-density result given by (6.5.12), using the value of Δ_0/R_s computed by Zlotnick and Newman. Clearly at $\vartheta_s = 5°$ the agreement is excellent. However, at $\vartheta_s = 32°$ we see that not only is u/u_s no longer linear, but that it is appreciably different from its value at the stagnation point.

Figure 6–26 is a summary of the body surface distributions of pressure, velocity, Mach number, and of the detachment distance for $M_\infty = 10$ and $\gamma = 1.2$ and 1.4, presented as a function of the shock angle. A comparison between the two cases again shows how the thin shock layer assumptions are more closely satisfied for $\gamma = 1.2$.

On the graphs of Fig. 6–26 we have also plotted the modified Newton-Busemann pressure distribution for a spherical shock in terms of the shock angle (cf. (6.3.36)).

$$\frac{p_b}{p_0} = 1 - \tfrac{4}{3}\sin^2\vartheta_s\,. \tag{6.5.13}$$

With ϵ small this result is close to that obtained by Lighthill [3] (cf. (4.4.11)) from the constant-density analysis for a spherical shock. Clearly, the agreement of the surface pressure distribution with (6.5.13) is better for $\gamma = 1.2$ than for $\gamma = 1.4$.

In Fig. 6–27 we show the stand-off distance at the axis of a sphere with $\gamma = 1.4$ as computed by Van Dyke [6]. Also plotted on this graph are experimental points taken from a compilation in Van Dyke's paper, and also the constant-density result obtained from a numerical solution of (4.4.8). Van Dyke's results were obtained with his inverse method in which the body shape is unknown in advance. From the results in Van Dyke and Gordon [1], his bodies are seen to very closely approximate spheres, at least up to a region somewhat beyond the sonic point on the body. In order to plot the constant-density result on this graph ϵ was taken equal to the value of the density ratio across a normal shock corresponding to the Mach number indicated. Of course the constant-density solution has been extrapolated far beyond its limit of validity. In spite of this, it is seen to give a reasonable approximation over most of the range of supersonic Mach numbers. A more rational comparison with the constant-density theory should be made on the basis of shock shape rather than body shape. This we have done in Fig. 6–28 where we have plotted the results of Van Dyke and Gordon [1] for a sphere with different values of γ, the Lighthill constant-density result of (4.4.8), and the simple local constant-density solution given by (4.4.16). Again, the transformation of the numerical results obtained from the inverse calculations to functions of ϵ was carried out, as in Fig. 6–27. We have drawn in the locus of the calculated points of Van Dyke and Gordon corresponding to $M_\infty = \infty$. The agreement with (4.4.16) is seen to be excellent, from which we conclude that this relation is empirically valid over a wide range of ϵ provided the effect of nonconstant ϵ can be neglected.

In Fig. 6–29 we have shown the corresponding results for a cylinder. The inverse calculations were taken from Fuller [1]. The constant-density result obtained from a numerical solution of (4.3.13) is shown, as is the asymptotic

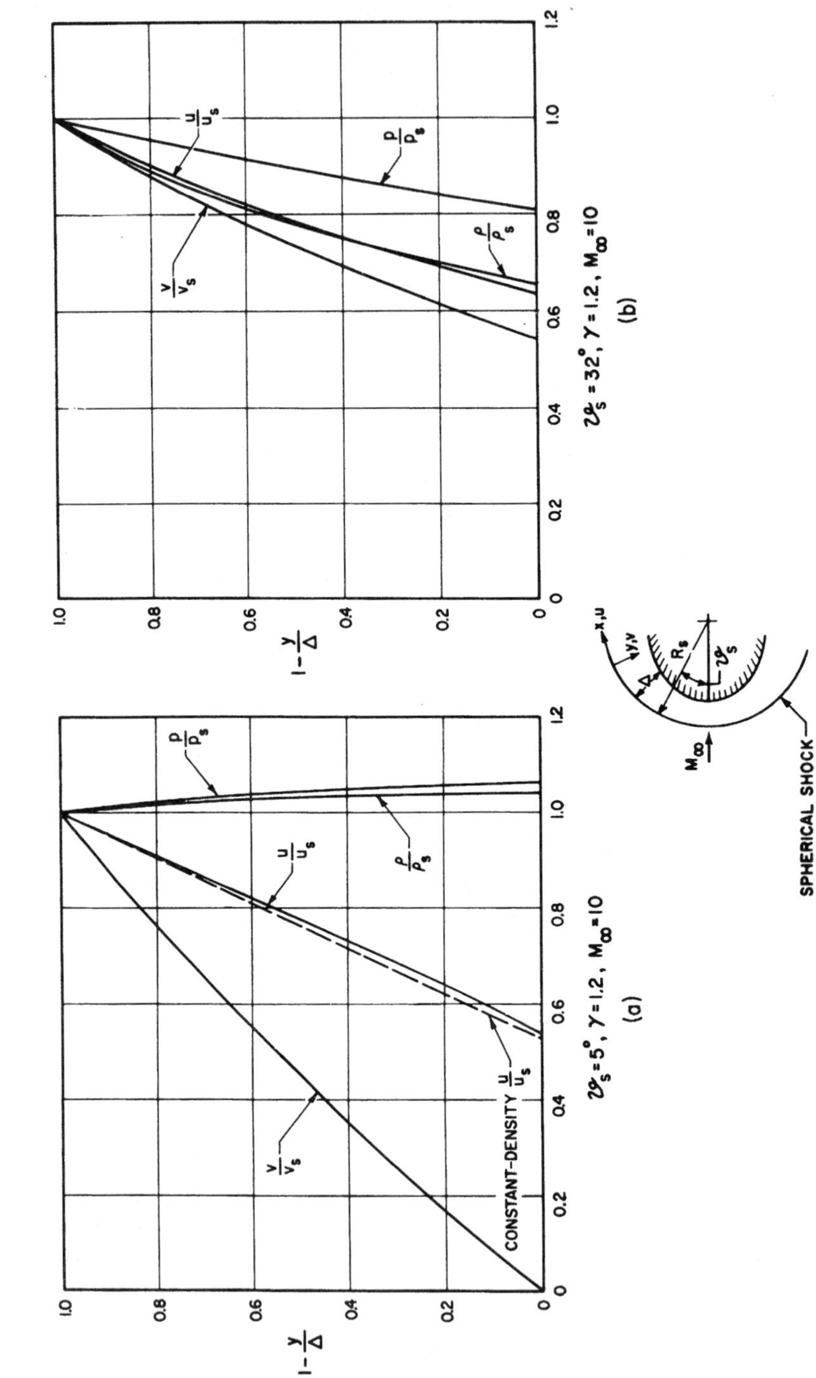

FIG. 6–25. Shock layer profiles for a spherical shock wave, $\gamma = 1.2$, $M_\infty = 10$ (Zlotnick and Newman, unpublished). (a) $\vartheta_s = 5°$. (b) $\vartheta_s = 32°$.

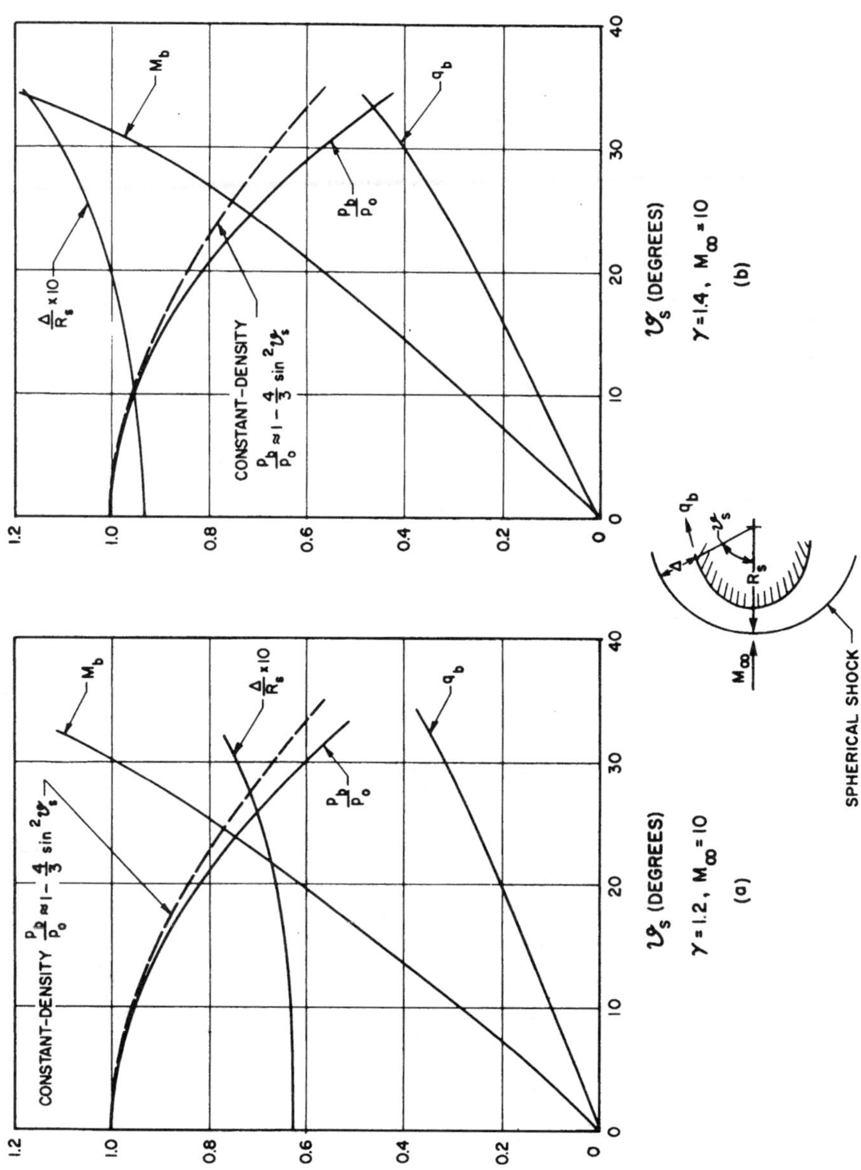

FIG. 6–26. Distribution of shock layer thickness and flow variables at body surface for a spherical shock wave. (a) $\gamma = 1.2$, $M_\infty = 10$ (Zlotnick and Newman, unpublished). (b) $\gamma = 1.4$, $M_\infty = 10$ (Zlotnick and Newman[1]).

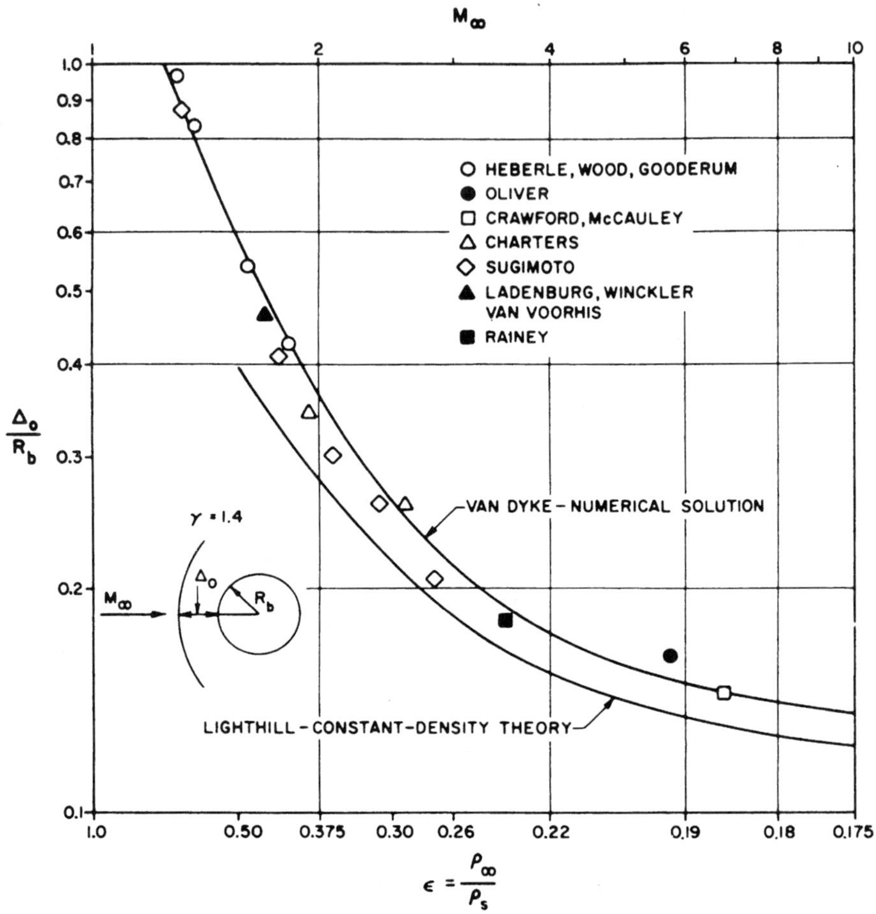

FIG. 6–27. Stand-off distance for a sphere (Van Dyke [6]), and comparison with constant-density result (Lighthill [3]) and experiment.

form of this result for small ϵ given by (4.3.14a). The agreement of this asymptotic result with the locus of $M_\infty = \infty$ points is good though, as is to be expected, not as good as in the axisymmetric case. Also plotted in this figure is the relation

$$\frac{\Delta_0}{R_s} = \frac{\epsilon}{2} \ln \frac{4}{3\epsilon} + \frac{\epsilon^2}{4}\left(\ln \frac{4}{3\epsilon} - 1\right). \tag{6.5.14}$$

This is the circular cylinder result given by (4.3.41), obtained by setting $\beta = 0$ (no noncircularity), $\mu = 0$ (constant ϵ), and $\gamma_s = 1$ (perfect gas compressibility). Although inverse calculations for $\gamma = 1$ were not given in Fuller [1], it is evident from this result that the principal departure from the constant-density result

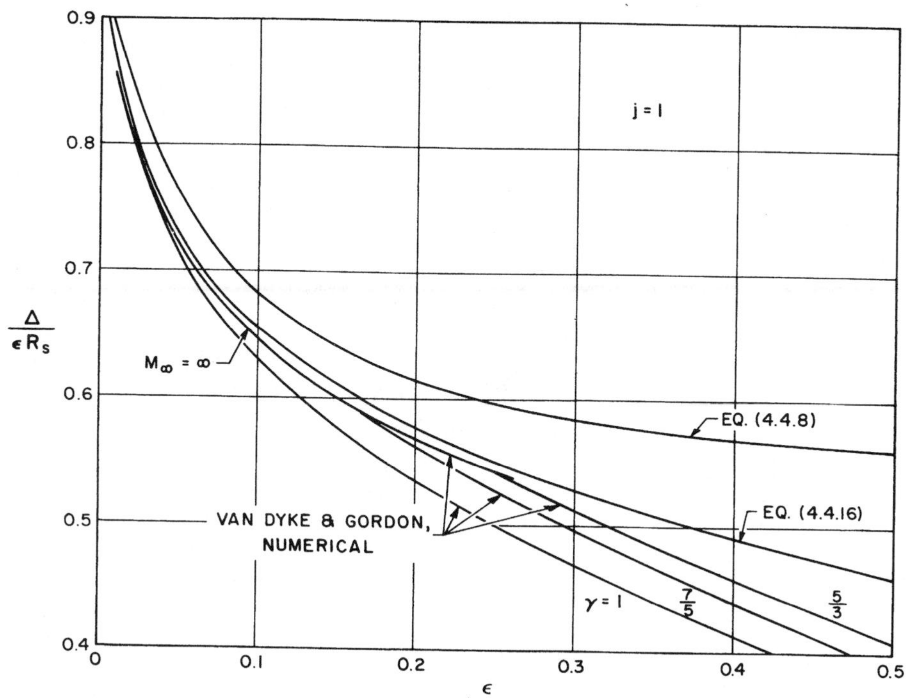

FIG. 6–28. Stand-off distance for a sphere (Van Dyke and Gordon [1]), and comparison with constant-density results (Lighthill [3] and (4.4.16)).

is due to compressibility and that the exact result is very closely represented by (4.3.41).

In connection with the inverse problem we now discuss the approach developed by Garabedian [1] and Garabedian and Lieberstein [1] for the symmetric problem. Since an initial value problem is in general inappropriate to the elliptic system of equations governing the subsonic domain, they suggest introducing an analytic continuation into the complex domain by an alteration of one of the independent variables. This then transforms the elliptic system into a hyperbolic system, for which an initial value technique is proper.

The basic equation for the flow field in terms of the stream function and the cartesian coordinates x and y (or cylindrical coordinates with y read as r for axisymmetric flow) is (6.4.2). This equation may be written in the form

$$a\psi_{xx} + 2b\psi_{xy} + c\psi_{yy} + d = 0, \tag{6.5.15}$$

where a, b, c, and d are known functions of ψ, ψ_x, ψ_y, and y. We note that this equation is elliptic for $b^2 - ac < 0$ and hyperbolic for $b^2 - ac > 0$. With the shock expressed as a known analytic curve $x = x(y)$, both ψ and ψ_x along the shock are also known analytic functions of the variable y (from the shock

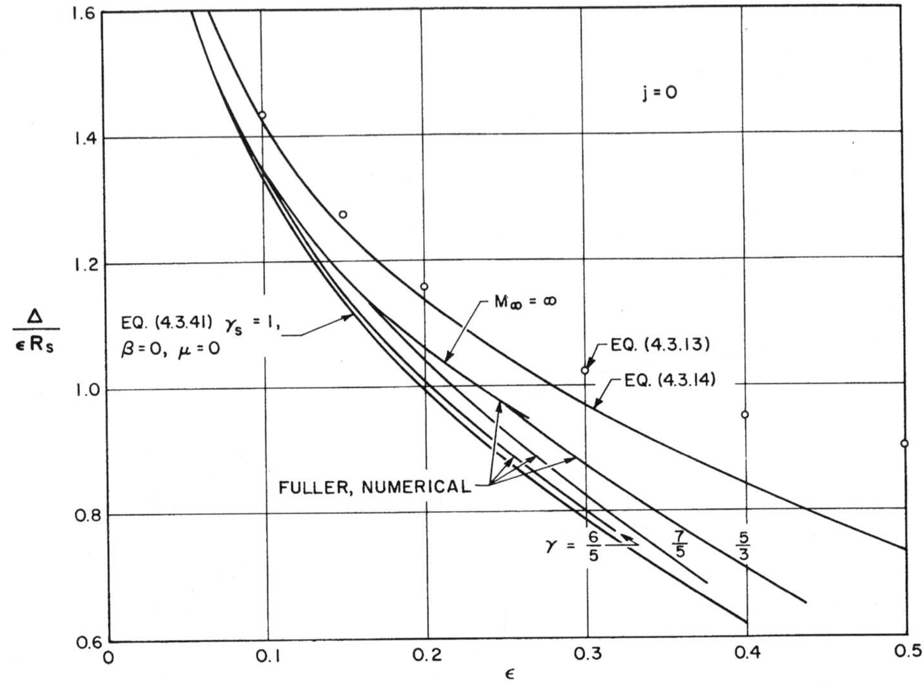

FIG. 6–29. Stand-off distance for a cylinder (Fuller [1]) and comparison with constant-density results ((4.3.13), (4.3.14a)) and (4.3.41) ($\beta = 0$, $\gamma_s^{-1} = 1$, $\mu = 0$).

relations), and the solution of (6.5.15) subject to these initial conditions constitutes a Cauchy problem for the determination of the stream function ψ.

To illustrate the Garabedian-Lieberstein method in the simplest manner possible, instead of dealing with (6.5.15) let us again consider the simpler Laplace equation (6.5.1) subject to the initial conditions

$$\phi(0, y) = f(y); \qquad \phi_x(0, y) = g(y). \tag{6.5.16}$$

Following the idea used by these authors, we consider that we have obtained the analytic continuation of f and g with respect to the variable y into the complex domain, leaving x as a real parameter; we then confine our attention to purely imaginary changes in y, with the real part of y held constant. We set $y = y_1 + iy_2$; then

$$\phi(0, y) = f(y) = f(y_1 + iy_2), \tag{6.5.17a}$$

$$\phi_x(0, y) = g(y) = g(y_1 + iy_2). \tag{6.5.17b}$$

In this case, for each value of y_1, $\phi(x, y)$ satisfies the wave equation

$$\frac{\partial^2\phi}{\partial x^2} - \frac{\partial^2\phi}{\partial y_2^2} = 0. \tag{6.5.18}$$

The function ϕ is therefore a complex-valued function $\phi(x, y_1 + iy_2)$ of three real arguments x, y_1, and y_2, which can be thought of as coordinates of a point in a cartesian three-dimensional space. In particular we are interested in the solution $\phi(x, y_1 + iy_2)$ for $y_2 = 0$ (that is, in the (x, y_1) plane) satisfying the initial conditions (6.5.17). This solution can be found for each value of y_1 from the hyperbolic equation (6.5.18) in the (x, y_2) plane rather than from the original elliptic equation (6.5.1) in the (x, y_1) plane. Thus, although the solution is determined in a three-dimensional region, we are interested only in the portion of this region which intersects the plane $y_2 = 0$. Clearly, for every value of the parameter y_1 we are dealing only with an initial value problem in two independent variables. This same idea of solution can also be applied to the original equation (6.5.15); we must depend here on a mathematical equivalence of the elliptic equation (6.5.15) to an equation in which the second-derivative terms are in the form of a Laplacian.

From the above reasoning it appears that the instability of the original elliptic problem has been avoided by this scheme, since we deal with a hyperbolic equation which can always be treated in a stable manner by the numerical method of characteristics. As a natural reaction to this rather ingenious idea we may ask: can a simple change in variable alter an essential instability in a calculation procedure? The answer to this question is definitely no. But, as we shall see, the unstable step may be avoided if an analytic closed-form procedure can be substituted for the unstable part of the numerical procedure. In fact, any sound mathematical method of analytic continuation appropriate to the mathematical expression describing the given shock shape may be used.

To understand these previous statements somewhat better we note, as both Garabedian [1] and C. C. Lin [1] have pointed out, that a small change in the initial data in the real domain can result in a large change in this same initial data in the complex domain. This is because the latter is obtained by an analytic continuation of the former. As Lin observed, the analytic continuation of the initial conditions $f(y)$ and $g(y)$ (see (6.5.16)) from real values of y into the complex domain is itself the solution of an initial value problem for Laplace's equation.

To show this, let us write

$$g(y_1 + iy_2) = u(y_1, y_2) + iv(y_1, y_2). \tag{6.5.19}$$

In this case u and v each satisfy Laplace's equation with y_1 and y_2 as independent variables with the initial conditions

$$u(y_1, 0) = g(y_1), \qquad u_{y_2}(y_1, 0) = 0; \tag{6.5.20a}$$

$$v(y_1, 0) = 0, \qquad v_{y_2}(y_1, 0) = g'(y_1). \tag{6.5.20b}$$

Returning to Hadamard's example (6.5.2), for which $f(y) = 0$ and $g(y) = A_n \sin ny$, we find that analytic continuation of $g(y)$ gives

(6.5.21a) $$u(y_1, y_2) = A_n \sin ny_1 \cosh ny_2,$$

(6.5.21b) $$v(y_1, y_2) = A_n \cos ny_1 \sinh ny_2.$$

It is clear that the same instability as discussed previously with the original Laplace equation applies in the analytical continuation as well.

In our example using Laplace's equation this method has merely managed to substitute a new but essentially equivalent problem involving Laplace's equation again. In this instance there is no gain from the method. However, the essential feature of the Garabedian-Lieberstein method is that the original elliptic differential equation need not be Laplace's equation but may be a much more complicated one. The part of the problem involving analytic continuation is always equivalent to the much simpler Laplace's equation. We may thus transform a complicated elliptic Cauchy problem such as our blunt-body inverse problem into a simple elliptic problem equivalent to a Cauchy problem for Laplace's equation, plus a complicated hyperbolic problem. We may also investigate the fundamental singularities appearing in the analytic continuation or choose shock shapes yielding specified singularities. The question remains as to whether we can carry out the analytic continuation.

If we were to choose simple closed-form analytic shock shapes such as hyperbolas or parabolas for our hypersonic blunt-body problem, permitting the initial conditions to be continued *exactly* into the complex domain, the type of instability which we have illustrated would not occur. Nevertheless, our example does show clearly the manner in which small alterations in the shape of the shock in the physical plane can result in very large changes in the shape of the body as a result of the unstable process of analytic continuation if carried out numerically. This therefore places a restriction, as with any inverse scheme, on the general applicability of the method. However, the method of Garabedian and Lieberstein may be applied with initial data given numerically if some reasonable smoothing process such as we have discussed earlier is applied.

Garabedian and Lieberstein formulated their procedure for solving the detached shock problem by reducing the second-order equation (6.5.15) following a standard procedure to a system of five first-order partial differential equations, through the introduction of characteristic coordinates α and β. The slope dz/dr of an α-increasing (β constant) line is σ^+, and that of a β-increasing (α constant) line is σ^-, where

(6.5.22) $$\sigma^+ = \frac{b + \sqrt{b^2 - ac}}{a}, \qquad \sigma^- = \frac{b - \sqrt{b^2 - ac}}{a}.$$

Here $b^2 - ac < 0$ for the elliptic region, and σ^+ and σ^- are complex quantities. Note that there is a certain degree of arbitrariness in the definition of α and β which must be resolved in the calculation procedure. By making the additional transformation

$$X = \frac{\alpha + \beta}{2}, \qquad Y = \frac{\alpha - \beta}{2i}, \tag{6.5.23}$$

the system of differential equations in the (X, Y) plane can then be written in the canonical matrix form

(6.5.24)

$$\begin{bmatrix} y_X \\ x_X \\ (\psi_y)_X \\ (\psi_x)_X \\ \psi_X \end{bmatrix} = \frac{1}{\sqrt{ac - b^2}} \begin{bmatrix} b & -c & 0 & 0 & 0 \\ a & -b & 0 & 0 & 0 \\ 0 & d & b & a & 0 \\ -d & 0 & -c & -b & 0 \\ b\psi_y + a\psi_x & -(c\psi_y + b\psi_x) & 0 & 0 & 0 \end{bmatrix} \begin{bmatrix} y_Y \\ x_Y \\ (\psi_y)_Y \\ (\psi_x)_Y \\ \psi_Y \end{bmatrix},$$

or in matrix notation

$$J_X = BJ_Y, \tag{6.5.25}$$

with J denoting the column vector of the unknown functions. In this formulation real solutions generate real flows in the subsonic region $b^2 - ac < 0$.

These authors set up initial conditions for an analytic shock curve of the form

$$(x_s + x_0)^2 = x_0^2 + g(y^2). \tag{6.5.26}$$

With $\psi_s = \rho_\infty U y_s^{1+j}/(1+j)$ both ψ_x and ψ_y at the shock can be computed explicitly in terms of y and ψ. These authors also choose the Y axis as the initial curve in the (X, Y) plane using the explicit relation

$$y = F(Y) = E_1 Y + E_2 Y^3 + \frac{E_3 Y}{Y^2 + E_4} \tag{6.5.27}$$

between y and Y, where the E's are arbitrary real parameters which are chosen for convenience in the numerical procedure.

In accordance with the scheme outlined previously we now let $Y = Y_1 + iY_2$, in which case (6.5.25) can be written as

$$J_X = \frac{B}{i} J_{Y_2}. \tag{6.5.28}$$

The initial conditions are then imposed at $X = 0$, and analytically continued in closed form into the complex Y plane. We now restrict our attention to the plane $Y_1 = constant$, in which X and Y_2 are rectangular coordinates. In the (X, Y_2) plane (6.5.28) is a hyperbolic system having the three families of characteristic curves $X + Y_2 = constant$, $X - Y_2 = constant$, and $Y_2 = constant$. In order to find the subsonic flow, the initial value problem (6.5.28) is solved, for a suitable set of values of the parameter Y_1, in a manner similar to that described for Laplace's equation. The solution of each initial value problem for a particular choice of the parameter Y_1 then yields values of ψ, ψ_y, and ψ_x along a curve in the (x, y) plane whose shape depends on the function $F(Y)$ in (6.5.27).

We may ask the general question as to how the sonic line and the supersonic part of the flow field are related to the subsonic part of the flow field treated by this method. The basic method breaks down near the sonic line because $b^2 - ac = 0$ there, but the location of the sonic line and flow properties there may be obtained by extrapolation. With a hypersonic axisymmetric sonic line (see Fig. 6–1(c)) the entire supersonic flow field and the sonic line can also be obtained by the method of characteristics with the original equations of motion. Lieberstein (private communication) has checked the location of the sonic line by these two procedures in the example shown in Fig. 6–30(b) and obtained excellent agreement. If a part of the body bounds a part of the transonic zone, as in Fig. 6–1(a) or Fig. 6–1(b), the solution in that part must be obtained starting with the sonic line determined by the solution in the subsonic region.

If the entire shock shape is an analytic shape its complete course is determined, including the part of the shock behind which the flow is supersonic. Implicit in this assumption that the shock shape is analytic is the fact that the entire flow field is determined, including the transonic zone. Therefore, no question arises as to the nature of the influence of the transonic zone on the subsonic flow. If the analytic shape is limited by the presence of a singularity or branch point, the situation may become much more complicated and has not yet been clarified.

Garabedian and Lieberstein, employing the method just described, have carried out machine calculations for several shock shapes in axisymmetric flow. In order to obtain different types of body shapes, the foci of the conic sections chosen to represent the shock were placed in different positions. Thus to obtain the flow past a flat-nosed body of revolution, an ellipse was chosen whose foci are located off the axis of symmetry near the points where the shoulder intersects the meridian plane. To obtain the flow past a blunt-nosed cone, a hyperbola was chosen for the shock with its focus just behind the stagnation point.

In Fig. 6–30 we have shown the body and sonic line shapes computed by these authors for two different hyperbolic shocks at $M_\infty = 5.8$ with $\gamma = 1.4$, and computed by Lieberstein [1] for a hyperbolic shock at $M_\infty = 20$ and $\gamma_s = 1.17$ ($\gamma = 1.4$). We have also indicated on these figures the ratio of the stand-off distance at the axis to the body radius of curvature at the vertex.

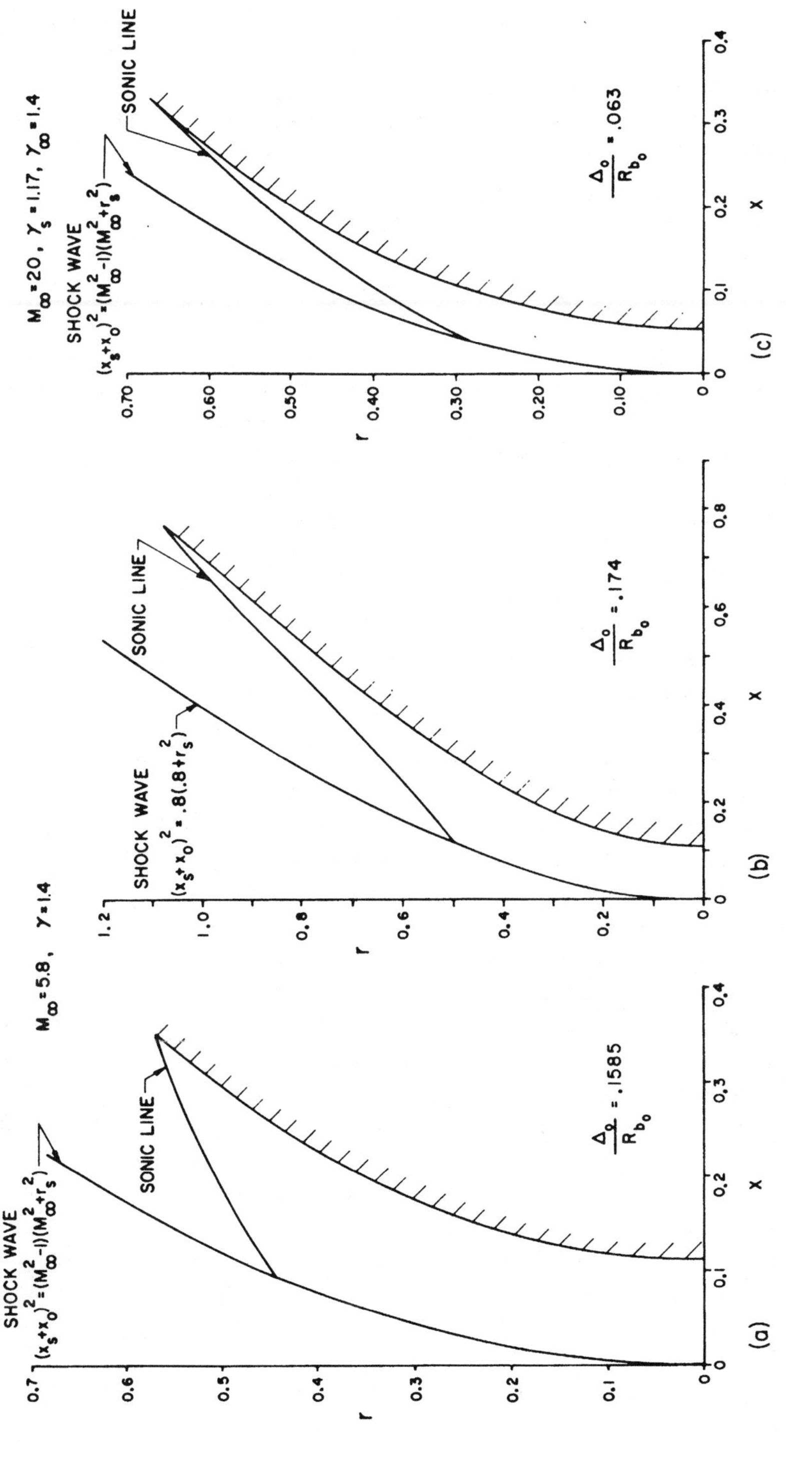

FIG. 6–30. Body and sonic line shapes for axisymmetric hyperbolic shock waves. (a) $(x_s + x_0)^2 = (5.8^2 - 1)(5.8^2 + r_s^2)$; $M_\infty = 5.8$, $\gamma = 1.4$ (Garabedian and Lieberstein [1]). (b) $(x_s + x_0)^2 = .8(.8 + r_s^2)$; $M_\infty = 5.8$, $\gamma = 1.4$ (Garabedian and Lieberstein [1]). (c) $(x_s + x_0)^2 = (20^2 - 1)(20^2 + r_s^2)$; $M_\infty = 20$, $\gamma_s = 1.17$ (Lieberstein [1]).

For all the bodies of Fig. 6–30 we again observe the hypersonic axisymmetric sonic line. The bodies obtained in Figs. 6–30(a) and (c) are roughly spherical, while that obtained in Fig. 6–30(b) approximates a blunt-nosed cone. An interesting result of the flow pattern in Fig. 6–30(b) is the shape of the sonic line, which is slightly convex with respect to the subsonic region. Such a behavior may be deduced from the results of Section 6.1 applied to a very blunt cone with a quasi-spherical capped nose. We showed there that the distance from the axis to the sonic point on the shock is roughly proportional to the radius of curvature of the shock at the axis times the cosine of the shock angle at the sonic point. Therefore the location of the sonic point on the shock is roughly fixed by the nose. In addition, as we pointed out, at high Mach numbers the sonic line will have a positive inclination angle with respect to the free stream direction. On the other hand, if the cone angle is sufficiently blunt the asymptotic flow near the surface far downstream will have remained subsonic. The flow will only become sonic on the body at the point where the cone terminates. Assuming a monotonic variation in the slope of the sonic line, this must lead to the shape convex to the subsonic region shown in Fig. 6–30(b). A further discussion of such behavior of sonic line shapes, discussed with respect to dependence upon geometry and flow conditions, may be found in Traugott [2].

A very interesting picture of the flow on a slightly concave truncated cylinder, which Garabedian and Lieberstein obtained by assuming an elliptic shock shape, is shown in Fig. 6–31. Here, we have indicated directly on the body the ratio of surface to stagnation point pressure. They have not presented the continuation of this solution into the supersonic region.

In concluding this discussion of the Garabedian and Lieberstein approach we would point out that although an accurate solution is obtainable by this scheme, the computational time on a digital machine may be very large. Van Dyke [7] points out that an accurate solution for a typical hemispherical-type nose shape required $3\frac{1}{2}$ hours of electronic computer time as compared with about one minute with a straightforward marching-ahead procedure.

The problem of applying an inverse technique to problems with asymmetry was considered by Vaglio-Laurin and Ferri [1]. They treat bodies of revolution at small angles of incidence by a perturbation in the radial and the azimuthal direction. Only first-order terms in the perturbation quantities are retained. The resulting perturbation equations are transformed to the coordinate system used in the symmetric problem. In their coordinate system the stream function is used as one of the independent variables. The perturbation equations in finite difference form are obtained at each point in the network of the original axisymmetric calculation. The perturbations are then calculated by marching ahead from the shock to the body. No asymmetric results have as yet been reported using this method.

Another method for treating the inverse problem with asymmetry has been developed by Swigart [3; 4]. Two distinct approximations are involved in his procedure. One of these is the approximation inherent in the series-expansion technique described at the beginning of this section. Swigart uses an expansion in distance from the axis of symmetry of the shock. The other approximation is the same as that used by Vaglio-Laurin and Ferri. It involves the assumption of small incidence of the free stream with respect to the axis of symmetry of the shock, and leads to a perturbation analysis in the angle of incidence. The first of these expansions involves an arbitrary truncation. As we have noted previously this scheme is thus similar to a successive approximation scheme. The result of the two expansions is to reduce the problem to one in a single independent variable rather than in three independent variables.

Swigart uses the same coordinate system and dependent variables as does Van Dyke [6]. This involves using a modified stream function and the density as dependent variables, and an orthogonal coordinate system in which the shock is a coordinate surface. The shock waves are taken to be conic sections. The azimuthal angle is introduced as the appropriate third orthogonal coordinate. Swigart [3; 4] presents several examples for both plane and axisymmetric flow, for different Mach numbers and values of γ. Comparison of one case with an experimental pressure distribution in Swigart [4] shows good agreement with the calculated results. Table 6–1 in Section 6.1 presents some of Swigart's results with respect to the stagnation streamline, which he clearly shows does not pass through the normal point on the shock. A discussion and comparison of his results has already been given in Section 6.1.

In view of the truncation procedure used it is likely that a better choice for the dependent variables would be the actual flow variables such as pressure and velocity, as were used by Zlotnick and Newman. Such a choice was found suitable in the analogous viscous layer problem discussed in Section 10.2 of the first edition of this book. Finally, the empirical suggestion of Van Dyke [7] regarding the appropriate expansion parameter should also be considered.

In general, once the inverse problem has been reduced to one in two independent variables it can always be treated by methods analogous to those described in connection with the symmetric case. In order to avoid the restriction to small incidence angles, such a reduction might be more appropriately carried out by a Fourier series expansion in the azimuth angle, as was described in connection with the direct integral relations method.

In summary, it would appear that the inverse methods described may afford a relatively simple way of finding exact numerical solutions to the hypersonic flow past blunt bodies, to an accuracy consonant with the smoothing requirements necessary to avoid the instability difficulty. Of course, the usual problem of interest is the direct problem, to which an inverse technique can only be applied through a method of trial and error. With the aid of modern automatic computing machines a "catalogue" of blunt-body solutions will undoubtedly be built up, including the effects of asymmetry, and this will certainly be of some value. The point of view that such a catalogue of solutions can take the place of a suitable method for the direct problem represents a dangerous philosophy, one that has appreciably delayed scientific advances in the past in various fields of science, including aerodynamics.

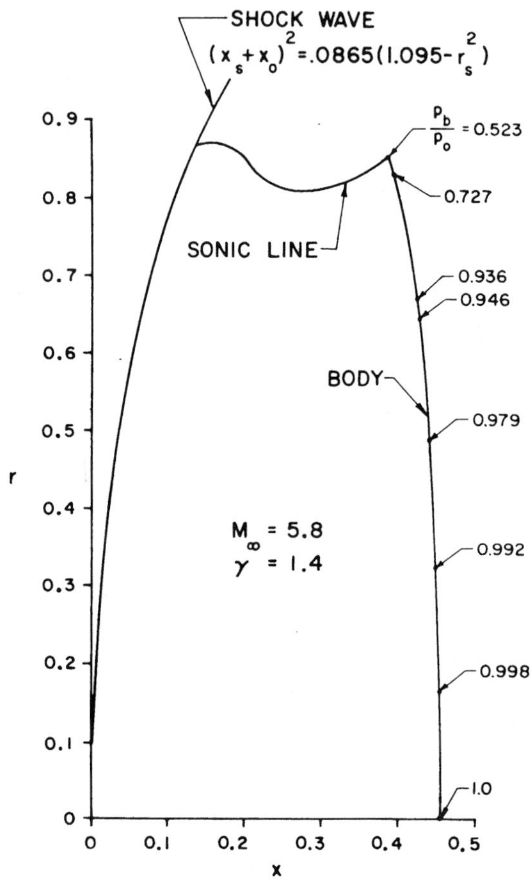

FIG. 6–31. Body and sonic line shape for oblate ellipsoidal shock wave (Garabedian and Lieberstein [1]).

6. Procedures with nonequilibrium

All of the numerical methods described in this chapter can be readily adapted to account for nonequilibrium effects. It is assumed that the physics and chemistry of the relevant nonequilibrium effects are sufficiently well known that the appropriate rate-determining equations may be specified in terms of the flow variables. In this case, any increased difficulty in obtaining a solution generally results from numerical features of the problem rather than from any new conceptual features. The main differences in numerical procedure arise when a characteristic reaction or relaxation time is small, so that the flow is near to equilibrium with respect to a reaction or energy adjustment. As we have noted in Section 1.8, we encounter in such a case a singular perturbation problem, which for the blunt-body problem may be characterized by a thin region of nonequilibrium flow behind the detached shock. In this region rapid changes in the flow variables can take place and any numerical technique generally must be altered. A discussion of procedures and difficulties encountered in the numerical integration of chemical and vibrational rate equations under conditions close to equilibrium can be found in Curtiss and Hirschfelder [1] and Emanuel [1].

As in the previous sections the discussion is intended to be illustrative rather than detailed. Only the symmetric problem is treated. We consider first the inverse method in the real physical plane, for which the type of modification required to take into account nonequilibrium effects is most easily demonstrated. The system (6.5.5) of two momentum equations and the overall continuity equation remains unchanged. In place of the energy equation for a perfect gas, we use the more general form

$$h + \tfrac{1}{2}q^2 = h_\infty + \tfrac{1}{2}U^2 = H. \tag{6.6.1}$$

To this must be added the equation of state and the rate-determining equations for the physical or chemical processes.

As an example, consider the simple vibrational relaxation model of a perfect diatomic gas presented in Section 2.11. For this model the enthalpy is given from (2.11.3) by

$$h = \tfrac{5}{2}\mathscr{R}T + e_{\text{vib}} + \frac{p}{\rho}, \tag{6.6.2}$$

where T is the temperature corresponding to the translational and rotational modes, which are assumed in equilibrium. The gas is taken to behave as a perfect gas with respect to the temperature T. The system of equations is therefore completed by the relaxation equation (2.11.5). In shock-oriented orthogonal curvilinear coordinates this equation is

$$\frac{u}{\mathscr{H}}\frac{\partial e_{\text{vib}}}{\partial x} + v\frac{\partial e_{\text{vib}}}{\partial y} = \frac{e_{\text{vib},e} - e_{\text{vib}}}{t_{\text{vib}}}. \tag{6.6.3}$$

One relation for the relaxation time t_{vib} is given by the Landau-Teller value (2.11.6). However, the model may be simplified still further, as was done in

Section 5.10, by considering the relaxation time to be a constant and the vibrational energy to be a linear function of T.

With the above relations the inverse procedure may be carried out in the same manner as for the perfect gas case illustrated in Section 6.5. Equations (6.5.8), (6.5.10), and (6.5.11) for u_y, ρ_y, and p_y remain unchanged. It is only necessary to alter (6.5.9) for v_y to take into account the modified energy equation resulting from (6.6.2), and to add the additional relaxation equation (6.6.3) expressed as an equation for the y derivative of e_{vib}.

Assuming a constant relaxation time Sedney [1] has calculated the initial gradients behind a curved two-dimensional shock wave. The calculations were made on the basis of an equilibrium flow ahead of the shock, while behind the shock the flow variables were computed from the frozen relations with constant specific heats, e_{vib} remaining unchanged across the shock. These calculations are the analogue, with vibrational relaxation downstream of the curved shock, of the perfect gas calculations presented in Section 6.1. The effect of the relaxation process is to increase the streamline curvature and to decrease the velocity gradient.

When there is more than one component present, as with chemical reactions, then to the system of equations must be added the species conservation equations

$$\nabla \cdot (\rho_i \mathbf{q}) = w_i , \tag{6.6.4}$$

where w_i is the mass rate of formation of the ith species per unit volume and time. In order that mass be conserved in any chemical change we require $\Sigma\, w_i = 0$. Summing over all components of the gas and using this condition gives the overall continuity equation $\nabla \cdot (\rho \mathbf{q}) = 0$. With the aid of this equation the continuity equation for each species can be rewritten in the form

$$\frac{Dc_i}{Dt} = (\mathbf{q} \cdot \nabla)c_i = \frac{w_i}{\rho} . \tag{6.6.5}$$

Here D/Dt indicates a total time derivative along a streamline and $c_i = \rho_i/\rho$ is the mass fraction of the ith component, with $\Sigma\, c_i = 1$.

A simple model of an ideal dissociating diatomic gas is the one due to Lighthill [3] and Freeman [2], which was discussed in Section 2.11. For such a gas the conservation of species equations reduces to a single rate-determining equation. If α is the fraction of diatomic molecules dissociated, we write

$$\frac{u}{\mathscr{H}} \frac{\partial \alpha}{\partial x} + v \frac{\partial \alpha}{\partial y} = \frac{d\alpha}{dt} \tag{6.6.6}$$

in the coordinate system of Section 5.1, with $d\alpha/dt$ obtained from (2.11.17). For each mole of the original diatomic gas there are $(1 - \alpha)$ moles of diatomic gas remaining and 2α moles of monatomic gas. Thus there are $(1 + \alpha)$ moles of atom-molecule mixture for each original mole of gas. The usual equation of state is therefore given by

$$p = \rho(1 + \alpha)\mathscr{R}T. \tag{6.6.7}$$

The system is completed with the definition of the enthalpy, which is equal to p/ρ plus the energy in the translational and internal degrees of freedom plus the energy in dissociation. For this model gas the translational and rotational modes are in equilibrium and the vibrational modes are only half excited, so that

$$h = 3\mathscr{R}T + D\alpha/2m + p/\rho, \tag{6.6.8}$$

where D is the dissociation energy per molecule and m the atomic mass. It is clear that a discussion of the solution procedure for this case would parallel the one for vibrational relaxation, with the difference that the y derivative of α is calculated in place of the y derivative of e_{vib}.

Neglecting ionization, we may consider a more general multicomponent model. In this case the rate equations are specified by (6.6.5), with the convective operator the same as that used in the preceding rate equations. If each component of the fluid is assumed to behave as a perfect gas in contributing to the total pressure then $p_i = \rho_i\mathscr{R}_iT$. Consequently for the mixture

$$p = \rho\bar{\mathscr{R}}T, \tag{6.6.9}$$

where

$$\bar{\mathscr{R}} = \sum c_i\mathscr{R}_i. \tag{6.6.10}$$

The specific enthalpy is given by

$$h = \sum c_i h_i, \tag{6.6.11}$$

where we write

$$h_i = \tfrac{3}{2}\mathscr{R}_iT + (e_{\text{int}})_i + \frac{p_i}{\rho_i} + h_i^0. \tag{6.6.12}$$

Here $(e_{\text{int}})_i$ is the specific internal energy of the ith species associated with the rotational and vibrational modes, if any, and h_i^0 is the specific enthalpy of formation of the species extrapolated to absolute zero and defined according to some consistent standard.

The simplest assumption that may be made with respect to the internal energy modes is that they are in equilibrium at the local temperature. Normally the rotational modes are fully equilibrated for the components of air undergoing dissociative reactions in hypersonic flows of the type considered. However, the vibrational modes may not be equilibrated and it may be necessary to take into account the vibrational relaxation, for example through a relation of the form (6.6.3). But even this model could prove to be too simplified, since it does not couple the vibrational excitation to the molecular dissociation. Under certain conditions dissociation and recombination directly affect the average vibrational energy of the molecules. Such coupling between vibration and dissociation would require the addition to the right-hand side of the vibrational rate equation terms involving both the species concentration and the dissociation and recombination rates. In addition, the chemical rates, which we have left unspecified, may also be coupled through a dependence on species concentrations. The inclusion of ionization only tends to complicate still further the state and rate equations. Here again, a simple model for a pure monatomic gas, such as is suggested by (2.11.24) and (2.11.25) may serve to illustrate the phenomena. The actual situation can, however, generally be expected to be considerably different. We mention all of these complications only to emphasize again the physical-chemical limitations of the simple models which have been chosen to illustrate the fluid mechanics of a relaxing and reacting gas in a hypersonic blunt-body flow.

Lick [1] has carried out inverse calculations for an axisymmetric shock and a pure diatomic gas model simulating a mixture of oxygen and nitrogen in the same ratio as that of atmospheric air. The calculations were based on a free stream density and temperature corresponding to atmospheric conditions at 100,000 ft altitude ($T_\infty = 227°\text{K}$, $\rho_\infty = 1.71 \times 10^{-5}\ \text{gm/cm}^3$) with a Mach number $M_\infty = 14$. The resulting temperatures in the shock layer are high enough that the dissociation of oxygen is important but low enough that the dissociation of nitrogen is sensibly negligible (see Section 2.11). The oxygen therefore determines the rate-controlling dissociative reaction. In these calculations vibrational equilibration was assumed. The shock shape chosen was the catenary $R/R_s = \cosh(r/R_s) - 1$, with R distance measured from the shock vertex along the symmetry axis and R_s the shock radius of curvature at the vertex.

The state equation and enthalpy relation used by Lick are analogous to the relations (6.6.9) to (6.6.12). The appropriate y derivatives of c_i, u, v, p, and ρ are solved for exactly as in the simpler cases discussed earlier and the complete expressions may be found in Lick's paper. For the examples calculated, the solution is shown to depend on the parameter

$$\Lambda = \frac{R_s}{U}\left(\frac{w_{O_2}}{\rho}\right)_0, \tag{6.6.13}$$

where the subscript 0 indicates the rate constant is evaluated at the shock vertex. This parameter is essentially the one introduced by Freeman [2] and represents the ratio of the characteristic macroscopic flow time to the characteristic time associated with the oxygen nonequilibrium process. The value of Λ was taken over a range from the frozen condition ($\Lambda = 0$) to a condition near equilibrium ($\Lambda = 100$). No difficulties in the marching-ahead procedure were reported for this latter case, despite its singular nature.

The numerical results bear out the fact that finite rates of reaction have little effect on the pressure and velocity, especially in the vicinity of the axis. In the stagnation region the behavior of the flow with increasing Λ generally follows the classification outlined in Table 5–2, although the inner layer details cannot be discerned from the numerical solutions. Figure 6–32 shows the density and temperature profiles along the stagnation streamline as a function of Λ. For Λ large the flow is near to equilibrium and there is a large density increase and temperature drop close to the shock due to the energy absorption by the molecular dissociation. The increase in density manifests itself in a corresponding decrease in detachment distance. The general behavior of the temperature and density profiles parallels that for a normal shock wave, in terms of the time variable discussed in Section 2.10. Downstream along the body surface, however, for an intermediate value of $\Lambda = 5$ the calculations indicate a frozen flow near the surface. This is a consequence of the fact that the expansion around the body produces a temperature and density drop which locally reduces the chemical reaction rate. Equilibration is thereby prevented and a frozen state results since the lagging mode cannot adjust to the local equilibrium. The lagging mode therefore "freezes out" at a constant level above the state corresponding to equilibrium. This sudden freezing out was noted by Bray [1] in connection with expanding dissociated nozzle flows.

An extension of Lick's inverse calculations to include more realistic chemical kinetics of high temperature air has been carried out by Hall, Eschenroeder, and Marrone [1]. A Runge-Kutta integration procedure was used along the y coordinate normal to the shock. Again no numerical difficulties were reported for conditions close to equilibrium, despite the singular nature of the almost-equilibrium case.

As noted at the beginning of this section, all of the methods of the present chapter can be extended to nonequilibrium flows. Extensive calculations for the direct problem have been performed using the method of integral relations discussed in Section 6.3. The appropriate equations for the method with uncoupled vibrational relaxation and dissociation were written down by Belotserkovskii [5]. In the version of the method used with equilibrium flow it proves convenient to use the entropy equation and utilize the fact that the entropy is constant along streamlines and equal to its value just behind the shock (cf. (6.3.12) and (6.3.14)). For nonequilibrium flows, however, the adiabatic condition and rate equations are written along the streamlines. In so doing ordinary differential equations are obtained for the change in the rate dependent variables

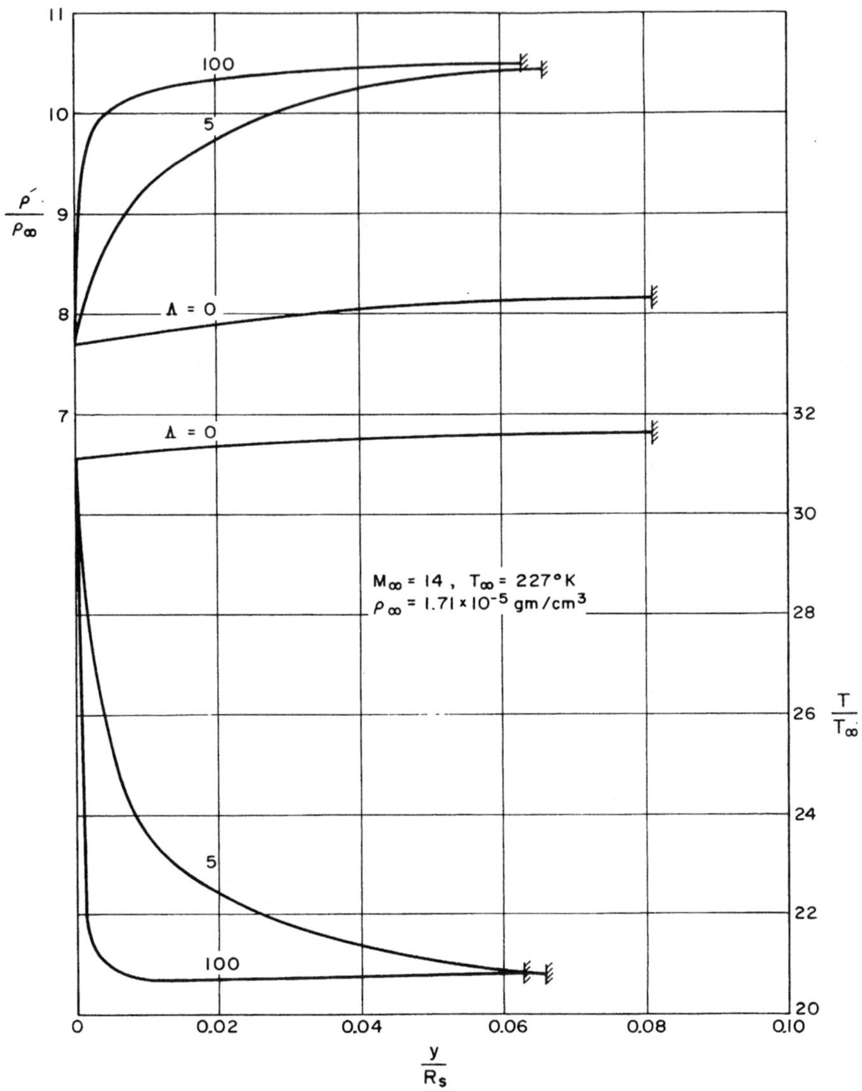

FIG. 6–32. Density and temperature along stagnation streamline for an axisymmetric catenary shock wave and dissociating oxygen in an air-like mixture of O_2 and N_2 (Lick [1]).

such as e_{vib} or c_i. It follows that the change in these quantities is determined exactly along the body.

Belotserkovskii and Dushin [1] in their application of the integral relations method to nonequilibrium flows chose the method of approximating and integrating the variables along the direction of the surface, with strips normal to rather than parallel to the body. The resulting set of ordinary differential equations are then solved in the coordinate normal to the body. These authors chose this approach because of the rapid changes in pressure and temperature which take place in the direction normal to the shock in a dissociating flow near to equilibrium. As a result of the procedure used, no particular numerical difficulties were encountered in their examples, which were carried out for dissociative relaxation with the internal modes equilibrated.

Calculations with the strips chosen parallel to the body surface have been reported, for example, for chemically coupled dissociating air by Shih and Baron [1] and for a dissociating diatomic gas by Lun'kin and Popov [1]. Shih and Baron considered only the one-strip problem. The exact form of the momentum equation in the direction of the surface and the exact rate equations are used in their procedure. No essential difficulties are encountered and no additional assumptions are required beyond those for the one-strip perfect gas formulation. This results from the fact that in the one-strip formulation the differential equations are required only along the body, thereby permitting the use of the exact form of the rate equations and momentum equation. However, the procedure then gives c_i only along the stagnation streamline on the body surface and behind the shock, but not in the shock layer itself. To determine the shock layer distributions of c_i an additional scheme is required (see Shih and Baron [1]). Shih and Baron did indicate that their system of equations was unstable when the conditions were such as to be near to equilibrium behind the shock. In order to maintain numerical stability a variable integration step size was required so that the step size could be decreased when required.

Lun'kin and Popov extended the parallel strip approach to two strips. In this case, difficulties are immediately encountered since the intermediate line is not a streamline and since the rate equations and tangential momentum equation must be applied in their complete form on the strip boundaries. It is necessary therefore to write these equations on the strip boundaries in terms of the unknown derivatives normal to the boundaries. These derivatives must then be determined by an appropriate finite difference scheme. The reader is referred to the original paper for the details of the procedure which was used.

Other extensions to nonequilibrium flow of the numerical methods discussed in this chapter have been reported. The streamtube-continuity approach of Section 6.2, for example, has been extended to nonequilibrium flows by Langelo [1]. The unsteady approach method of Section 6.4 has also been extended by Bohachevsky, Rubin, and Mates [1].

CHAPTER VII

OTHER METHODS FOR LOCALLY SUPERSONIC FLOWS

1. Method of characteristics

In the preceding chapters we have discussed various means of obtaining approximate solutions to inviscid hypersonic flow problems. In addition we have presented various schemes for obtaining exact solutions for the flow past blunt bodies. We next turn to the problem in which the flow field is completely supersonic. We can in this case carry out an exact numerical determination of the inviscid rotational supersonic flow by the method of characteristics, provided we are supplied with sufficient initial data. The problem involves the solution by finite differences of a system of first-order partial differential equations of hyperbolic type. In what follows, we shall assume that the reader is acquainted to some extent with the philosophy of this method or with the details of the method for a two-dimensional or axisymmetric rotational supersonic gas flow in a perfect gas (see, for example, Ferri [2; 3], Isenberg and Lin [1] and Rockett and Hayes [1], Meyer [1], or Belotserkovskii and Chushkin [2]). Our main purpose will be to point out those features which are characteristic of hypersonic flows.

The method of characteristics in two independent variables is well known and well established in the theory of supersonic flow. The extension to three dimensions is less familiar, since a simple direct extension of the approach used with two variables is not possible. We treat the method of characteristics in this book for two reasons. The first is that we shall need the equations for the exact theory in order to describe and to evaluate the approximate methods given later in this chapter. The second is that the method of characteristics as usually presented in the literature includes the unnecessary restriction to a perfect gas of constant γ, and greater generality is desired.

Since any result of characteristic theory must be able to serve as an exact solution with which we may compare approximate theories (e.g. those of Chapter II or of the other sections in this chapter), we cannot properly utilize any of the simplifying hypersonic flow assumptions discussed in Section 1.3. This being the case, the essential new feature which arises in a hypersonic characteristics calculation is the marked deviation in gas properties from those of a perfect gas resulting from the high stagnation temperatures associated with extreme flight speeds.

Our considerations will at first be restricted to the two limiting cases in which the flow is either frozen or in thermodynamic equilibrium. The extension to nonequilibrium will be considered separately in the last section of this chapter. In general, for a frozen flow the gas is taken to have a composition appropriate to the equilibrium composition at some specified state on the downstream side of the attached or detached shock wave. We may, however, define more than one type of frozen flow (cf. Section 1.8).

We begin our discussion by first deriving the characteristic equations for either two-dimensional or axisymmetric flow, deferring until afterwards the extension to three dimensions. It is convenient to write the hydrodynamic equations again in terms of the intrinsic or natural orthogonal coordinates defined by the streamlines and their normals. We let s denote distance measured along a streamline and n denote distance measured perpendicular to it (see Fig. 7–1). It must be emphasized that such intrinsic coordinates do not form a

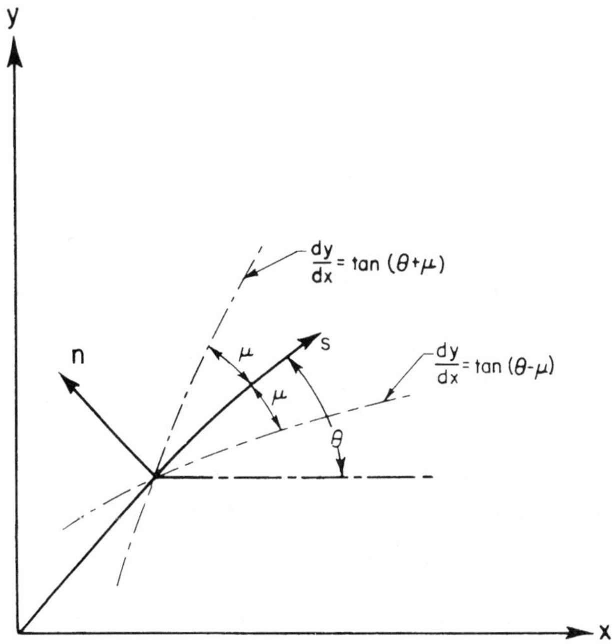

FIG. 7–1. Intrinsic coordinates.

coordinate system in the usual sense. Streamlines are not curves for which a quantity n is constant, and their orthogonal trajectories are not curves for which a quantity s is constant. The quantities s and n are not independent variables in their own right, but are used only in terms of their differentials, which are not perfect ones. We may not take second derivatives with respect to intrinsic coordinates without the introduction of additional terms involving the streamline geometry. Any result obtained in terms of intrinsic coordinates can equally be expressed in terms of a curvilinear orthogonal coordinate system aligned with the streamlines, in terms of a metric which is necessarily unknown if the streamline geometry is unknown. We use intrinsic coordinates in this chapter and in Sections 6.1 and 6.2 primarily because of the simplicity of form which they afford.

The local relation between intrinsic coordinates and fixed cartesian coordinates is expressed by the differentiation formulas

$$\frac{\partial}{\partial s} = \cos\theta \frac{\partial}{\partial x} + \sin\theta \frac{\partial}{\partial y}, \tag{7.1.1a}$$

$$\frac{\partial}{\partial n} = -\sin\theta \frac{\partial}{\partial x} + \cos\theta \frac{\partial}{\partial y}, \tag{7.1.1b}$$

where θ denotes the angle between the streamlines and the free stream direction x (see Fig. 7–1). As before, for the case of axisymmetric flow y is the radial cylindrical coordinate and is read as r.

The continuity equation in intrinsic coordinates assumes the form

$$\frac{\partial \rho q y^j}{\partial s} + \rho q y^j \frac{\partial \theta}{\partial n} = 0, \tag{7.1.2}$$

while the corresponding momentum equations along and normal to the streamlines are

$$\rho q \frac{\partial q}{\partial s} + \frac{\partial p}{\partial s} = 0, \tag{7.1.3}$$

$$\rho q^2 \frac{\partial \theta}{\partial s} + \frac{\partial p}{\partial n} = 0. \tag{7.1.4}$$

The assumption of an inviscid adiabatic flow which is in thermodynamic equilibrium (or frozen) leads to the condition of constancy of entropy along a streamline

$$\frac{\partial S}{\partial s} = 0. \tag{7.1.5}$$

The above relations are supplemented by the equation of state (1.4.3), from which the pressure may be expressed as a function of density and entropy. Finally, since we will consider only problems in which the total enthalpy is a constant throughout the flow, we have

$$H = h + \tfrac{1}{2}q^2 = h_\infty + \tfrac{1}{2}U^2. \tag{7.1.6}$$

A characteristic is a line in the field on which an ordinary differential equation may be written. Such an equation must be written as a relation connecting total differentials and in which partial derivatives do not appear. One such characteristic line which we may obtain immediately is the streamline, for which $dn = 0$. For this characteristic (7.1.3) and (7.1.5) are already in characteristic form, as no partial derivatives normal to the streamline appear. Since we are treating isoenergetic flow, we use (7.1.3) in its integrated form (7.1.6) rather than as a characteristic equation. But our entire development would hold essentially without change for flows which are not isoenergetic. From (7.1.5) we obtain the essential characteristic equation

$$dS = 0, \quad \text{on} \quad dn = 0. \tag{7.1.7}$$

We next eliminate $\partial q/\partial s$ from (7.1.2) by use of (7.1.3), substitute $\sin\theta$ for $\partial y/\partial s$, substitute $a^{-2}\,\partial p/\partial s$ for $\partial \rho/\partial s$, and replace q/a by M. The result is an altered form of the continuity equation,

$$\frac{M^2 - 1}{\rho q^2}\frac{\partial p}{\partial s} + \frac{\partial \theta}{\partial n} = -\frac{j}{y}\sin\theta. \tag{7.1.8}$$

Since we wish to obtain relations involving total differentials, we write the expressions for dp and $d\theta$

$$dp = \frac{\partial p}{\partial s}\,ds + \frac{\partial p}{\partial n}\,dn, \tag{7.1.9a}$$

$$d\theta = \frac{\partial \theta}{\partial s}\,ds + \frac{\partial \theta}{\partial n}\,dn. \tag{7.1.9b}$$

We now have four equations (7.1.4), (7.1.8), and (7.1.9) which involve the four partial derivatives of p and θ with respect to s and n. A characteristic equation must be completely independent of these partial derivatives, and the determinant of the coefficients of the partial derivatives in our four equations must vanish. This condition determines two characteristic directions, besides the one already found. Since the four equations are not homogeneous in the partial derivatives another condition must be satisfied in order that solutions to the system may exist. This condition gives us two characteristic equations, each corresponding to one of the characteristic directions. These characteristic equations are found to be

$$dp \pm \frac{\rho q^2}{\sqrt{M^2 - 1}}\,d\theta + \frac{j\rho q^2 \sin\theta}{M^2 - 1}\frac{ds}{y} = 0, \tag{7.1.10}$$

$$\text{on} \quad \frac{dn}{ds} = \pm\frac{1}{\sqrt{M^2 - 1}}.$$

Transforming the characteristic equations and directions (7.1.7) and (7.1.10) to cartesian coordinates, we obtain

$$dS = 0, \qquad \text{on} \quad \frac{dy}{dx} = \tan\theta, \tag{7.1.11}$$

and

$$\frac{dp}{\rho q^2 \tan\mu} \pm d\theta + \frac{j \sin\theta \sin\mu}{\sin(\theta \pm \mu)} \frac{dy}{y} = 0, \tag{7.1.12}$$

$$\text{on} \quad \frac{dy}{dx} = \tan(\theta \pm \mu),$$

where μ is the local Mach angle $\sin^{-1}(1/M)$ (see Fig. 7–1). Thus the characteristic directions are the streamlines and Mach lines in the flow field. We note that (7.1.12) has exactly the same form as in potential flow, for which the pressure is a function of the velocity alone. However, in a general rotational flow it is necessary to determine the relation between pressure and velocity at each point through the equation of state, the isoenergetic relation (7.1.6), and the condition of constancy of entropy along a streamline (7.1.7).

If the flow is two-dimensional and isentropic (irrotational) an explicit integral may be obtained for both characteristic equations (7.1.10) or (7.1.12). If the flow is axisymmetric or rotational such an integral may not be obtained, and a stepwise numerical integration procedure is required to obtain the solution.

There are many developments of the method of characteristics for two-dimensional or axisymmetric flow in the literature and in standard texts (e.g. in the references cited at the beginning of this section). We do not consider it worthwhile in this book to repeat the details of various numerical procedures, and the reader is referred to the works cited for such details. A few points should be emphasized, however. In a general fluid the thermodynamic behavior of the one-dimensional steady isentropic flow along each streamline must be known. Since the flow is assumed to be isocompositional and isoenergetic these one-dimensional flows form a one-parameter family, with the entropy as the parameter. The simplicity of form of the characteristic equations as expressed in terms of dp and $d\theta$ is also to be emphasized. If another variable is introduced in place of p the characteristic equations become significantly more complicated. That the quantities p and θ should be considered as the basic variables is related to the fact that it is these quantities which must be matched across a contact discontinuity or slipstream.

In a calculation it may be more convenient to replace the quantity $dp/\rho q^2$ appearing in the first term of (7.1.12) by $dp/\gamma_e M^2 p$, where γ_e is defined in (1.4.22). A chart of γ_e for air is available in Moeckel [2] or in Hansen [1, Fig. 5].

The explicit integral available for two-dimensional isentropic flow is equivalent to the integral describing Prandtl-Meyer flow in a general fluid. We shall next consider Prandtl-Meyer flow, the results for which are used in the shock-expansion method of the next section. It is possible for hypersonic flow problems to arise in which the flow is approximately or strictly two-dimensional and isentropic, and in such a problem the explicit integral may be used directly. If the flow is isentropic, then $dS = 0$ throughout the flow field, and the Bernoulli relation $q\,dq = -dp/\rho$ is valid in any direction. From (7.1.12) we then obtain the usual form of the characteristic relations for two-dimensional, isentropic, supersonic flow:

$$d\nu \mp d\theta = 0; \qquad \nu \mp \theta = constant, \qquad \text{on} \quad \frac{dy}{dx} = \tan(\theta \pm \mu), \tag{7.1.13}$$

where ν is the Prandtl-Meyer turning angle defined by

$$d\nu = \cot \mu \frac{dq}{q} = \sqrt{M^2 - 1}\,\frac{dq}{q}. \tag{7.1.14}$$

We define ν to be zero at $M = 1$. With this convention ν is a function $\nu(H, S, p)$ of the total enthalpy and the thermodynamic state.

It is at this point that the problem in a general fluid differs from the conventional perfect gas treatment. In a general fluid no simple explicit relation exists for the speed of sound as a function of the velocity. Instead this relation is an implicit one expressed through the equation of state and the isoenergetic relation (7.1.6). Thus with the aid of (7.1.6), we have from (7.1.14)

$$\nu = \frac{1}{2}\int \left[\frac{2(H-h)}{a^2} - 1\right]^{1/2} \frac{d(H-h)}{H-h}, \tag{7.1.15}$$

with the integration constant chosen by convention so that $\nu = 0$ at $M = \sqrt{2(H-h)/a^2} = 1$. It is to be emphasized that this integration is taken along an isentrope. To integrate (7.1.15) it is necessary to specify the equation of state, to obtain the function $a^2(h, S)$.

Let us now first consider the case of an equilibrium flow for which the state relation is defined by a Mollier diagram such as is shown in Fig. 1–4, or given in the charts of Feldman [1] or Chance Vought [1]. Before carrying out the integration, however, it is also necessary to specify both the total enthalpy and the entropy. Under such circumstances, it is evident that a characteristics diagram, such as is employed for a perfect gas, can be constructed for a general fluid only for a given choice of total enthalpy and the entropy. For a given fluid, the characteristic diagrams form a two-parameter family.

Although (7.1.15) is in a convenient form for numerical integration, since values of the speed of sound are available from charts or may be obtained from the equation of state, it does not permit a simple interpretation of how a Prandtl-Meyer flow in a general fluid differs from the usual perfect gas solution. We present here two alternative formulations of Prandtl-Meyer flow which show clearly such an interpretation. In order to do so we must use quantities involving an isentropic derivative of a^2, even though such a derivative does not appear in (7.1.15).

We turn our attention from ν to the quantity $\nu + \frac{1}{2}\pi - \mu$, with μ the Mach angle $\sin^{-1}(1/M)$. This quantity is also zero at $M = q/a = 1$, and its differential may be expressed as

$$d\nu - d\mu = \frac{1}{\sqrt{M^2 - 1}}\left[(M^2 - 1)\frac{dq}{q} + \frac{dq}{q} - \frac{da}{a}\right] \tag{7.1.16}$$

$$= \frac{q\,dq - a\,da}{a\sqrt{q^2 - a^2}}.$$

Since the process involved is always isentropic we may write partial derivatives as ratios of differentials, and using (7.1.6) may write

$$\frac{da}{a} = a_h\frac{dh}{a} = -aa_hM^2\frac{dq}{q}, \tag{7.1.17}$$

with $a_h = (\partial a/\partial h)_S$. We then obtain

$$\frac{dM}{M(1 + aa_hM^2)} = \frac{dq}{q} = \frac{dM}{M} + \frac{da}{a}, \tag{7.1.18}$$

and may substitute this expression for dq and for da into (7.1.16). The resulting equation may be put in the form

$$\nu + \tfrac{1}{2}\pi - \mu = \int_1 \left(\frac{1 + aa_h}{1 + aa_hM^2}\right)\frac{M\,dM}{\sqrt{M^2 - 1}}, \tag{7.1.19}$$

with the integral defined to be zero at $M = 1$. Noting that $\frac{1}{2}\pi - \mu = \cos^{-1}(1/M) = \tan^{-1}\sqrt{M^2 - 1}$, we can arrive at the following formally simple expression for the turning angle

$$\nu = \int_0 \frac{d\beta}{1 + (F_*\beta)^2} - \tan^{-1}\beta, \tag{7.1.20}$$

where $\beta = \sqrt{M^2 - 1}$, and

$$F_*^2 = \frac{aa_h}{1 + aa_h} = \frac{\left(\dfrac{\partial a^2}{\partial h}\right)_S}{2 + \left(\dfrac{\partial a^2}{\partial h}\right)_S}. \tag{7.1.21}$$

The significance of the function F_* is made clearer by recognizing that it may be expressed in terms of the quantity γ_* defined in (1.4.24). The quantity γ_* may be re-expressed as

$$\gamma_* = 1 + \left(\frac{\partial a^2}{\partial h}\right)_S, \tag{7.1.22}$$

and we obtain immediately

$$F_*^2 = \frac{\gamma_* - 1}{\gamma_* + 1}. \tag{7.1.23}$$

For a perfect gas of constant γ, $a^2 = (\gamma - 1)h$, $\gamma = \gamma_*$, and $F_*^2 = (\gamma - 1)/(\gamma + 1)$. In this case (7.1.20) may be integrated directly to give the conventional form of the Prandtl-Meyer relation,

$$\nu = F^{-1} \tan^{-1} (\beta F) - \tan^{-1} \beta, \tag{7.1.24}$$

in which the subscript $*$ on F has been dropped.

The second formulation for the Prandtl-Meyer flow is due to Heims [1]. The formulation just presented and that of Heims were developed independently, and we have slightly changed Heims' analysis to fit the development of this section. Heims defines a quantity η, which we here replace by the square root of its inverse

$$F_e = \frac{a}{\sqrt{a^2 + 2h}}. \tag{7.1.25}$$

It must be noted that this quantity requires the assignment of a reference value of the enthalpy before it is uniquely defined. Although any assignment of reference enthalpy will serve, an arbitrary assignment may lead to complex values of F_e or to a zero value of F_e at absolute zero temperature. To avoid such mathematical behavior and to provide the most significant interpretation of the results, we must assign to the enthalpy h the value zero at absolute zero temperature.

With the enthalpy zero at absolute zero temperature the total enthalpy of the stream H is equal to $\frac{1}{2}q_{\max}^2$, where $q_{\max}$ is the maximum velocity of the steady flow. The energy equation then takes the form

$$q^2 = q_{\max}^2 - 2h. \tag{7.1.26}$$

We may now rewrite (7.1.16) in the form

$$d\nu - d\mu = -\frac{a\,da + dh}{F_e \sqrt{q_{\max}^2 - a^2 - 2h}\sqrt{a^2 + 2h}} \tag{7.1.27}$$

$$= F_e^{-1}\, d \cos^{-1}\left(\frac{a}{q_{\max} F_e}\right).$$

Next we introduce the angle φ (in Heims' notation ψ), defined by the relation

$$\varphi = \cos^{-1}\left(\frac{a}{q_{\max} F_e}\right), \tag{7.1.28}$$

which takes the value zero at $M = 1$. We may now express the Prandtl-Meyer turning angle ν as

$$\nu = \int_0 F_e^{-1}\, d\varphi - \cos^{-1}(1/M), \tag{7.1.29}$$

in terms of what is essentially a Stieltje's integral.

The radial component of velocity in a Prandtl-Meyer flow is $\sqrt{q^2 - a^2}$, and is expressed in terms of φ as

$$\sqrt{q^2 - a^2} = q_{\max} \sin \varphi. \tag{7.1.30}$$

The velocity of sound is simply

$$a = F_e\, q_{\max} \cos \varphi, \tag{7.1.31}$$

and the quantity $\beta = \sqrt{M^2 - 1}$ is

$$\beta = F_e^{-1} \tan \varphi. \tag{7.1.32}$$

In a perfect gas of constant γ, we have $F_e^2 = (\gamma - 1)/(\gamma + 1)$ is constant, and we again obtain (7.1.24) with the subscript e dropped. The quantity F_e^2 is expressible in general in terms of the γ_ϵ and γ_e defined in (1.4.20) and (1.4.22) as

$$F_e^2 = \frac{\gamma_e(\gamma_\epsilon - 1)}{\gamma_\epsilon \gamma_e + 2\gamma_\epsilon - \gamma_e}. \tag{7.1.33}$$

In comparing these two formulations we find no tremendous distinction. Both involve derivatives of the speed of sound a, in the definition of γ_* in the first, and in the differential of φ in (7.1.29) in the second. Both require a calculation of M as a function of thermodynamic state along the isentrope. The first formulation is closer in form to the usual one for a perfect gas, and is expressed in terms of the quantity $\beta = \sqrt{M^2 - 1}$. It is perhaps somewhat more closely related to our characteristics calculation. The second formulation shows more clearly the relation between the velocity components and the maximum velocity $q_{\max}$, provided the correct zero point has been assigned to h.

In general it is not possible to carry out the quadrature (7.1.20) or (7.1.29) in closed form. Charts or tables of F_* or of F_e are needed, and should be prepared for air under conditions of practical interest. Information on the quantity F_e

would be needed more accurately than on the quantity F_*, as F_e is differentiated in the process of obtaining the quadrature (7.1.29). One empirical approximation would be to take F_* or F_e equal to some constant average value, corresponding to some fictitious perfect gas of constant γ. Another would be to replace F_* by $(\gamma_e - 1)/(\gamma_e + 1)$. Results of two sample calculations using this latter empirical approximation are given in Fig. 7–2, and compared with the results for a perfect gas and with one exact calculation according to (7.1.15). The empirical approximation $\gamma_* = \gamma_e$ appears to be reasonably accurate for air at moderate altitudes.

Equilibrium calculations for Prandtl-Meyer flow, including the effect of partially excited vibrational modes, were reported by Eggers, Syvertson, and

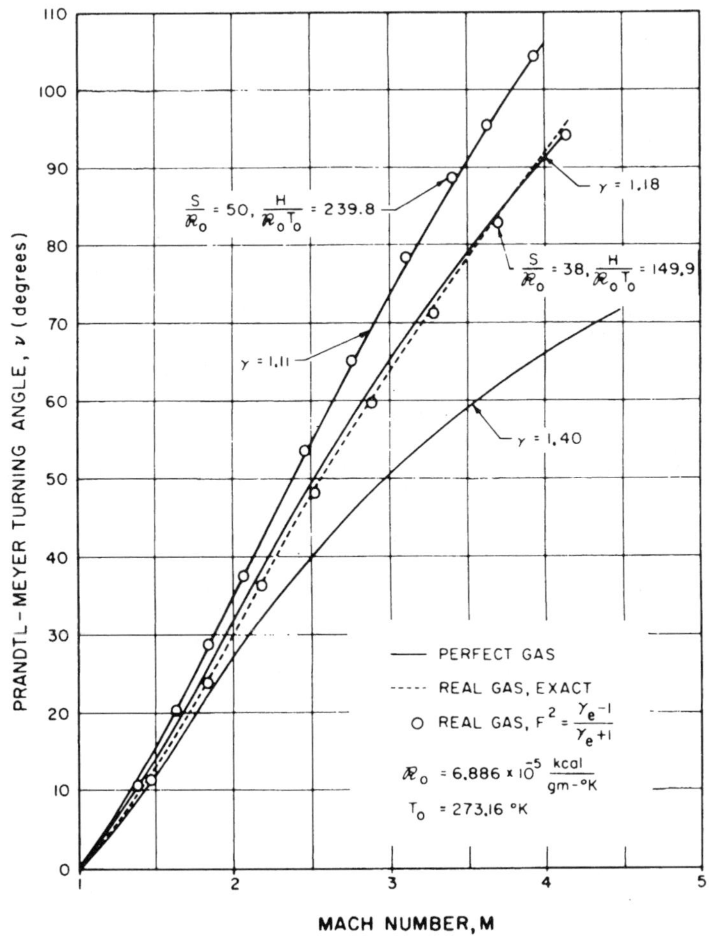

FIG. 7–2. Prandtl-Meyer turning angle.

Kraus [1]. Calculations for dissociated oxygen, nitrogen, air, and ionized argon, covering a range typical of shock tube operating conditions, have been made by Glass and Kawada [1]. Frozen flow calculations discussed below are also given there.

We now discuss the case of frozen equilibrium, in which the gas is assumed to have a fixed composition corresponding to a specified reference state. In contrast to the case of thermodynamic equilibrium, this kind of flow can be treated with relative simplicity. To show this, we first write the usual equation of state for the pressure in terms of the "compressibility factor" $Z(p, T)$ as

$$p = \rho \mathscr{R} T Z, \tag{7.1.34}$$

where $\mathscr{R}$ is the undissociated perfect gas constant.

We concern ourselves with air, which when undissociated is essentially entirely composed of diatomic molecules, and let α be the fraction of original molecules dissociated. Following the argument leading to (6.6.6) we see that $Z = 1 + \alpha$, and that the gas law may be written as

$$p = \rho(1 + \alpha)\mathscr{R} T = \rho \mathscr{R}_f T, \tag{7.1.35}$$

where $\mathscr{R}_f$ is the gas constant for the frozen state. With this form of the gas law, we can express all the frozen flow relationships in the same form as for a perfect gas by the use of the modified $\mathscr{R}_f$ instead of $\mathscr{R}$, and an effective γ_f instead of γ. All of the quantities for a frozen state are then functions of α, which is assumed constant during the frozen flow process.

Two frozen states would appear to be of particular interest. The first is one in which any chemical reaction rates involved are sufficiently slow that no change in composition occurs in the process, and also the vibrational relaxation times are sufficiently long that no change in the vibrational energy levels occurs. In this frozen state we assume that the rotational energy of the molecules is always in equilibrium. From the fact that the specific heat at constant pressure equals $7\mathscr{R}/2$ for a diatomic gas and $5\mathscr{R}/2$ for a monatomic gas, we may calculate

$$c_{p_f} = \frac{7 + 3\alpha}{2} \mathscr{R} = \frac{7 + 3\alpha}{2(1 + \alpha)} \mathscr{R}_f . \tag{7.1.36}$$

But with $\mathscr{R}_f = c_{p_f} - c_{v_f}$ and $\gamma_f = c_{p_f}/c_{v_f}$, we have

$$\gamma_f = \frac{7 + 3\alpha}{5 + \alpha} . \tag{7.1.37}$$

The isentropic exponent for the frozen state γ_f varies from $7/5$ at $\alpha = 0$ to $5/3$ at $\alpha = 1$ for an originally diatomic gas.

We define the other frozen state of interest as one in which the relaxation time for chemical recombination is assumed to be much longer than the relaxation time for the vibrational degrees of freedom. If the temperatures are sufficiently high in a diatomic gas that the vibrational mode is fully excited, it contributes the classical amount $\mathscr{R}T$ to the heat capacities. It follows that for a frozen state with vibrational equilibrium

$$c_{p_f} = \frac{9+\alpha}{2}\mathscr{R} = \frac{9+\alpha}{2(1+\alpha)}\mathscr{R}_f\,, \tag{7.1.38}$$

and

$$\gamma_f = \frac{9+\alpha}{7-\alpha}\,. \tag{7.1.39}$$

Here γ_f varies from 9/7 at $\alpha = 0$ to 5/3 at $\alpha = 1$ for an originally diatomic gas. Clearly, there are other frozen states which can be defined, and a fuller discussion of this point is given by Feldman [2].

From the preceding discussion it can be seen that the usual form of the Prandtl-Meyer relation (7.1.24) is valid for the frozen case provided the correct effective value of γ is utilized. In order to know just which value to employ, we need to know the appropriate relaxation times for the gas. Once the frozen state is specified, however, it is only necessary to determine α from a Mollier diagram to complete the problem. For air at high temperatures the effective value of γ_f lies between the limits $9/7 \leqslant \gamma_f \leqslant 5/3$.

In using the Prandtl-Meyer relation it is often useful to have it in inverted form, in terms of the function $M(\nu)$. In the general case this inversion must be done numerically, but for a perfect gas with $\gamma = 5/4$ or 5/3 it may be done analytically (see Probstein [1]). If $\gamma = 5/3$ the inverted relation is

$$M = \frac{1 + (\tan \nu/2)^{2/3}}{1 - (\tan \nu/2)^{2/3}}\,. \tag{7.1.40}$$

The inverted relation for $\gamma = 5/4$ is much more complicated in form.

In the three-dimensional steady case the part of the analysis corresponding to streamline characteristics is the same as in the two-dimensional case. Thus equations (7.1.3), (7.1.5), (7.1.6), and (7.1.7) given above are valid in the same context in three-dimensional flows, with $dn = 0$ meaning "along a streamline." In the remainder of this section we consider only the wave characteristics and not the streamline characteristics. We wish to find the three-dimensional analogues to the equations of the type of (7.1.10).

Thus we turn now to the question of the method of characteristics in three-dimensional flows. It has been mentioned that a simple straightforward extension to three-dimensional cases of the method used with two independent variables is not available. The presence of the additional independent variable naturally entails additional complication. However, there is a three-dimensional method of characteristics, and it can be discussed in such a way that the analogies with the two-dimensional method are most apparent.

In steady supersonic two-dimensional flow, a characteristic direction is one which is at an angle with respect to the flow direction equal to the Mach angle. In three-dimensional flow the two-dimensional characteristic is replaced by two entities. A bicharacteristic direction in the three-dimensional case is one which is at an angle with respect to the flow direction equal to the Mach angle. A characteristic surface orientation is one for which the angle between the normal and the flow direction is the complement of the Mach angle (see Fig. 7–3).

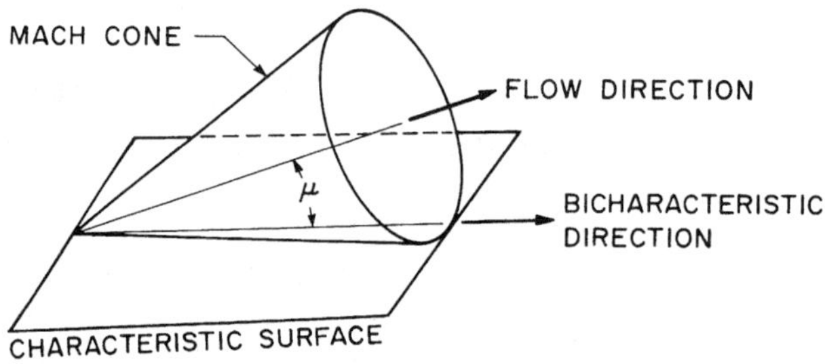

FIG. 7–3. Bicharacteristic direction and corresponding characteristic surface orientation.

A general bicharacteristic line is an integral curve for the bicharacteristic direction. A characteristic surface is one whose orientation is everywhere the characteristic orientation. A general bicharacteristic line through a given point is far from unique. A characteristic surface through a given smooth line in space (with angle relative to the flow direction greater than the Mach angle) is locally unique. Any envelope of characteristic surfaces is also a characteristic surface. See, for example, Fig. 7–4.

The two concepts are related in the following way. To a given bicharacteristic direction at a point corresponds a unique characteristic orientation, for which the bicharacteristic direction is tangent to the characteristic surface. And to a given characteristic orientation at a point corresponds a unique bicharacteristic direction, corresponding to the same criterion. As a consequence, through a given nonsingular point in a given characteristic surface passes a unique bicharacteristic line. In this sense singular points include points on an envelope of bicharacteristic lines in a characteristic surface. The term bicharacteristic generally refers to such a bicharacteristic line without singular points.

With each point is associated a particular characteristic surface termed the characteristic conoid for that point. This surface is tangent to the Mach cone at the point in question, and is thus singular there. Physically, it represents the boundary of the zone of action for the point in question. The characteristic surfaces generated by a given generating line may be interpreted as the envelope of the characteristic conoids for the points on the line, as indicated in Fig. 7–4. Conversely, the characteristic conoid for a point may be interpreted as the envelope of a suitably chosen family of characteristic surfaces passing through the point.

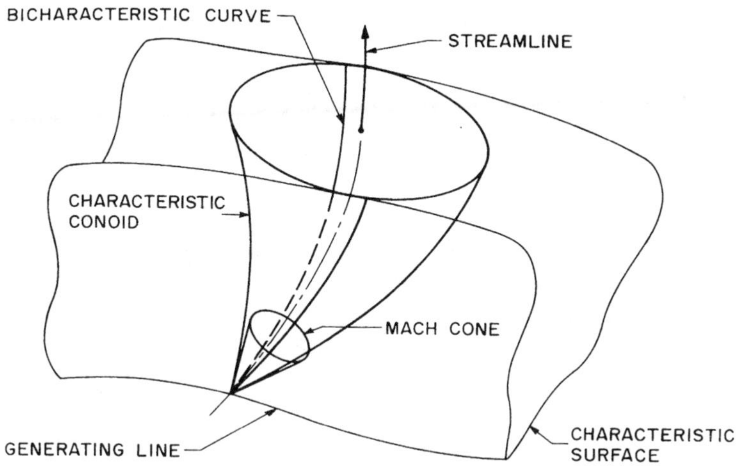

FIG. 7–4. Characteristic conoid.

Although bicharacteristic lines in the general sense defined are far from unique, the bicharacteristic (in a characteristic surface without singular points) enjoys a different status. The bicharacteristic through a point with given initial direction is unique, at least locally. The characteristic conoid through a point may be interpreted as the locus of bicharacteristics through the point.

In the two-dimensional case, the characteristic equation is a one-dimensional (ordinary) differential equation in a characteristic line. In the three-dimensional case, the analogous equation is a two-dimensional differential equation in a characteristic surface, termed the bicharacteristic equation. It turns out that the principal terms in this two-dimensional equation are made up of one-dimensional derivatives along the bicharacteristic in the surface. It is the presence of the terms involving derivatives not in the bicharacteristic direction which defeats the direct application of the two-dimensional method.

In the two-dimensional case we have two characteristic directions and corresponding equations. In the three-dimensional case there are an infinite number of bicharacteristic directions (or characteristic orientations) and corresponding equations. In the various schemes different suggested choices are made as to the equations used. Two equations are insufficient. Three equations are sufficient, and may consist of three characteristic equations or of two characteristic equations and one partial derivative equation in three variables. If desired, more than three equations may be used, with the resulting over-determination providing a check on the accuracy of the scheme.

To obtain the equations for a method of characteristics in three-dimensional steady flow we may extend the two-dimensional analysis given above. The intrinsic variable ds is as given before, and corresponds to distance along a streamline. The intrinsic variable dn is replaced by a two-vector $d\mathbf{n}$ normal to the streamline, and the operator $\partial/\partial n$ is replaced by the operator ∇. This operator may be defined

(7.1.41) $$\nabla = \mathbf{e}_1 \frac{\partial}{\partial n_1} + \mathbf{e}_2 \frac{\partial}{\partial n_2},$$

where the $\mathbf{e}_i$ are unit vectors orthogonal to each other and to the streamline directions, and $\partial/\partial n_i$ are the intrinsic spatial derivatives in those directions. For further details of three-dimensional intrinsic equations the reader is referred to Truesdell [2] and to the references cited therein.

The velocity vector in the three-dimensional space is expressed

(7.1.42) $$\mathbf{q} = q\mathbf{t},$$

where $\mathbf{t}$ is a unit three-vector having the direction of a streamline. The continuity equation may be written

(7.1.43) $$\frac{\partial \rho q}{\partial s} + \rho q \nabla \cdot \mathbf{t} = 0.$$

The quantity $\nabla \cdot \mathbf{t}$ is the one denoted Θ by Truesdell, and may be interpreted as the logarithmic derivative $\partial A/A\, \partial s$ of the area of a streamtube. Equations (7.1.3) and (7.1.5) above hold without change, and the same derivation used earlier leads to

(7.1.44) $$\frac{M^2 - 1}{\rho q^2} \frac{\partial p}{\partial s} + \nabla \cdot \mathbf{t} = 0$$

in place of (7.1.8).

The $\mathbf{n}$ components of the momentum equation take the form

(7.1.45) $$\rho q^2 \frac{\partial \mathbf{t}}{\partial s} + \nabla p = 0.$$

This equation replaces (7.1.4) above. The two-vector $\partial \mathbf{t}/\partial s$ is the curvature of the streamline. Equations (7.1.9) are replaced by

(7.1.46a) $$dp = \frac{\partial p}{\partial s} ds + d\mathbf{n} \cdot \nabla p,$$

(7.1.46b) $$d\mathbf{t} = \frac{\partial \mathbf{t}}{\partial s} ds + d\mathbf{n} \cdot \nabla \mathbf{t}.$$

Equations (7.1.44), (7.1.45), and (7.1.46) serve as the basis for three-dimensional characteristic equations.

We digress slightly to discuss the nature of the two-tensor or two-dyadic $\nabla \mathbf{t}$ which appears in (7.1.46b). Its trace is the $\nabla \cdot \mathbf{t}$ which enters the continuity equation. Its antisymmetric part is essentially the abnormality

$$\Omega = \mathbf{t} \cdot \nabla \times \mathbf{t}, \tag{7.1.47}$$

which measures the "swirl" of the streamline pattern. The parameter Ω plays a key role in the study of Beltrami flows. The remaining two independent components measure the way in which the cross section of a streamtube is distorted as we pass along the tube.

The bicharacteristic equation we are seeking is one which may involve total derivatives in the bicharacteristic direction, but which, as we have noted, generally involves partial derivatives in the characteristic surface. The basic property is that derivatives across the surface may not enter.

In the present context, we may consider an arbitrary direction (ds, $d\mathbf{n}$) and impose the condition that only derivatives in a surface may enter. Alternatively, we may introduce the correct bicharacteristic directions and verify that the condition is met. We choose this latter simpler course.

From our understanding of the physical problem, we recognize the characteristic conoid locally as the Mach cone (as was pointed out earlier). An arbitrary unit vector $\mathbf{n}_i$ is chosen normal to $\mathbf{t}$, and a bicharacteristic direction defined by

$$d\mathbf{n} = \frac{\mathbf{n}_i\, ds}{\sqrt{M^2 - 1}}. \tag{7.1.48}$$

A combination of dp and $d\mathbf{t}$ in which derivatives across the Mach cone or tangent Mach plane do not appear is obtained by analogy with (7.1.10). The resulting bicharacteristic equation is

$$dp + \frac{\rho q^2}{\sqrt{M^2 - 1}} \mathbf{n}_i \cdot d\mathbf{t} + \frac{\rho q^2}{M^2 - 1} \{\nabla \cdot \mathbf{t} - \mathbf{n}_i \cdot \nabla \mathbf{t} \cdot \mathbf{n}_i\}\, ds = 0 \tag{7.1.49}$$

This equation may be compared with (7.1.10).

To verify that the bicharacteristic condition is met, let $i = 1$ and define $\mathbf{n}_2$ as the unit vector orthogonal to both $\mathbf{t}$ and $\mathbf{n}_1$. The vector $\mathbf{n}_2$ lies in the characteristic plane associated with the bicharacteristic. See Fig. 7–5. The quantity $\nabla \cdot \mathbf{t} = \mathbf{n}_1 \cdot \partial \mathbf{t}/\partial n_1 + \mathbf{n}_2 \cdot \partial \mathbf{t}/\partial n_2$, while the quantity $\mathbf{n}_1 \cdot \nabla \mathbf{t} \cdot \mathbf{n}_1 = \mathbf{n}_1 \cdot \partial \mathbf{t}/\partial n_1$. Thus the bracket in (7.1.49) may be written

$$\{\nabla \cdot \mathbf{t} - \mathbf{n}_1 \cdot \nabla \mathbf{t} \cdot \mathbf{n}_1\} = \mathbf{n}_2 \cdot \frac{\partial \mathbf{t}}{\partial n_2}, \tag{7.1.50}$$

and involves only a partial derivative in the characteristic plane.

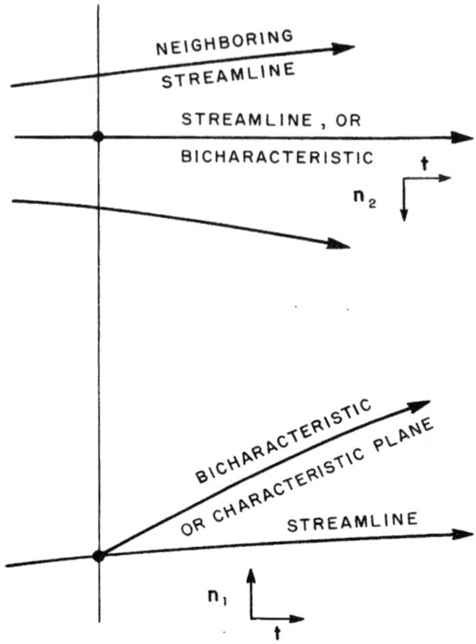

FIG. 7–5. Bicharacteristic direction with intrinsic coordinates.

This term $\mathbf{n}_2 \cdot \partial\mathbf{t}/\partial n_2$ may be interpreted as the apparent divergence of the streamlines passing through a line from the point of interest passing in the $\mathbf{n}_2$ direction, as seen in projection on the $(\mathbf{t}, \mathbf{n}_2)$ plane. In the axisymmetric case this divergence is the inverse of the distance upstream to the point of intersection of the tangents to the streamlines. This distance may readily be seen to be $r/\sin\theta$, so that (7.1.10) may be recognized as a special case of (7.1.49).

To solve a physical problem using these equations we must use three independent values of $\mathbf{n}_i$, because we have three dependent variables. These three variables are p and the two independent components of $\mathbf{t}$. As in the two-dimensional and axisymmetric cases we must keep track of S, and of H if it is variable. If only two bicharacteristic equations are used, another equation obtained from the equations of motion must be used.

Many schemes have been described for carrying out numerical calculations with the aid of bicharacteristic equations. An example of a method in which difference relations are applied along three bicharacteristics intersecting each new point to be located in space is to be found in one of the approaches of Thornhill [1]. Coburn and Dolph [1] suggested a method developed further by Holt [1], in which difference relations are set up along two intersecting bicharacteristics and a nonbicharacteristic curve through the intersection. This curve is taken to correspond to the intersection of two characteristic surfaces through the two bicharacteristics. Related approaches with the bicharacteristic equations in intrinsic form have been suggested by Sauer [1] and Bruhn and Haack [1]. In still another approach to the problem Butler [1] has found that through a particular point the bicharacteristic equations may be linearly combined to give two independent relations involving only derivatives in bicharacteristic directions at the point. Another nonbicharacteristic curve to be used in this difference scheme is fixed by the form of the bicharacteristic conditions he uses. Butler notes that this curve always lies inside the characteristic conoid for the point being considered; this property is in contrast to that in the Coburn and Dolph method. For further discussion of the details involved in these procedures and additional references the reader is referred to the works mentioned.

2. Shock-expansion theory

Although the method of characteristics serves to supply an exact scheme for computing a supersonic flow field, the lengthiness of such a calculation makes its application to hypersonic flows quite onerous. One fairly accurate technique for finding simple approximate hypersonic flow solutions for two-dimensional sharp-nosed airfoils, for which the shock is attached at the leading edge and the flow behind the shock is supersonic, is the shock-expansion method first used by Epstein [1]. Epstein considered principally polygonal profiles for which the reflected waves neglected in the shock-expansion method do not hit the body. In the shock-expansion method, the airfoil characteristics are computed by assuming that the flow behind the leading edge shock is the same as an isentropic Prandtl-Meyer expansion, with only a single family of characteristics taken into account. The obvious calculational advantage of the method is that while it does take into account the change in entropy through the strong leading edge shock, it yields a result for the pressure on the body which is dependent only upon the flow inclination angle. In addition, none of the basic hypersonic assumptions discussed in Section 1.3 need be applied. But we must examine with care the inherent limitations on the accuracy of the method.

The concept introduced by Epstein for the calculation of surface pressures was extended by Eggers and Syvertson [1] in their "generalized shock-expansion method" (see also, Eggers, Syvertson, and Kraus [1]) to include an approximate determination of the shock shape and of the entire flow field. In this procedure we still only consider a single family of principal characteristics, with reflections from the shock wave and from the vortex lines in the flow neglected. The field is not taken to be isentropic, however, and the Mach lines are not taken to be straight as in a Prandtl-Meyer flow. This allows the development of a procedure similar to the method of characteristics but much simpler than it. The scheme is illustrated in Fig. 7–6, in which no reflected characteristics are shown. We use the established condition that the entropy is constant along streamlines downstream of the shock wave. An additional assumption is now needed to permit a calculation of the flow field. We may, for example, assume that the pressure is constant along the principal characteristics, and thereby calculate the corresponding deflection angle θ, the Mach angle μ, and the location of the

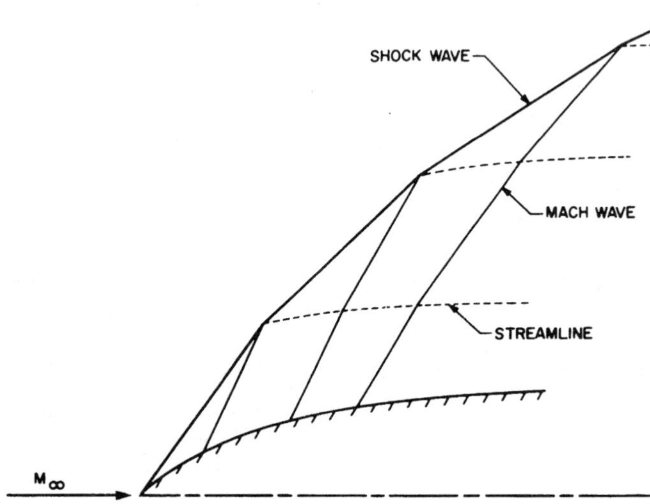

FIG. 7–6. Sketch of flow field in shock-expansion method.

principal characteristics and streamlines in the field. An alternative assumption would be that the flow deflection angle θ is constant along principal characteristics, with the pressure and the Mach angle calculated in the flow field. The reader will observe that the method as thus proposed has a large degree of arbitrariness. Since it is assumed that there are no reflected waves, the pressure on the body is not affected by the choice of method used for the flow field. It was pointed out by Eggers, Syvertson, and Kraus [1] that the choice of method affects the primary geometry of the flow field but slightly. But the arbitrariness prevents any rational estimate of the change in pressure along a principal characteristic. This deficiency can be partially corrected and a closer approximation to the true flow geometry obtained by simply averaging the shock wave angle determined by assuming constant flow deflection with that obtained by assuming constant pressure.

We present a rough analysis to justify this averaging procedure in two-dimensional flow. Referring to Fig. 7–7, the characteristic equations (with $j = 0$) applied along the principal characteristics with a difference scheme give

$$p_1 - p_0 = -\Gamma_0 p_0(\theta_1 - \theta_0), \tag{7.2.1a}$$

$$p_3 - p_2 = -\Gamma_0 p_0(\theta_3 - \theta_2), \tag{7.2.1b}$$

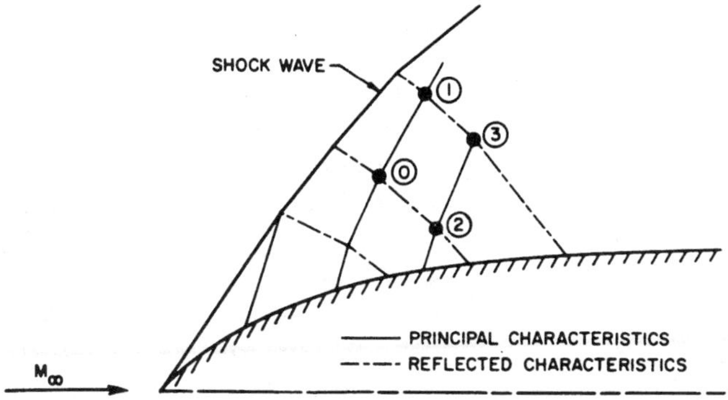

FIG. 7–7. Interaction of principal and reflected characteristics.

in which Γp is used for $\rho q^2 \tan \mu$ in (7.1.12) and

$$\Gamma = \frac{\gamma_e M^2}{\sqrt{M^2 - 1}}. \tag{7.2.2}$$

The differences appearing in (7.2.1) are small, and the change in Γp is neglected. The characteristic equations applied along the reflected characteristics give

$$p_2 - p_0 = \Gamma_0 p_0(\theta_2 - \theta_0), \tag{7.2.3a}$$

$$p_3 - p_1 = \Gamma_0 p_0(\theta_3 - \theta_1) + \delta(\Gamma p)(\theta_2 - \theta_0), \tag{7.2.3b}$$

in which the change in Γp between point 1 and point 0 is designated $\delta(\Gamma p)$. In the shock-expansion method (7.2.1) is not used but is replaced by an assumed relation.

If we now assume constant pressure along principal characteristics, we obtain

$$p_3 - p_2 = p_1 - p_0 = 0, \tag{7.2.4a}$$

$$\Gamma_0 p_0(\theta_3 - \theta_2) = \Gamma_0 p_0(\theta_1 - \theta_0) - \delta(\Gamma p)(\theta_2 - \theta_0). \tag{7.2.4b}$$

If, on the other hand, we assume constant flow deflection, we obtain

$$p_3 - p_2 = p_1 - p_0 + \delta(\Gamma p)(\theta_2 - \theta_0), \tag{7.2.5a}$$

$$\Gamma_0 p_0(\theta_3 - \theta_2) = \Gamma_0 p_0(\theta_1 - \theta_0) = 0. \tag{7.2.5b}$$

By using both sets of characteristic equations (7.2.1) and (7.2.3) we can obtain without further assumption the results

$$p_3 - p_2 = p_1 - p_0 + \tfrac{1}{2}\,\delta(\Gamma p)(\theta_2 - \theta_0), \tag{7.2.6a}$$

$$\Gamma_0 p_0(\theta_3 - \theta_2) = \Gamma_0 p_0(\theta_1 - \theta_0) - \tfrac{1}{2}\,\delta(\Gamma p)(\theta_2 - \theta_0). \tag{7.2.6b}$$

If we now assume that the strength of the reflected wave is zero along the first principal characteristic and that $p_1 = p_0$ and $\theta_1 = \theta_0$ we see that the complete result may be obtained by taking the arithmetic mean of the constant pressure result and the constant deflection result. If we had assumed that both p and θ were constant along the principal characteristic preceding (0–1), so that $p_1 - p_0$ and $\theta_1 - \theta_0$ could be obtained by the averaging process, the same would be again true of $p_3 - p_2$ and $\theta_3 - \theta_2$. Thus the argument may be continued by induction along the reflected characteristic.

This argument is based on the assumption that we may take p and θ both constant along a first principal characteristic (negligible reflection from the shock), and on the assumption that the strength of the reflected waves remains small compared with that of the principal waves (negligible reflection from the vorticity). These assumptions are examined below. In any case the choice of the arithmetic mean of the constant pressure procedure and the constant deflection procedure is not irrational and removes the arbitrariness of the method.

From the description of the shock-expansion method we see that it depends upon two basic assumptions: First, the reflections of the Mach waves from the curved shock are weak, and second, the reflections of the Mach waves from the streamlines are also weak.

Let us consider first the problem of determining the strength of the Mach waves reflected from the shock. The problem of the ratio of shock curvature to body curvature at the nose of a two-dimensional ogive is a closely related problem, treated by Crocco [1] and others (see references in Kraus [1]). Munk and Prim [1] carried out calculations of the ratio of the exact surface pressure gradient to that obtained by the use of shock-expansion over a range of free stream Mach numbers and leading edge deflection angles in a perfect gas.

The calculation for the strength of a wave reflected from a shock front in terms of the incident wave has been carried out by Lighthill [1], Chu [1], Eggers and Syvertson [1], Eggers, Savin, and Syvertson [1], and Waldman and Probstein [1].

To compute the value of this reflection coefficient, let us consider the incidence of an expansion wave on an oblique shock (see Fig. 7–8). If we denote the change in pressure across the incident wave as δp_1 and that across the reflected wave as δp_2, then from the characteristic relations for a simple wave flow we have

$$\delta p_1 = \rho_s q_s^2 \tan \mu_s \, \delta\theta_1 , \tag{7.2.7}$$

$$\delta p_1 + \delta p_2 = \rho_s q_s^2 \tan \mu_s \, (\delta\theta_1 - \delta\theta_2). \tag{7.2.8}$$

Here we neglect the change in the reference quantity $\rho_s q_s^2 \tan \mu_s$, as this change is itself of the order of $\delta\theta_1$. On the other hand, from the oblique shock relations

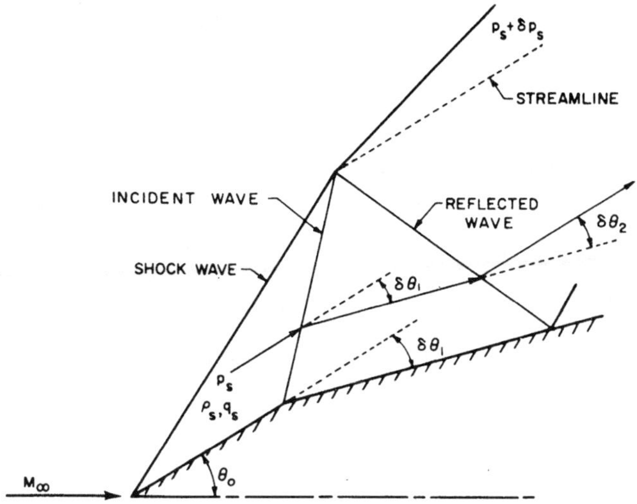

FIG. 7–8. Wave reflection from a shock wave.

we may formally write the change in pressure associated with a small change in deflection angle as

$$\delta p_s = \rho_s q_s^2 \tan \varphi \, \delta\theta_s = \rho_s q_s^2 \tan \varphi \, (\delta\theta_1 + \delta\theta_2), \tag{7.2.9}$$

where $\tan \varphi$ is some function of the free stream conditions and the shock strength. Since the pressure in the region separated from the shock by the slipstream must equal the pressure directly behind the shock, we find on equating (7.2.8) and (7.2.9) that

$$\delta\theta_2 = \left[\frac{\tan \mu_s - \tan \varphi}{\tan \mu_s + \tan \varphi}\right] \delta\theta_1 \, . \tag{7.2.10}$$

If a reflection coefficient $\mathfrak{R}_s$ is now defined as the change in pressure across the reflected wave divided by the change in pressure across the incident wave, we obtain

$$\mathfrak{R}_s = \frac{\tan \varphi - \tan \mu_s}{\tan \varphi + \tan \mu_s} \, . \tag{7.2.11}$$

This is the form given for this quantity by both Lighthill and Chu.

In principle, one can now calculate the change in the pressure behind the shock associated with a small change in deflection, and in this manner determine $\tan \varphi$. By way of example, we consider the special case in which we may make the assumption that ϵ is constant along the shock, and that the shock inclination angle σ is small. These assumptions are the same as those underlying the similar solutions of Section 2.7. From the oblique shock relation (1.4.5b) the pressure jump across the shock is

$$p_s - p_\infty = \rho_\infty U^2 \sin^2 \sigma \,(1 - \epsilon). \tag{7.2.12}$$

From (1.5.1) and (1.5.2a) we obtain

$$q_s^2 = U^2(\cos^2 \sigma + \epsilon^2 \sin^2 \sigma) \approx U^2. \tag{7.2.13}$$

Differentiation of (7.2.12) gives the result

$$\frac{\delta p_s}{\rho_s q_s^2} = 2\epsilon(1 - \epsilon)\sigma \,\delta\sigma, \tag{7.2.14}$$

in which $\sin \sigma$ has been replaced by σ and $\cos \sigma$ by one. We now only need to differentiate the relation

$$\theta_s = \sigma(1 - \epsilon) \tag{7.2.15}$$

obtainable from (4.1.2), in order to obtain $\delta\sigma$ as a function of $\delta\theta_s$. Then, from the definition of $\tan \varphi$ in (7.2.9) we obtain immediately

$$\tan \varphi = \frac{2\epsilon}{1 - \epsilon}\,\theta_s. \tag{7.2.16}$$

The local Mach angle behind the shock can be found by employing the formula $a_s^2 = \gamma_s p_s/\rho_s$ from (1.4.22), with the effective γ_s taking either a frozen or equilibrium value. Thus, with the strong shock assumption ($p_\infty = 0$) and (7.2.12) we find

$$M_s^2 = \frac{1 - \epsilon}{\gamma_s \epsilon \theta_s^2}. \tag{7.2.17}$$

Since $\tan \mu_s \approx 1/M_s$, from (7.2.16) and (7.2.17) the result for this limiting value of the reflection coefficient is

$$\Re_s = \frac{2\sqrt{\epsilon} - \sqrt{\gamma_s(1 - \epsilon)}}{2\sqrt{\epsilon} + \sqrt{\gamma_s(1 - \epsilon)}} \tag{7.2.18}$$

(cf. Chernyi [3, p. 189]).

One of the most interesting features of this result is that the limiting value of the reflection coefficient is independent of the deflection angle and dependent only on the state of the gas behind the shock. In fact, when expressed in terms of $\Re_s$, the calculations of Eggers, Syvertson, and Kraus [1] (see Waldman and

Probstein [1]) show this conclusion to be quite generally accurate, practically up to the detachment point angle. Furthermore, we see from (7.2.18) that as $\epsilon \to 0$, the reflection coefficient approaches -1. In this extreme case an expansion wave will reflect from the shock as a compression wave of undiminished strength.

In order to give some idea of the magnitude of the reflection, we have presented in Table 7–1, for different values of γ, the limiting small-disturbance value of the reflection coefficient in a perfect gas. Even in this extreme limiting case of $M_\infty \to \infty$, except for ϵ very close to zero, and except near the detachment point, the absolute value of the reflection coefficient is never large. This implies that the disturbances are absorbed to a great extent by the shock wave and are only weakly reflected.

TABLE 7–1

Reflection coefficient at shock for $M_\infty = \infty$ in a perfect gas according to small-disturbance theory

γ	1.4	1.3	1.2	1.1	1.05	1
$\mathfrak{R}_s$	-0.14	-0.19	-0.27	-0.40	-0.53	-1.0

As we have pointed out, the reflection coefficient is not the entire story, because of the interactions which take place between the waves from the airfoil surface and the vorticity or entropy layers in the flow. An estimate of the magnitude of this effect can be found by calculating the reflection of a simple wave from an idealized supersonic shear layer (see Fig. 7–9). By using the form of the characteristic relations as given by (7.2.7) and (7.2.8), with the pressure and flow deflection on both sides of the shear layer taken to be constant, one can show that the ratio of the pressure rise across the reflected wave to that across the incident wave is given by

$$\mathfrak{R}_v = \frac{\Gamma_2 - \Gamma_1}{\Gamma_2 + \Gamma_1} \tag{7.2.19}$$

where the quantity Γ is the same as the one defined by (7.2.2).

When the shear layer is very weak, $M_2 \to M_1$ and $\Gamma_2 \to \Gamma_1$, with the result that the strength of the reflected wave tends to zero. Let us consider, on the other hand, the case of a strong shear layer with the conditions $M_2 \gg M_1$ and M_1 not too close to unity $[(M_1^2 - 1)^{-1/2} \ll M_2]$. It follows in this case that $\Gamma_2 \gg \Gamma_1$ and $\mathfrak{R}_v \approx 1$. In such a case, the change in pressure across the reflected wave is almost equal to the change in pressure across the incident

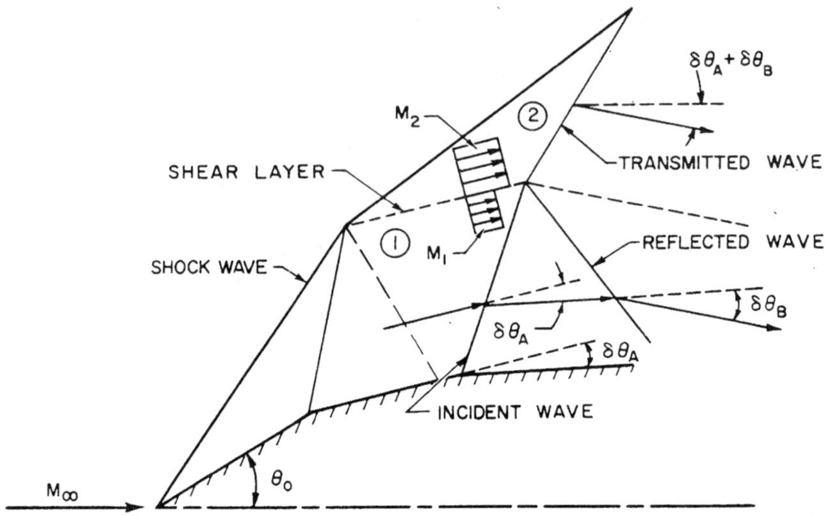

FIG. 7–9. Wave reflection from a shear layer.

wave, so that a wave reflects from the shear layer as if from a solid wall. Therefore, if such a shear layer were present in the flow field the shock-expansion method could not be applied. This extreme example points out that vorticity reflections may be equally important to or more important than the reflections from the shock.

In considering the applicability of the shock-expansion method, we must not be misled into treating the influence on surface pressure of either of the reflection processes which we have discussed as if it by itself necessarily characterized the accuracy of the method. An important feature of these processes is a tendency for them to cancel each other in their effect on the surface pressure. Thus, an expansion wave reflects from a shear layer as an expansion, but reflects from the shock as a compression. This annulment would tend to make the shock-expansion method more accurate for the surface pressure distribution than a consideration of either one of the two inherent sources of error would indicate. A similar conclusion has been reached by Mahony [1], and Mahony and Skeat [1] through a somewhat different approach. The conclusion that these reflections tend to cancel depends upon $\mathfrak{R}_s$ and $\mathfrak{R}_v$ having opposite signs, in particular on $\mathfrak{R}_s$ being negative and $\mathfrak{R}_v$ being positive. Changes of sign can appear in either quantity, such as near the detachment point and at low Mach numbers (see Waldman and Probstein [1]).

A factor which we have not yet discussed is the geometry of the streamlines and Mach waves. In Epstein's original studies for polygonal bodies the first outgoing wave starts at a shoulder of the body well aft of the leading edge, and its reflection from the shock may be assumed to miss the body completely at hypersonic velocities. In general, on a body with a wedge-shaped forebody shock-expansion theory should be either very accurate or should be exact in giving the pressure on the body. On the other hand, on a body with large curvatures very near the nose or with a blunted nose there will be a strong concentration of vorticity in the streamlines very near the body (the entropy wake discussed in Sections 1.2 and 2.7), and the shock-expansion method is subject to serious inaccuracy. Thus the shape of the forebody is extremely important, particularly at hypersonic speeds.

Stocker [1] has made a careful analysis of shock-expansion theory for a perfect gas. He points out that considerable error may arise in a region if the body curvature is small compared with the body curvature at the nose, and emphasizes the role of the parameter γ. If $\gamma = 1.4$ shock-expansion theory can yield a good estimate, while at $\gamma = 1.2$ the accuracy is already appreciably reduced.

Figure 7–10(a) presents the results of calculations by Eggers, Syvertson, and Kraus [1] of the shock wave shape and surface pressure distribution by the characteristics method and by the generalized shock-expansion method in a perfect gas with $\gamma = 1.4$ and $M_\infty = \infty$. The body shown is a 10 per cent thick parabolic arc biconvex airfoil at zero incidence. In Fig. 7–10(b) surface pressure distribution calculations are shown for the same body in the same flow but with $\gamma = 1.05$. Although shock-expansion checks well with the characteristics result for $\gamma = 1.4$ and would be even better for a lower Mach number, its accuracy is seen to be considerably reduced for $\gamma = 1.05$ (see Eggers, Syvertson, and Kraus [1]).

Following the development of Mahony [1] and Mahony and Skeat [1], Meyer [2] formulated the generalized shock-expansion method analytically by using the stream function and flow deflection angle as independent variables. In principle, by this method the solution can be built up in a simple rectangular network in the plane of the independent variables. The determination of the shock shape in this approach is based on matching the flow deflection for the generalized simple wave with the flow deflection given by the shock equation. The result given for the shape of the shock is expressed in an integral form which can be evaluated by quadrature.

We shall now discuss some relatively simple procedures for improving the accuracy of shock-expansion theory in precisely those cases where we have noted the method is limited. To be of advantage the procedures must necessarily be simpler than the method of characteristics. Such approaches as are suggested are to some extent empirically based. Following these considerations we shall briefly discuss rational successive approximation schemes, which have also been devised for the purpose of improving upon the basic shock-expansion results.

We have seen that any modification of the shock-expansion method should

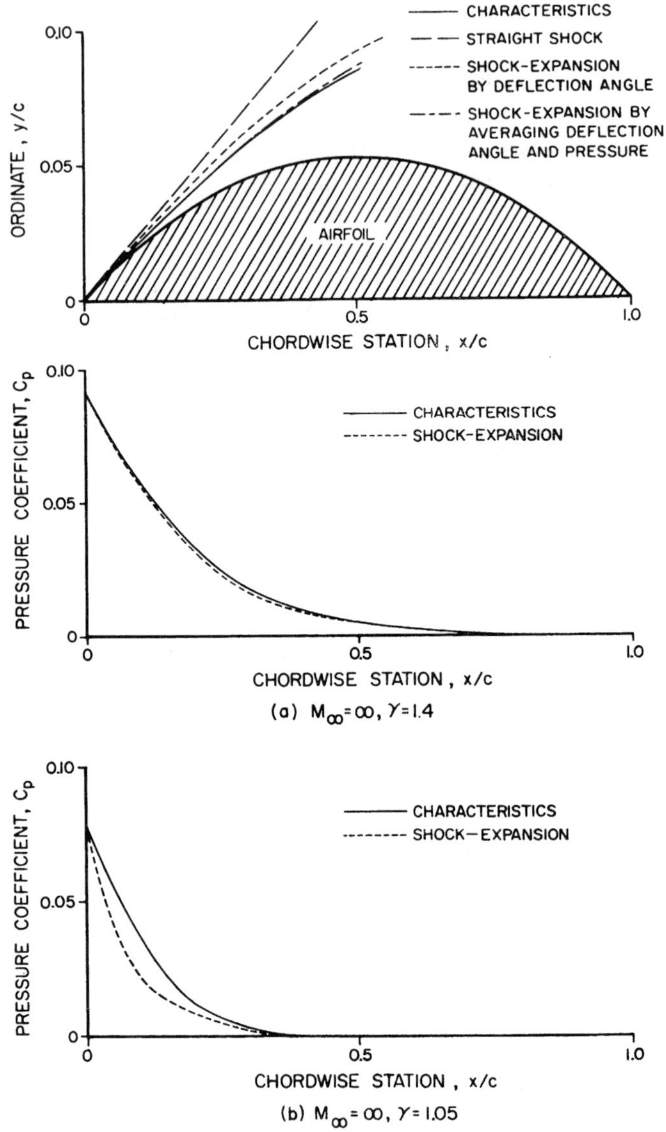

FIG. 7–10. Comparison of shock-expansion method and characteristics theory for a 10 per cent thick biconvex airfoil (Eggers, Syvertson, and Kraus [1]). (a) Pressure distribution and shock shape, $M_\infty = \infty$, $\gamma = 1.4$. (b) Pressure distribution, $M_\infty = \infty$, $\gamma = 1.05$.

suitably take into account the reflections from the shock and from the vorticity. We should not only have the strength of any reflection correct but also the location on the body of where the reflection returns to influence the pressure on the body. The proper expression for the pressure gradient on the body must include the curvature term $\Gamma_b p_b \, d\theta_b/ds$, where s is distance from the leading edge along the surface. However, it must also include an influence integral giving the total effect at the point of interest of reflection of waves emitted from the body upstream. The key element here is the delay as measured in distance downstream between the issuing of a wave from the body and the return to the body of reflections from that wave.

This problem may be attacked by several methods. One involves setting up the influence integrals directly. One method, explored by Chu [1] (see also, Chernyi [3]) involves expressing the body shape in a power series as a perturbation of a wedge flow. A third method, available if the delay is small, involves establishing a relaxation-type equation. Finally, if the delay may be considered negligible, we can neglect it and attempt to take multiple reflections into account. The delay may be expected to be small very near the leading edge of a sharp ogive. The delay may be expected to be smaller with a thinner shock layer, in general.

We first look at the problem of multiple reflections with delay neglected. The approach we present essentially follows that given in Scheuing *et al.* [1] and may also be found summarized in Scheuing [1]. To obtain a distributed reflection coefficient from the shock analogous to (7.2.11) we require a geometrical parameter, which measures the ratio of the spacing of the waves incident on the shock to those reflected from it. This parameter is defined as the ratio of the lengths l_1/l_2 identified in Fig. 7–11. The result for this parameter is

$$r = \frac{l_1}{l_2} = \frac{1 - \epsilon\sqrt{M_s^2 - 1}\,\tan\sigma}{1 + \epsilon\sqrt{M_s^2 - 1}\,\tan\sigma}, \tag{7.2.20}$$

where we have used (4.1.1) to replace $\tan(\sigma - \theta_s)$. This quantity is the same as the quantity $1/\Gamma$ of Chu [1] or the k of Chernyi [3, p. 183].

For the limiting case in which ϵ is constant along the shock and the shock inclination angle σ is small a reduction analogous to that used in obtaining (7.2.18) may be applied with the result (cf. Chernyi [3, p. 189])

$$r = \frac{\sqrt{\gamma_s(1-\epsilon)} - \sqrt{\epsilon}}{\sqrt{\gamma_s(1-\epsilon)} + \sqrt{\epsilon}}. \tag{7.2.21}$$

In the limit $\epsilon \to 0$, the reflection distance ratio approaches 1. To give an indication of the magnitude of the reflection distance ratio we have shown in Table

7–2 the limiting value of r according to (7.2.21) for different values of γ with $M_\infty \to \infty$ in a perfect gas. In this limiting case this quantity is seen to lie approximately between $\frac{1}{2}$ and 1.

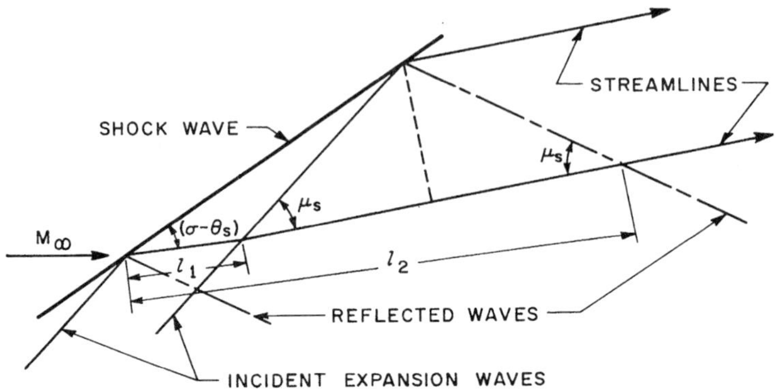

FIG. 7–11. Reflection distances for distributed reflections from a shock wave.

TABLE 7–2

Reflection distance ratio for $M_\infty = \infty$ in a perfect gas according to small-disturbance theory

γ	1.4	1.3	1.2	1.1	1.05	1
r	0.45	0.50	0.55	0.65	0.73	1.0

We now define a distributed reflection coefficient at the shock as the differential change in pressure across the reflected wave system divided by the differential pressure change across the incident wave system. This ratio is simply r times the single wave reflection coefficient $\Re_s$, and gives

$$\pi_- = r\Re_s\pi_+ , \tag{7.2.22}$$

where π_+ and π_- denote the strengths of the incident and reflected waves expressed in terms of pressure gradient. The boundary condition at the wall is

$$\pi_+ = \pi_- + \Gamma_b p_b \frac{d\theta_b}{ds} , \tag{7.2.23}$$

while the surface pressure gradient is by definition

$$\frac{dp_b}{ds} = \pi_+ + \pi_- . \tag{7.2.24}$$

Solving (7.2.22) to (7.2.24) we obtain, for constant curvature,

$$\frac{dp_b}{ds} = \left[\frac{1 + r\mathfrak{R}_s}{1 - r\mathfrak{R}_s}\right] \Gamma_b p_b \frac{d\theta_b}{ds} . \tag{7.2.25}$$

This relation determines the initial pressure gradient at the nose of a two-dimensional ogive as a function of its curvature. This result is the same as that obtained by retaining only the first term of the series obtained by Chu and by Chernyi from a perturbation of the wedge flow solution.

Although we have neglected the effect of the delay we must still take into account the multiple reflections arising from the distributed vorticity in the shock layer due to the shock curvature. This effect is not present in (7.2.25).

With the subscript 1 denoting the principal characteristic direction and 2 the corresponding reflected direction, the characteristic equations are written

$$\frac{dp}{ds_1} = -\Gamma p \frac{d\theta}{ds_1} , \tag{7.2.26a}$$

$$\frac{dp}{ds_2} = \Gamma p \frac{d\theta}{ds_2} . \tag{7.2.26b}$$

By cross differentiation we can obtain

$$\frac{d}{ds_1}\left(\frac{dp}{ds_2}\right) = \frac{1}{2}\frac{d\theta}{ds_2}\frac{d(\Gamma p)}{ds_1} - \frac{1}{2}\frac{d\theta}{ds_1}\frac{d(\Gamma p)}{ds_2} . \tag{7.2.27}$$

The two derivatives of Γp are of the same order of magnitude, while $d\theta/ds_1$ is much smaller than $d\theta/ds_2$ in the shock-expansion theory. Thus in the shock-expansion theory we neglect the second term on the right-hand side. The approximate relation resulting is a differential form of (7.2.6a). Inserting the resulting relation into (7.2.26b) and integrating along the principal characteristic from the body gives

$$\frac{dp/ds_2}{(dp/ds_2)_b} = \left[\frac{\Gamma p}{(\Gamma p)_b}\right]^{1/2} . \tag{7.2.28}$$

This expression relates the strength of the wave (dp/ds_2) at any position along s_1 to its strength at the body surface.

Referring to Fig. 7–12, we can now define an average reflection coefficient for the incident principal characteristic passing through the region of varying vorticity in direct analogy to the reflection coefficient defined for the wave reflection from a single shear layer. In this case we define the coefficient as the ratio of the total pressure rise across all the waves reflected from the vorticity

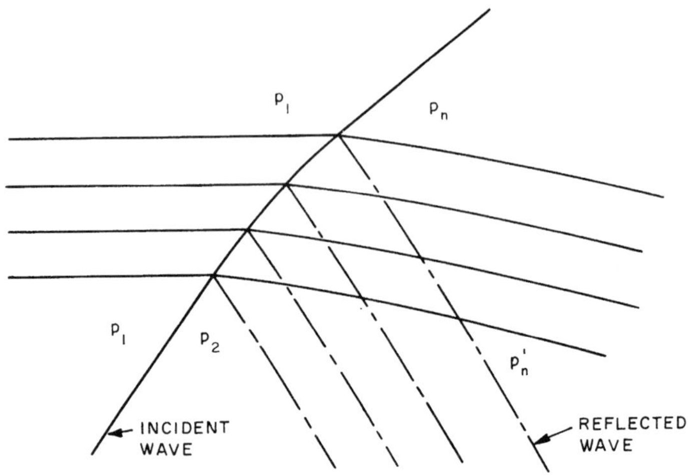

FIG. 7–12. Wave reflection from a region of vorticity.

region to the pressure rise at the surface across the initial incident wave. The coefficient is given by

$$\bar{\mathfrak{R}}_v = \frac{p'_n - p_2}{p_2 - p_1}. \tag{7.2.29}$$

With the assumption that the flow is a parallel flow of constant pressure which is perturbed by a single weak wave crossing an isolated region of vorticity, we can apply (7.2.19) to the global solution. If we do this, we are including with the transmitted wave the second, fourth, etc., reflections; with the reflected wave we are including the third, fifth, etc., reflections as well as the primary ones.

If we are interested only in the primary transmitted wave under these linearizing assumptions, we may apply (7.2.28) with the factor p dropped. We obtain

$$\frac{p_n - p_1}{p_2 - p_1} = \left(\frac{\Gamma_n}{\Gamma_1}\right)^{1/2}. \tag{7.2.30}$$

The reflected wave is modified by an analogous factor in passing from the point at which it is created to below the region of vorticity (in Fig. 7–12). Thus we write

$$dp'_n = \left(\frac{\Gamma_1}{\Gamma_n}\right)^{1/2} dp_n. \tag{7.2.31}$$

We may now integrate dp'_n to calculate $\bar{\mathfrak{R}}_v$, and obtain

$$\bar{\mathfrak{R}}_v = \tfrac{1}{2} \ln \frac{\Gamma_n}{\Gamma_1}. \tag{7.2.32}$$

This expression may be expanded in powers of $(\Gamma_n - \Gamma_1)/(\Gamma_n + \Gamma_1)$, to show that (7.2.19) and (7.2.32) agree to terms of second order in this variable. Note that it is third and higher order reflections which are neglected in obtaining (7.2.32).

We may thus consider the whole region of varying vorticity as a discontinuous shear layer. Whether we include only the primary transmitted and reflected waves or the entire system including repeated reflections (with their delays), the reflection process may be looked upon as a single reflection phenomena. If we in fact make this simplifying assumption, then it is a relatively simple matter to combine these effects of shock and vorticity reflection.

Figure 7–13 is a sketch of the wave geometry involving the combined re-

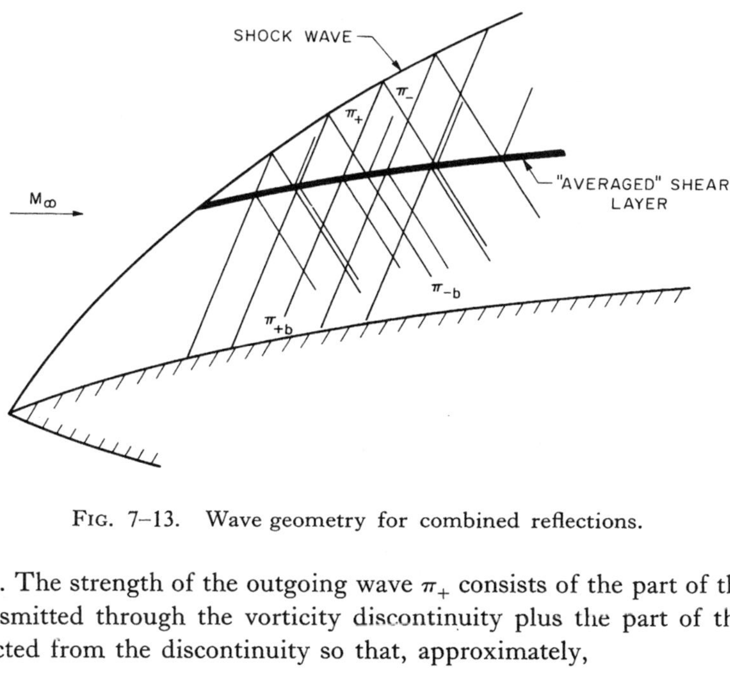

FIG. 7–13. Wave geometry for combined reflections.

flections. The strength of the outgoing wave π_+ consists of the part of the wave π_{+b} transmitted through the vorticity discontinuity plus the part of the wave π_- reflected from the discontinuity so that, approximately,

$$\pi_+ = \pi_{+b}(1 - \overline{\mathfrak{R}}_v) + \pi_-(-\overline{\mathfrak{R}}_v). \tag{7.2.33}$$

Similarly

$$\pi_{-b} = \pi_-[1 - (-\overline{\mathfrak{R}}_v)] + \pi_{+b}\overline{\mathfrak{R}}_v\,. \tag{7.2.34}$$

From the above relations and (7.2.22) we find the combined reflection coefficient $\mathfrak{R}$ to be

$$\mathfrak{R} = \frac{\pi_{-b}}{\pi_{+b}} = \frac{\overline{\mathfrak{R}}_v + r\mathfrak{R}_s}{1 + r\mathfrak{R}_s\overline{\mathfrak{R}}_v}\,. \tag{7.2.35}$$

In a manner analogous to the shock reflection calculation the surface pressure gradient is readily shown to be

$$\frac{dp_b}{ds} = \frac{1 + \mathfrak{R}}{1 - \mathfrak{R}} \Gamma_b p_b \frac{d\theta_b}{ds} . \tag{7.2.36}$$

This relation is an evident generalization of (7.2.25), and may be solved by a step-by-step numerical integration.

Calculations using (7.2.36) for the body and conditions of Fig. 7–10(b) have been carried out (Scheuing [1]). For simplicity $\mathfrak{R}$ and Γ_b were taken constant. The results indicate good agreement with the characteristics calculation up to 10 per cent of the chord, beyond which point the underlying approximations could not be expected to be valid. Other calculations reported by Scheuing *et al.* [1] using (7.2.36) show clearly the effect of cancellation of shock reflections by vorticity reflections, as mentioned earlier in the section. A numerical determination under hypersonic conditions of distributed shock reflection coefficients and average vorticity reflection coefficients for specific configurations has been given by Bird [1].

We next show how the delay effect associated with the issuing of a wave from the body and the return of reflections from that wave may be taken into account in a semiempirical manner if the effect is small. We consider an equation of the form

$$\frac{dp_b}{ds} = \Gamma_b p_b \frac{d\theta_b}{ds} + \frac{2\mathfrak{R}}{1 - \mathfrak{R}} \Gamma_b p_b \frac{d\theta_b}{ds} . \tag{7.2.37}$$

We recognize this equation as equivalent to (7.2.36). The integral from the leading edge is assumed known, and of the form $p = p_{\text{asymp}}(s)$, with the subscript "asymp" denoting asymptotic. We now consider that the argument of the second term in the right-hand side is $s - \delta s$ instead of s. The quantity δs is an estimate of the average delay. We replace the second term by its value at s less δs times its derivative with respect to s. Treating δs as a constant, we can integrate to obtain

$$p_b - p_{\text{asymp}} = -\delta s \frac{2\mathfrak{R}}{1 - \mathfrak{R}} \Gamma_b p_b \frac{d\theta_b}{ds} . \tag{7.2.38}$$

A combination of the two equations above gives

$$\frac{dp_b}{ds} - \Gamma_b p_b \frac{d\theta_b}{ds} = -\frac{1}{\delta s} (p_b - p_{\text{asymp}}). \tag{7.2.39}$$

This equation, with δs appropriately estimated, serves to estimate the effect of the reflection delay. It is of classical relaxation form. A comparison of this approach with characteristics calculations would be worthwhile.

A different group of problems appears farther downstream, where the reflections from the shock and the bulk of the vorticity have a delay which is large enough to make the reflections negligible. The foremost of these comes from the fact that the pressure on the body can change by an order of magnitude, and no procedure based upon linearization may be applied. Here appears a prime virtue of the shock-expansion theory. The quantity $\Gamma_b p_b$ is a function of p_b alone, and the shock-expansion (or Prandtl-Meyer) result $\theta_b = \int dp_b/\Gamma_b p_b$ applies for large changes in θ_b and p_b. One problem in extending the shock-expansion theory in various directions is the problem of preserving this nonlinear property.

In the case of a thin entropy wake next to the body, a theory of Sullivan, Donaldson, and Hayes [1] is available. The theory is linear, and predicts the pressure changes on a plate associated with a sudden change in its angle. With an entropy profile that decreases monotonically from the body to a free stream value outside the entropy wake the solution is dominated by an exponential decay to the Prandtl-Meyer solution based upon exterior flow conditions.

This result suggests again a semiempirical relaxation equation of the form

$$\frac{dp_b}{ds} = (p_b\Gamma_b)_0 \frac{d\theta_b}{ds} + \kappa(p_{\text{ext}} - p_b), \tag{7.2.40}$$

where $(p_b\Gamma_b)_0$ and κ are slowly varying and functions of p_b. The theory yields values for $(p_b\Gamma_b)_0$ and κ, and offers the approximation

$$\kappa^{-1} = \Gamma_{\text{ext}} \int \frac{\sqrt{M^2 - 1}\, dy}{\Gamma} . \tag{7.2.41}$$

In a semiempirical approach, it is simpler to take $\Gamma_b p_n$ in place of $(\Gamma_b p_b)_0$. What κ should then be is a somewhat open question, and perhaps should be determined on an empirical basis rather than by an estimate such as (7.2.41). The virtue of not being restricted to almost linear problems is present in this method.

This theory and the closely related relaxation equation have been applied by Donaldson, Sullivan, and Hayes [1], to a number of hypersonic aerodynamic problems and good agreement with experiment was indicated.

The approaches so far suggested for the improvement of the shock-expansion method cannot be considered completely rational. We shall therefore briefly discuss those procedures which have been devised for systematically improving the results of shock-expansion theory by successive approximation. Such approaches are limited practically by the fact that they often become as lengthy as the method of characteristics. Nevertheless, they offer a mathematically consistent treatment of the improvement and at the same time are capable of providing general analytic results and an indication as to when the simple approximate scheme breaks down.

Mahony [1], for example, seeks to improve on and investigate the accuracy of the shock-expansion method in a perfect gas. His method is to solve the unsteady one-dimensional piston problem through an iteration technique starting with the isentropic shock-expansion solution. He carries his results over to steady two-dimensional flow by using the equivalence principle described in Section 2.1. Recognizing the dominant role played by the principal characteristics in contrast to the secondary role of the reflected characteristics, he introduces two new independent variables. One is a parameter which is defined so that it is constant along reflected characteristics, and which serves as a coordinate. The other is the characteristic parameter for the principal characteristics, which would be constant along principal characteristics only if the flow were isentropic. Using these variables he establishes an approximate basic solution including a curved shock and straight principal characteristics. He then applies an iterative procedure in which the independent variables are held fixed, but an entropy distribution appears at the shock and is made to propagate along the streamlines. This then permits the corrected pressure distribution at the airfoil to be obtained by integrating back along the reflected characteristics from the shock. His results show, as we have already seen, that for $\gamma = 7/5$ and $\gamma = 5/3$ the predictions of shock-expansion theory are very accurate in the hypersonic similarity range. Mahony makes the observation that this is partly due to a tendency of the inherent errors to cancel, an observation which we made previously on the basis of general considerations of the wave reflection process. He also finds, in agreement with our previous discussion, that the pressure distribution predictions for $\gamma = 1$ are in great error, and that the shock-expansion method is completely unsatisfactory with γ near one.

One of the difficulties in the analysis of Mahony is that it applies only to the region of hypersonic similarity and hence to slender bodies in hypersonic flow. Mahony and Skeat [1] and Waldman and Probstein [1] extended the approach of Mahony to general supersonic two-dimensional flows in a perfect gas, without the restrictions inherent in using the equivalence principle. The analysis of Mahony and Skeat followed the original analysis of Mahony very closely. Waldman and Probstein obtain subsequent approximations beyond the lowest isentropic one by expanding in powers of the shock reflection coefficient $\mathfrak{R}_s$. Through this technique they find a general expression to second order for the pressure distribution on two-dimensional airfoils in rotational flow, valid for relatively thick bodies. Evidently, these techniques can always be continued to any order desired, although beyond a first correction to shock-expansion the algebra becomes rather extensive. The results of such second-order calculations, however, are found to be in excellent agreement with characteristic calculations.

Guiraud [5] (cf. Guiraud [1]) has also suggested an iterative method for improving upon shock-expansion theory. The basic independent variables used are essentially the stream function and distance along the outgoing characteristic. A general solution is expressed through a Volterra integral equation in which the normally known functions are expressed in terms of a parameter which is essentially the distributed shock reflection coefficient $r\mathfrak{R}_s$. The reflection coefficient depends, however, upon the solution. A formal iterative scheme starting with shock-expansion is indicated. Guiraud [1] notes that the principal characteristics must make a relatively small angle with the shock wave if shock-expansion is to be valid.

A successive approximation scheme closely related to those mentioned, though not iterating upon shock-expansion, is due to Kogan [1–3]. In his solution Kogan takes as the zeroth-order approximation the flow behind a plane shock wave. He compares [3] his results with the shock-expansion method and with iterations upon the method.

In the light of the practical success of the shock-expansion method in treating two-dimensional hypersonic flows past smooth bodies whose curvature is only slowly varying and with the isentropic exponent γ not too close to one, it is reasonable to inquire whether such an approach could be adopted in three-dimensional problems, even though there exists no simple general solution analogous to the Prandtl-Meyer flow. With the basic assumption of shock-expansion theory that the disturbances reflected from the leading edge shock can be neglected, Sheppard and Thomson [1; 2] (see also Vincenti and Fisher [1]) have shown how it is possible to generalize the method to a three-dimensional wing whose surface is a developable ruled surface, provided the generators of the surface are supersonic. A ruled surface is one which can be considered as the locus of a family of straight lines; the straight lines are termed the rulings or generators of the surface. A developable surface is a ruled surface whose generators have an envelope which is a curved line in space, and has the property of being developable into a plane. By supersonic generators is meant that the local velocity component normal to each generator is supersonic. An example of such a wing is shown in Fig. 7–14, where we have also indicated how the actual cross section is replaced by a polygonal one corresponding to the division of the wing into facets bounded by the rulings.

At a facet junction it is clear that the flow at the surface normal to the generator defining the junction is locally cylindrical, provided we are outside the zones of action of the apex and tip. The velocity component parallel to the generator remains constant, while the remaining velocity components are the same as those in a two-dimensional flow. The flow across the junction behind the leading edge shock is therefore a Prandtl-Meyer expansion of the normal component of the flow. Similarly, if we neglect the reflected characteristics from the shock wave and vorticity layers then the flow at the surface of the facet also behaves locally like a two-dimensional isentropic flow, and the concepts of shock-expansion theory apply directly. The flow over each facet is assumed to be uniform. At the trailing edge of a facet the flow is considered to be decomposed

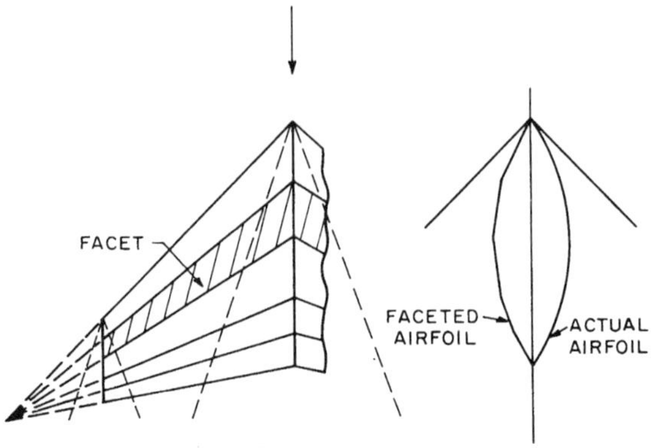

FIG. 7–14. Ruled surface wing.

into normal and parallel components for the determination of the flow conditions over the next facet.

The simple result which appears in two-dimensional shock-expansion theory, $p = f(\theta_{le}, \theta)$ with the subscript "le" denoting the leading edge, does not hold here. The pressure is obtained by a quadrature which depends upon the orientation of the generators across which each angle change $d\theta$ takes place.

This approach does not apply within the zones of action of the apex and tip of a wing such as that shown in Fig. 7–14. These regions can be handled by an approach which will be discussed later in this section.

The rather limited application of shock-expansion theory to the three-dimensional configurations just described is perhaps its most logical extension. The limitations involved are of a quite different nature from those inherent in strip theory. However, it is also clear that the shock-expansion method can be extended to three-dimensional flows whenever the concepts of strip theory apply (cf. Sections 1.7, 2.4, and 2.9). In this case the flow may be considered to be approximately two-dimensional locally.

On the basis of the above discussion we may argue that for values of the hypersonic similarity parameter $K = M_\infty \tau > 1$ the pressure distribution can be calculated approximately on bodies of revolution by assuming a conical shock wave at the nose followed by a two-dimensional Prandtl-Meyer expansion. We can justify such an approach by considering the behavior of the axisymmetric term in the characteristic equations (7.1.12). This term can be expressed as $r^{-1} \sin \theta \sin \mu \, ds_{1,2}$, where $ds_{1,2}$ are elements of length along the first and second family characteristics respectively. Since $\sin \mu = 1/M$, we can write the coefficient of $ds_{1,2}$ for small deflections as $\theta^2/r(M\theta)$. We see then that the axisymmetric term will indeed be small away from the axis in comparison with the other terms in the characteristic equations if θ is small and $M\theta$ is of the order of one or greater. With θ fixed, the larger the value of the local similarity parameter $M\theta$, the better will be the approximation of neglecting the axisymmetric contribution.

Following Eggers and Savin [1; 2] (see also Eggers, Savin, and Syvertson [1]), these considerations can be extended to general three-dimensional flows past slender bodies by noting that at high local Mach numbers the shock layer is quite thin, so that we may confine our attention to the neighborhood of the surface. Such an assumption is the basic one employed in the strip theory similitude of Section 2.4, for example.

Justifications of the application of the shock-expansion method to axisymmetric and other three-dimensional flows have appeared in the references cited. If the flow inclination angles are small and the local Mach numbers are large a justification is afforded by the arguments of Section 2.4. In axisymmetric flow where these conditions are not met a correction (described below) may be used to reduce the errors inherent in such a direct application. For other three-dimensional bodies such as yawed bodies of revolution the justifications are somewhat clouded and we should consider the application of the direct shock-expansion method, to be discussed further below, as primarily empirically based.

In Fig. 7–15 taken from Eggers, Savin, and Syvertson [1] the surface pressure distributions computed by shock-expansion for an unyawed ogive of fineness ratio 3 are compared with results obtained by the method of characteristics and experiment. The good agreement for values of the hypersonic similarity parameter K greater than one is clearly evident, although for K of the order of one the comparison is somewhat less favorable.

We have seen that the application of the shock-expansion method to the three-dimensional case requires the use of two-dimensional techniques in a three-dimensional flow, so that the continuity equation is only approximately satisfied. Syvertson and Dennis [1] have attempted to increase the accuracy of the method for the axisymmetric case by adjusting the streamtube thickness to satisfy the continuity requirement.

In the approach of Syvertson and Dennis, their correction to the shock-expansion method for axisymmetric flow is based on representing the body of arbitrary contour by a series of tangents. The surface pressure distribution around the corners of the tangent-body is then found by the Prandtl-Meyer relation. The problem is to determine the change in surface pressure along the straight line elements, for which no pressure change is predicted by shock-expansion.

On the basis of a quasi-rational argument based on continuity the authors conclude that the pressure on such a straight line element should have an approximate behavior given by

$$p = b + ce^{-as}, \tag{7.2.42}$$

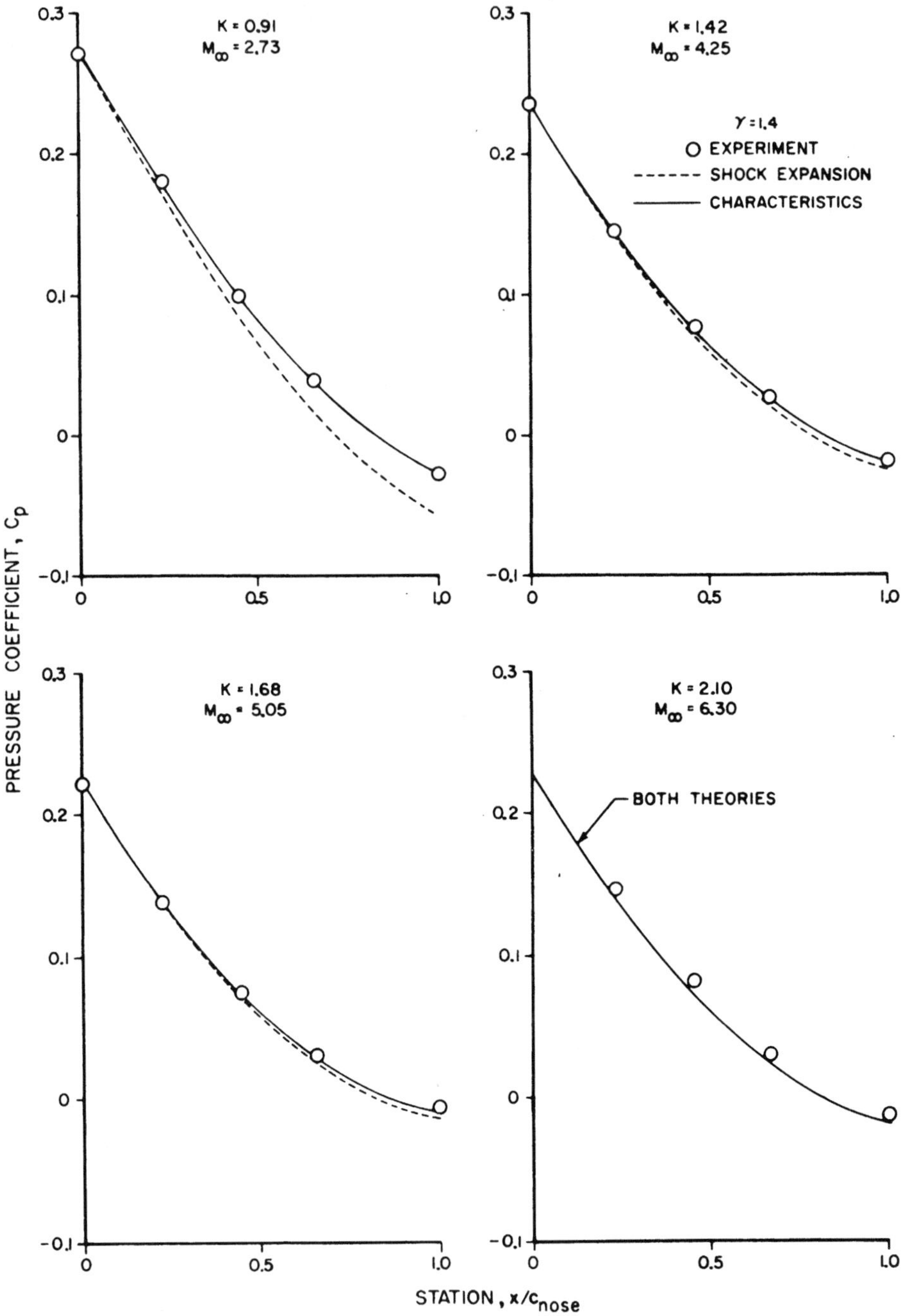

FIG. 7–15. Pressure distributions on an ogive of fineness ratio 3 at zero incidence (Eggers, Savin, and Syvertson [1]).

where s is distance along the element taken from the corner and a, b, and c are constants. This relation is seen to be of the relaxation form discussed earlier in connection with the two-dimensional delay problem. The pressure immediately behind the corner is $b + c$ and is obtained from the local Prandtl-Meyer expansion. The pressure gradient immediately behind the corner $\partial p/\partial s$ is needed, and gives the quantity ac. If the line element were extended to infinity the pressure should approach asymptotically the value on a cone of the same inclination angle, and this fact is used as a condition to evaluate b. It is difficult to see how the surface pressure of (7.2.42) could be far wrong, since it has the correct initial value and initial derivative and approaches a correct limit. However, we must consider this result as primarily empirically based.

It should be noted at this point that the effects of axial symmetry could be represented by the interaction of the waves generated at the surface with an area change. With this model a reflection coefficient could be defined, and a procedure analogous to that used in the two-dimensional case could be set up to take into account both reflections and a delay process. In this regard, the relaxation form arrived at in (7.2.42) is of interest. Bird [1] has defined an appropriate axisymmetric reflection coefficient and numerically determined its value for specific configurations.

The evaluation of the pressure gradient on a straight line element immediately after a Prandtl-Meyer expansion is needed in the evaluation of the constants in (7.2.42). Determinations of the exact value of this pressure gradient for a perfect gas have been given by Johannsen and Meyer [1] and Ferrari [1], with numerical calculations carried out for $\gamma = 1.4$. An approximate evaluation of this pressure gradient has been made by Syvertson and Dennis [1], who calculated the streamtube areas by a one-dimensional analysis, and obtained a simplified analytical result. We refer the reader to the original references for detailed results.

It is evident that an approximate distribution for the surface pressure along the tangent-body can be determined by a step-by-step process using (7.2.42). Calculations of Syvertson and Dennis which are shown in Fig. 7–16 illustrate for $K = 1$ the improved accuracy which can be obtained for unyawed bodies of revolution with this correction. Other numerical results by these same authors indicate equally good comparisons for K as low as 0.4, while for $K > 1.5$ their calculations show that the correction becomes small, and the results of this method approach those of the generalized shock-expansion method, as they should. Syvertson and Dennis have extended this technique to the evaluation of normal-force derivatives and to the location of centers of pressure at zero angle of attack.

It is clear from our discussion of the method of characteristics in Section 7.1 that any generalization of the shock-expansion method to three dimensions should impose a requirement on the bicharacteristics, their number and choice.

We recall that for the complete method of characteristics we required at least three bicharacteristic equations, or two bicharacteristic equations and an additional equation obtained from the equations of motion. We may expect, therefore, that a complete extension to three dimensions would involve conditions on two instead of three bicharacteristics. Eggers and Savin [1; 2] reduce this requirement

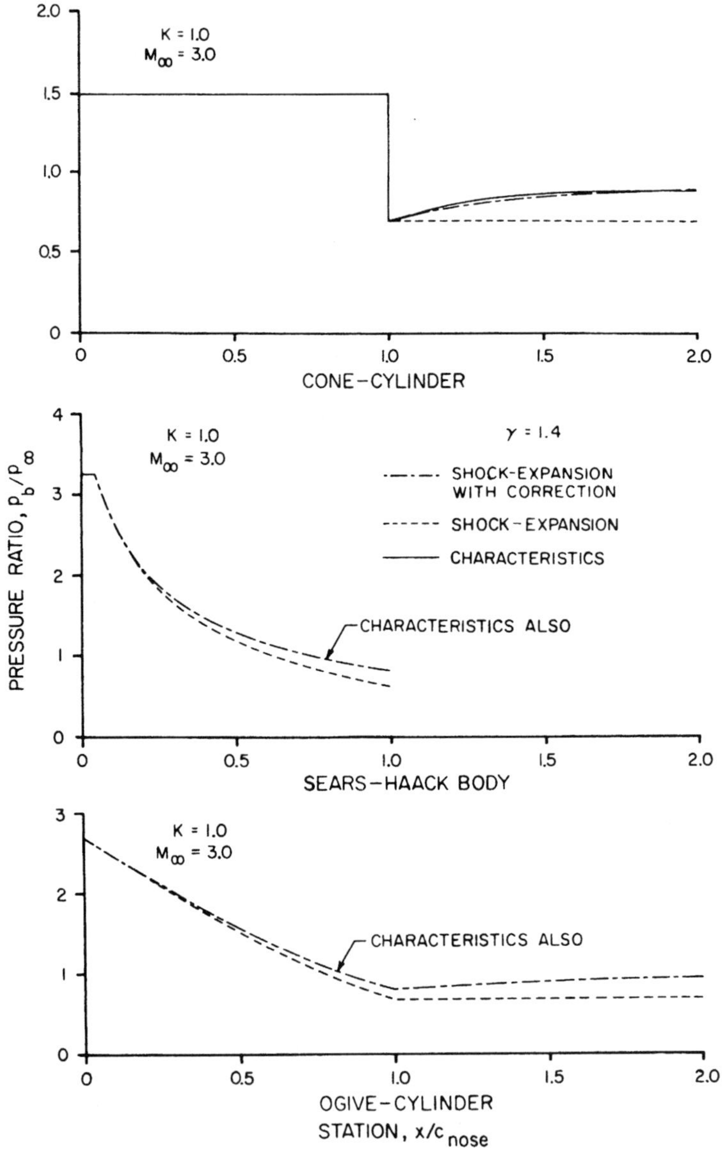

FIG. 7–16. Pressure distributions on axisymmetric bodies (Syvertson and Dennis [1]).

again by one, by assuming the streamlines are geodesics of the surface and known in advance. As we have already noted, the last term in the bicharacteristic equation (7.1.49) represents the apparent divergence of the streamlines. This is the term dropped by Eggers and Savin, thereby implicitly introducing a strip theory requirement. With $\mathbf{n}_i$ in (7.1.49) taken as the principal normal to a streamline, then $\mathbf{n}_i \cdot d\mathbf{t}$ is simply the principal curvature of the streamline times the change in distance along the streamline. The problem is thus readily solved once the surface geodesics are determined.

One method is available for three-dimensional bodies, which appears to overcome quite successfully the shortcomings of the simpler approach which assumes geodesic streamlines. This method was proposed and applied by Sheppard and Thomson [2], and in different form by Mauger [2]. The method may be considered to be based upon a physical conception of the nature of the flow near the surface.

We conceive of the wave system as comprising two systems of waves. The first system is generated by the body shape in a manner similar to that with the ruled developable body. Reflections from this system are neglected, this assumption corresponding to the standard shock-expansion assumption. The second system of waves consists of essentially two-dimensional waves of both characteristic families propagated tangent to the body. This conception is essentially that of Sheppard and Thomson [2].

To express the method in more analytical terms, we use the three-dimensional intrinsic coordinates of Section 7.1, and choose $\mathbf{n}_1$ normal to the body (outward) and $\mathbf{n}_2$ tangential to the body and normal to the surface streamline. The pressure equation (7.1.44) takes the form

$$\frac{\partial p}{\partial s} + \frac{\rho q^2}{M^2 - 1}\left(\mathbf{n}_1 \cdot \frac{\partial \mathbf{t}}{\partial n_1} + \mathbf{n}_2 \cdot \frac{\partial \mathbf{t}}{\partial n_2}\right) = 0. \tag{7.2.43}$$

The two-dimensional wave system tangential to the body affects dp and $\mathbf{n}_2 \cdot d\mathbf{t}$ but does not affect $\mathbf{n}_1 \cdot d\mathbf{t}$. The shock-expansion assumption is then chosen to be that $\mathbf{n}_1 \cdot d\mathbf{t} = 0$ along bicharacteristics corresponding to $\mathbf{n}_1$, or that

$$\mathbf{n}_1 \cdot \left(\frac{\partial \mathbf{t}}{\partial s} + \frac{1}{\sqrt{M^2 - 1}} \frac{\partial \mathbf{t}}{\partial n_1}\right) = 0. \tag{7.2.44}$$

The substitution of (7.2.44) into (7.2.43) leads to

$$\frac{\partial p}{\partial s} + \frac{\rho q^2}{M^2 - 1} \mathbf{n}_2 \cdot \frac{\partial \mathbf{t}}{\partial n_2} = \frac{\rho q^2}{\sqrt{M^2 - 1}} \mathbf{n}_1 \cdot \frac{\partial \mathbf{t}}{\partial s}. \tag{7.2.45}$$

This equation is essentially the basic equation of Mauger [2]. The geometric term $\mathbf{n}_1 \cdot \partial \mathbf{t}/\partial s$ is the component of the streamline curvature normal to the surface, equal to $-\mathbf{t} \cdot \mathfrak{K} \cdot \mathbf{t}$ in terms of the curvature tensor defined in Section 3.5.

The left-hand side of (7.2.45) is the same as that of (7.1.8), apart from the factor $\rho q^2/(M^2 - 1)$. The characteristics equations for the second system are then the same as those of (7.1.10), with two essential differences. One difference is, of course, that the third term of (7.1.10) is replaced by a term coming from the right-hand side of (7.2.45). The second difference is that the calculation is to be carried out in the surface of the body and not in a two-dimensional cartesian space. For this reason the form (7.1.12) is not applicable.

The weaknesses and dangers in the method are obvious from the assumptions made. Both Sheppard and Thomson [2] and Mauger [2] report encouraging results from calculations using the method, and it clearly offers a significant improvement over the simple method proposed earlier by Eggers and Savin. A critical and rational analysis of the method with a view to expressing satisfactory conditions for validity would be highly desirable.

3. Tangent-wedge and tangent-cone

Two other approximate inviscid methods for obtaining surface pressure distributions on moderately slender bodies at hypersonic speeds are the "tangent-wedge" approximation for two-dimensional bodies and the "tangent-cone" approximation for bodies of revolution. Although these empirical methods are extremely simple to apply, they give no information on the structure of the shock layer and they neglect centrifugal effects as usually applied. Yet because of the simple relation they afford between surface pressure and local streamline inclination, we feel that some discussion of them is warranted, even though they must be considered as empirically based in the range in which they are usually used.

In the tangent-wedge approximation for two-dimensional flow the surface pressure at any point on a body at an arbitrary angle of attack is taken to be equal to the pressure on a wedge whose half-angle equals the local inclination angle of the body with respect to the flight direction. Physically the approximation relies on the fact that at hypersonic speeds the shock layer is sufficiently thin that there is little change in either flow inclination or pressure along a normal to the airfoil. Hence the surface values are approximately the same as those at the shock.

The extreme simplicity of the tangent-wedge approximation lies in the fact that a single function of body angle is used to estimate the pressure. With such simplicity it is evident that in general the tangent-wedge approximation should be less accurate than the result of the shock-expansion method. However, for ϵ very small it may be better. We shall present the pressure law given by the tangent-wedge approximation in some special cases and shall present a method for improving it if ϵ is very small.

In the special case of a slender body in a hypersonic flow of a perfect gas with constant γ we have available the result of Linnell [1] given in (2.5.2). It is convenient to rewrite (2.5.2) in a different form, in which $K = M_\infty \theta_b$ is used to replace θ_b. In this form the result for the pressure may be expressed

$$\frac{p_b}{p_\infty} = 1 + \gamma K^2 \left[\sqrt{\left(\frac{\gamma+1}{4}\right)^2 + \frac{1}{K^2}} + \frac{\gamma+1}{4} \right]. \tag{7.3.1}$$

For K small this result may be rewritten as a power series in K,

$$\frac{p_b}{p_\infty} = 1 + \gamma K + \frac{\gamma(\gamma+1)}{4} K^2 + \frac{\gamma(\gamma+1)^2}{32} K^3 + O(K^5), \tag{7.3.2}$$

which represents the hypersonic limit of the well known Busemann expansion for the pressure behind an oblique shock (cf. (2.10.1) for an isentropic flow).

Another special case is that of a gas which is a perfect gas behind the shock of fixed $\gamma_e = \gamma_s$, with $\epsilon_{\lim} = (\gamma_s - 1)/(\gamma_s + 1)$, but for which $\gamma_s \neq \gamma$. Here we assume not only that the body is a slender one in a hypersonic flow but that K is large. We take the approximation afforded by (1.4.14) and (1.5.4), and obtain

$$\frac{\sigma - \theta_b}{\sigma} = \epsilon = \epsilon_{\lim} \left[1 + \frac{\epsilon_\infty^{-1} - \epsilon_{\lim}}{1 - \epsilon_{\lim}} \frac{1}{\gamma M_\infty^2 \sigma^2} \right]. \tag{7.3.3}$$

Solving this equation for σ/θ_b yields for p_b/p_∞, from (1.4.12) or (4.1.13), the result

$$\frac{p_b}{p_\infty} = 1 + \gamma K^2 \left(\frac{\sigma}{\theta_b}\right) = 1 + \frac{\gamma K^2}{1 - \epsilon_{\lim}} \left[1 + \frac{\epsilon_{\lim}(\epsilon_\infty^{-1} - \epsilon_{\lim})}{\gamma K^2} \right], \tag{7.3.4}$$

with terms of higher order in K^{-2} dropped because terms of order K^{-4} have been neglected in (7.3.3). Since (7.3.4) is only valid for large K, no expansion analogous to (7.3.2) may be made.

As was pointed out by Lees [1], the tangent-wedge approximation is based on the constancy of both pressure and flow inclination angle across the shock layer (along the same path), and will be in error for at least two reasons. First, centrifugal force effects arising from the curvature of the body cause a pressure gradient across the shock layer. Second, the pressure gradient along the shock layer causes a divergence or convergence of the streamlines, and the flow inclination angle has a gradient across the shock layer. The only completely rational theory for a correction of these errors must be made for ϵ very small. With ϵ very small the correction for the change in flow inclination angle is small in comparison with the correction from the centrifugal effects, and must be neglected within the accuracy to which we are able to estimate the latter. Within this order of accuracy the correction for the centrifugal effects is the same as that appearing in the Newtonian theory treated in Chapter III.

Within hypersonic slender body theory the pressure immediately behind the shock is given by (1.4.12) or (4.1.13) as

$$\frac{p_s - p_\infty}{\gamma p_\infty M_\infty^2} = (1 - \epsilon)\sigma^2 = \frac{\theta_s^2}{1 - \epsilon} . \tag{7.3.5}$$

The centrifugal pressure correction across the shock layer with ϵ small may be approximated from (7.1.4) with $d\theta/ds = -1/R_b$ and $\rho(y_s - y_b) \approx \rho_\infty y_b$ by

$$p_s - p_b = \rho U^2 \left(\frac{y_s - y_b}{R_b}\right) = \gamma p_\infty M_\infty^2 \left(\frac{y_b}{R_b}\right). \tag{7.3.6}$$

As before R_b is the radius of curvature of the body, taken positive for a convex body.

Taking $\theta_b = \theta_s$, we obtain for the corrected pressure on the body

$$\frac{p_b}{p_\infty} = 1 + \frac{\gamma K^2}{1 - \epsilon} - \gamma M_\infty^2 \frac{y_b}{R_b} . \tag{7.3.7}$$

The third term on the right-hand side of (7.3.7) is the centrifugal pressure correction. Since we have assumed that ϵ is small in calculating the pressure correction, we should ask whether the factor $(1 - \epsilon)^{-1}$ in the second term should not also consistently be dropped. If the quantity $y_b/R_b\theta_b^2$ is of order 1 this question should be answered yes, and we are left simply with the result of Newtonian theory for a slender body. In order for the result (7.3.7) to give us something new, we must assume that $y_b/R_b\theta_b^2$ is small compared with one. We may apply a similar centrifugal correction if θ_b is finite instead of small, provided $\cos\theta_b$ is not small and provided the other conditions discussed above are met.

One problem to which this correction may be immediately applied is to the calculation of the initial pressure gradient on a two-dimensional ogive. Here we must take the Newtonian limit in order to retain consistency. We take $d\theta_b/ds = -1/R_b$ and $dy_b/ds = \theta_b$, where s is distance along the body surface, and obtain

$$\frac{dC_{p_b}}{ds} = -6\frac{\theta_b}{R_b} . \tag{7.3.8}$$

This result is consistent with the Newtonian result shown on Fig. 2–5. Had we omitted the correction we would have obtained the same result with the incorrect factor 4 in place of 6.

The centrifugal pressure correction which appears in (7.3.7) was suggested and calculated for a perfect gas by Lees [1]. In addition to calculating this correction he also included a correction for the change in flow deflection angle across the shock layer. However, as we have pointed out, the order of magnitude of this deflection angle correction is of the same order of magnitude as that of the error in estimating the centrifugal pressure correction.

The corresponding procedure for bodies of revolution is the so-called tangent-cone method, which has been applied to the calculation of surface pressure distributions (e.g., Fowler and French [1]). According to this approximation, the pressure at any point on the surface of a slender body of revolution at arbitrary angles of pitch and yaw is identical with the Taylor-Maccoll value at the same Mach number on a semi-infinite unyawed circular cone of half-angle equal to the local inclination of the body with respect to the flight direction. Of course, this method will be in error for the same reasons as those given in the two-dimensional case.

We can in a very simple manner carry out a centrifugal correction for the body of revolution similar to the one which was made for two-dimensional flow, with the same assumptions and under the same conditions as those for the tangent-wedge correction.

For the cone, (7.3.6) is repeated but with a factor $\frac{1}{2}$ arising from the relation $\rho(r_s - r_b) \approx \frac{1}{2}\rho_\infty r_b$, and the analogue of (7.3.5) may be obtained from (4.2.26). The final result for the pressure on a slender body is

$$\frac{p_b}{p_\infty} = 1 + \frac{\gamma K^2}{1 - \frac{1}{4}\epsilon} - \frac{\gamma M_\infty^2}{2}\frac{r_b}{R_b}. \tag{7.3.9}$$

In order to justifiably keep the factor $1 - \frac{1}{4}\epsilon$ above we must again impose the condition that $r_b/R_b\theta_b^2$ is small, and again a similar correction for axisymmetric flow may be applied if θ_b is finite under the suitable restrictions discussed previously. In calculating the initial pressure gradient on an ogive of revolution (7.3.8) is again obtained, but with a factor 5 instead of 6, consistent with Fig. 2–9. Had the correction been omitted, the incorrect factor 4 would have appeared.

Some additional support for the corrected tangent-wedge and tangent-cone approximations is provided by the results of Section 5.5. There the lowest-order result for a quasi wedge or quasi cone was the tangent-wedge or tangent-cone formulas for the shock shape, with the pressure corrected by the Busemann centrifugal correction based on the body curvature. Thus the tangent-wedge and tangent-cone formulas with correction have a rational basis when applied to a somewhat restricted class of flows. The limitations of the theory of Section 5.5. should be kept in mind: The parameter ϵ must be small; the body shape upstream of the point of interest must approximate that of a wedge or cone; and the body curvature must be small enough that the centrifugal pressure correction is small compared with the basic pressure $\rho_\infty U^2 \sin^2 \theta_b$. The tangent-wedge solutions of Sections 5.3 and 5.4 also lend support. The extension of the results beyond these limitations can be defended only empirically.

Despite the lack of an unassailable theoretical justification, the tangent-wedge and tangent-cone approximations are useful, particularly when other approximate theories such as shock-expansion break down. Thus, the tangent-wedge approximation might be especially valuable for an estimate of the pressure on a blunted wedge or on a two-dimensional ogive which is highly curved near the nose.

4. Conical flows

The definition of conical flow has been given in Section 3.7. Most of the concepts of conical flow given there and in Section 5.6 apply to general conical flows. But additional concepts are needed in the general case, primarily to take into account phenomena connected with the speed of sound. For this purpose the classical concept of the zone of action of a point in a compressible flow and the closely related concept of the characteristic conoid are essential.

The standard proof that the flow on a conical body is conical under appropriate conditions is based upon the scale transformation. An infinite conical body is invariant under a scale transformation with the vertex as origin. If the appropriate determining conditions on the flow at infinity are invariant under the transformation the flow is conical, provided only the flow solution is unique. It is necessary that the fluid be nondissipative.

If the body is finite, only part of the flow field can be conical. This part of the flow must be completely determined by boundary conditions on the part of the body which is conical. The conical part of the flow field must exclude the zones of action of the nonconical parts or cut-off ends of the body. Thus if the body is finite a necessary condition for the flow field of interest to be conical is that the flow be everywhere supersonic. If the body is infinite or if the flow is only approximately conical the flow field may be partly subsonic. In some cases the flow on a part of a finite body may be approximated by that on an infinite body.

We outline briefly the proof that a supersonic flow on a finite conical body is conical, with the aid of Fig. 7–17. Body A is a finite conical body in a uniform supersonic flow, with the flow everywhere supersonic. It is transformed into body B by a scale transformation, with point 1 in the flow field of A transformed into point 1 in the flow field of B. The velocity vector and all thermodynamic quantities are the same at these two points. Upstream of the zone of action of the end of the shorter body the two bodies are identical and support the same flow fields if uniqueness of the solution is accepted. Thereby the velocity vector and all thermodynamic quantities are the same at point 2 for body A as at point 1 for body B. Thus these quantities are the same at points 1 and 2 for

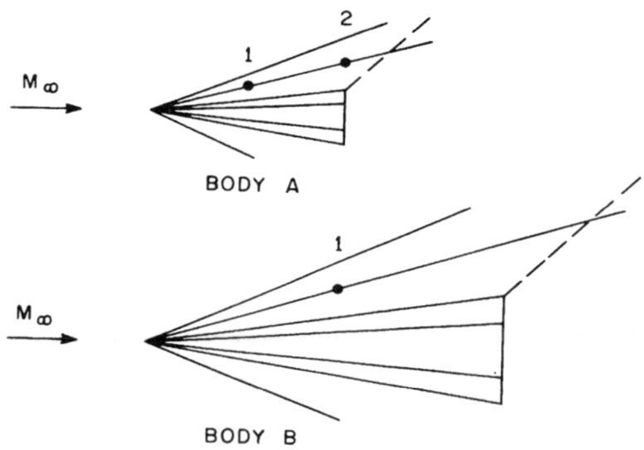

Fig. 7–17. Correspondence of flows on conical bodies.

body A and, since the scale transformation may be applied with any stretching parameter, are constant along the entire ray from the vertex to the zone of action of the end of the body.

If the vertex is missing from a finite conical body the part of the flow field which is conical must lie outside the zone of action of the vertex. An example of a body which is not conical at the vertex but which does have a conical section supporting a conical flow is shown in Fig. 7–18. A cylindrical flow on a cylindrical

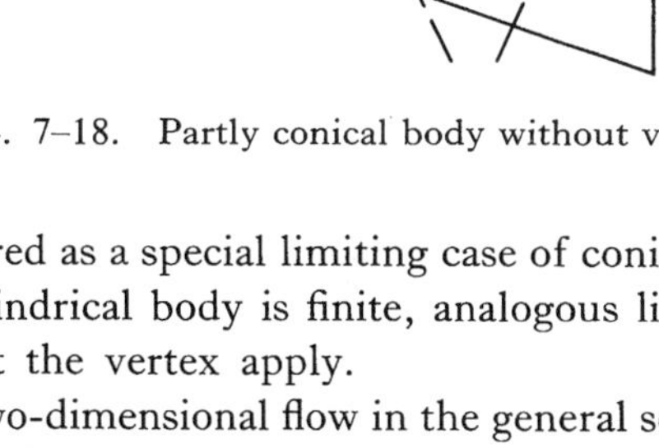

Fig. 7–18. Partly conical body without vertex.

body may be considered as a special limiting case of conical flow with the vertex at infinity. If the cylindrical body is finite, analogous limitations to those for a conical body without the vertex apply.

Conical flow is a two-dimensional flow in the general sense, with two coordinates identifying a ray from the vertex as an essential independent variable and with distance along a ray from the vertex an inessential variable. Conical flow problems are similar in many ways to classical two-dimensional and axisymmetric flow problems, and essentially all the approaches we have used in these cases are applicable. The most significant difference is that in conical flow the continuity equation cannot be put into a divergence form in the two essential coordinates, so that a stream function of the conventional type does not exist. To describe the flow field at least two dependent variables are needed.

The most natural coordinate systems for the study of conical flow are those termed conical. In these one coordinate is the distance R from the vertex, and is constant on concentric spheres. The other two are coordinates determining rays through the vertex, and may be considered as coordinates on a unit sphere. Of the possible choices polar coordinates on the unit sphere ϑ, ϕ are usually the most convenient, with the three-dimensional coordinate system the standard spherical coordinate system (R, ϑ, ϕ). Other choices of conical coordinate systems have been used to advantage with conical flow problems.

Although no stream function of the convential type exists, it is possible and convenient to express the continuity condition through a pair of stream functions. The appropriate form is

$$\rho \mathbf{q} = \tfrac{1}{2} \nabla R^2 \chi \times \nabla \psi, \tag{7.4.1}$$

where ψ and χ are functions on the unit sphere, functions of ϑ and ϕ alone with spherical coordinates. Writing gradients on the unit sphere with the subscript 2, we have, for example

$$\nabla \psi = \frac{1}{R} \nabla_2 \psi. \tag{7.4.2}$$

Then we may write, from (7.4.1),

$$\rho \mathbf{q} = [\chi \mathbf{e}_R + \tfrac{1}{2} \nabla_2 \chi] \times \nabla_2 \psi, \tag{7.4.3}$$

where $\mathbf{e}_R$ is the unit vector in the radial direction. Decomposing $\mathbf{q}$ into a radial and a cross flow component through

$$\mathbf{q} = q_R \mathbf{e}_R + \mathbf{q}_2 , \tag{7.4.4}$$

we have

$$\rho q_R \mathbf{e}_R = \tfrac{1}{2} \nabla_2 \chi \times \nabla_2 \psi, \tag{7.4.5}$$

$$\rho \mathbf{q}_2 = \mathbf{e}_R \times \chi \nabla_2 \psi. \tag{7.4.6}$$

The functions ψ and χ are not unique. The transformation

$$\bar{\psi} = f(\psi), \tag{7.4.7a}$$

$$\bar{\chi} = \frac{\chi}{f'(\psi)} , \tag{7.4.7b}$$

where f is any monotonic differentiable function, yields a new pair $(\bar{\psi}, \bar{\chi})$ with the same properties as (ψ, χ). But ψ and $\bar{\psi}$ are both constant on the same lines on the unit sphere. These lines are the trajectories of the cross velocity $\mathbf{q}_2$ on the unit sphere and are termed the cross streamlines. They are the intercept on the unit sphere of the conical streamsurfaces defined by the condition $\psi = constant$ in space. Such a streamsurface is illustrated in Fig. 7–19.

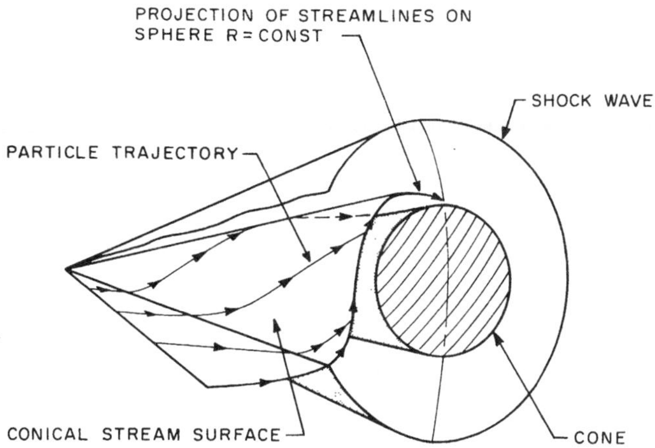

FIG. 7–19. Conical streamsurface formed by conical flow streamlines.

The entropy S is independent of R because it is a thermodynamic variable. The basic condition $\mathbf{q} \cdot \nabla S = 0$ from (5.1.10) or (7.1.5) leads to the equivalent conditions

$$\mathbf{q}_2 \cdot \nabla_2 S = 0, \tag{7.4.8a}$$

$$\nabla_2 \psi \times \nabla_2 S = 0. \tag{7.4.8b}$$

The physical meaning of these conditions is that the entropy is constant on cross streamlines on the unit sphere or on conical streamsurfaces in space, and is a function $S(\psi)$ of the stream function ψ alone.

We shall not develop the equations of motion for conical flow in detail. The reader is referred to the literature for these. Besides the original papers of Busemann [1; 3] on general conical flow, there are the studies of Ferri [1; 3], Ferrari [2], and Maslen [1]. Later references are mentioned below.

In the equations of motion we may treat the three velocity components as distinct dependent variables. In general, it is far more convenient to use the variables ψ and χ as dependent variables. In any case, the resulting conical equations of motion show essentially all the characteristics of the more familiar two-dimensional equations, and differ primarily by being more complicated and requiring two dependent variables instead of one. Boundary conditions at infinity are that the flow is uniform, with this property expressed in conical terms.

Boundary conditions on the body are simply that $\mathbf{q}_2 \cdot \mathbf{n}$ be zero there. Shock waves are conical in shape and governed by the classical Hugoniot conditions in terms of the cross flow alone. Contact discontinuities may appear and must lie on cross streamlines (or conical streamsurfaces).

The equations governing the conical flow may be either elliptic or hyperbolic in nature. Thus the unit sphere is divided into elliptic and hyperbolic domains, with the boundaries between them shock waves, contact discontinuities, or parabolic lines. Elliptic regions are characterized by the condition $q^2 - q_R^2 = q_2^2 < a^2$, parabolic lines by $q_2^2 = a^2$, and hyperbolic regions by $q_2^2 > a^2$. Conical characteristics and conical zones of action on the unit sphere may be defined in full analogy with two-dimensional flow. The flow upstream of a shock must be (conically) hyperbolic, while that downstream of a shock may be of either type.

Elliptic and hyperbolic regions are regions of subsonic and supersonic cross flow, respectively. The distinction may also be characterized in terms of the three-dimensional flow field. If the three-dimensional flow at a point is subsonic or if the Mach cone at the point contains the ray from the vertex the flow is elliptic. If the Mach cone at a point does not contain the ray the flow is hyperbolic. If the ray lies on the surface of the Mach cone the point is parabolic.

The characteristics on the unit sphere are the integral curves of the direction field making an angle $\mu_2 = \sin^{-1}(a/q_2)$ with the cross streamlines. A method of characteristics for the hyperbolic regions may be established, as was done by Maslen [1]. For the general case in which the flow is not irrotational the choice of dependent variable should probably be the pressure p and a measure of the angular orientation of $\mathbf{q}_2$, say $\tan^{-1}(q_\phi/q_\vartheta)$ in spherical coordinates. These are the two quantities which are necessarily continuous across a contact discontinuity.

Based upon the conical zones of action, limiting characteristics and pseudo-elliptic zones may be defined in a manner analogous to that in two-dimensional or axisymmetric flow. Much of the discussion of the first two pages of Section 6.1 applies in this case. One important difference appears, however. In the two-dimensional and axisymmetric cases, the pseudoelliptic or transonic zones encountered are of the accelerating type in general, with a parabolic or sonic line as a boundary upstream and a limiting characteristic as a boundary downstream. In the conical case the pseudoelliptic zones encountered may be either of the accelerating type or of the decelerating type. In the latter case the zone has a limiting line as a boundary upstream in the cross flow and a parabolic line as a boundary downstream. This difference may be explained through the physical idea that nature "abhors" a compression in which the pressure increases and the flow passes from supersonic to subsonic continuously without a shock. With this idea decelerating pseudoelliptic zones are ruled out for two-dimensional and axisymmetric flows. In conical flow there is no cross-flow Bernoulli equation which connects pressure with cross velocity or cross Mach number. The cross flow can decelerate without an increase in pressure, and transitions from hyperbolic to elliptic are not necessarily compressive.

A limiting characteristic can contact a parabolic line only at a physical boundary, a shock, or at a point of tangency. In any case the streamline is normal to the limiting characteristic there. A point on the parabolic line for which the streamline is normal to it and from which characteristics do leave is generally a tangency point for a limiting characteristic and a point across which the limiting characteristic changes the family to which it belongs.

Figure 7–20 will serve to illustrate these comments and a possible local

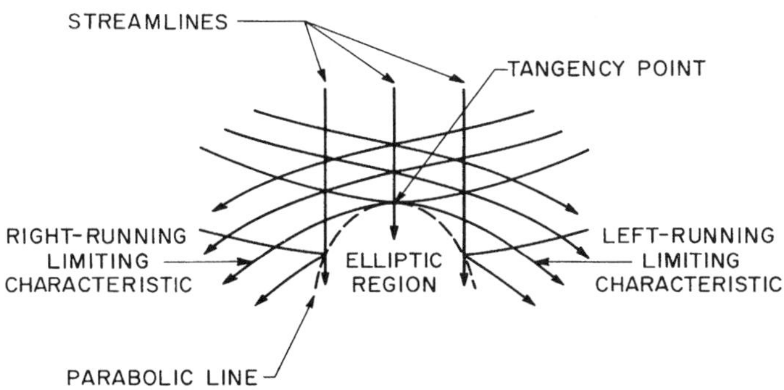

FIG. 7–20. Limiting characteristic tangent to parabolic line.

solution, with the limiting characteristics decelerating. The analogy, as well as the points of difference, with the mixed flow past a blunt body is apparent (cf. Fig. 6–1).

To illustrate further the nature of conical flows, we sketch in Fig. 7–21 the solution for a flat delta wing with supersonic leading edges, taken from Maslen [1] (with minor corrections). In this example the top and bottom surfaces are independent since the leading edges are supersonic. The flow field on the upper surface contains two embedded shocks, one on each side of the central elliptic region. Only the part of the flow which has not passed through either of these shocks is irrotational. On the lower surface, the entire flow passes through a shock. The portion of this shock between a leading edge and the corresponding parabolic point is planar, and the flow which has passed through such a portion of the shock is irrotational.

Our discussion so far has brought out the fact that a conical flow when transformed to spherical coordinates has many features of a two-dimensional mixed blunt-body flow. This analogy is most complete in what we have termed in Chapters IV and V the outer layer, although it breaks down in the inner layer.

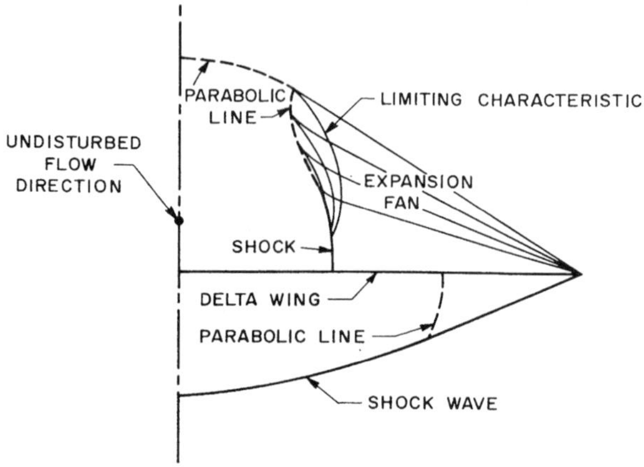

FIG. 7–21. Solution for delta wing (Maslen [1]).

The reason for this may be seen from the cross-flow continuity equation. This equation may be expressed in the form

$$R\nabla \cdot (\rho \mathbf{q}) = \nabla_2 \cdot (\rho \mathbf{q}_2) + 2\rho q_R = 0. \tag{7.4.9}$$

The term in q_R acts as a distribution of sinks (or sources) which negates some of the classical continuity concepts with respect to the cross flow. Therefore, as has been indicated in Sections 3.7 and 5.6, cross streamlines may terminate in the middle of the flow field or on the body. The streamline directions on the sphere surface representing the independent conical variables form a vector direction field on this surface and the termination points are singularities of this direction field. These singular points occur where $\mathbf{q}_2$ is zero and necessarily lie in an elliptic region. They are points for which the pressure p is stationary.

The elementary singular points which appear in two-dimensional continuous vector fields are classified into source-like nodes, source-like foci, sink-like nodes, sink-like foci, and saddle points. Physically, the case $q_R < 0$ can exist at a singular point only for the source on the unit sphere representing the uniform flow coming from upstream infinity. With this case excluded the vector field $\mathbf{q}_2$ can have as elementary singular points only sink-like nodes, sink-like foci, and saddle points.

With a fixed boundary in the two-dimensional space and the boundary condition of tangency, we must add to the classification sink-like half-nodes and half-saddles which lie on the boundary. These are illustrated in Fig. 7–22. The cross flow is generally well behaved at a half-saddle, which is similar to a stagnation point. The entropy, which is constant along cross streamlines, is well defined there. However, the entropy is not defined at a half-node (sink-like), or at a sink-like node or focus in the flow field. In the neighborhood of such singularities

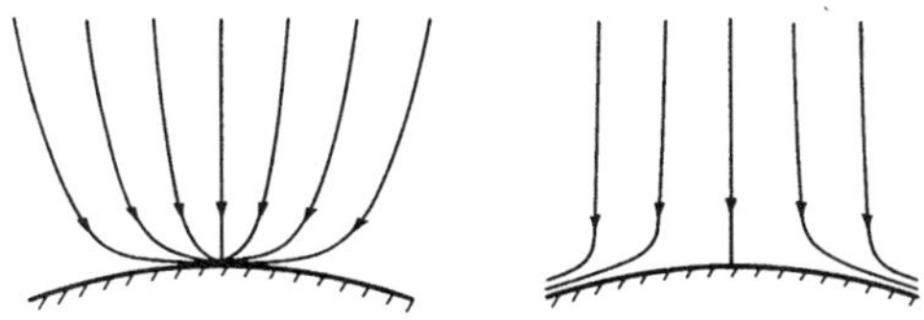

FIG. 7–22. Elementary singular points on a boundary.

the entropy gradients generally become very large. Such a singularity with nonconstant entropy is the "vortical singularity" of Ferri [1; 3, Art. 15]. The nature of vortical singularities is fairly well understood (see Melnik [1; 2]). Following Ferri [3], these singularities are illustrated in Fig. 7–23 for an elliptic cone at small and at large incidence. It can be seen that the half-node which is on the upper surface of the cone at small incidence lifts off the cone surface to become an isolated node at large incidence.

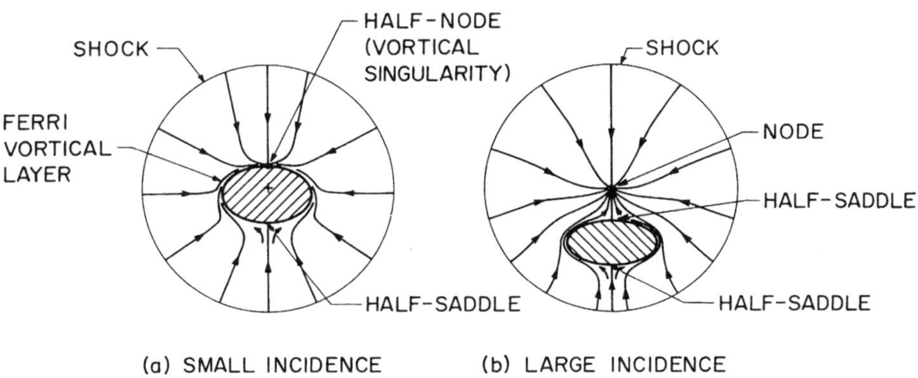

FIG. 7–23. Singularities in solutions for a cone.

A topological law based on the theory of Poincaré indices relates the numbers of elementary singular points provided only such points occur. This law may be expressed

$$N + F - S + O = 2 \tag{7.4.10}$$

for the entire unit sphere, with N the number of nodes, F the number of foci, S the number of saddle points, and O the number of obstacles contributing boundaries of the type discussed. In this formula the source representing the approaching uniform flow counts as a node, a half-node counts as half a node, and a half-saddle counts as half a saddle point. For details see Scheuing *et al.* [1].

For a circular cone at zero incidence the entire boundary on the unit sphere consists of degenerate singular points. In other words, the entire boundary is made up of points for which $\mathbf{q}_2 = 0$ and which are neither half-nodes nor half-saddles. The limiting direction of the cross streamlines is normal to the body. If now the cone is slightly perturbed the degenerate singular points all disappear. Half-nodes and half-saddles appear at isolated points on the boundary. At other points on the boundary $\mathbf{q}_2$ is not zero, and is tangent to the boundary. The perturbation is weak, and in the flow field not far from the boundary the streamlines are approximately normal to the body. The adjustment of the outer flow to the new conditions at the boundary occurs in a thin layer near the surface. This layer is termed a Ferri vortical layer (see Ferri [1; 3, Art. 15]). This layer has almost constant pressure, a concentration of the radial component of vorticity, but only a very small effect on the outer part of the flow field. Its presence may be discerned in Fig. 7–23(a).

There are actually two layers which can be distinguished. One is a layer characterized by the property that q_ϕ and q_ϑ are of the same order of magnitude. This layer has a thickness of the order of α, where α is a suitably defined perturbation parameter (say angle of attack divided by cone half-angle). This layer is analogous to the transitional layer part of the Newtonian vortical layer discussed in Section 5.6, and can be correctly described through a straightforward perturbation analysis on the zero-incidence cone solution.

The other layer is a singular region in which a straightforward perturbation analysis fails. This layer is analogous to the inner layer part of the Newtonian vortical layer (Section 5.6), and has a thickness which is of the order of the exponential of minus a function of position on the boundary times α^{-1}. For further details the reader is referred to Ferri [1; 3, Art. 15], Willett [1], Bulakh [1], Woods [1], and Munson [1]. The phenomenon of the Ferri vortical layer is not characteristically hypersonic and we do not treat it in detail here.

Many numerical schemes have been developed for the analysis of nonlinear conical flows. All of these methods are closely related to those described in Chapter VI or make use, where appropriate, of a method of characteristics analogous to that presented in Section 7.1. The outer conical field may have much the same nature and present the same difficulties in solution as do detached shock blunt-body flow fields. Thus in a direct problem the shock location is unknown in advance, and the flow may be of mixed type. In addition, the presence of an inner vortical layer can introduce additional complications in any numerical approach. In addition to the method of characteristics developed for the hyperbolic regions, Maslen developed a relaxation method for handling the corresponding elliptic regions; as an example he treated the flat supersonic delta wing (Fig. 7–21). For treating the same problem Babaev [1; 2] presents a difference scheme somewhat modified from that of Maslen. In [3] Babaev compares calculations on the compression side of the wing at $M_\infty = \infty$ and $\gamma = 1.4$

for a 45 degree sweepback angle with the conical Newtonian flow results of Gonor [2].

Chushkin and Schennikov [1] have applied the integral relations method to the direct problem of the unyawed elliptic cone and the yawed right circular cone. Their calculations were restricted to cross flows in which the condition $q_2^2 < a^2$ was satisfied. Brook [1] reports that the extension of the method with one-strip and linear profiles to the mixed cross-flow problem leads to an indeterminate situation. He suggests the introduction of an inner shock to overcome this difficulty. He also notes that placing an elliptic cone at an angle of attack introduces the complication of unknown stagnation point locations. The problem faced here is analogous to that with the blunt body at angle of attack, considered earlier in Section 6.3.

Of inverse methods applied to conical flows we mention those of Briggs [1; 2], Mauger [1], Leavitt [1], and Stocker and Mauger [1]. The approach of Briggs has its analogue in the inverse methods of Section 6.5, which are carried out in the physical plane. The approach of Mauger makes use of the complex plane and is the analogue of the Garabedian-Lieberstein method, while that of Leavitt employs a double power series expansion in the manner of the Richtmeyer-Lewis procedure. All of these authors started with elliptic conical shocks and all encountered calculational difficulties on reaching the vortical layers of the bodies generated by such shocks. The calculations were stopped in the region of the inner layer, where the body was taken to be located. Where comparison is possible, the agreement is good among the body shapes and surface pressure distributions obtained by the different methods. The failure of these outer layer approaches in the inner layer is to be expected, though there should be little effect on the surface pressure distribution or body shape when a reasonably behaved solution can be obtained.

Stocker and Mauger [1] also examine an inverse approach in the physical plane, but attempt to account for the presence of the vortical layer by their choice of independent variables. They replace the physical variables ϑ and ϕ by the two stream functions ψ and χ defined by (7.4.1). As independent variables they choose ψ and a variable η defined by

$$\eta = \ln\left(\frac{\chi_s(\psi)}{\chi}\right), \tag{7.4.11}$$

where $\chi_s(\psi)$ is the value of χ at the shock for a given value of ψ. The variable ψ is defined to satisfy $\psi = \phi$ at the shock. The variable η is chosen in order that the body not be reached in a finite number of steps. The vortical layer and the vortical singularity are thus approached slowly, with the singular behavior modified. In the vortical layer (inner layer) two points which are physically close together in the unit sphere may still be separated by appreciable intervals in both ψ and η. The fact that the computations give distances from the body and pressures which are also very close together is cited by the authors as a reason for believing their computations are very accurate. The obvious limitation is that details of the vortical layers and singularities are not described; as we have indicated, these are not important for the main flow field. In computing various cases they encountered some computational difficulties in those cases in which difficulties should be expected, as with large incidence of a cone or slight shock concavity. In accord with our discussion in Section 6.5 and elsewhere, they found a small change in shock shape produced a large change in body shape.

5. Nonequilibrium flows

In this last section we extend our treatment of purely supersonic flows to include the effects of lack of equilibrium, restricting ourselves for simplicity to two-dimensional and axisymmetric flows. The principal part of our discussion will be concerned with the generalization to nonequilibrium of the method of characteristics and topics related to those considered in the first section of this chapter.

In deriving the characteristic equations we again use the intrinsic coordinates of Section 7.1. Alternative derivations using intrinsic coordinates may be found, for example, in Chu [2] or Clarke and McChesney [1], and in cartesian coordinates in Katskova and Kraiko [1]. The overall continuity and momentum equations (7.1.2)–(7.1.4) remain unchanged in the absence of equilibrium, while the relationship (7.1.5) for constancy of entropy along a streamline now no longer holds.

As in Section 6.6, to this system must be added suitable equations of state and the appropriate rate-determining equations for the physical or chemical processes involved. In place of the energy equation (7.1.6) we use the form

$$\frac{Dh}{Dt} = \frac{1}{\rho}\frac{Dp}{Dt}, \tag{7.5.1}$$

with D/Dt again denoting a time derivative along a streamline.

We may write the rate equations in a form which is sufficiently general to represent the type of relaxation processes considered previously. Thus the relation

$$\frac{DC_i}{Dt} = W_i(p, \rho, C_j) \tag{7.5.2}$$

may be considered as a representative rate relation for a chemically or vibrationally relaxing system, for example. Here we have used the notation C_j to denote the finite set of relaxing parameters C_1, C_2, ... These parameters may be species concentrations, temperatures of internal modes, or any other quantities which are considered constant during a sudden volume change. The defining relation for the enthalpy may be expressed functionally by

$$h = h(p, \rho, C_j), \tag{7.5.3}$$

while the usual engineering equation of state is written

$$p = p(\rho, T, C_j). \tag{7.5.4}$$

The above equations as presented parallel the equivalent more specific forms given in Section 6.6.

As in the equilibrium (or frozen) case the streamlines and Mach lines are the characteristic directions in the flow field. As a necessary preliminary to finding what the new characteristic equations are it is necessary to determine the value of the local sound speed on which the characteristic Mach angle is based. As noted, the streamline $dn = 0$ is a characteristic line and for this line (7.5.1) and (7.5.2) are already in characteristic form. These relations may be written

$$q\,dC_i = W_i\,ds, \tag{7.5.5}$$

$$(h_p - \rho^{-1})\,dp + h_\rho\,d\rho + \sum h_i\,dC_i = 0, \qquad \text{on} \quad dn = 0, \tag{7.5.6}$$

to which can be added the total enthalpy relation (7.1.6). We have here used the notation

$$h_p = \left(\frac{\partial h}{\partial p}\right)_{\rho, C_j}, \qquad h_\rho = \left(\frac{\partial h}{\partial \rho}\right)_{p, C_j}, \qquad h_i = \left(\frac{\partial h}{\partial C_i}\right)_{p, \rho, C_{j(\neq i)}}. \tag{7.5.7}$$

The square of the frozen speed of sound is the ratio $dp/d\rho$ with the C_i's all constant and may be obtained directly from (7.5.6). We obtain

$$a_f^2 = \left(\frac{\partial p}{\partial \rho}\right)_{S, C_j} = \frac{-h_\rho}{h_p - \rho^{-1}}. \tag{7.5.8}$$

In deriving the characteristic equations on the Mach lines it is again convenient to obtain an altered form of the continuity equation analogous to (7.1.8). We follow the same procedure as in deriving (7.1.8) except that $\partial\rho/\partial s$ is eliminated by means of (7.5.5) and (7.5.6). Using the definition (7.5.8) this procedure gives

$$\frac{M_f^2 - 1}{\rho q^2}\frac{\partial p}{\partial s} + \frac{\partial \theta}{\partial n} = \frac{-j}{y}\sin\theta + \frac{\sum h_i W_i}{h_\rho \rho q}. \tag{7.5.9}$$

Here q/a_f has been replaced by the Mach number M_f based on the frozen speed of sound a_f.

Using (7.5.9) the derivation of the characteristic equations analogous to (7.1.10) proceeds exactly as for that case, with the result

$$(7.5.10)\quad dp \pm \frac{\rho q^2}{\sqrt{M_f^2 - 1}}\, d\theta + \frac{\rho q^2}{M_f^2 - 1}\left\{\frac{j \sin\theta}{y} - \frac{\sum h_i W_i}{h_\rho \rho q}\right\} ds = 0,$$

$$\text{on}\quad \frac{dn}{ds} = \pm \frac{1}{\sqrt{M_f^2 - 1}}.$$

Transformed to cartesian coordinates the streamline characteristic equations are

$$(7.5.11)\qquad q\cos\theta\, dC_i = W_i\, dx, \qquad \text{on} \quad \frac{dy}{dx} = \tan\theta,$$

along with (7.5.6) and the energy relation (7.1.6). For the Mach line characteristics in cartesian coordinates

$$(7.5.12)\quad \frac{dp}{\rho q^2 \tan\mu_f} \pm d\theta + \left\{\frac{j\sin\theta}{y} - \frac{\sum h_i W_i}{h_\rho \rho q}\right\} \frac{\sin\mu_f}{\sin(\theta \pm \mu_f)}\, dy = 0,$$

$$\text{on}\quad \frac{dy}{dx} = \tan(\theta \pm \mu_f),$$

where μ_f is the local Mach angle $\sin^{-1}(1/M_f)$ based on the frozen sound speed.

From the preceding relations we see that appropriate characteristic directions are those defined in terms of the frozen speed of sound in nonequilibrium flow. The result that the frozen speed of sound is the correct one to be used for nonequilibrium characteristics was noted early by Broer [1; 2] in connection with sound propagation in a gas with lagging internal modes, and by Brinkley and Richardson [1] in connection with the propagation of a rarefaction wave in a chemically reacting medium. Later Wood and Kirkwood [1] verified this result in their characteristic analysis of nonequilibrium one-dimensional unsteady flow, as did Chu [2].

The square of the equilibrium sound speed is defined as the ratio $dp/d\rho$ provided the C_i's always have their equilibrium values

$$(7.5.13)\qquad C_{i,e} = C_{i,e}(p, \rho).$$

From (7.5.6) we may write

$$(7.5.14)\qquad a_e^2 = \frac{-(h_\rho + \sum h_i C_{i,e,\rho})}{h_p + \sum h_i C_{i,e,p} - \rho^{-1}}.$$

Note that following Broer [2] we have avoided the use of the entropy in our formulation. The definitions of the two speeds of sound may be shown to be equivalent to the standard ones involving isentropic derivatives. An alternative formulation with entropy as a basic variable is feasible but less convenient.

A basic inequality connects the frozen and equilibrium sound speeds. This relation is

$$a_f > a_e \,. \tag{7.5.15}$$

The inequality follows directly from Le Chatelier's principle (cf. Landau and Lifshitz [1, Eq. (78.11)]). According to this principle, if the volume of a system originally in equilibrium is reduced adiabatically so that its pressure is increased, subsequent internal processes which tend to bring the system again to equilibrium must act to reduce the pressure. The initial pressure change for a sudden volume change is proportional to a_f^2, the subsequent pressure change to $a_e^2 - a_f^2$. This latter quantity must be negative. If the inequality $a_f < a_e$ were to hold, this would entail various anomalous phenomena according to standard theories. These anomalous phenomena would include sound waves with negative absorption or a bulk viscosity which is negative.

We may properly ask why the frozen sound speed is always the correct characteristic speed even in a flow which may be very close to equilibrium, in spite of the fact that for complete equilibrium the correct sound speed must change discontinuously to the equilibrium value a_e. Further, if the correct sound speed is in general the frozen one, then what significance does the equilibrium sound speed have?

We may attach two separate but related interpretations to the significance of the different sound speeds. The first is that with finite reaction rates the head or forerunner of any (linearized) sound signal travels with the frozen sound speed. To see this we consider a weak discontinuity in p and ρ. In the characteristic equations (7.5.10) or their analogue in unsteady flow dp/ds and $d\theta/ds$ (or du/ds) are infinite. With W_i finite the propagation is governed by the leading terms in terms of the quantity a_f. The frozen sound speed is the speed characteristic of the propagation of a weak discontinuity. It is a maximum propagation speed beyond which no other signal can propagate. In a steady frame of reference the frozen characteristics, therefore, yield the largest domain of dependence and zone of action of a point. Thus they are always the correct characteristics in the absence of equilibrium. Intuitively, if all of the characteristic relaxation times are considered to approach zero we expect that $C_i \rightarrow C_{i,e}$; for an infinitely fast reaction local equilibrium must be established instantaneously. However, this result is only true provided that in the equilibrium limit W_i is infinite if Dp/Dt or $D\rho/Dt$ is infinite. As pointed out in Section 1.8, we have in this case an example of a singular perturbation problem.

To explain the singular nature of the perturbation problem, we note first that we can derive pseudocharacteristic equations similar to (7.5.10) but based upon the equilibrium speed of sound a_e. In these equations the term in W_i is replaced by one in $W_i - DC_{i,e}/Dt$. Since the derivatives of $C_{i,e}$ are not in the equilibrium characteristic direction these are not characteristic equations. Relaxation equations have the general form

$$C_i - C_{i,e} = \sum \tau_{ij} W_i, \tag{7.5.16}$$

where the τ_{ij} are functions of state. Eigenvalues of the matrix τ_{ij} may be interpreted as the characteristic relaxation times for the system. If the τ_{ij}'s approach zero, so do the relaxation times. In the limit the pseudocharacteristic equations become characteristic. But since the τ_{ij} in the equations in which $DC_{i,e}/Dt$ has been eliminated through (7.5.16) multiply essential derivatives, the limiting process is a singular one.

As a consequence of the singular perturbation nature of the almost-equilibrium case, we expect the occurrence of thin layers analogous to boundary layers aligned along equilibrium characteristics. These are termed relaxation zones, and also occur behind shock waves. With the relaxation times small but finite, these boundary layers (in the general sense) provide the mechanism by which the C_i's do not have discontinuities where p and ρ do. For such boundary layer phenomena the characteristic wave speed is clearly that relevant to the singular perturbation problem, that is, the equilibrium sound speed. We may associate such a singular perturbation problem either with the relaxation times approaching zero and the macroscopic scale fixed or with the relaxation times fixed and the macroscopic scale becoming infinite. The essentials of this argument on the singular perturbation nature of the near-equilibrium problem was first given by Chu [2]. Chu has also shown by examining the waves produced by a piston in a reacting mixture that in an almost-equilibrium the bulk of the signal propagates at the equilibrium speed even though the wave front moves with the frozen sound speed.

The second interpretation of the significance of the frozen and equilibrium sound speeds follows that given by Broer [2]. Broer shows by a standard linear analysis of plane progressive sound waves in a medium with a single relaxation time that the phase velocity takes on two limiting values depending on whether the relaxation time t_r times the sound frequency ω is zero or infinity. In the limiting case of frozen flow in which $\omega t_r \to \infty$, the phase velocity is shown to be the frozen speed of sound. In the limit of equilibrium flow in which $\omega t_r = 0$, the phase velocity is the equilibrium speed of sound. The frozen speed of sound is interpretable as the phase velocity of very short waves and the equilibrium speed of sound as the phase velocity of very long waves. For intermediate times the phase velocity is complex, which indicates that there is sound absorption. A similar analysis is given by Landau and Lifshitz [1, Sect. 78].

In our discussion so far we have considered nonequilibrium flows which are everywhere hyperbolic, that is, $M_f > 1$. However, since $a_f > a_e$ conditions may exist locally for which $M_f < 1$ and $M_e > 1$. In this case the flow has an elliptic character with respect to the frozen sound speed but a hyperbolic character with respect to the equilibrium sound speed. We observe that when going from a finite to a zero relaxation time the flow type must change discontinuously from elliptic to hyperbolic. This shift in type is another evidence of the singular perturbation nature of almost-equilibrium flow. We do not consider this interesting case further but consider only the purely hyperbolic problem of flows with $M_f > 1$.

We turn next to the numerical application of the method of characteristics and inquire as to how successful we may expect a method to be which is always based on frozen characteristics even though the flow may be arbitrarily close to equilibrium. In principle, the use of frozen characteristics is permissible even in an equilibrium flow. It is essential that the zone of action of a point inherent in the computational scheme include the physical zone of action of the point. The main effect would be errors in a computation with finite differences.

Sedney, South, and Gerber [1] have applied the method of characteristics based on frozen characteristics to the calculation of flow on a wedge with vibrational relaxation (see Fig. 7–24). They found that the standard difference

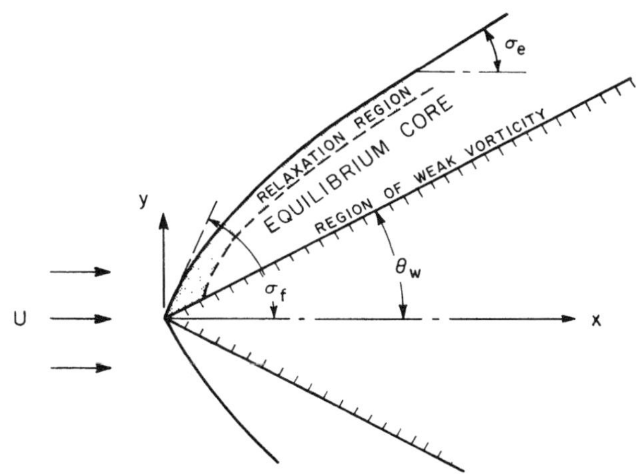

FIG. 7–24. Nonequilibrium flow on a wedge.

technique using two input points to determine conditions at a third point was not accurate enough, and that a method based on using information at three input points was needed to improve the accuracy. These calculations were carried out using the characteristic equations in intrinsic coordinates. However, they found a scheme which incorporated a transformation of the characteristic equations to characteristic coordinates more satisfactory. It was also found impracticable to use the entropy as an explicit dependent variable. Capiaux and Washington [1] extended these calculations to include chemical reactions, where the rate equation is based on the ideal dissociating gas discussed in Section 2.11. They found that satisfactory accuracy was achieved using as coordinates the stream function and the cartesian coordinate normal to the free stream.

Sedney, South, and Gerber [1] used the simple relaxation model (2.11.5) with a modified form of (2.11.6) for t_{vib}. Their results are described in Fig. 7–24. The shock angle at the leading edge of the wedge corresponds to frozen conditions. Far downstream of the tip the shock approaches the equilibrium shock angle. However, with the exception of the pressure far downstream, the values of the flow variables at the wedge surface are different from the equilibrium shock values. There is a variation in these quantities across the shock layer far downstream though the pressure is essentially constant. Near the shock surface there is a characteristic relaxation region, behind which the flow is entirely in equilibrium. In the equilibrium flow there is a region of nonuniform entropy and temperature near the surface, termed an entropy layer by some authors. This region of vorticity near the surface results firstly from the curved shock at the tip, and secondly from the fact that the entropy increase due to the relaxation process along the streamlines near the wedge is greater than that for the streamlines which are farther away. This results from the fact that the flow nearest the surface is the furthest out of equilibrium behind the shock. Sedney, South, and Gerber note that in the case they studied in detail about three-quarters of the entropy difference results from the shock curvature at the leading edge. In general, however, these effects are very weak ones. It would appear most inappropriate to associate the term entropy wake or entropy layer with this region of vorticity near the surface. These terms generally describe much stronger effects associated with a blunt nose (see Section 2.7). The layer arising from the relaxation effects here is properly described as a region of vorticity or very weak entropy wake analogous to that created in equilibrium flow by a wedge which is slightly curved near its tip.

A similar study for a cone has been made by Sedney and Gerber [1]. In this case the approximate result of Chapman reproduced at the end of Section 5.10 was available to the authors for comparison purposes. This comparison between the simple formula (5.10.36) with $j = 1$ interpreted in terms of shock angle and the results of Sedney and Gerber's characteristic calculations is shown in Fig. 7–25. Here t_{vib_0} is the characteristic time of the relaxation law evaluated at the nose.

In our discussion of nonequilibrium wedge flow we observed that there was a relatively larger entropy increase near the surface due to the shock curvature and to the dissipative relaxation process. We can indicate the nature of the second of these effects alone by considering the example of a Prandtl-Meyer

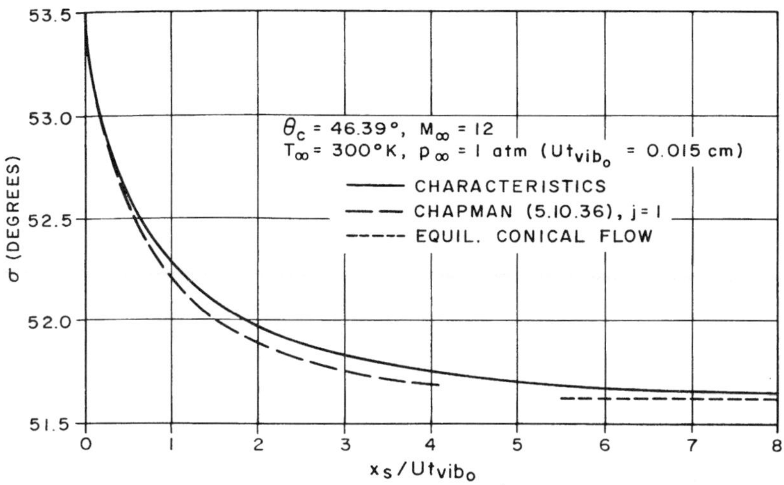

FIG. 7–25. Shock wave angle on a cone versus distance along shock for vibrationally relaxing nitrogen (Sedney and Gerber [1]).

expansion. With nonequilibrium flow there is no simple general solution analogous to the Prandtl-Meyer flow with equilibrium discussed in Section 7.1. This is evident from (7.5.12), where the presence of the reaction rate term in W_i prevents a simple solution in much the same way as does the axial symmetry term in general. The added nonequilibrium contribution may be considered to represent the effect of heat release and the changing state of the relaxing quantity on the production of pressure waves. We shall return to this question below.

An example of a Prandtl-Meyer calculation in a relaxing gas may be found in the paper of Appleton [1], who considers the case of an ideal dissociating gas. Other and earlier solutions may be found as, for example, the analogous unsteady centered expansion wave solution of Wood and Parker [1] with a relaxing internal degree of freedom. In his solution Appleton numerically solved (7.5.11), (7.5.12) and the energy relation on a characteristic net in the x, y plane. His calculations were carried out for a particular set of initial conditions with several expansion angles.

Appleton's results bear out in most details the picture of a Prandtl-Meyer flow shown in Fig. 7–26. The flow which starts out initially in a uniform state undergoes a frozen expansion in the immediate vicinity of the corner, while at large radial distances the flow undergoes essentially an isentropic equilibrium Prandtl-Meyer expansion. Downstream of this equilibrium expansion the flow approaches the uniform flow appropriate to the equilibrium case. Far downstream near the surface the flow is also in equilibrium at the same constant pressure as in the outer uniform flow. In this layer near the surface the tem-

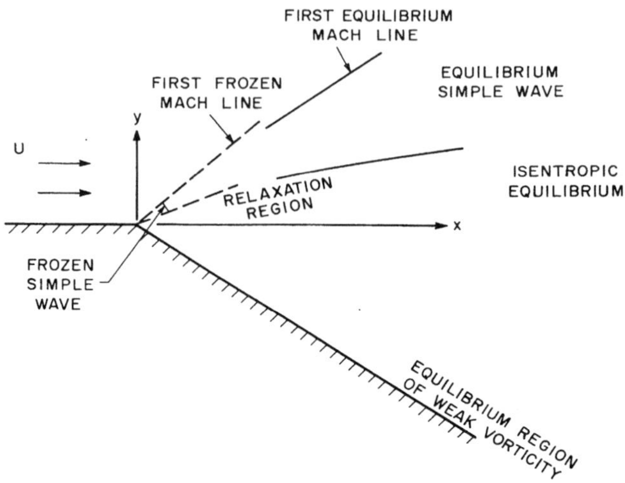

FIG. 7–26. Nonequilibrium Prandtl-Meyer flow.

perature and entropy are slightly higher than in the uniform flow region. The calculations were unable to show this increased temperature, partly because of computational limitations but mostly because the effect is so weak. This observation reinforces the point made earlier regarding the weakness of the vortical region introduced by the relaxation. The remaining zone, which is the main body of the flow in the wave system, represents the relaxation region itself. It is interesting to note that the calculations indicate that the wave reflections between the vortical layers in the expansion and the wall can cause pressure oscillations following the primary expansion.

The nature of the relaxation term as it appears in the characteristic equations and of its wave producing effect suggests that a modified shock-expansion method might be developable for nonequilibrium flows. For almost-frozen flows a shock-expansion method could be sought of a type similar to that discussed for equilibrium flows in Section 7.2. The near-equilibrium case is however, a somewhat more difficult problem, and any shock-expansion procedure must involve modifications analogous to those which would be required for the use of the method with axial symmetry. In general, it would be necessary to estimate the effect of the waves sent towards the surface from the effective heat release from the reactions. Since these are induced originally by a wave from the body this effect may be looked at as a reflection analogous to that from area (radial coordinate r) changes in axisymmetric flow. The procedure should be more feasible in almost-frozen flow than in almost-equilibrium flow. In the latter case, however, the concept of the thin (but weak) entropy wake on the surface is the essential one. Such extensions should be worthwhile exploring.

CITED REFERENCES

In the references cited we have used, as far as possible, the abbreviations for journals and reports used by *Applied Mechanics Reviews*. A list of these abbreviations may be found through any recent index number of *Applied Mechanics Reviews*.

Transliteration of Russian names has essentially followed the system adopted by the Library of Congress, but with no distinction between е and ё or between и and й, and with yu used for ю and ya for я. Russian titles have been translated into English, but where a translation is indicated the title given is that appearing on the translated version. A source of an English translation for all cited Russian references has been given whenever known to the authors. Translations listed as "M. D. Friedman *Transl.*" may be obtained from the Clearinghouse for Federal Scientific and Technical Information, Springfield, Virginia. The sources of all other translations are self-explanatory.

Except where a distinguishing reference to a superseded more informal report was desired, we have consistently referenced the final published version of a paper if available.

Anderson, W. A.
See Xerikos and Anderson.

Antonov, A. M., and Hayes, W. D.
[1] Calculation of the hypersonic flow about blunt bodies, *J. Appl. Math. Mech.* (*Prikl. Mat. Mekh.*, pp. 347–352) **30**, no. 2 (1966), to appear.

Appleton, J. P.
[1] Structure of a Prandtl-Meyer expansion in an ideal dissociating gas, *Physics of Fluids* **6**, 1057–1062 (1963).
[2] Aerodynamic pitching derivatives of a wedge in hypersonic flow, *AIAA J.* **2**, 2034–2036 (1964).

ARO
[1] Mollier diagram for equilibrium air, ARO, Inc., Arnold Air Force Station, Tenn., 1964.

Ashley, H.
See Zartarian, Hsu, and Ashley.

Ashley, H., and Zartarian, G.
[1] Piston theory—a new aerodynamic tool for the aeroelastician, *J. Aero. Sci.* **23**, 1109–1118 (1956).

Babaev, D. A.
[1] Numerical solution of the problem of flow round the upper surface of a triangular wing by a supersonic stream, *Zh. Vychislitel'noi Mat. i Mat. Fiz.* **2**, 278–289 (1962). Transl. in *U.S.S.R. Comput. Math. and Math. Phys.*, 296–308.

[2] The numerical solution of the problem of the flow about the lower surface of a triangular wing in a supersonic stream, *Zh. Vychislitel'noi Mat. i Mat. Fiz.* **2**, 1086–1101 (1962). Transl. in *U.S.S.R. Comput. Math. and Math. Phys.*, 1305–1324. Also *AIAA J.* **1**, 2224–2231 (1963).

[3] Flow about a triangular wing for large values of *M*, *Zh. Vychislitel'noi Mat. i Mat. Fiz.* **3**, 397–400 (1963). Transl. in *U.S.S.R. Comput. Math. and Math. Phys.*, 528–532.

Babenko, K. I., Voskresenskii, G. P., Lyubimov, A. N., and Rusanov, V. V.

[1] *Three Dimensional Flows of an Ideal Gas Past Smooth Bodies.* Izdat. Nauka, Moscow, 1964.

Bam-Zelikovich, G. M., Bunimovich, A. I., and Mikhailova, M. P.

[1] The motion of slender bodies at hypersonic speeds, *Collection of Papers No.* 4, *Teoretich. Gidromekhanika.* Oborongiz, Moscow, 1949. See also *Izv. Akad. Nauk SSSR, Otd. Tekhn. Nauk, Mekh. i Mashinostr.*, 1959, no. 1, 34-40.

Baron, J. R.

See Shih and Baron.

Bartos, J.

See Gerber and Bartos.

Batchelder, R. A.

[1] An inverse method for inviscid ideal gas flow fields behind analytic shock shapes, *Rep.* SM-42588, Santa Monica Div., Douglas Aircraft Co., Inc., Santa Monica, Calif., 1963.

Bates, D. R. (ed.)

[1] *Atomic and Molecular Processes.* Academic Press, New York, 1962.

Bausch, H.

[1] Die Singularitätsstelle in der Grenzschichtmethode der Hyperschallgasdynamik, *ZAMM* **42**, 557–564 (1962). See also *Vestnik Leningrad. Univ.* 1963, no. 1, 86–89.

Bazzhin, A. P.

[1] On the calculation of hypersonic gas flow past a flat plate with a detached shock wave, *Inzhen. Zh., Moscow* **3**, 222–227 (1963).

Bazzhin, A. P., and Gladkov, A. A.

[1] On a solution of the inverse problem by the method of series expansion, *Inzhen. Zh., Moscow* **3**, 517-518 (1963).

Bechert, K.

[1] Über die Differentialgleichungen der Wellenausbreitungen in Gasen, *Ann. Physik* **39**, 357–372 (1941).

Bedder, D.

See Freeman, Cash, and Bedder.

Belotserkovskii, O. M.

[1] Flow past a circular cylinder with a detached shock, *Dokladi Akad. Nauk SSSR* **113**, 509–512 (1957). M. D. Friedman *Transl.* no. B-131.

[2] Flow with a detached shock wave about a symmetrical profile, *J. Appl. Math. Mech.* (*Prikl. Mat. Mekh.*) **22**, 279-296 (1958).

[3] Flow past a circular cylinder with a detached shock wave, *Vychislitel'noi Mat.*, 1958, no. 3, 149–185. Transl. as *Tech. Memo.* no. RAD-9-TM-59-66, Research and Advanced Dev. Div., Avco Corp., Wilmington, Mass., 1959.

[4] *The Calculation of Flows Past Axisymmetric Bodies with Detached Shock Waves* (*Calculation Formulas and Flow Field Tables*). Vychislitel'nyi Tsentr Akad.

Nauk SSSR, Moscow, 1961. Transl. as *Tech. Memo.* no. RAD-TM-62-64, Research and Advanced Dev. Div., Avco Corp., Wilmington, Mass., 1962. See also *J. Appl. Math. Mech.* (*Prikl. Mat. Mekh.*) **24**, 744–755 (1960).

[5] Symmetric flow about blunt bodies in a supersonic stream of a perfect and real gas, *Zh. Vychislitel'noi Mat. i Mat. Fiz.* **2**, 1062–1085 (1962). Transl. in *U.S.S.R. Comput. Math. and Math. Phys.*, 1272–1304.

Belotserkovskii, O. M., and Chushkin, P. I.

[1] Supersonic flow past blunt bodies, *Arch. Mech. Stos.* **14**, 461–490 (1962).

[2] The numerical solution of problems in gas dynamics, *Basic Developments in Fluid Dynamics*, Vol. I (M. Holt, ed.), pp. 1–126. Academic Press, New York, 1965.

Belotserkovskii, O. M., and Dushin, V. K.

[1] The supersonic flow round blunt bodies of an unbalanced (nonequilibrium) gas stream, *Zh. Vychislitel'noi Mat. i Mat. Fiz.* **4**, 61–77 (1964). Transl. in *U.S.S.R. Comput. Math. and Math. Phys.*, 83–104.

Berger, S. A.

[1] Hypersonic flow over cones, *Rep.* no. 523, Dept. of Aero. Eng., Princeton Univ., Princeton, N. J., 1960.

Bird, G. A.

[1] Effect of wave interactions on pressure distributions in supersonic and hypersonic flow, *AIAA J.* **1**, 634–639 (1963).

Bird, R. B.

See Hirschfelder, Curtiss, and Bird.

Bloor, M. I. G.

[1] Determination of the sonic line in hypersonic flow past a blunt body, *J. Fluid Mech.* **21**, 495–501 (1965).

Blythe, P. A.

[1] The effects of vibrational relaxation on hypersonic flow past blunt bodies, *Aeronaut. Quart.* **14**, 357–373 (1963).

Bogacheva, A. A., and Ladyzhenskii, M. D.

[1] Hypersonic flow past slender blunted elliptic cones, *Inzhen. Zh., Moscow* **2**, no. 3, 14–20 (1962).

Bogdonoff, S. M.

See Vas, Bogdonoff, and Hammitt.

Bohachevsky, I. O., Rubin, E. L., and Mates, R. E.

[1] A direct method for computation of nonequilibrium flows with detached shock waves. I: Two-dimensional flows (Bohachevsky and Rubin), *AIAA J.* **4**, 600–607 (1966); II: Axisymmetric blunt-body at an angle of attack (Bohachevsky and Mates), *AIAA J.* **4**, 776–782 (1966).

Bowen, E. N.

See Ivey, Klunker, and Bowen.

Brainerd, J. J.

See Cole and Brainerd.

Bray, K. N. C.

[1] Atomic recombination in a hypersonic wind-tunnel nozzle, *J. Fluid Mech.* **6**, 1–32 (1959).

[2] Electron-ion recombination in argon flowing through a supersonic nozzle, *The High Temperature Aspects of Hypersonic Flow* (W. C. Nelson, ed.), pp. 67-87. Pergamon, New York, 1964.

Briggs, B. R.
[1] Calculation of supersonic flow past bodies supporting shock waves shaped like elliptic cones, *NASA Tech. Note* no. D-24 (1959).
[2] The numerical calculation of flow past conical bodies supporting elliptic conical shock waves at finite angles of incidence, *NASA Tech. Note* no. D-340 (1960).

Brinkley, S. R., Jr., and Richardson, J. M.
[1] On the structure of plane detonation waves with finite reaction velocity, *Fourth Symposium (Int.) on Combustion*, pp. 450–457. Williams & Wilkins, Baltimore, Md., 1953.

Brocher, E. F.
[1] Comments on the behavior of Sedov's blast-wave solution as $\gamma \to 1$, *J. Aero/Space Sci.* **27**, 955–956 (1960).
[2] On similar solutions for strong blast waves and their application to steady hypersonic flow, *J. Aerospace Sci.* **29**, 694-701 (1962).

Broer, L. J. F.
[1] On the influence of acoustic relaxation on compressible flow, *Appl. Sci. Res. (A)* **2**, 447–468 (1951).
[2] Characteristics of the equations of motion of a reacting gas, *J. Fluid Mech.* **4**, 276–282 (1958).

Brong, E. A., and Leigh, D. C.
[1] Method of Belotserkovskii for asymmetric blunt-body flows, *AIAA J.* **2**, 1852–1853 (1964).

Brook, J. W.
[1] The calculation of nonlinear supersonic conical flows by the method of integral relations, *Rep.* no. FDL-TDR-64-7, AF Flight Dynamics Lab., Res. and Tech. Div., Wright-Patterson Air Force Base, Ohio, 1964.
See Scheuing *et al.*

Bruhn, G., and Haack, W.
[1] Ein Charakteristikenverfahren für dreidimensionale instationäre Gasströmungen, *ZAMP* **96**, 173–190 (1958).

Bulakh, B. M.
[1] Supersonic flow around an inclined circular cone, *J. Appl. Math. Mech. (Prikl. Mat. Mekh.)* **26**, 430–440 (1962).
[2] On the nonsymmetric hypersonic flow around a circular cone, *J. Appl. Math. Mech. (Prikl. Mat. Mekh.)* **26**, 1473–1479 (1962).

Bunimovich, A. I.
See Bam-Zelikovich, Bunimovich, and Mikhailova.

Burhop, E. H. S.
See Massey and Burhop.

Burke, A. F.
[1] Blunt nose and real fluid effects in hypersonic aerodynamics, *IAS Paper* no. 59-114, Inst. of Aero. Sci., New York, N. Y., 1959.

Burstein, S. Z.
[1] Numerical methods in multidimensional shocked flows, *AIAA J.* **2**, 2111–2117 (1964).

Busemann, A.
[1] Drucke auf Kegelformige Spitzen bei Bewegung mit Überschallgeschwindigkeit, *ZAMM* **9**, 496–498 (1929).

[2] Flüssigkeits- und Gasbewegung, Handwörterbuch der Naturwissenschaften, Vol. IV, 2nd Edition, pp. 244–279. Gustav Fischer, Jena, 1933.

[3] Aerodynamischer Auftrieb bei Überschallgeschwindigkeit, *Luftfahrtforschung* **12**, 210–220 (1935).

Butler, D. S.

[1] The numerical solution of hyperbolic systems of partial differential equations in three independent variables, *Proc. Roy. Soc. London. Ser. A* **255**, 232–252 (1960).

Byron, S.

See Petschek and Byron.

Cabannes, H.

[1] Détermination théorique de l'écoulement d'un fluide derrière une onde de choc détachée, *ONERA Note Tech.* no. 5 (1951).

[2] Tables pour la détermination des ondes de choc détachées, *Rech. Aéronaut.* no. 49, 11–15 (1956).

Capiaux, R., and Washington, M.

[1] Nonequilibrium flow past a wedge, *AIAA J.* **1**, 650–660 (1963).

Cash, R. F.

See Freeman, Cash, and Bedder.

Chance Vought

[1] Thermodynamic properties of high temperature air, *Rep.* no. RE-1R-14, Chance Vought Research Center, Dallas, Texas, 1961.

Cheng, H. K.

[1] Similitude of hypersonic real-gas flows over slender bodies with blunted noses, *J. Aero/Space Sci.* **26**, 575–585 (1959).

[2] Hypersonic flows past a yawed circular cone and other pointed bodies, *J. Fluid Mech.* **12**, 169–191 (1962).

Cheng, H. K., Hall, J. G., Golian, T. C., and Hertzberg, A.

[1] Boundary-layer displacement and leading-edge bluntness effects in high-temperature hypersonic flow, *J. Aerospace Sci.* **28**, 353–381, 410 (1961).

Cheng, H. K., and Pallone, A. J.

[1] Inviscid leading-edge effect in hypersonic flow, *J. Aero. Sci.* **23**, 700–702 (1956).

Chernyi, G. G.

[1] Hypersonic flow around a body by an ideal gas, *Izv. Akad. Nauk SSSR, Otd. Tekhn. Nauk*, 1957, no. 6, 77–85. M. D. Friedman *Transl.* no. C-111. See also *Dokladi Akad. Nauk SSSR* **107**, 221–224 (1956). M. D. Friedman *Transl.* no. C-106.

[2] Effect of slight blunting of leading edge of an immersed body on the flow around it at hypersonic speeds, *Izv. Akad. Nauk SSSR, Otd. Tekhn. Nauk*, 1958, no. 6, 54–66. *NASA Tech. Transl.* no. F-35 (1960).

[3] *Introduction to Hypersonic Flow.* Fizmatgiz, Moscow, 1959. English transl. (R. F. Probstein, transl. and ed.), Academic Press, New York, 1961.

[4] Application of integral relationships in problems of propagation of strong shock waves, *J. Appl. Math. Mech.* (*Prikl. Mat. Mekh.*) **24**, 159–165 (1960).

[5] Integral methods for the calculation of gas flows with strong shock waves, *J. Appl. Math. Mech.* (*Prikl. Mat. Mekh.*) **25**, 138-147 (1961).

See Gonor and Chernyi.

Chester, W.

[1] Supersonic flow past a bluff body with a detached shock. Part I. Two-dimensional body, *J. Fluid Mech.* **1**, 353–365 (1956).

[2] Supersonic flow past a bluff body with a detached shock. Part II. Axisymmetrical body, *J. Fluid Mech.* **1**, 490–496 (1956).

Chisnell, R. F.

[1] The motion of a shock wave in a channel, with applications to cylindrical and spherical shock waves, *J. Fluid Mech.* **2**, 286–298 (1957).

Chu, B. T.

[1] On weak interaction of strong shock and Mach waves generated downstream of the shock, *J. Aero. Sci.* **19**, 433–446 (1952).

[2] Wave propagation and the method of characteristics in reacting gas mixtures with applications to hypersonic flow, *WADC Tech. Note* no. 57-213, Div. of Eng., Brown Univ., Providence, R. I., 1957. See also *Preprints 1958 Heat Transfer and Fluid Mech. Inst.*, pp. 80-90. Stanford Univ. Press, Stanford, 1958.

Chushkin, P. I.

[1] Investigation of flow round blunt bodies of revolution at supersonic velocities, *Zh. Vychislitel'noi Mat. i Mat. Fiz.* **2**, 255–277 (1962). Transl. in *U.S.S.R. Comput. Math. and Math. Phys.*, 272–295.

See Belotserkovskii and Chushkin.

Chushkin, P. I., and Shchennikov, V. V.

[1] The calculation of certain conical flows without axial symmetry, *Inzhen.—Fiz. Zh.* **3**, no. 7, 88–94 (1960). Transl. as *Rep.* no. TR-20, Res. Dept., Grumman Aircraft Eng. Corp., Bethpage, N. Y., 1962.

Chushkin, P. I., and Shulishnina, N. P.

[1] *Tables of Supersonic Flow About Blunted Cones.* Vychislitel'nyi Tsentr Akad. Nauk SSSR, Moscow, 1961. Transl. as *Tech. Memo.* no. RAD-TM-62-63, Research and Advanced Dev. Div., Avco Corp., Wilmington, Mass., 1962.

Clarke, J. F., and McChesney, M.

[1] *The Dynamics of Real Gases.* Butterworths, Washington, 1964.

Cline, C. W.

See Ivey and Cline.

Coburn, N., and Dolph, C. L.

[1] The method of characteristics in the three-dimensional stationary supersonic flow of a gas, *Proc. of Symposia in Appl. Math.*, Vol. I, pp. 55–66. Am. Math. Soc., New York, 1949.

Codd, J.

See Maccoll and Codd.

Cole, J. D.

[1] Newtonian flow theory for slender bodies, *J. Aero. Sci.* **24**, 448–455 (1957).

[2] Note on the lift of slender nose shapes according to Newtonian theory, *J. Aero. Sci.* **25**, 399 (1958).

Cole, J. D., and Brainerd, J. J.

[1] Slender wings at high angles of attack in hypersonic flows, *Hypersonic Flow Research* (F. R. Riddell, ed.), pp. 321–343. Academic Press, New York, 1962.

Cottrell, T. L., and McCoubrey, J. C.

[1] *Molecular Energy Transfer in Gases.* Butterworths, London, 1961.

Courant, R., and Hilbert, D.
[1] *Methods of Mathematical Physics*, Vol. I. Wiley (Interscience), New York, 1953.

Cranz, C.
[1] *Lehrbuch der Ballistik*, Vol. I (*Äussere Ballistik*), 5th Edition. Julius Springer, Berlin, 1925. Reprinted by Edwards Bros., Ann Arbor, 1943.
[2] *Lehrbuch der Ballistik, Ergänzungsband.* Julius Springer, Berlin, 1936. Reprinted by Edwards Bros., Ann Arbor, 1943.

Crocco, L.
[1] Singolarita della corrente gassosa iperacustica nell'intorno di una prora a diedro, *Aerotecnica* **17**, 519–534 (1937).

Culler, G. J., and Fried, B. D.
[1] The propagation of shock waves. I, *Rep.* no. ARL-6-20, Aero. Res. Lab., The Ramo-Wooldridge Corp., Los Angeles, Calif., 1956.

Curtiss, C. F.
See Hirschfelder, Curtiss, and Bird.

Curtiss, C. F., and Hirschfelder, J. O.
[1] Integration of stiff equations, *Proc. Nat. Acad. Sci.* **38**, 235–243 (1952).

Daskin, W., and Feldman, L.
[1] The characteristics of two-dimensional sails in hypersonic flow, *J. Aero. Sci.* **25**, 53–55 (1958).

Dennis, D. H.
See Eggers, Resnikoff, and Dennis.
See Syvertson and Dennis.

Dolph, C. L.
See Coburn and Dolph.

Donaldson, C. duP.
See Scheuing *et al.*
See Sullivan, Donaldson, and Hayes.

Donaldson, C. duP., Sullivan, R. D., and Hayes, W. D.
[1] The effect of surface perturbations on the pressure distributions about blunt two-dimensional bodies at high Mach numbers, *Rep.* no. FDL-TDR-64-17, AF Flight Dynamics Lab., Res. and Tech. Div., Wright-Patterson Air Force Base, Ohio, 1964.

Dorodnitsyn, A. A.
[1] On a method of numerical solution of some nonlinear problems of aero-hydrodynamics, *Proc. 9th Internatl. Congr. Appl. Mech.*, Vol. I, p. 485. Univ. of Brussels, Brussels, 1957.
[2] A contribution to the solution of mixed problems of transonic aerodynamics, *Advances in Aeronautical Sciences*, Vol. 2 (Th. von Kármán, chmn. ed. comm.), 832–843. Pergamon, New York, 1959.
[3] Numerical methods in gasdynamics, *Arch. Mech. Stos.* **12**, 13–27 (1960).
[4] General method of integral relations and its application to boundary layer theory, *Advances in Aeronautical Sciences*, Vol. 3 (Th. von Kármán, chmn. ed. comm.), 207–219. Macmillan, New York, 1962.

Dorrance, W. H.
[1] *Viscous Hypersonic Flow.* McGraw-Hill, New York, 1962.

Drebinger, J. W.
[1] Detached shock waves, *Ph.D. Thesis*, Faculty of Arts and Sci., Harvard Univ., Cambridge, Mass., 1950.

Dushin, V. K.
See Belotserkovskii and Dushin.

East, R. A.
[1] A theoretical and experimental study of oscillating wedge shaped airfoils in hypersonic flow, *A.A.S.U. Rep.* no. 228, Dept. of Aero. and Astro., Univ. of Southampton, Hampshire, England, 1962.

Edelfelt, I. H.
See Gravalos, Edelfelt, and Emmons.

Eggers, A. J., Jr., Resnikoff, M. M., and Dennis, D. H.
[1] Bodies of revolution having minimum drag at high supersonic airspeeds, *NACA Rep.* no. 1306 (1957). Supersedes *Tech. Note* no. 3666 (1956).

Eggers, A. J., Jr., and Savin, R. C.
[1] Approximate methods for calculating the flow about non-lifting bodies of revolution at high supersonic airspeeds, *NACA Tech. Note* no. 2579 (1951).
[2] A unified two-dimensional approach to the calculation of three-dimensional hypersonic flows, with application to bodies of revolution, *NACA Rep.* no. 1249 (1955). Supersedes *Tech. Note* no. 2811 (1952).

Eggers, A. J., Jr., Savin, R. C., and Syvertson, C. A.
[1] The generalized shock-expansion method and its application to bodies traveling at high supersonic air speeds, *J. Aero. Sci.* **22**, 231-238, 248 (1955).

Eggers, A. J., Jr., and Syvertson, C. A.
[1] Inviscid flow about airfoils at high supersonic speeds, *NACA Tech. Note* no. 2646 (1952). Superseded by Eggers, Syvertson, and Kraus [1].

Eggers, A. J., Jr., Syvertson, C. A., and Kraus, S.
[1] A study of inviscid flow about airfoils at high supersonic speeds, *NACA Rep.* no. 1123 (1953). Supersedes Eggers and Syvertson [1] and Kraus [1].

Emanuel, G.
[1] Problems underlying the numerical integration of the chemical and vibrational rate equations in a near-equilibrium flow, *Rep.* no. AEDC-TDR 63-82, Arnold Eng. Dev. Center, AF Systems Command, Arnold Air Force Station, Tenn., 1963.

Emmons, H. W.
See Gravalos, Edelfelt, and Emmons.

Epstein, P. S.
[1] On the air resistance of projectiles, *Proc. Nat. Acad. Sci.* **17**, 532–547 (1931).

Eschenroeder, A. Q.
See Hall, Eschenroeder, and Marrone.

Evans, M. E.
See Mangler and Evans.

Evans, M. W., and Harlow, F. H.
[1] The particle-in-cell method for hydrodynamic calculations, *Rep.* no. LA-2139, Los Alamos Sci. Lab., Los Alamos, New Mexico, 1957.
[2] Calculation of supersonic flow past an axially symmetric cylinder, *J. Aero. Sci.* **25**, 269–270 (1958).

[3] Calculation of unsteady supersonic flow past a circular cylinder, *ARS J.* **29**, 46–48 (1959).

Feldman, L.
See Daskin and Feldman.

Feldman, S.
[1] *Hypersonic gas dynamic charts for equilibrium air.* Avco Research Lab., Everett, Mass., 1957.
[2] Some shock tube experiments on the chemical kinetics of air at high temperatures, *J. Fluid Mech.* **3**, 225–242 (1957).
[3] Hypersonic conical shocks for dissociated air in thermodynamic equilibrium, *Jet Propulsion* **27**, 1253–1255 (1957).

Ferrari, C.
[1] Campo aerodinamico a velocita iperacustica attorno a un solido di rivoluzione a proro acuminata, *Aerotecnica* **16**, 121–130 (1936). *Transl.* no. A9-T-18, Grad. Div. Appl. Math., Brown Univ., Providence, R. I., 1948.
[2] Sui moti conici rotazionali, *Aerotecnica* **31**, 64–66 (1951).

Ferri, A.
[1] Supersonic flow around circular cones at angles of attack, *NACA Rep.* no. 1045 (1951). Supersedes *Tech. Note* no. 2236 (1950).
[2] The method of characteristics, *General Theory of High Speed Aerodynamics* (W. R. Sears, ed.), Sect. G, pp. 583-669 (Vol. VI of *High Speed Aerodynamics and Jet Propulsion*). Princeton Univ. Press, Princeton, 1954.
[3] Supersonic flows with shock waves, *General Theory of High Speed Aerodynamics* (W. R. Sears, ed.), Sect. H, pp. 670–747 (Vol. VI of *High Speed Aerodynamics and Jet Propulsion*). Princeton Univ. Press, Princeton, 1954.
See Vaglio-Laurin and Ferri.

Fisher, N. H.
See Vincenti and Fisher.

Flemming, D. P.
[1] A numerical method for solving the inverse blunt body problem in hypersonic flow, *Tech. Rep.* no. 480, Canadian Armament Res. and Dev. Establish., Valcartier, Quebec, 1964.

Fowler, J. E., and French, D. M.
[1] A graphical correlation based on the Taylor-Maccoll results of the air conditions just outside the boundary layer of a cone as related to the conditions of the undisturbed air, *Data Folder* no. 85501, General Electric Company, Schenectady, N. Y., 1946.

Fraenkel, L. E.
[1] The hypersonic flow of a polyatomic gas past bodies of finite thickness, *Proc. 9th Internatl. Congr. Appl. Mech.*, Vol. II, pp. 255–265. Univ. of Brussels, Brussels, 1957.

Freeman, N. C.
[1] On the theory of hypersonic flow past plane and axially symmetric bluff bodies, *J. Fluid Mech.* **1**, 366–387 (1956).
[2] Non-equilibrium flow of an ideal dissociating gas, *J. Fluid Mech.* **4**, 407–425 (1958).
[3] On the Newtonian theory of hypersonic flow for a blunt body, *Rep.* no. 467, Dept. of Aero. Eng., Princeton Univ., Princeton, N. J., 1959.

[4] A note on the explosion solution of Sedov with application to the Newtonian theory of unsteady hypersonic flow, *J. Aero/Space Sci.* **27**, 77-78, 956 (1960).

[5] On a singular point in the Newtonian theory of hypersonic flow, *J. Fluid Mech.* **8**, 109–122 (1960).

[6] Newtonian theory of hypersonic flow at large distances from bluff axially symmetric bodies, *Hypersonic Flow Research* (F. R. Riddell, ed.), pp. 345–377. Academic Press, New York, 1962.

[7] Asymptotic solutions in hypersonic flow: (an approach to second-order solutions of hypersonic small-disturbance theory), *Research Frontiers in Fluid Dynamics* (R. J. Seegar and G. Temple, eds.), pp. 284–307. Wiley (Interscience), New York, 1965.

Freeman, N. C., Cash, R. F., and Bedder, D.

[1] An experimental investigation of asymptotic hypersonic flows, *J. Fluid Mech.* **18**, 379–384 (1964).

French, D. M.

See Fowler and French.

Fried, B. D.

See Culler and Fried.

Fuller, F. B.

[1] Numerical solutions for supersonic flow of an ideal gas around blunt two-dimensional bodies, *NASA Tech. Note* no. D-791 (1961).

Garabedian, P. R.

[1] Numerical construction of detached shock waves, *J. Math. Phys.* **36**, 192–205 (1957).

Garabedian, P. R., and Lieberstein, H. M.

[1] On the numerical calculation of detached bow shock waves in hypersonic flow, *J. Aero. Sci.* **25**, 109–118 (1958).

Geiger, R. E.

See Li and Geiger.

Gerber, N.

See Sedney and Gerber.

See Sedney, South, and Gerber.

Gerber, N., and Bartos, J.

[1] Calculation of flow-variable gradients behind curved shock waves, *J. Aerospace Sci.* **27**, 958–959 (1960).

Gibson, W. E., and Marrone, P. V.

[1] Correspondence between normal-shock and blunt-body flows, *Physics of Fluids* **5**, 1649–1656 (1962).

[2] A similitude for non-equilibrium phenomena in hypersonic flight, *in The High Temperature Aspects of Hypersonic Flow* (W. C. Nelson, ed.), pp. 105–129. Pergamon, New York, 1964.

Gilinskii, S. M., Telenin, G. F., and Tinyakov, G. P.

[1] A method for computing supersonic flow around blunt bodies, accompanied by a detached shock wave, *Izv. Akad. Nauk SSSR, Otd. Tekhn. Nauk, Mekh. i Mashinostr.*, 1964, no. 4, 9–28. *NASA Tech. Transl.* no. F-297 (1965).

Gladkov, A. A.

See Bazzhin and Gladkov.

Glass, I. I., and Kawada, H.
[1] Prandtl-Meyer flow of dissociated and ionized gases, *UTIA Rep.* no. 85, Inst. of Aerophysics, Univ. of Toronto, Toronto, Canada, 1962.

Godunov, S. K.
[1] A difference method for the numerical calculation of discontinuous solutions of the equations of hydrodynamics, *Mat. Sb.* **47**, 271–306 (1959).

Godunov, S. K., Zabrodin, A. V., and Prokopov, G. P.
[1] A computational scheme for two-dimensional nonstationary problems of gas dynamics and calculation of the flow from a shock wave approaching a stationary state, *Zh. Vychislitel'noi Mat. i Mat. Fiz.* **1**, 1020–1050 (1961). Transl. in *U.S.S.R. Comput. Math. and Math. Phys.*, 1187–1219.

Gold, R., and Holt, M.
[1] Calculation of supersonic flow past a flat-headed cylinder by Belotserkovskii's method, *AFOSR Tech. Note* no. 59–199, Div. of Appl. Math., Brown Univ., Providence, R. I., 1959.

Goldsworthy, F. A.
[1] Two-dimensional rotational flow at high Mach number past thin airfoils, *Quart. J. Mech. Appl. Math.* **5**, 54–63 (1952).

Golian, T. C.
See Cheng, Hall, Golian, and Hertzberg.

Gonor, A. L.
[1] Hypersonic flow around a cone at an angle of attack, *Izv. Akad. Nauk SSSR, Otd. Tekhn. Nauk*, 1958, no. 7, 102–105. M. D. Friedman *Transl.* no. G-154.
[2] Hypersonic gas flow around conical bodies, *Izv. Akad. Nauk SSSR, Otd. Tekhn. Nauk, Mekh. i Mashinostr.*, 1959, no. 1, 34–40. M. D. Friedman *Transl.* no. G-163.
[3] Location of frontal wave in asymmetrical flow of gas at high supersonic speed over a pointed body, *Izv. Akad. Nauk SSSR, Otd. Tekhn. Nauk, Mekh. i Mashinostr.*, 1959, no. 5, 117–118. Transl. *ARS J.* **30**, 841–842 (1960).
[4] Determination of the shape of a body of minimum drag at hypersonic speed, *J. Appl. Math. Mech.* (*Prikl. Mat. Mekh.*) **24**, 1628–1635 (1960).
[5] Exact solution of the problem of supersonic flow of gas past some three-dimensional bodies, *J. Appl. Math. Mech. (Prikl. Mat. Mekh.)* **28**, 1178–1181 (1964).

Gonor, A. L., and Chernyi, G. G.
[1] On minimum drag bodies at hypersonic speeds, *Izv. Akad. Nauk SSSR, Otd. Tekhn. Nauk*, 1957, no. 7, 89–93. M. D. Friedman *Transl.* no. G-141.

Gordon, H. D.
See Van Dyke and Gordon.

Gravalos, F. G.
[1] Supersonic flow about blunt bodies of revolution, *Advances in Astronautical Sciences*, Vol. 2, *Paper* no. 5. Plenum Press, New York, 1958.

Gravalos, F. G., Edelfelt, I. H., and Emmons, H. W.
[1] The supersonic flow about a blunt body of revolution for gases at chemical equilibrium, *Proc. 9th Internatl. Astronautical Cong., Amsterdam 1958*, Vol. I, pp. 312–332. Springer, Vienna, 1959.

Gray, K. E.
See Scheuing *et al.*

Green, J. R., and Southwell, R. V.

[1] Relaxation methods applied to engineering problems, IX. High-speed flow of compressible fluid through a two-dimensional nozzle, *Phil. Trans. Roy. Soc. London. Ser. A* **239**, 367–386 (1944).

Grimminger, G., Williams, E. P., and Young, G. B. W.

[1] Lift on inclined bodies of revolution in hypersonic flow, *J. Aero. Sci.* **17**, 675–690 (1950).

Grodzovskii, G. L.

[1] Certain peculiarities of the flow around bodies at high supersonic velocities, *Izv. Akad. Nauk SSSR, Otd. Tekhn. Nauk*, 1957, no. 6, 86–92. M. D. Friedman *Transl.* no. G-140.

Guderley, G.

[1] Starke kugelige und zylindrische Verdichtungstösse in der Nähe des Kugelmittelpunktes bzw. der Zylinderachse, *Luftfahrtforschung* **19**, 302–312 (1942).

Guiraud, J. P.

[1] Sur la méthode de choc-détente, *Compt. Rend. Acad. Sci. Paris* **245**, 1778–1780 (1957).

[2] Écoulement hypersonique d'un fluide parfait sur une plaque plane comportant un bord d'attaque d'épaisseur finie, *Compt. Rend. Acad. Sci. Paris* **246**, 2842–2845 (1958).

[3] Lignes de courant d'un écoulement newtonien, 736–738; Écoulement newtonien sur une surface, 775–778, *Compt. Rend. Acad. Sci. Paris* **247** (1958). See also *Rech. Aéronaut.*, no. 71, 11–28 (1959).

[4] Newtonian flow over a surface—theory and applications, *Hypersonic Flow* (A. R. Collar and J. Tinkler, eds.), pp. 253–296. Butterworths, London, 1960.

[5] Topics in hypersonic flow theory, *SUDAER Rep.* no. 154, Dept. of Aero. and Astro., Stanford Univ., Stanford, Calif., 1963.

[6] Effet d'émoussement en hypersonique, *Arch. Mech. Stos.* **16**, 244–271 (1964). See also Asymptotic theory in hypersonic flow, *ONERA T. P.* no. 132 (1964). Also in *Proc. Internatl. Symp. on Fundamental Phenomena in Hypersonic Flow* (J. G. Hall, ed.), pp. 70–84. Cornell Univ. Press, Ithaca, New York, 1966.

Guiraud, J. P., Vallee, D., and Zolver, R.

[1] Bluntness effects in hypersonic small disturbance theory, *Basic Developments in Fluid Dynamics*, Vol. I (M. Holt, ed.), pp. 127–247. Academic Press, New York, 1965.

Haack, W.

See Bruhn and Haack.

Hadamard, J.

[1] *Lectures on Cauchy's Problem in Linear Partial Differential Equations*. Yale Univ. Press, New Haven, 1923. Reprinted by Dover, New York, 1952.

Hall, J. G.

See Cheng, Hall, Golian, and Hertzberg.

Hall, J. G., Eschenroeder, A. Q., and Marrone, P. V.

[1] Blunt-nose inviscid airflows with coupled nonequilibrium processes, *J. Aerospace Sci.* **29**, 1038–1051 (1962).

Hamaker, F. M.

[1] Numerical solution of the flow of a perfect gas over a circular cylinder at infinite Mach number, *NASA Tech. Memo.* no. 2-25-59A (1959).

Hamaker, F. M., Neice, S. E., and Wong, T. J.

[1] The similarity law for hypersonic flow and requirements for dynamic similarity of related bodies in free flight, *NACA Rep.* no. 1147 (1953). Supersedes Hamaker and Wong [1] and *Tech. Note* no. 2443 (1951).

Hamaker, F. M., and Wong, T. J.

[1] The similarity law for nonsteady hypersonic flows and requirements for the dynamical similarity of related bodies in free flight, *NACA Tech. Note* no. 2631 (1952). Superseded by Hamaker, Neice, and Wong [1].

Hammitt, A. G.

See Vas, Bogdonoff, and Hammitt.

Hammitt, A. G., and Murthy, K. R. A.

[1] Approximate solutions for supersonic flow over wedges and cones, *J. Aero/Space Sci.* **27**, 71–73 (1960).

Hansen, C. F.

[1] Approximations for the thermodynamic and transport properties of high-temperature air, *NACA Tech. Note* no. 4150 (1958).

Harlow, F. H.

[1] The particle-in-cell computing method for fluid dynamics, *Methods in Computational Physics*, Vol. 3 (B. Alder, S. Fernbach, M. Rotenberg, eds.), pp. 319–343. Academic Press, New York, 1964.

See Evans and Harlow.

Harwell, K. E., and Jahn, R. G.

[1] Initial ionization rates in shock-heated argon, krypton, and xenon, *Physics of Fluids* **7**, 214–222 (1964).

Hasimoto, Z.

[1] Some local properties of plane flow behind a curved shock wave, Dept. Math. and Phys. *Rep.*, Ritumeikan Univ., Kyoto, Japan, 1956.

Hayes, W. D.

[1] On hypersonic similitude, *Quart. Appl. Math.* **5**, 105–106 (1947).

[2] Pseudotransonic similitude and first-order wave structure, *J. Aero. Sci.* **21**, 721-730 (1954).

[3] Some aspects of hypersonic flow, The Ramo-Wooldridge Corp. *Rep.*, Los Angeles, Calif., 1955.

[4] Hypersonic flow fields at small density ratio (and its continuation), The Ramo-Wooldridge Corp. *Rep.*, Los Angeles, Calif., 1955.

[5] The vorticity jump across a gasdynamic discontinuity, *J. Fluid Mech.* **2**, 595–600 (1957).

[6] The basic theory of gasdynamic discontinuities, *Fundamentals of Gas Dynamics* (H. W. Emmons, ed.), Sect. D, pp. 416–481 (Vol. III of *High Speed Aerodynamics and Jet Propulsion*). Princeton Univ. Press, Princeton, 1958.

[7] Rotational stagnation point flow, *J. Fluid Mech.* **19**, 366–374 (1964). See also *J. Appl. Math. Mech. (Prikl. Mat. Mekh.)* **28**, 840–843 (1964).

[8] Inviscid hypersonic flows on blunt bodies, *Proc. 12th Internatl. Congr. Appl. Mech.* Munich, 1964. Springer-Verlag, Berlin, to appear.

[9] Hypersonic stagnation-region flows with asymmetry, *Arch. Mech. Stos.*, to appear.

See Antonov and Hayes.

See Donaldson, Sullivan, and Hayes.

See Rockett and Hayes.

See Scheuing *et al.*
See Sullivan, Donaldson, and Hayes.
See Wecker and Hayes.

Hayes, W. D., and Probstein, R. F.
[1] Viscous hypersonic similitude, *J. Aero/Space Sci.* **26**, 815–824 (1959).

Heims, S. P.
[1] Prandtl-Meyer expansion of chemically reacting gases in local chemical and thermodynamic equilibrium, *NACA Tech. Note* no. 4230 (1958).

Hertzberg, A.
See Cheng, Hall, Golian, and Hertzberg.

Herzfeld, K. F., and Litovitz, T. A.
[1] *Absorption and Dispersion of Ultrasonic Waves.* Academic Press, New York, 1959.

Hida, K.
[1] An approximate study on the detached shock wave in front of a circular cylinder and a sphere, *J. Phys. Soc. Japan* **8**, 740–745 (1953).
[2] Blunt body theory for hypersonic flow, *AFOSR Rep.* no. 204, Guggenheim Aero. Lab., Calif. Inst. Tech., Pasadena, Calif., 1961.
[3] Thickness effects on the force of slender delta wings in hypersonic flow, *AIAA J.* **3**, 427–433 (1965).

Hilbert, D.
See Courant and Hilbert.

Hirschfelder, J. O.
See Curtiss and Hirschfelder.

Hirschfelder, J. O., Curtiss, C. F., and Bird, R. B.
[1] *Molecular Theory of Gases and Liquids.* Wiley, New York, 1954.

Holt, M.
[1] The method of characteristics for steady supersonic rotational flow in three dimensions, *J. Fluid Mech.* **1**, 409–423 (1956).
See Gold and Holt.

Hord, R. A.
[1] An approximate solution for axially symmetric flow over a cone with an attached shock wave, *NACA Tech. Note* no. 3485 (1955).

Hsu, P. T.
See Zartarian, Hsu, and Ashley.

Huckel, V.
See Morgan, Runyan, and Huckel.

Inger, G. R.
[1] Similitude of hypersonic flows over slender bodies in nonequilibrium dissociated gases, *AIAA J.* **1**, 46–53 (1963).

Il'yushin, A. A.
[1] The law of plane sections in the aerodynamics of high supersonic speeds, *Prikl. Mat. Mekh.* **20**, 733–755 (1956).

Inouye, M., and Lomax, H.
[1] Comparison of experimental and numerical results for the flow of a perfect gas about blunt-nosed bodies, *NASA Tech. Note* no. D-1246 (1962).

Isenberg, J. S., and Lin, C. C.

[1] The method of characteristics in compressible flow, Part I (Steady supersonic flow), *Tech. Rep.* no. F-TR-1173A-ND (GDAM A9-M 11/1), Air Materiel Command, Wright-Patterson Air Force Base, Ohio, 1947. See also Part IA (Tables and Charts), and Part IB (Numerical Examples), 1947.

Ivey, H. R., and Cline, C. W.

[1] Effect of heat-capacity lag on the flow through oblique shock waves, *NACA Tech. Note* no. 2196 (1950).

Ivey, H. R., Klunker, E. B., and Bowen, E. N.

[1] A method for determining the aerodynamic characteristics of two- and three-dimensional shapes at hypersonic speeds, *NACA Tech. Note* no. 1613 (1948).

Ivey, H. R., and Morrissette, R. R.

[1] An approximate determination of the lift of slender cylindrical bodies and wing-body combinations at very high supersonic speeds, *NACA Tech. Note* no. 1740 (1948).

Jahn, R. G.

See Harwell and Jahn.

Johannsen, N. H., and Meyer, R. E.

[1] Axially-symmetrical supersonic flow near the center of an expansion, *Aeronaut. Quart.* **2**, 127–142 (1950).

Kao, H. C.

[1] A new technique for the direct calculation of blunt-body flow fields, *AIAA J.* **3**, 161–163 (1965).

Katskova, O. N., and Kraiko, A. N.

[1] *The Calculation of Plane and Axisymmetric Supersonic Flows in the Presence of Nonequilibrium Processes.* Vychislitel'nyi Tsentr Akad. Nauk SSSR, Moscow, 1964.

Kawada, H.

See Glass and Kawada.

Kendall, J. M., Jr.

[1] Experiments on supersonic blunt-body flows, *Progress Rep.* no. 20-372, Jet Prop. Lab., Calif. Inst. Tech., Pasadena, Calif., 1959.

Kholyavko, V. I.

[1] Flow past plates at high supersonic speeds, *Izv. Akad. Nauk SSSR, Otd. Tekhn. Nauk, Mekh. i Mashinostr.*, 1962, no. 5, 26–31.

Kim, C. S.

[1] Experimental studies of supersonic flow past a circular cylinder, *J. Phys. Soc. Japan* **11**, 439–445 (1956).

Kirkwood, J. G.

See Wood and Kirkwood.

Klunker, E. B.

See Ivey, Klunker, and Bowen.

Kochina, N. N., and Mel'nikova, N. S.

[1] Strong point-blasts in a compressible medium, *J. Appl. Math. Mech.* (*Prikl. Mat. Mekh.*) **22**, 1–19 (1958).

[2] On the properties of the solution of the problem of a point explosion in compressible media, *Soviet Physics—Doklady* (*Dokladi Akad. Nauk SSSR*) **6**, 380–383 (1961).

Kogan, A.

[1] On inviscid flow near an airfoil leading edge or an ogive tip at high supersonic Mach numbers, *J. Aero. Sci.* **23**, 794–795 (1956).

[2] An application of Crocco's stream function to the study of rotational supersonic flow past airfoils, *Quart. J. Mech. Appl. Math.* **11**, 1–23 (1958).

[3] On supersonic flow past thick airfoils, *J. Aero/Space Sci.* **27**, 504–508, 516 (1960).

Kopal, Z.

[1] Tables of supersonic flow around cones, *Tech. Rep.* No. 1, Dept. of Elect. Eng., Mass. Inst. Tech., Cambridge, Mass., 1947.

Korobeinikov, V. P., Mel'nikova, N. S., and Ryazanov, E. V.

[1] *Theory of the Point Explosion.* Fizmatgiz, Moscow, 1961.

Kraiko, A. N.

[1] The determination of minimal drag bodies by Newton's and Busemann's drag laws, *J. Appl. Math. Mech.* (*Prikl. Mat. Mekh.*) **27**, 723–739 (1963).

See Katskova and Kraiko.

Krasheninnikova, N. L.

[1] On the unsteady motion of a gas displaced by a piston, *Izv. Akad. Nauk SSSR, Otd. Tekhn. Nauk*, 1955, no. 8, 22–36.

Kraus, S.

[1] An analysis of supersonic flow in the region of the leading edge of curved airfoils, including charts for determining surface-pressure gradient and shock-wave curvature, *NACA Tech. Note* no. 2729 (1952). Superseded by Eggers, Syvertson, and Kraus [1].

Kruger, C. H., Jr.

See Vincenti and Kruger.

Kubota, T.

[1] Investigation of flow around simple bodies in hypersonic flow, *Memo.* no. 40, Hypersonic Research Project, Guggenheim Aero. Lab., California Inst. of Tech., Pasadena, Calif., 1957.

See Lees and Kubota.

Kuehn, D. M.

[1] Experimental and theoretical pressures on blunt cylinders for equilibrium and nonequilibrium air at hypersonic speeds, *NASA Tech. Note* no. D-1979 (1963).

Ladyzhenskii, M. D.

[1] On hypersonic flow past slender blunted bodies, *Izv. Akad. Nauk SSSR, Otd. Tekhn. Nauk, Mekh. i Mashinostr.*, 1961, no. 1, 150–151.

[2] Hypersonic area rule, *Inzhen. Zh., Moscow* **1**, no. 1, 159–163 (1961). Transl. *AIAA J.* **1**, 2696–2698 (1963).

[3] Generalized hypersonic area rule, *Izv. Akad. Nauk SSSR, Otd. Tekhn. Nauk, Mekh. i Mashinostr.*, 1961, no. 3, 188–189.

See Bogacheva and Ladyzhenskii.

Landahl, M. T.

[1] Unsteady flow around thin wings at high Mach numbers, *J. Aero. Sci.* **24**, 33–38 (1957).

Landau, L. D., and Lifshitz, E. M.

[1] *Fluid Mechanics.* Addison-Wesley, Reading, Mass., 1959.

Langelo, V. A.

[1] The inviscid reacting flow field about hypersonic bodies, *Rep.* no. R635D90,

Space Sci. Lab., Missile and Space Div., General Electric Co., King of Prussia, Pa., 1963.

Latter, R.

[1] Similarity solution for a spherical shock wave, *J. Appl. Phys.* **26**, 954-960 (1955).

Laval, P.

[1] Écoulements newtoniens sur des surfaces coniques en incidence, *Rech. Aéronaut.*, no. 73, 5–16 (1959).

[2] Écoulements hypersoniques sur des surfaces coniques, *ONERA Publ.* no. 106 (1962).

[3] Écoulement supersonique au voisinage d'une surface conique de forme arbitraire, *Rech. Aerospatiale*, no. 97, 11–21 (1963).

[4] Méthode P.L.K. et problème inverse en hypersonique pour un choc en x^n, *ONERA T.P.* no. 176 (1964).

Lax, P. D.

[1] Weak solutions of nonlinear hyperbolic equations and their numerical computation, *Commun. Pure Appl. Math.* **7**, 159–193 (1954).

Lax, P. D., and Wendroff, B.

[1] System of conservation laws, *Commun. Pure Appl. Math.* **13**, 217–237 (1960).

Leavitt, J. A.

[1] A power series solution for compressible flow past a conical shock wave, *Rep.* no. NYO-10, 432, Inst. of Math. Sci., New York Univ., New York, 1963.

Lee, J. T.

[1] Inviscid hypersonic flow for power-law shock waves, *Rep.* no. 9813-6003-KU 000, Space Technology Labs., Inc., Redondo Beach, Calif., 1963.

Lees, L.

[1] Hypersonic flow, *Proc. Fifth Internatl. Aeronautical Conf., Los Angeles*, pp. 241–275. Inst. Aero. Sci., New York, 1955.

Lees, L., and Kubota, T.

[1] Inviscid hypersonic flow over blunt-nosed slender bodies, *J. Aero. Sci.* **24**, 195–202 (1957).

Leigh, D. C.

See Brong and Leigh.

Lewis, C. H.

[1] Plane, cylindrical, and spherical blast waves based upon Oshima's quasi-similarity model, *AEDC Tech. Note* no. 61-157, Arnold Eng. Dev. Center, ARO, Inc., Arnold Air Force Station, Tenn., 1961.

[2] The blast-hypersonic flow analogy based upon Oshima's quasi-similarity model, *AEDC Tech. Note* no. 61-158, Arnold Eng. Dev. Center, ARO, Inc., Arnold Air Force Station, Tenn., 1961.

Lewis, G.

[1] Two methods using power series for solving analytic initial value problems, *Rep.* no. NYO-2881, Inst. of Math. Sci., New York Univ., New York, 1960.

Li, T. Y., and Geiger, R. E.

[1] Stagnation point of a blunt body in hypersonic flow, *J. Aero. Sci.* **24**, 25–32 (1957).

Lick, W.

[1] Inviscid flow of a reacting mixture of gases around a blunt body, *J. Fluid Mech.* **7**, 128–144 (1960).

Lieberstein, H. M.
[1] Further numerical data on blunt bodies, *J. Aero/Space Sci.* **25**, 660–661 (1958).
See Garabedian and Lieberstein.

Lifshitz, E. M.
See Landau and Lifshitz.

Lighthill, M. J.
[1] The flow behind a stationary shock, *Phil. Mag.* **40**, 214–220 (1949).
[2] Oscillating airfoils at high Mach number, *J. Aero. Sci.* **20**, 402–406 (1953).
[3] Dynamics of a dissociating gas. Part I. Equilibrium flow, *J. Fluid Mech.* **2**, 1–32 (1957).

Lin, C. C.
[1] Note on Garabedian's paper "Numerical construction of detached shock waves," *J. Math. Phys.* **36**, 206–209 (1957).
See Isenberg and Lin.
See Shen and Lin.

Lin, C. C., Reissner, E., and Tsien, H. S.
[1] On two-dimensional non-steady motion of a slender body in a compressible fluid, *J. Math. Phys.* **27**, 220–231 (1948).

Lin, C. C., and Shen, S. F.
[1] An analytic determination of the flow behind a symmetrical curved shock in a uniform stream, *NACA Tech. Note* no. 2506 (1951).

Lin, C. C., and Rubinov, S. I.
[1] On the flow behind curved shocks, *J. Math. Phys.* **27**, 105–129 (1948).

Lin, S. C.
[1] Cylindrical shock waves produced by instantaneous energy release, *J. Appl. Phys.* **25**, 54–57 (1954).

Lin, S. C., and Teare, J. D.
[1] Rate of ionization behind shock waves in air. II. Theoretical interpretations, *Physics of Fluids* **6**, 355–375 (1963).

Linnell, R. D.
[1] Two-dimensional airfoils in hypersonic flows, *J. Aero. Sci.* **16**, 22–30 (1949).

Litovitz, T. A.
See Herzfeld and Litovitz.

Lomax, H.
See Inouye and Lomax.

Lukasiewicz, J.
[1] Blast-hypersonic flow analogy theory and application, *ARS J.* **32**, 1341–1346 (1962).

Lunev, V. V.
[1] Hypersonic flows around thin blunted power-law shapes, *J. Appl. Math. Mech.* (*Prikl. Mat. Mekh.*) **26**, 572–575 (1962).

Lun'kin, U. P., and Popov, F. D.
[1] The effect of nonequilibrium dissociation on supersonic flow past blunted bodies, *Zh. Vychislitel'noi Mat. i Mat. Fiz.* **4**, 896–904 (1964).

Lyubimov, A. N.
See Babenko, Voskresenskii, Lyubimov, and Rusanov.

Maccoll, J. W.
See Taylor and Maccoll.

Maccoll, J. W., and Codd, J.
[1] Theoretical investigations of the flow around various bodies in the sonic region of velocities, *Theor. Res. Rep.* no. 17/45, Armament Res. Dept., Ministry of Supply, Fort Halstead, Kent, 1945.

Mahony, J. J.
[1] A critique of shock-expansion theory, *J. Aero. Sci.* **22**, 673–680, 720 (1955).

Mahony, J. J., and Skeat, P. R.
[1] The flow around a supersonic airfoil, *Aero. Note* no. 147, Res. and Dev. Branch, Aero. Res. Lab., Melbourne, Australia, 1955.

Maikapar, G. I.
[1] Calculation of the influence of centrifugal forces on surface pressure for a body of arbitrary shape in hypersonic flow, *J. Appl. Math. Mech.* (*Prikl. Mat. Mekh.*) **23**, 83–91 (1959).
[2] On the wave drag of nonaxisymmetric bodies at supersonic speeds, *J. Appl. Math. Mech.* (*Prikl. Mat. Mekh.*) **23**, 528–531 (1959).

Mangler, K. W.
[1] The calculation of the flow field between a blunt body and the bow wave, *Hypersonic Flow* (A. R. Collar and J. Tinkler, eds.), pp. 219–237. Butterworths, London, 1960.

Mangler, K. W., and Evans, M. E.
[1] The calculation of the inviscid flow between a detached bow wave and a body, *RAE Tech. Note Aero.* no. 2536, Royal Aircraft Establish. (Gt. Brit.), Farnborough, 1957.

Marrone, P. V.
See Gibson and Marrone.
See Hall, Eschenroeder, and Marrone.

Maslen, S. H.
[1] Supersonic conical flow, *NACA Tech. Note* no. 2651 (1952).
[2] Inviscid hypersonic flow past smooth symmetric bodies, *AIAA J.* **2**, 1055–1061 (1964).

Maslen, S. H., and Moeckel, W. E.
[1] Inviscid hypersonic flow past blunt bodies, *J. Aero. Sci.* **24**, 683–693 (1957).

Massey, H. S. W., and Burhop, E. H. S.
[1] *Electronic and Ionic Impact Phenomena.* Oxford, London, 1952.

Mates, R. E.
See Bohachevsky, Rubin, and Mates.

Mauger, F. E.
[1] Steady supersonic flow past conical bodies, *A.R.D.E. Rep.* no. (B) 3/60, Gt. Brit. Armament Res. and Dev. Establish., Fort Halstead, Kent, 1960.
[2] The extension of shock-expansion theory to flows past three-dimensional bodies, *R.A.R.D.E. Rep.* no. (B) 10/63, Gt. Brit. Royal Armament Res. and Dev. Establish., Fort Halstead, Kent, 1963.
See Stocker and Mauger.

McCall, F.
See Mitchell and McCall.

McChesney, M.
See Clarke and McChesney.

McCoubrey, J. C.
See Cottrell and McCoubrey.

McIntosh, S. C., Jr.
[1] Hypersonic flow over an oscillating wedge, *AIAA J.* **3**, 433–440 (1965).

Mead, H. R.
See Scheuing *et al.*

Melnik, R. E.
[1] Newtonian entropy layer in the vicinity of a conical symmetry plane, *AIAA J.* **3**, 520–522 (1965).
[2] A conical thin-shock-layer theory uniformly valid in the entropy layer, *Rep.* no. FDL-TDR-64-82, AF Flight Dynamics Lab., Res. and Tech. Div., USAF, Wright-Patterson Air Force Base, Ohio, 1965.
See Scheuing *et al.*

Melnik, R. E., and Scheuing, R. A.
[1] Shock layer structure and entropy layers in hypersonic conical flows, *Hypersonic Flow Research* (F. R. Riddell, ed.), pp. 379–420. Academic Press, New York, 1962. See also *Rep. no. RE-149*, Res. Dept., Grumman Aircraft Eng. Corp., Bethpage, N. Y., 1961.

Mel'nikova, N. S.
See Kochina and Mel'nikova.
See Korobeinikov, Mel'nikova, and Ryazanov.

Messiter, A. F.
[1] Lift of slender delta wings according to Newtonian theory, *AIAA J.* **1**, 794–802 (1963).

Meyer, R. E.
[1] The method of characteristics, *Modern Developments in Fluid Dynamics, High Speed Flow* (L. Howarth, ed.), Vol. I, pp. 71–104. Oxford, London, 1953.
[2] On supersonic flow behind a curved shock, *Quart. Appl. Math.* **14**, 433–436 (1957).
See Johannsen and Meyer.

Miele, A.
[1] A study of slender shapes of minimum drag using the Newton-Busemann pressure coefficient law, *AIAA J.* **1**, 168–178, 195 (1963).

Mikhailova, M. P.
See Bam-Zelikovich, Bunimovich, and Mikhailova.

Miles, J. W.
[1] *The Potential Theory of Unsteady Flow.* Cambridge Univ. Press, Cambridge, 1959.
[2] Unsteady flow at hypersonic speeds, *Hypersonic Flow* (A. R. Collar, and J. Tinkler, eds.), pp. 185–197. Butterworths, London, 1960.

Milne-Thomson, L. M.
[1] *Theoretical Hydrodynamics*, 3rd Edition. Macmillan, New York, 1955.

Minailos, A. N.
[1] Calculation of flow over blunt bodies of revolution at an angle of incidence in supersonic gas flow, *Zh. Vychislitel'noi Mat. i Mat. Fiz.* **4**, 171–177 (1964). Transl. in *U.S.S.R. Comput. Math. and Math. Phys.*, 238–248.

Mirels, H.

[1] Approximate analytical solutions for hypersonic flow over slender power law bodies, *NASA Tech. Rep.* no. R-15 (1959).

[2] Hypersonic flow over slender bodies associated with power-law shocks, *Advances in Applied Mechanics*, Vol. VII, pp. 1–54, 317–319. Academic Press, New York, 1962.

Mirels, H., and Mullen, J. F.

[1] Shock-bounded, self-similar flows with volumetric mass, momentum, and energy addition, *Rep.* no. TDR-69 (2230-01) TR-2; DCAS-TDR-62-135, Aerospace Corp., El Segundo, Calif., 1962.

Mirels, H., and Thornton, P. R.

[1] Effect of body perturbations on hypersonic flow over slender power law bodies, *NASA Tech. Rep.* no. R-45 (1959).

Mitchell, A. R.

[1] Application of relaxation to the rotational field of flow behind a bow shock wave, *Quart. J. Mech. Appl. Math.* **4**, 371–383 (1951).

Mitchell, A. R., and McCall, F.

[1] The rotational field behind a bow shock wave in axially symmetric flow using relaxation methods, *Proc. Roy. Soc. Edinburgh, Sect. A* **53**, 371–380 (1952).

Moeckel, W. E.

[1] Approximate method for predicting form and location of detached shock waves, *NACA Tech. Note* no. 1921 (1949).

[2] Oblique-shock relations at hypersonic speeds for air in chemical equilibrium, *NACA Tech. Note* no. 3895 (1957).

See Maslen and Moeckel.

Morgan, H. G., Runyan, R. L., and Huckel, V.

[1] Theoretical considerations of flutter at high Mach numbers, *J. Aero. Sci.* **25**, 371–381 (1958).

Morrissette, R. R.

See Ivey and Morrissette.

Muggia, A.

[1] Regione di arresto per un profilo a bordo di attaco quasi-circolare in corrente ipersonica non simmetrica, *Atti Accad. Sci., Torino* **94**, 836–854 (1960).

[2] Sul flusso ipersonico di un fluido non viscoso nell'intorno del punto di arresto per un solido a prora arrotondata, *Aerotecnica* **40**, 112–118 (1960).

Mullen, J. F.

See Mirels and Mullen.

Munk, M. M., and Prim, R. C.

[1] Surface-pressure gradient and shock-front curvature at the edge of a plane ogive with attached shock front, *J. Aero. Sci.* **15**, 691–695 (1948).

Munson, A. G.

[1] The vortical layer on an inclined cone, *J. Fluid Mech.* **20**, 625–643 (1964).

Murthy, K. R. A.

See Hammitt and Murthy.

Neice, S. E.

See Hamaker, Neice, and Wong.

Newman, D. J.

See Zlotnick and Newman.

Newton, I.

[1] *Mathematical principles of natural philosophy* (*Philosophiae naturalis principia mathematica*), transl. by A. Motte (1729), revised by A. Cajori. Univ. of California Press, Berkeley, 1934. Reprinted 1946.

Nikol'skii, A. A.

[1] Some unsteady gas motions and their steady hypersonic analogues, *Inzhen. Zh., Moscow* **2**, no. 2, 246–253 (1962).

Nonweiler, T. R. F.

[1] Aerodynamic problems of manned space vehicles, *J. Roy. Aeronaut. Soc.* **63**, 521–528 (1959).

Oliver, R. E.

[1] An experimental investigation of flow over simple blunt bodies at a nominal Mach number of 5.8, *Memo* no. 26, Hyp. Wind Tunnel, Guggenheim Aero. Lab., Calif. Inst. Tech., Pasadena, Calif., 1955.

Oshima, K.

[1] Blast waves produced by exploding wire, *Rep.* no. 358, *Aeronaut. Res. Inst., Univ. Tokyo*, 137–193 (1960).

Oswatitsch, K.

[1] Similarity laws for hypersonic flow, *Tech. Note* no. 16, Institutionen för Flygteknik, Kungl. Tekniska Högskolan, Stockholm, 1950.

[2] *Ähnlichkeitsgesetze für Hyperschallströmung*, *ZAMP* **2**, 249–264 (1951).

[3] Similarity and equivalence in compressible flow, *Advances in Applied Mechanics*, Vol. VI, pp. 153–271. Academic Press, New York, 1960.

Pallone, A. J.

See Cheng and Pallone.

Parker, F. R.

See Wood and Parker.

Peckham, D. H.

[1] On three-dimensional bodies of delta planform which can support plane attached shock waves, *Aeronaut. Res. Council, London, Curr. Pap.* no. 640 (1963).

Penner, S. S.

[1] *Introduction to the Study of Chemical Reactions in Flow Systems*. Butterworths, London, 1955.

Petschek, H., and Byron, S.

[1] Approach to equilibrium ionization behind strong shock waves in argon, *Ann. Phys.* **1**, 270–315 (1957).

Pleshanov, A. S.

See Predvoditelev *et al.*

Popov, F. D.

See Lun'kin and Popov.

Predvoditelev, A. S., *et al.*

[1] Predvoditelev, A. S., Stupochenko, E. V., Pleshanov, A. S., Samuilov, E. V., and Rozhdestvenskii, I. B. *Thermodynamic Functions of Air* (*for Temperatures of* 12,000 *to* 20,000°*K and Pressures of* 0.001 *to* 1000 *atm.*). Izdat. Akad. Nauk SSSR, Moscow, 1959. English transl., Assoc. Technical Services, Inc., Glen Ridge, N. J., 1962.

[2] Predvoditelev, A. S., Stupochenko, E. V., Rozhdestvenskii, I. B., Samuilov, E. V.,

and Pleshanov, A. *Tables of Gasdynamic and Thermodynamic Quantities*. Vychislitel'nyi Tsentr Akad. Nauk SSSR, Moscow, 1962.

Prim, R. C.
See Munk and Prim.

Probstein, R. F.
[1] Inversion of the Prandtl-Meyer relation for specific heat ratios of 5/3 and 5/4, *J. Aero. Sci.* **24**, 316–317, 632 (1957).
[2] On the nature of the sonic line for supersonic and hypersonic flow over blunt bodies, *WADC Tech. Note* no. 57–349, Div. of Eng., Brown Univ., Providence, R. I., 1957.
See Hayes and Probstein.
See Waldman and Probstein.

Prokopov, G. P.
See Godunov, Zabrodin, and Prokopov.

Raizer, Yu. P.
See Zel'dovich and Raizer.

Reissner, E.
See Lin, Reissner, and Tsien.

Resnikoff, M. M.
See Eggers, Resnikoff, and Dennis.

Richardson, J. M.
See Brinkley and Richardson.

Richtmeyer, R. D.
[1] Power series solution by machine of a non-linear problem in two-dimensional fluid flow, *Ann. New York Acad. Sci.* **86**, 828–843 (1960).
See von Neumann and Richtmeyer.

Rockett, J. A., and Hayes, W. D.
[1] The method of characteristics in compressible flow, Part IC (Two-dimensional flow with large entropy changes), *Tech. Rep.* no. 102-AC 49/6-100, Air Materiel Command, Wright-Patterson Air Force Base, Ohio, 1949.

Romig, M. F.
[1] Conical flow parameters for air in dissociation equilibrium: final results, *Res. Rep.* no. 7, Convair Scientific Res. Lab., San Diego, Calif., 1960. Supersedes *Res. Note* no. 14, 1958.

Rossini, F. D. (ed.)
[1] *Thermodynamics and Physics of Matter*. (Vol. I of *High Speed Aerodynamics and Jet Propulsion*). Princeton Univ. Press, Princeton, 1955.

Rozhdestvenskii, I. B.
See Predvoditelev *et al.*

Rubin, E. L.
See Bohachevsky, Rubin, and Mates.

Rubinov, S. I.
See Lin and Rubinov.

Runyan, R. L.
See Morgan, Runyan, and Huckel.

Rusanov, V. V.
See Babenko, Voskresenskii, Lyubimov, and Rusanov.

Ryazanov, E. V.
See Korobeinikov, Mel'nikova, and Ryazanov.

Sakurai, A.
[1] On the propagation and structure of the blast wave, I, *J. Phys. Soc. Japan* **8**, 662–669 (1953).
[2] On the propagation and structure of the blast wave, II, *J. Phys. Soc. Japan* **9**, 256–266 (1954).

Samuilov, E. V.
See Predvoditelev *et al.*

Sänger, E.
[1] *Raketenflugtechnik*. R. Oldenburg, Munich, 1933.

Sapunkov, Ya. G.
[1] Hypersonic flow past a circular cone at angle of attack, *J. Appl. Math. Mech.* (*Prikl. Mat. Mekh.*) **27**, 281–285 (1963).
[2] Hypersonic flow past a circular cone at an angle of attack, *J. Appl. Math. Mech.* (*Prikl. Mat. Mekh.*) **27**, 1422–1436 (1963).

Sauer, R.
[1] Dreidimensionale Probleme der Charakteristikentheorie partieller Differentialgleichungen, *ZAMM* **30**, 347–356 (1950).

Savin, R. C.
See Eggers and Savin.
See Eggers, Savin, and Syvertson.

Scheuing, R. A.
[1] Outer inviscid hypersonic flow with attached shock waves, *ARS J.* **31**, 486–505 (1961).
See Melnik and Scheuing.

Scheuing, R. A. *et al.*
[1] Scheuing, R. A., Mead, H. R., Brook, J. W., Melnik, R. E., Hayes, W. D., Gray, K. E., Donaldson, C. duP., and Sullivan, R. D. Theoretical prediction of pressures in hypersonic flow with special reference to configurations having attached leading-edge shock. Part I. Theoretical investigation, *ASD Tech. Rep.* no. 61–60, Part I. Flight Control Lab., Aeronaut. Systems Div., USAF, Wright-Patterson Air Force Base, Ohio, 1962.

Sedney, R.
[1] Some aspects of nonequilibrium flows, *J. Aerospace Sci.* **28**, 189–196, 208 (1961).

Sedney, R., and Gerber, N.
[1] Nonequilibrium flow over a cone, *AIAA J.* **1**, 2482–2486 (1963).

Sedney, R., South, J. C., and Gerber, N.
[1] Characteristic calculation of non-equilibrium flows, *The High Temperature Aspects of Hypersonic Flow* (W. C. Nelson, ed.), pp. 89–103. Pergamon Press, New York, 1964.

Sedov, L. I.
[1] On certain unsteady motions of a compressible fluid, *Prikl. Mat. Mekh.* **9**, 293–311 (1945). *Transl.* no. T-57, The Rand Corp., Santa Monica, Calif., 1956. See also, On unsteady motions of a compressible fluid, *Dokladi Akad. Nauk SSSR* **47**, 91–93 (1945) in English.
[2] Propagation of strong blast waves, *Prikl. Mat. Mekh.* **10**, 241–250 (1946). See

also, Le mouvement d'air en cas d'une forte explosion, *Dokladi Akad. Nauk SSSR* **52**, 17–20 (1946).

[3] *Similarity and Dimensional Methods in Mechanics*. Gostekhizdat, Moscow, 4th edition, 1957. English transl. (M. Holt, ed.), Academic Press, New York, 1959.

Serbin, H.

[1] The high speed flow of gas around blunt bodies, *Aeronaut. Quart.* **9**, 313–330 (1958). Supersedes *Res. Memos.* nos. RM-1713 and RM-1772, The Rand Corp., Santa Monica, Calif., 1956.

Shchennikov, V. V.

See Chushkin and Shchennikov.

Shen, S. F.

See Lin and Shen.

Shen, S. F., and Lin, C. C.

[1] On the attached curved shock in front of a sharp-nosed axially symmetric body placed in a uniform stream, *NACA Tech. Note* no. 2505 (1951).

Sheppard, L. M., and Thomson, K. D.

[1] The application of shock-expansion theory to the flow outside regions influenced by the apex or tip on wings with supersonic edges, *ARL Tech. Note* no. HSA TN 11, Aeronaut. Res. Lab., Australia, 1954.

[2] A shock-expansion theory applicable to wings with attached shock waves, *Aeronaut. Res. Council, London, Curr. Pap.* no. 392 (1958).

Shih, W. C. L., and Baron, J. R.

[1] Nonequilibrium blunt-body flows using the method of integral relations, *AIAA J.* **2**, 1062–1071 (1964).

Shmyglevskii, U. D.

[1] Supersonic profiles with minimum drag, *J. Appl. Math. Mech.* (*Prikl. Mat. Mekh.*) **22**, 368–374 (1958).

[2] *Some Variational Problems in Gas Dynamics*. Vychislitel'nyi Tsentr Akad. Nauk SSSR, Moscow, 1963.

Shulishnina, N. P.

See Chushkin and Shulishnina.

Skeat, P. R.

See Mahony and Skeat.

South, J. C.

See Sedney, South, and Gerber.

Southwell, R. V.

See Green and Southwell.

Stanyukovich, K. P.

[1] On automodel solutions of equations of hydrodynamics possessing central symmetry, *Dokladi Akad. Nauk SSSR* **48**, 310–312 (1945).

[2] *Unsteady Motion of Continuous Media*. Gostekhizdat, Moscow, 1955. English transl. (M. Holt, ed.), Academic Press, New York, 1960.

Stephenson, J. D.

[1] A technique for determining relaxation times by free-flight tests of low-fineness-ratio cones; with experimental results for air at equilibrium temperatures up to 3440°K, *NASA Tech. Note* no. D-327 (1960).

Stocker, P. M.

[1] Hypersonic flow. Part I. General considerations and two-dimensional inviscid flow theory, *Rep.* no. (B) 22–55, Gt. Brit. Armament Res. and Dev. Establish., Fort Halstead, Kent, 1955.

Stocker, P. M., and Mauger, F. E.

[1] Supersonic flow past cones of general cross-section, *J. Fluid Mech.* **13**, 383–399 (1962).

Stupochenko, E. V.

See Predvoditelev *et al.*

Sullivan, R. D.

See Donaldson, Sullivan, and Hayes.

See Scheuing *et al.*

Sullivan, R. D., Donaldson, C. duP., and Hayes, W. D.

[1] Linearized pressure distributions with strong entropy layers, *J. Fluid Mech.* **16**, 481–496 (1963).

Swigart, R. J.

[1] Third-order blast wave theory and its application to hypersonic flow past blunt-nosed cylinders, *J. Fluid Mech.* **9**, 613–620 (1960).

[2] On the shock shape and pressure distribution about blunt-nosed cylinders using blast-wave theory, *J. Aerospace Sci.* **28**, 828–829 (1961).

[3] A theory of asymmetric hypersonic blunt-body flows, *AIAA J.* **1**, 1034–1042 (1963).

[4] Hypersonic blunt-body flow fields at angle of attack, *AIAA J.* **2**, 115–117 (1964).

Sychev, V. V.

[1] Three-dimensional hypersonic gas flow past slender bodies at high angles of attack, *J. Appl. Math. Mech.* (*Prikl. Mat. Mekh.*) **24**, 296–306 (1960).

[2] On the theory of hypersonic gas flow with a power-law shock wave, *J. Appl. Math. Mech.* (*Prikl. Mat. Mekh.*) **24**, 756–764 (1960).

[3] On the theory of hypersonic flow over blunt-nosed slender bodies, *Advances in Aeronautical Sciences*, Vol. 3 (Th. von Kármán, chmn. ed. comm.), pp. 87–102. Macmillan, New York, 1962.

[4] On the method of small disturbances in the problem of hypersonic gas flow over thin blunted bodies, *Zh. Prikl. Mekh. Tekh. Fiz.* (*PMTF*), 1962, no. 6, 50–59. Transl. as *Rep.* no. TR-25, Grumman Aircraft Eng. Corp., Bethpage, N. Y., 1963.

Syvertson, C. A.

See Eggers, Savin, and Syvertson.

See Eggers and Syvertson.

See Eggers, Syvertson, and Kraus.

Syvertson, C. A., and Dennis, D. H.

[1] A second-order shock-expansion method applicable to bodies of revolution near zero lift, *NACA Rep.* no. 1328 (1957). Supersedes *Tech. Note* no. 3527 (1956).

Taylor, G. I.

[1] The formation of a blast wave by a very intense explosion, *Proc. Roy. Soc. London, Ser. A* **201**, 159–186 (1950).

Taylor, G. I., and Maccoll, J. W.

[1] The air pressure on a cone moving at high speeds, *Proc. Roy. Soc. London, Ser. A* **139**, 278–311 (1933).

Teare, J. D.
See Lin and Teare.

Telenin, G. F.
See Gilinskii, Telenin, and Tinyakov.

Telenin, G. F., and Tinyakov, G. P.
[1] A method for calculating three-dimensional flows about bodies with detached shock waves, *Dokladi Akad. Nauk SSSR* **154**, 1056–1058 (1964).

Thomson, K. D.
See Sheppard and Thomson.

Thornhill, C. K.
[1] The numerical method of characteristics for hyperbolic problems in three independent variables, *Aeronaut. Res. Council, London, Rep. Memor.* no. 2615 (1952).

Thornton, P. R.
See Mirels and Thornton.

Tinyakov, G. P.
See Gilinskii, Telenin, and Tinyakov.
See Telenin and Tinyakov.

Traugott, S. C.
[1] Shock generated vorticity and sonic line slope on the surface of blunt bodies in supersonic and hypersonic flight, *Res. Rep.* no. 14, The Martin Co., Baltimore, Md., 1959.
[2] Some features of supersonic and hypersonic flow about blunted cones, *Res. Memo.* no. 64, The Martin Co., Baltimore, Md., 1960.

Truesdell, C.
[1] On curved shocks in steady plane flow of an ideal fluid, *J. Aero. Sci.* **19**, 826–828 (1952).
[2] Intrinsic equations of spatial gas flow, *ZAMM* **40**, 9–14 (1960).

Tsien, H. S.
[1] Similarity laws of hypersonic flows, *J. Math. Phys.* **25**, 247–251 (1946).
See Lin, Reissner, and Tsien.

Uchida, S., and Yasuhara, M.
[1] The rotational field behind a curved shock wave calculated by the method of flux analysis, *J. Aero. Sci.* **23**, 830–845 (1956).

Vaglio-Laurin, R.
[1] On the PLK method and the supersonic blunt-body problem, *J. Aerospace Sci.* **29**, 185–206, 248 (1962).
[2] Asymptotic flow pattern of a hypersonic body, *PIBAL Rep.* no. 805, Dept. of Aerospace Eng. and Appl. Mech., Polytechnic Inst. of Brooklyn, Brooklyn, N. Y., 1964.

Vaglio-Laurin, R., and Ferri, A.
[1] Theoretical investigation of the flow field about blunt-nosed bodies in supersonic flight, *J. Aero/Space Sci.* **25**, 761–770 (1958).

Vallee, D.
See Guiraud, Vallee, and Zolver.

Van Dyke, M. D.
[1] The combined supersonic-hypersonic similarity rule, *J. Aero. Sci.* **18**, 499–500 (1951).

[2] A study of hypersonic small-disturbance theory, *NACA Rep.* no. 1194 (1954). Supersedes *Tech. Note* no. 3173 (1954).

[3] Applications of hypersonic small-disturbance theory, *J. Aero. Sci.* **21**, 179–186 (1954).

[4] Supersonic flow past oscillating airfoils including nonlinear thickness effects, *NACA Rep.* no. 1183 (1954). Supersedes *Tech. Note* no. 2982 (1953).

[5] A model of supersonic flow past blunt axisymmetric bodies, with application to Chester's solution, *J. Fluid Mech.* **3**, 515–522 (1958).

[6] The supersonic blunt body problem—review and extension, *J. Aero/Space Sci.* **25**, 485–496 (1958).

[7] The blunt-body problem revisited, *Proc. Internatl. Symp. on Fundamental Phenomena in Hypersonic Flow* (J. G. Hall, ed.), pp. 52–65. Cornell Univ. Press, Ithaca, New York, 1966.

Van Dyke, M. D., and Gordon, H. D.

[1] Supersonic flow past a family of blunt axisymmetric bodies, *NASA Tech. Rep.* no. R-1 (1959).

Van Hise, V.

[1] Analytic study of induced pressure on long bodies of revolution with varying nose bluntness at hypersonic speeds, *NASA Tech. Rep.* no. R-78 (1961).

Van Tuyl, A. H.

[1] The use of rational approximations in the calculation of flows with detached shocks, *J. Aero/Space Sci.* **27**, 559–560 (1960).

Vas, I. E., Bogdonoff, S. M., and Hammitt, A. G.

[1] An experimental investigation of the flow over simple two-dimensional and axial symmetric bodies at hypersonic speeds, *Jet Propulsion* **28**, 97–104 (1958).

Vincenti, W. G., and Fisher, N. H.

[1] Calculation of the supersonic pressure distribution on a single-curved tapered wing in regions not influenced by the root or tip, *NACA Tech. Note* no. 3499 (1955).

Vincenti, W. G., and Kruger, C. H., Jr.

[1] *Introduction to Physical Gas Dynamics.* Wiley, New York, 1965.

Volterra, V.

[1] *Theory of Functionals and of Integral and Integro-Differential Equations.* Dover, New York, 1959.

von Kármán, T.

[1] The problem of resistance in compressible fluids, *Proc. 5th Volta Cong., Rome*, pp. 222–277. Accad. d'Italia Roma, Fondazione Allessandro Volta, Rome, 1935.

von Neumann, J., and Richtmeyer, R. D.

[1] A method for the numerical calculation of hydrodynamic shocks, *J. Appl. Phys.* **21**, 232–237 (1950).

Voskresenskii, G. P.

See Babenko, Voskresenskii, Lyubimov, and Rusanov.

Wagner, R. D., Jr., and Watson, R.

[1] Induced pressures and shock shapes on blunt cones in hypersonic flow, *NASA Tech. Note* no. D-2182 (1964).

Waldman, G. D.

[1] Extended hypersonic small-disturbance theory, *WADC Tech. Note* no. 59-173, Div. of Eng., Brown Univ., Providence, R. I., 1959.

[2] Integral approach to the yawed blunt body problem, *Paper* no. 65-28, AIAA, New York, N. Y., 1965.

Waldman, G. D., and Probstein, R. F.
[1] An analytic extension of the shock-expansion method, *J. Aerospace Sci.* 28, 119–132 (1961). Supersedes WADC Tech. Note no. 57-214, Div. of Eng., Brown Univ., Providence, R. I., 1957.

Washington, M.
See Capiaux and Washington.

Watson, R.
See Wagner and Watson.

Wecker, M. S., and Hayes, W. D.
[1] Self-similar fluids, *Rep.* no. 528, Dept. of Aero. Eng., Princeton Univ., Princeton, N. J., 1960.

Wendroff, B.
See Lax and Wendroff.

Whitham, G. B.
[1] A note on the stand-off distance of the shock in high speed flow past a circular cylinder, *Commun. Pure Appl. Math.* **10**, 531–535 (1957).

Willett, J. E.
[1] Supersonic flow at the surface of a circular cone at angle of attack, *J. Aerospace Sci.* **27**, 907–912, 920 (1960).

Williams, E. P.
See Grimminger, Williams, and Young.

Wong, T. J.
See Hamaker, Neice, and Wong.
See Hamaker and Wong.

Wood, W. W., and Kirkwood, J. G.
[1] Hydrodynamics of a reacting and relaxing fluid, *J. Appl. Phys.* **28**, 395–398 (1957).

Wood, W. W., and Parker, F. R.
[1] Structure of a centered rarefaction wave in a relaxing gas, *Physics of Fluids* **1**, 230–241 (1958).

Woods, B. A.
[1] The flow close to the surface of a circular cone at incidence to a supersonic stream, *Aeronaut. Quart.* **23**, 115–128 (1962).

Wray, K. L.
[1] Chemical kinetics of high temperature air, *Hypersonic Flow Research* (F. R. Riddell, ed.), pp. 181–204. Academic Press, New York, 1962.

Xerikos, J., and Anderson, W. A.
[1] A critical study of the direct blunt body integral method, *Rep.* SM-42603, Santa Monica Div., Douglas Aircraft Co., Inc., Santa Monica, Calif., 1962.

Yakura, J. K.
[1] Theory of entropy layers and nose bluntness in hypersonic flow, *Hypersonic Flow Research* (F. R. Riddell, ed.), pp. 421–470. Academic Press, New York, 1962.

Yasuhara, M.
See Uchida and Yasuhara.

Young, G. B. W.
See Grimminger, Williams, and Young.

Zabrodin, A. V.
See Godunov, Zabrodin, and Prokopov.

Zadoff, L. N.
[1] Axisymmetric shocks in the Newtonian limit for nonrigid boundaries, *Physics of Fluids* **5**, 831–839 (1962).

Zartarian, G.
See Ashley and Zartarian.

Zartarian, G., Hsu, P. T., and Ashley, H.
[1] Dynamic airloads and aeroelastic problems at entry Mach numbers, *J. Aerospace Sci.* **28**, 209–222 (1961).

Zel'dovich, Ya. B., and Raizer, Yu. P.
[1] *The Physics of Shock Waves and High Temperature Hydrodynamic Phenomena.* Fizmatlit, Moscow, 1963. English transl. (W. D. Hayes and R. F. Probstein, eds.), Academic Press, New York, 1966.

Zienkiewicz, H. K.
[1] Flow about cones at very high speeds, *Aeronaut. Quart.* **8**, 384–394 (1957).

Zlotnick, M., and Newman, D. J.
[1] Theoretical calculation of the flow on blunt-nosed axisymmetric bodies in a hypersonic stream, *Tech. Rep.* no. 2-57-29, Research and Advanced Dev. Div., Avco Mfg. Corp., Lawrence, Mass., 1957.

Zolver, R.
See Guiraud, Vallee, and Zolver.

SYMBOL INDEX

Throughout the book we have tried wherever possible to use standard aerodynamic and mathematical notation. In order to maintain some degree of self-consistency and in order to avoid confusing repetition, it was necessary to use a few symbols different from those which are commonly accepted. We have also tried to avoid duplication within the obvious limitations imposed by common usage and the limited number of symbols available to us in the Roman and Greek alphabets. However, in symbols used only locally frequent repetition is to be found. In order to avoid as far as possible the use of dummy variables, we have consistently used the integral sign with one limit to represent an indefinite integral evaluated as zero at that limit.

Except as indicated all standard mathematical symbols are taken to have their usual meanings. No attempt has been made to include every symbol used in the book. In particular, we have generally not included symbols which are only employed locally and which are not referred to in other parts of the book.

In this index we refer primarily to the page where the symbol is introduced and defined; as a rule, we include also a reference to where the symbol is reintroduced in a different context. A verbal definition is usually included, and in addition where appropriate an equation number, mathematical definition, or both are given.

f	shock shape function in Newtonian separation analysis, (5.9.49b), 377
$f(w^*)$	function of modified stream function in blunt-faced body analysis, 301
$f(\eta)$	detachment function, (4.1.5), (4.2.19), 218, 227
f_i	function representing dependent variable in perturbation of similar power-law solutions with numerical subscript ($i = 0, 1, 2$) denoting order of the variable, 89, 92
F	integrand in calculus of variations, (3.4.8), (3.4.40), 153, 164
$\mathbf{F}$	total force on a surface in Newtonian flow, 189
$F(x)$	shape function for slender body of revolution at small incidence, (3.8.20), 205
$F(\mathscr{K})$	function in constant-density general stagnation point solution, (4.5.12), 257
F_*	thermodynamic function in Prandtl-Meyer solution, (7.1.21), (7.1.23), 486
F_e	thermodynamic function in Prandtl-Meyer solution, (7.1.25), (7.1.33), 487
g	shock wave shape function, (2.1.7), (5.7.13), 35, 346
g	dimensionless pressure gradient in Newtonian vortical layer analysis, (5.6.38), 336
$g(z)$	function of coordinate parallel to surface in blunt-faced body analysis, 301
$G(\lambda)$	function in constant-density general stagnation point solution, (4.5.13), 257
h	specific enthalpy, 11
H	total enthalpy, $h + \frac{1}{2}q^2$, 11, 273
$\mathscr{H}$	metric variable in boundary layer coordinate system, (5.1.1), 267
$\mathbf{i}, \mathbf{j}, \mathbf{k}$	unit vectors in x, y, and z directions, 104, 128, 181, 183
I	dimensionless total momentum in similar power-law solutions, (2.6.20b), 61
I_N	concentrated lateral impulse due to blunt nose or leading edge, 360
$\mathscr{I}$	total momentum per unit area ($j = 0$), per unit angle times depth ($j = 1$), or per unit solid angle ($j = 2$) in similar power-law solutions, (2.6.18b), 61
j	integer specifying number of dimensions, $= 0$ planar or two-dimensional case, $= 1$ cylindrical or axisymmetric case, $= 2$ spherical case, 56
$\mathbf{j}$	see $\mathbf{i}, \mathbf{j}, \mathbf{k}$, 104, 181, 183
k	exponent specifying power-law body or shock shape, r or y proportional to x^k or t^k, 56, 141
k	wave number of oscillating plate, 212
$\mathbf{k}$	see $\mathbf{i}, \mathbf{j}, \mathbf{k}$, 104, 183
k_i	exponent specifying axial variation of zeroth-order dependent variable in perturbation of similar power-law solutions, 89
k_0	exponent specifying power-law perturbation of power-law body shape, (2.7.37), 85
K	hypersonic similarity parameter, (2.2.11), (2.9.9), 42, 106
K	streamline curvature, (3.5.17), (5.6.39), 178, 337
$K(x)$	curvature of reference surface in thin shock layer analysis, curvature of shock in Newtonian separation analysis and inverse method, 267, 367, 454
$K(y)$	curvature function, (3.6.8), 183
K'	unified supersonic-hypersonic similarity parameter, $\beta\tau$, 45
K_{N_j}	hypersonic similarity parameter for slight blunting in two-dimensional ($j = 0$) or axisymmetric ($j = 1$) flow, (2.8.5), 96, 97, 107, 110
$K_0(\phi)$	curvature in unit sphere of curve describing conical body shape, (3.7.6), 191
K_0	curvature of body in Newtonian separation analysis, (5.9.5a), 367

y — cartesian boundary layer coordinate for blunt-faced bodies in direction normal to reference surface, 343

y, z — cartesian boundary layer coordinate system for blunt-faced bodies, y distance in free stream direction, 299

y' — reduced cartesian coordinate in y direction for yawed wing, (2.9.14b), 108

y^* — reduced cartesian coordinate in direction normal to reference surface in blunt-faced body analysis, (5.4.1), (5.5.11), (5.6.19), (5.7.5b), 299, 322, 327, 345

z — see x, y, z and y, z

z — dimensionless square of speed of sound, (2.6.6), 57

z — cartesian coordinate parallel to body in asymmetric constant-density stagnation solution, 252

z — coordinate normal to body in constant-density general stagnation point solution, (4.5.6), 256

z — variable dependent on shock angle in constant-streamtube-area analysis, (5.3.23), 286

z — cartesian boundary layer coordinate for blunt-faced bodies in direction of flow across body, 343

z — reduced axial coordinate in Newtonian separation analysis, (5.9.49a), 377

z' — reduced lateral cartesian coordinate, (2.4.4), 48

z^* — reduced cartesian coordinate in direction of flow across body in blunt-faced body analysis, (5.7.5c), 345

Z — compressibility factor in equation of state $p = \rho \mathscr{R} TZ$, 490

α — angle of attack or incidence, 27, 43, 104, 166, 201, 219

α — small parameter characterizing perturbation of similar power-law solutions, 89

α — fraction of original diatomic molecules dissociated, 121, 474, 490

α — inclination of conical body surface to latitude line of constant ϑ, (3.7.3), 191, 325

α — exponent in constant-streamtube-area ogival body solution, (5.3.43), 292

α — angle of incidence of coordinate reference surface from free stream direction, 320, 342, 386

$\alpha_{1,2}$ — parameters in constant-density general stagnation point solution, (4.5.5), 256

β — Prandtl-Glauert parameter, $\sqrt{M_\infty^2 - 1}$, 45

β — ratio of body thickness to forebody thickness, (3.4.26), 162

β — parameter governing noncircularity effect on constant-density solutions, (4.3.31), 239, 288, 310, 314

β — inclination of streamline relative to a conical ray, dependent variable in Newtonian vortical layer analysis, (5.6.37), 336

β — local Prandtl-Glauert parameter, $\sqrt{M^2 - 1}$, 486

γ — isentropic exponent or effective ratio of specific heats in free stream (actual specific heat ratio for perfect gas), $\rho_\infty a_\infty^2 / p_\infty$, 13

γ_e — isentropic exponent or effective ratio of specific heats, $\rho a^2/p$, (1.4.22), 13

γ_f — isentropic exponent for frozen state, c_{p_f}/c_{v_f}, 490

γ_s — γ_e immediately behind shock, inverse of parameter governing compressibility effect on constant-density solutions, 239

γ_ϵ — ratio of enthalpy to internal energy, (1.4.20), 18

γ_* — thermodynamic coefficient analogous to γ, (1.4.24), 18

Γ — thermodynamic parameter, $\frac{1}{2}(\gamma_* + 1)$, (1.4.26), 20

SUBSCRIPTS

SUPERSCRIPTS

—	dimensionless variables in method of integral relations, 411, 412

MATHEMATICAL AND SPECIAL SYMBOLS

Æ	aspect ratio, 26
Æ′	aspect ratio for Æ $\gg$ 1, 342
D	differential operator in analysis of blunt-faced body similitudes, (5.7.7), 345
$\tilde{D}$	differential operator in inner layer analysis of blunt-faced body similitudes, 351
D/Dt	total time derivative along a streamline, 382, 474, 536
D/Dx	partial derivative with respect to x with ψ held constant, 268
$O(\ )$	of order of or of smaller order than, in a limiting sense, 14
$\approx$	approximately equal to (requires dimensional consistency), 30
$\propto$	proportional to, 87
[ ]	difference between the quantity inside the brackets after the shock and the same quantity before the shock, 35

AUTHOR INDEX

SUBJECT INDEX